To Rosalie and our children and grandchildren
S.K.

To Ania, Joseph, and Kamil
T.J.K.

To my parents
K.P.

Samuel Kotz
Tomasz J. Kozubowski
Krzysztof Podgórski

The Laplace Distribution and Generalizations

A Revisit with Applications to Communications, Economics, Engineering, and Finance

Birkhäuser
Boston • Basel • Berlin

Samuel Kotz
Department of Engineering Management
and Systems Engineering
George Washington University
Washington, DC 20052
U.S.A.

Tomaz J. Kozubowski
Department of Mathematics
University of Nevada
Reno, NV 89557
U.S.A.

Krzysztof Podgórski
Department of Mathematical Sciences
Indiana University–Purdue University
Indianapolis, IN 46202
U.S.A.

Library of Congress Cataloging-in-Publication Data

Kotz, Samuel.
The Laplace distribution and generalizations : a revisit with applications to communications, economics, engineering, and finance / Samuel Kotz, Tomasz J. Kozubowski, and Krzysztof Podgórski.
p. cm.
Includes bibliographical references and index.
ISBN 0-8176-4166-1 (alk. paper) – ISBN 3-7643-4166-1 (alk. paper)
1. Distribution (Probability theory) I. Kozubowski, Tomasz J., 1962- II. Podgórski, Krzysztof, 1963- III. Title.

QA273.6.K684 2001
519.2'4–dc21 00-068900

AMS Subject Classifications: 60-02, 60E05, 60E07, 60F05, 62-02, 62A10, 62E10, 62E15, 62E20, 62F03, 62F10, 62F12, 62F25, 62G30, 62H05, 62N10, 62P05, 62P10, 62P20, 62P99, 90A09, 94A05, 94A17

Printed on acid-free paper

Birkhäuser

ISBN 0-8176-4166-1 SPIN 10754106
ISBN 3-7643-4166-1

Reformatted from authors' files by John Spiegelman, Philadelphia, PA
Cover design by Jeff Cosloy, Newton, MA
Printed and bound by Hamilton Printing, Rensselaer, NY
Printed in the United States of America

9 8 7 6 5 4 3 2 1

Contents

Preface xi

Abbreviations xiii

Notation xv

I Univariate Distributions 1

1 Historical Background 3

2 Classical Symmetric Laplace Distribution 15

2.1 Definition and basic properties . . . 16

2.1.1 Density and distribution functions . . . 16

2.1.2 Characteristic and moment generating functions . . . 19

2.1.3 Moments and related parameters . . . 19

2.1.3.1 Cumulants . . . 19

2.1.3.2 Moments . . . 20

2.1.3.3 Mean deviation . . . 20

2.1.3.4 Coefficients of skewness and kurtosis . . . 21

2.1.3.5 Entropy . . . 21

2.1.3.6 Quartiles and quantiles . . . 21

2.2 Representations and characterizations . . . 22

2.2.1 Mixture of normal distributions . . . 22

2.2.2 Relation to exponential distribution . . . 23

2.2.3 Relation to the Pareto distribution . . . 24

2.2.4 Relation to 2×2 unit normal determinants . . . 25

2.2.5 An orthogonal representation . . . 25

2.2.6 Stability with respect to geometric summation . . . 27

2.2.7 Distributional limits of geometric sums . . . 30
2.2.8 Stability with respect to the ordinary summation . . . 32
2.2.9 Distributional limits of deterministic sums . . . 35
2.3 Functions of Laplace random variables . . . 35
2.3.1 The distribution of the sum of independent Laplace variates . . . 35
2.3.2 The distribution of the product of two independent Laplace variates . . . 40
2.3.3 The distribution of the ratio of two independent Laplace variates . . . 41
2.3.4 The t-statistic for a double exponential (Laplace) distribution . . . 43
2.4 Further properties . . . 46
2.4.1 Infinite divisibility . . . 46
2.4.2 Geometric infinite divisibility . . . 48
2.4.3 Self-decomposability . . . 49
2.4.4 Complete monotonicity . . . 49
2.4.5 Maximum entropy property . . . 51
2.5 Order statistics . . . 53
2.5.1 Distribution of a single order statistic . . . 53
2.5.1.1 The minimum . . . 54
2.5.1.2 The maximum . . . 55
2.5.1.3 The median . . . 55
2.5.2 Joint distributions of order statistics . . . 55
2.5.2.1 Range, midrange, sample median . . . 56
2.5.3 Moments of order statistics . . . 60
2.5.4 Representation of order statistics via sums of exponentials . . . 63
2.6 Statistical inference . . . 64
2.6.1 Point estimation . . . 65
2.6.1.1 Maximum likelihood estimation . . . 66
2.6.1.2 Maximum likelihood estimation under censoring . . . 77
2.6.1.3 Maximum likelihood estimation of monotone location parameters 78
2.6.1.4 The method of moments . . . 79
2.6.1.5 Linear estimation . . . 84
2.6.2 Interval estimation . . . 91
2.6.2.1 Confidence bands for the Laplace distribution function . . . 93
2.6.2.2 Conditional inference . . . 94
2.6.3 Tolerance intervals . . . 99
2.6.4 Testing hypothesis . . . 103
2.6.4.1 Testing the normal versus the Laplace . . . 103
2.6.4.2 Goodness-of-fit tests . . . 105
2.6.4.3 Neyman–Pearson test for location . . . 106
2.6.4.4 Asymptotic optimality of the Kolmogorov–Smirnov test . . . 110
2.6.4.5 Comparison of nonparametric tests of location . . . 110
2.7 Exercises . . . 112

3 Asymmetric Laplace Distributions **133**
3.1 Definition and basic properties . . . 136
3.1.1 An alternative parametrization and special cases . . . 136
3.1.2 Standardization . . . 137
3.1.3 Densities and their properties . . . 137
3.1.4 Moment and cumulant generating functions . . . 140
3.1.5 Moments and related parameters . . . 141

3.1.5.1 Cumulants . . . 141
3.1.5.2 Moments . . . 142
3.1.5.3 Absolute moments . . . 142
3.1.5.4 Mean deviation . . . 142
3.1.5.5 Coefficient of Variation . . . 143
3.1.5.6 Coefficients of skewness and kurtosis . . . 143
3.1.5.7 Quantiles . . . 143
3.2 Representations . . . 144
3.2.1 Mixture of normal distributions . . . 144
3.2.2 Convolution of exponential distributions . . . 146
3.2.3 Self-decomposability . . . 147
3.2.4 Relation to 2×2 normal determinants . . . 148
3.3 Simulation . . . 149
3.4 Characterizations and further properties . . . 150
3.4.1 Infinite divisibility . . . 150
3.4.2 Geometric infinite divisibility . . . 151
3.4.3 Distributional limits of geometric sums . . . 152
3.4.4 Stability with respect to geometric summation . . . 155
3.4.5 Maximum entropy property . . . 155
3.5 Estimation . . . 158
3.5.1 Maximum likelihood estimation . . . 158
3.5.1.1 Case 1: The values of κ and σ are known . . . 159
3.5.1.2 Case 2: The values of θ and κ are known . . . 161
3.5.1.3 Case 3: The values of θ and σ are known . . . 162
3.5.1.4 Case 4: The value of κ is known . . . 166
3.5.1.5 Case 5: The value of θ is known . . . 167
3.5.1.6 Case 6: The value of σ is known . . . 170
3.5.1.7 Case 7: The values of all three parameters are unknown . . . 172
3.6 Exercises . . . 174

4 Related Distributions **179**
4.1 Bessel function distribution . . . 179
4.1.1 Definition and parametrizations . . . 180
4.1.2 Representations . . . 181
4.1.2.1 Mixture of normal distributions . . . 181
4.1.2.2 Relation to gamma distribution . . . 183
4.1.3 Self-decomposability . . . 184
4.1.3.1 Relation to sample covariance . . . 186
4.1.4 Densities . . . 188
4.1.4.1 Asymmetric Laplace laws . . . 189
4.1.4.2 Symmetric case . . . 189
4.1.4.3 An integer value of τ . . . 191
4.1.5 Moments . . . 192
4.2 Laplace motion . . . 193
4.2.1 Symmetric Laplace motion . . . 193
4.2.2 Representations . . . 194
4.2.3 Asymmetric Laplace motion . . . 197
4.2.3.1 Subordinated Brownian motion . . . 198
4.2.3.2 Difference of gamma processes . . . 198

4.2.3.3 Compound Poisson approximation . . . 198
4.3 Linnik distribution . . . 199
4.3.1 Characterizations . . . 200
4.3.1.1 Stability with respect to geometric summation . . . 200
4.3.1.2 Distributional limits of geometric sums . . . 202
4.3.1.3 Stability with respect to deterministic summation . . . 203
4.3.2 Representations . . . 204
4.3.3 Densities and distribution functions . . . 206
4.3.3.1 Integral representations . . . 206
4.3.3.2 Series expansions . . . 208
4.3.4 Moments and tail behavior . . . 212
4.3.5 Properties . . . 213
4.3.5.1 Self-decomposability . . . 213
4.3.5.2 Infinite divisibility . . . 214
4.3.6 Simulation . . . 215
4.3.7 Estimation . . . 215
4.3.7.1 Method of moments type estimators . . . 216
4.3.7.2 Least-squares estimators . . . 216
4.3.7.3 Minimal distance method . . . 217
4.3.7.4 Fractional moment estimation . . . 218
4.3.8 Extensions . . . 218
4.4 Other cases . . . 219
4.4.1 Log-Laplace distribution . . . 219
4.4.2 Generalized Laplace distribution . . . 219
4.4.3 Sargan distribution . . . 220
4.4.4 Geometric stable laws . . . 220
4.4.5 ν-stable laws . . . 222
4.5 Exercises . . . 222

II Multivariate Distributions 227

Introduction 229

5 Symmetric Multivariate Laplace Distribution 231
5.1 Bivariate case . . . 231
5.1.1 Definition . . . 231
5.1.2 Moments . . . 232
5.1.3 Densities . . . 232
5.1.4 Simulation of bivariate Laplace variates . . . 233
5.2 General symmetric multivariate case . . . 234
5.2.1 Definition . . . 234
5.2.2 Moments and densities . . . 235
5.3 Exercises . . . 236

6 Asymmetric Multivariate Laplace Distribution 239
6.1 Bivariate case: Definition and basic properties . . . 240
6.1.1 Definition . . . 240
6.1.2 Moments . . . 241

6.1.3 Densities . . . 241
6.1.4 Simulation of bivariate asymmetric Laplace variates . . . 242
6.2 General multivariate asymmetric case . . . 243
6.2.1 Definition . . . 243
6.2.2 Special cases . . . 244
6.3 Representations . . . 246
6.3.1 Basic representation . . . 246
6.3.2 Polar representation . . . 247
6.3.3 Subordinated Brownian motion . . . 248
6.4 Simulation algorithm . . . 248
6.5 Moments and densities . . . 249
6.5.1 Mean vector and covariance matrix . . . 249
6.5.2 Densities in the general case . . . 249
6.5.3 Densities in the symmetric case . . . 250
6.5.4 Densities in the one-dimensional case . . . 250
6.5.5 Densities in the case of odd dimension . . . 251
6.6 Unimodality . . . 251
6.6.1 Unimodality . . . 251
6.6.2 A related representation . . . 252
6.7 Conditional distributions . . . 253
6.7.1 Conditional distributions . . . 253
6.7.2 Conditional mean and covariance matrix . . . 254
6.8 Linear transformations . . . 254
6.8.1 Linear combinations . . . 254
6.8.2 Linear regression . . . 255
6.9 Infinite divisibility properties . . . 256
6.9.1 Infinite divisibility . . . 256
6.9.2 Asymmetric Laplace motion . . . 257
6.9.3 Geometric infinite divisibility . . . 258
6.10 Stability properties . . . 258
6.10.1 Limits of random sums . . . 258
6.10.2 Stability under random summation . . . 259
6.10.3 Stability of deterministic sums . . . 260
6.11 Linear regression with Laplace errors . . . 261
6.11.1 Least-squares estimation . . . 261
6.11.2 Estimation of σ^2 . . . 262
6.11.3 The distributions of standard t and F statistics . . . 263
6.11.4 Inference from the estimated regression function . . . 264
6.11.4.1 Estimating the regression function at $\mathbf{x}_0$ . . . 264
6.11.4.2 Forecasting a new observation at $\mathbf{x}_0$ . . . 264
6.11.5 Maximum likelihood estimation . . . 265
6.11.6 Bayesian estimation . . . 267
6.12 Exercises . . . 268

III Applications 273

Introduction 275

7 Engineering Sciences 277
- 7.1 Detection in the presence of Laplace noise . . . 277
- 7.2 Encoding and decoding of analog signals . . . 280
- 7.3 Optimal quantizer in image and speech compression . . . 281
- 7.4 Fracture problems . . . 284
- 7.5 Wind shear data . . . 285
- 7.6 Error distributions in navigation . . . 286

8 Financial Data 289
- 8.1 Underreported data . . . 289
- 8.2 Interest rate data . . . 290
- 8.3 Currency exchange rates . . . 292
- 8.4 Share market return models . . . 294
  - 8.4.1 Introduction . . . 294
  - 8.4.2 Stock market returns . . . 294
- 8.5 Option pricing . . . 296
- 8.6 Stochastic variance Value-at-Risk models . . . 297
- 8.7 A jump diffusion model for asset pricing with Laplace distributed jump-sizes . . . 300
- 8.8 Price changes modeled by Laplace–Weibull mixtures . . . 302

9 Inventory Management and Quality Control 303
- 9.1 Demand during lead time . . . 303
- 9.2 Acceptance sampling for Laplace distributed quality characteristics . . . 304
- 9.3 Steam generator inspection . . . 306
- 9.4 Adjustment of statistical process control . . . 306
- 9.5 Duplicate check-sampling of the metallic content . . . 308

10 Astronomy and the Biological and Environmental Sciences 309
- 10.1 Sizes of sand particles, diamonds, and beans . . . 309
- 10.2 Pulses in long bright gamma-ray bursts . . . 310
- 10.3 Random fluctuations of response rate . . . 311
- 10.4 Modeling low dose responses . . . 312
- 10.5 Multivariate elliptically contoured distributions for repeated measurements . . . 312
- 10.6 ARMA models with Laplace noise in the environmental time series . . . 313

Appendix: Bessel Functions 315

References 319

Index 343

Preface

The aim of this monograph is quite modest: It attempts to be a systematic exposition of all that appeared in the literature and was known to us by the end of the 20th century about the Laplace distribution and its numerous generalizations and extensions. We have tried to cover both theoretical developments and applications. There were two main reasons for writing this book. The first was our conviction that the areas and situations where the Laplace distribution naturally occurs is so extensive that tracking the original sources is unfeasible. The second was our observation of the growing demand for statistical distributions having properties tangent to those exhibited by the Laplace laws. These two "necessary" conditions justified our efforts that led to this book.

Many details are arranged primarily for reference, such as inclusion of the most commonly used terminology and notation. In several cases, we have proposed unification to overcome the ambiguity of notions so often present in this area. Personal taste may have done some injustice to the subject matter by omitting or emphasizing certain topics due to space limitations. We trust that this feature does not constitute a serious drawback—in our literature search we tried to leave no stone unturned (we collected over 400 references).

Because we view this monograph as a textbook, the exposition in the earlier chapters proceeds at a rather pedestrian pace and each part of the book presupposes all earlier developments. A slightly more advanced approach is taken in the second part of the book, where quite a few of our results appear in print for the first time.

The exercises are supposed to be an integral part of the discussion, but a number of them are intended simply to aid in understanding the concepts employed.

The monograph should be read (and studied!) with the constant reminder that it aims to provide an alternative to the dominance of the "normal" law (the eponymous "Gaussian distribution") that reigned almost without opposition in statistical theory and applications for almost two centuries.

We have tried to make sufficiently precise statements while striving to keep the mathematical level of the book appealing to the widest possible readership—including users of distribution theory in various applied sciences. We hopefully did not overplay the simplicity card so popular among expositors of probabilistic and statistical concepts in the last two decades or so. The prerequisites are calculus, matrix algebra, and familiarity with the basic concepts of probability theory and statistical inference. As always, the most desirable prerequisites for books of this kind are ill-defined qualities of mathematical sophistication and understanding of the intricate nature of somewhat elusive probabilistic reasoning.

Since so much of this book is a synthesis of other people's work, the text and the extensive bibliography (which reflects the rich diversity of sources) must stand as an expression of our intellectual gratitude to the pioneers and contributors to the subject matter of the monograph. Special thanks are due to librarians at the George Washington University (first and foremost Mrs. Debra Bensazon), Indiana University–Purdue University, Indianapolis, the University of California at Santa Barbara, and the University of Nevada at Reno, who generously assisted us in digging out sources related to Laplace distributions. Modern communication technology has helped us overcome the problem of the "academic geography" among the authors located at opposite corners of the United States and at its geographical midpoint. We tender our very warm thanks to Ann Kostant and Tom Grasso, our editors at Birkhaüser Boston, for their efficient, expeditious, and meticulous handling of the production of this monograph.

We hope that this work will trigger additional theoretical research and provide tools that will generate further fruitful applications of the distributions presented in various branches of life and behavioral sciences. It is the applications that provide the special vitality to probabilistic laws that in our opinion are of permanent interest on their own from both mathematical and conceptual aspects. We wish our readers a pleasant and instructive journey as they travel leisurely or rapidly through the text.

S.K.
Washington, D.C.

T.J.K.
Reno, Nevada

K.P.
Indianapolis, Indiana

July, 2000

Abbreviations

AAI	average adjustment interval
ABLUE	asymptotically best linear unbiased estimator
AD	Anderson–Darling (test)
AL	asymmetric Laplace
AMP	asymptotically most powerful
ARE	asymptotic relative efficiency
BAL	bivariate asymmetric Laplace
BLUE	best linear unbiased estimator
c.d.f.	cumulative distribution function
ch.f.	characteristic function
CLT	central limit theorem
CvM	Cramér–von Mises (test)
EC	eliptically contoured
GAL	generalized asymmetric Laplace
GGC	generalized gamma convolution
GIG	generalized inverse Gaussian
GS	geometric stable
i.i.d.	independent, identically distributed

LLN	law of large numbers
LMP	locally most powerful
LSE	least-squares estimator
MLE	maximum likelihood estimator
MME	method of moments estimator
MR	midrange
MSD	mean squared deviation
MSE	mean squared error
OC	operating characteristic (curve)
p.d.f.	probability density function
r.v.	random variable (vector)
UMVU	uniformly minimum variance unbiased
UMP	uniformly most powerful (test)
VaR	Value-at-Risk

Notation

$\mathbf{A}'$	the transpose of a matrix $\mathbf{A}$
$\lvert\mathbf{A}\rvert$	the determinant of a square matrix $\mathbf{A}$
$\mathcal{AL}(\theta, \mu, \sigma)$	univariate AL law with mode at θ, mean $\theta + \mu$, and variance $\mu^2 + \sigma^2$
$\mathcal{AL}(\mu, \sigma)$	univariate AL law with mode at zero, mean μ, and variance $\mu^2 + \sigma^2$
$\mathcal{AL}(\mu)$	standard univariate AL law with mode at 0, mean μ, and variance $\mu^2 + 1$
$\mathcal{AL}^*(\theta, \kappa, \sigma)$	univariate AL law with mode at θ, skewness parameter κ, and scale parameter σ
$\mathcal{AL}^*(\kappa, \sigma)$	univariate AL law with mode at 0, skewness parameter κ, and scale parameter σ
$\mathcal{AL}^*(\kappa)$	standard univariate AL law with mode at 0, skewness parameter κ, and scale parameter 1
$\mathcal{AL}_d(\mathbf{m}, \mathbf{\Sigma})$	d-dimensional asymmetric Laplace distribution with mean $\mathbf{m}$ and variance-covariance matrix $\mathbf{\Sigma} + \mathbf{mm}'$
$\mathcal{ALM}(\mu, \sigma, \nu)$	asymmetric Laplace motion
$\mathcal{BAL}(m_1, m_2, \sigma_1, \sigma_2, \rho)$	bivariate asymmetric Laplace distribution
$\mathrm{Beta}(\alpha, \beta)$	Beta distribution with parameters α and β
$\mathcal{BSL}(\sigma_1, \sigma_2, \rho)$	bivariate symmetric Laplace distribution
$\mathcal{CL}(\theta, s)$	classical Laplace distribution with mean θ and scale parameter s
D_n	the Kolmogorov statistic

$D_n^{\pm}$	the Smirnov one-sided statistic
EX	the expected value of a random variable X
$E_1(x)$	the exponential integral function, $E_1(x) = \int_x^{\infty} \frac{e^{-t}}{t} dt,\ x > 0$
$EC_d(\mathbf{m}, \mathbf{\Sigma}, g)$	elliptically contoured distribution
$G(\alpha, \beta)$	gamma distribution with shape parameter α and scale parameter β
$G(\alpha)$	standard gamma distribution with scale parameter 1
$\mathcal{GAL}(\theta, \mu, \sigma, \tau)$	generalized asymmetric Laplace distribution (Bessel K-function distribution, variance-gamma distribution) with parameters $\theta, \mu, \sigma, \tau$
$\mathcal{GAL}(\mu, \tau)$	standard generalized asymmetric Laplace distribution (the $\mathcal{GAL}(\theta, \mu, \sigma, \tau)$ distribution with $\theta = 0$ and $\sigma = 1$)
$\mathcal{GAL}^*(\theta, \kappa, \sigma, \tau)$	generalized asymmetric Laplace distribution (Bessel K-function distribution, variance-gamma distribution) with parameters $\theta, \kappa, \sigma, \tau$
$\mathcal{GAL}^*(\kappa, \tau)$	standard generalized asymmetric Laplace distribution (the $\mathcal{GAL}^*(\theta, \kappa, \sigma, \tau)$ distribution with $\theta = 0$ and $\sigma = 1$)
$\mathcal{GAL}_d(\mathbf{m}, \mathbf{\Sigma}, s)$	d-dimensional generalized Laplace distribution
$GIG(\lambda, \chi, \psi)$	generalized inverse Gaussian distribution
$GS_\alpha(\sigma, \beta, \mu)$	geometric stable distribution with index α, scale parameter σ, skewness parameter β, and location parameter μ; in particular, $GS_\alpha(\sigma, 0, 0) = L_{\alpha,\sigma}$, $GS_2(s, 0, 0) = \mathcal{CL}(0, s)$, $GS_2(\sigma/\sqrt{2}, \beta, \mu) = \mathcal{AL}(0, \mu, \sigma)$
$H_d(\lambda, \alpha, \beta, \delta, \mu, \mathbf{\Sigma})$	d-dimensional generalized hyperbolic distribution
$\mathbf{I}_d$	d-dimensional identity matrix
$I(\theta)$	the Fisher information about θ
J_λ	the Bessel function of the first kind of order λ
K_λ	the modified Bessel function of the third kind with index λ
$\mathcal{L}(\theta, \sigma)$	Laplace distribution with mean θ and variance σ^2
$L_{\alpha,\sigma}$	Linnik distribution with index α and scale parameter σ
$\mathcal{LM}(\sigma, \nu)$	symmetric Laplace motion with space-scale parameter σ and time-scale parameter ν
log	natural logarithm
$\mathbb{N}$	the set of natural numbers
$N(\mu, \sigma^2)$	normal distribution with mean μ and variance σ^2
$N_d(\mathbf{m}, \mathbf{\Sigma})$	d-dimensional normal distribution with mean vector $\mathbf{m}$ and variance-covariance matrix $\mathbf{\Sigma}$

$o(g(x))$	$f(x) = o(g(x))$ as $x \to x_0$ means that $f(x)/g(x)$ converges to zero as $x \to x_0$
$o(1)$	$f(x) = o(1)$ if the function f converges to zero
$O(g(x))$	$f(x) = O(g(x))$ as $x \to x_0$ means that $\|f(x)/g(x)\|$ is bounded for x close to x_0
$O(1)$	$f(x) = O(1)$ if the function f is bounded
$\mathbb{R}$	the set of real numbers
$\mathbb{R}^d$	d-dimensional Eucledean space
$\mathrm{Re}(z)$	the real part of z
$\mathcal{SL}_d(\mathbf{\Sigma})$	d-dimensional symmetric Laplace distribution with mean zero and variance-covariance matrix $\mathbf{\Sigma}$
$\mathbf{s}'\mathbf{t}$	the inner product of the vectors $\mathbf{s}$ and $\mathbf{t}$
$\mathbb{S}_d$	unit sphere in $\mathbb{R}^d$: $\{\mathbf{s} \in \mathbb{R}^d : \|\|\mathbf{s}\|\| = 1\}$
$\mathrm{sign}(x)$	1 for $x > 0$, -1 for $x < 0$, 0 for $x = 0$
$\|\|\mathbf{t}\|\|$	$(\mathbf{t}'\mathbf{t})^{1/2}$ — the Euclidean norm of $\mathbf{t} \in \mathbb{R}^d$
$\mathbf{t}'$	the transpose of a column vector $\mathbf{t}$
$\mathbf{U}^{(d)}$	uniform distribution on $\mathbb{S}_d$
$\mathrm{Var}(X)$	the variance of a random variable X
$X \sim \mathcal{CL}(\theta, s)$	X has the distribution $\mathcal{CL}(\theta, s)$
$[[x]]$	the greatest integer less than or equal to x
$x_{k:n}$	the kth smallest of $x_1, x_2, \ldots, x_n$
x^+	x if $x \geq 0$, 0 if $x \leq 0$
x^-	$-x$ if $x \leq 0$, 0 if $x \geq 0$
$\mathbb{I}_A$	indicator function of the set A
$\stackrel{a.s.}{\to}$	convergence with probability one
$\stackrel{p}{\to}$	convergence in probability
$\stackrel{d}{\to}$	convergence in distribution
$\stackrel{d}{=}$	equality of distributions
γ_1	the coefficient of skewness
γ_2	the coefficient of kurtosis (excess kurtosis)
$\Gamma(\alpha)$	the gamma function, $\Gamma(\alpha) = \int_0^\infty x^{\alpha-1}e^{-x}dx$

$\kappa_n(X)$	the nth cumulant of a random variable X
χ^2	chi-square distribution
$\mu_n(X)$	the nth central moment of a random variable X
ν_p	geometric random variable with mean $1/p$
δ_{ij}	Kronecker's symbol: 1 if $i = j$; 0 if $i \neq j$

Part I

Univariate Distributions

1
Historical Background

Over 75 years ago in a paper that appeared in the 1923 issue of the *Journal of American Statistical Association* (pp. 841–852) entitled "First and Second Laws of Error," the late professor and head of vital statistics at the Harvard School of Public Health, Edwin Bidwell Wilson (1879–1964)[1] concurs with economics professor W.L. Crum's conclusions expressed in a paper published in the same journal in March 1923, entitled "The Use of the Median in Determining Seasonal Variation" (pp. 607–614) that "a good many series of data from economic sources probably may be better treated by the median than by the mean." These remarks may be viewed as revolutionary at the period of unquestionable dominance of the arithmetic mean and normal distribution in statistical theory. E.B. Wilson reminds us that the first two laws of error both originated with P.S. Laplace. The first law, presented in 1774, states that the frequency of an error could be expressed as an exponential function of the numerical magnitude of the error, disregarding sign, or equivalently that the logarithm of the frequency of an error (without regard to sign) is a linear function of the error.

The second law (proposed four years later in 1778) states that the frequency of the error is an exponential function of the *square* of the error, or equivalently that the logarithm of the frequency is a quadratic (parabolic) function of the error. See Figure 1.1.

The second Laplace law is usually called the normal distribution or the Gauss law. Wilson—among several other scholars—doubts the attribution of that law to Gauss and remarks that Gauss "in spite of his well-known precocity had probably not made his discovery before he was two years old." He notes that there are excellent mathematical reasons for the far greater attention that has been paid to the second law, since it involves the variable x^2 (if x is the error) and this is "subject to all the laws of elementary mathematical analysis," while the first law involving the absolute value of the error x is not an analytic function and presents considerable mathematical difficulty in its manipulation.

Next, however, E.B. Wilson states that the frequency we actually meet in everyday work in economics, biometrics, or vital statistics often fails to conform closely to the so-called normal

[1]Wilson's name is known to many statisticians in view of the so-called Wilson–Hilferty transformation (see Wilson, E.B. and Hilferty, M.M., *Proc. Nat. Acad. Sci.*, **17**, pp. 684–688)—a device that allows the use of a normal approximation for chi-square probabilities.

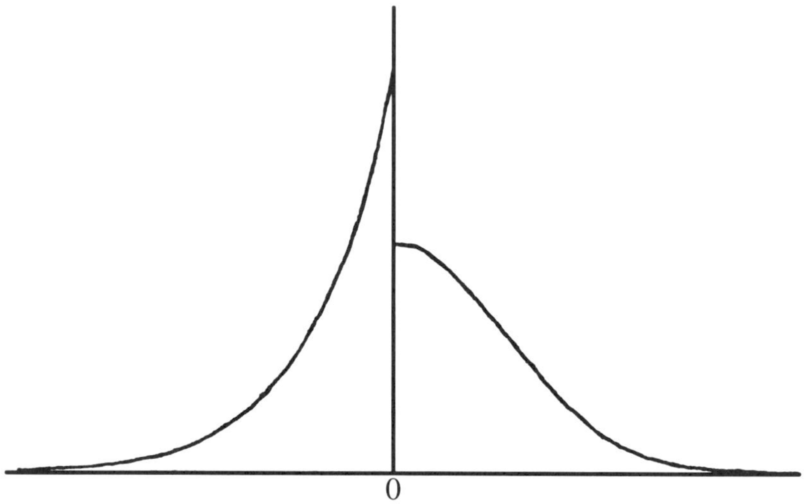

Figure 1.1: On the left, Laplace's first frequency curve $F = \frac{k}{2}e^{-k|x|}$. On the right, Laplace's second (Gauss's) frequency curve: $F = \frac{1}{\sqrt{2\pi}\sigma}e^{-x^2/2\sigma^2}$. Each curve should be reproduced symmetrically on the other side of the central vertical line. The figure is taken from Wilson's 1923 paper. Reprinted with permission from the *Journal of the American Statistical Association*. Copyright 1923 by the American Statistical Association. All rights reserved.

distribution. He points out that the fact that in extraordinarily precise measurements of astronomy of position the errors are dispersed about the mean in accordance with the Gauss law and that the dispersion of shots in artillery and small arms practice are covered very well by the generalization of this law are "no justification for attempting to force the (normal) law with its various generalizations upon the data for which it is not fitted." Wilson emphasizes that it is important to examine the actual data for the purpose of determining the proper statistical treatment, and "it is by no means safe to rush ahead and apply the second law of Laplace or the various extensions of it developed by the Scandinavian School on the one hand (Gram, Charlier) or the (British) Biometric School (Pearson, Yule) on the other." He analyzes the example provided by Crum (Table 1.1).

He also notes that for the normal distribution if e_i denotes a deviation[2] from a mean and S_1 denotes the mean deviation, S_2 denotes the mean square deviation, etc., namely,

$$nS_1 = \sum e_i, \qquad nS_2^2 = \sum e_i^2,$$

$$nS_3^3 = \sum e_i^3, \qquad nS_4^4 = \sum e_i^4,$$

the ratios S_i ought to satisfy

$$S_1 : S_2 : S_3 : S_4 = 1.000 : 1.253 : 1.465 : 1.645.$$

Commenting on these ratios, Wilson echoes and modifies Bertrand's famous dictum ("if these equalities are not satisfied — someone has retouched and altered the immediate results of experiment") and asserts that "when confronted with data that do not satisfy this continued proportion — it is very obvious that the data are not distributed in frequency according to the second law (with some

[2]Deviation here means absolute deviation.

Deviation	Frequency	Deviation	Frequency	Deviation	Frequency
*Over −30	2	−11	6	6	13
−30	1	−10	3	7	8
−29	1	−9	5	8	6
−28	1	−8	11	9	5
−24	1	−7	6	10	2
−23	1	−6	23	11	4
−22	1	−5	10	12	3
−21	2	−4	13	13	1
−20	1	−3	19	14	2
−19	2	−2	9	15	1
−18	2	−1	11	16	1
−17	2	0	28	17	1
−16	2	1	22	18	2
−15	1	2	22	23	1
−14	3	3	13	24	1
−13	6	4	19	28	1
−12	3	5	13	†Over 30	7

*−32, −37. †34, 35, 35, 41, 41, 42, 45.

Table 1.1: Crum's data: Frequencies of deviations from the medians ($N = 324$ total frequency).

latitude of departure from the straight proportion must be permitted)." Now for the data supplied by Crum, we have approximately

$$S_1 = 7.0,\ S_2 = 10.3,\ S_3 = 13.8,\ S_4 = 17.0;$$

thus the ratios are

$$1 : 1.5 : 2.0 : 2.4,$$

a far cry from those to be obeyed based on the normal distribution. The spread is just too wide, and no reasonable allowance for the behavior of probable errors can produce such great a spread.

On the other hand, applying the first law of Laplace, where the frequency varies as e^{-kd} (d is the numerical value of the deviation), we obtain after some "annoying" calculations involving calculus the theoretical values

$$S_1 : S_2 : S_3 : S_4 = 1.000 : 1.414 : 1.817 : 2.213.$$

Wilson justifiably asserts that the distribution in frequency of the data is much nearer Laplace's first law than the second, and it is no longer reasonable to maintain that the differences are within the presumptive errors due partly to the scarcity and irregularity of material.

However there is "a little evidence" that the observations are *more* dispersed than they would be even under the first law. To account for possible asymmetry Wilson suggests the classical graphical method representing the frequency law as

$$f = \frac{1}{2} N \kappa e^{-\kappa x},$$

where $N = 324$, the deviation is x and the number n of deviations beyond a given value x is

$$n = \int_x^\infty f dx = \frac{1}{2} N e^{-\kappa x}.$$

Hence,

$$\log_{10} n = \log_{10}\left(\frac{1}{2}N\right) - (\kappa \log_{10} e)x$$

plots as a straight line on the so-called arith-log paper with x as abscissa and n the ordinate. Since for the first law of Laplace $\kappa = 1/\theta$, where θ is the mean deviation, it is reasonable to choose $\theta = S_1 = 7.0$. (The values of θ calculated from the four S's are 7.0, 7.3, 7.5, and 7.7.) A fair representation of the distribution of the data is given by $f = 23e^{-x/7}$ (recall that we are using the absolute value of x — the numerical value of deviation) and the arith-log chart constructed for the first Laplacian law like the probability chart for the Gaussian law was also based on the total integrated frequency outside a certain limit. Figure 1.2 presents a probability plot — a chart in which the ordinates are the percentage of deviations that are *less* than (left scale) or *greater* than (right scale) a given deviation plotted as an abscissa — under the assumption of the Gaussian law, namely if the Gaussian law had been followed the line would have been straight.

Evidently the Gaussian fit is inadequate. The straight line fitted to the four central points results in no deviation in the observations greater than +20 and smaller than −19. For comparison the arith-log chart constructed for the first Laplacian law is presented in Figure 1.3. On this chart, the points $(\hat{n}, x)$ are represented for the number of empirical deviations $\hat{n}$ beyond x and compared to the graph of $\log_{10} n = \log_{10}(N/2) - (\kappa \log_{10} e)x$. Examining the chart, Wilson asserts: "This chart shows on arith-log paper the number of deviations as ordinates greater than the values given as abscissae. If Laplace's first law holds, the points should lie on a straight line. The lowest set of points and the lowest line are for the negative deviations (left scale), and for them the law holds as well as could be desired. The top line and set of points are for the positive deviations; the fit to the straight dotted line is bad (right scale). The middle line and set of points are for positive and negative deviations taken together (left scale) without regard to sign, and the fit is fair — better than for the (Gaussian) curve" (in Figure 1.2).

Wilson concludes by stating that these data give internal evidence of following Laplace's first law instead of his second law and should be fitted to that law.

In spite of the prestige of the journal in which the paper appeared and the prominence of the author, Wilson's plea remained a call in the wilderness for over five decades and only recently attention has been shifted to Laplace's first law, known as the *Laplace distribution* or occasionally *double exponential distribution*, as a candidate for fitting data in economics and health sciences.

For many years the Laplace distribution was a popular topic in probability theory due to the simplicity of its characteristic function and density, the curious phenomenon that a random variable with only slightly different characteristic function loses the simplicity of the density function and other numerous attractive probabilistic features enjoyed by this distribution.

Perhaps one of the earliest sources in which the Laplace distribution is discussed as a law of errors in the English language is the 1911 paper by the famous economist and probabilist J.M. Keynes in the *Journal of the Royal Statistical Society*, **74**, New Series, pp. 322–331.

With his usual lucidity, Keynes discusses the probability of a measurement x_q, assuming the real (actual) value to be a_s, as an algebraic function $f(x_q, a_s)$, the same function for all values of x_q and a_s "within the limits of the problem." The task is to find the value of a_s, namely x, which maximizes

$$\prod_{q=1}^{m} f(x_q, x).$$

This is equivalent to solving

$$\sum_{q=1}^{m} \frac{f'(x_q, x)}{f(x_q, x)} = 0$$

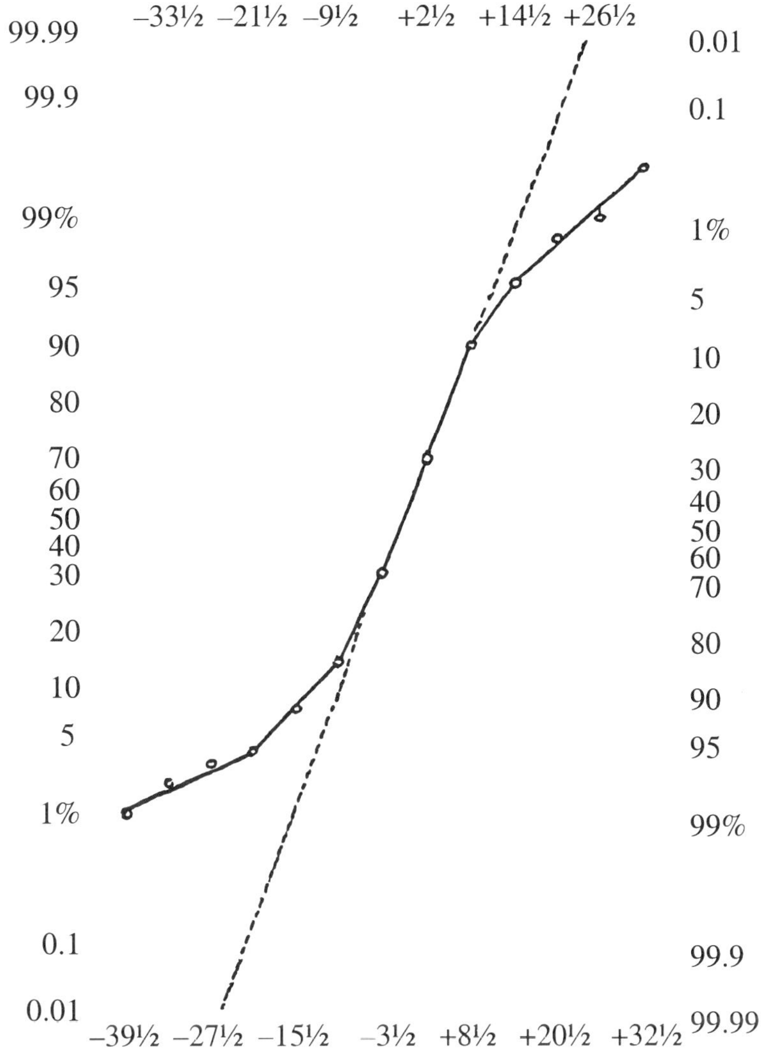

Figure 1.2: Probability plot for Crum's data discussed in Wilson's article. Reprinted with permission from the *Journal of the American Statistical Association*. Copyright 1923 by the American Statistical Association.

or $\sum f_q'/f_q = 0$ for brevity. Now, the law of errors determines the *form* of $f(x_q, x)$ and the form of $f(x_q, x)$ determines the algebraic relation $\sum f_q'/f_q = 0$ between the measurements and the most probable value. Keynes analyzes several situations.

1. If the most probable value of the quantity is equal to the arithmetic mean of measurements $\frac{1}{m}\sum_{q=1}^{m} x_q$, then $\sum f_q'/f_q = 0$ is equivalent to $\sum(x - x_q) = 0$. Thus, f_q'/f_q can be written as $\Phi''(x)(x - x_q)$, where $\Phi''(x)$ is a nonzero function independent of x_q. Integrating, we get

$$\log f_q = \Phi'(x)(x - x_q) - \Phi(x) + \Psi(x_q),$$

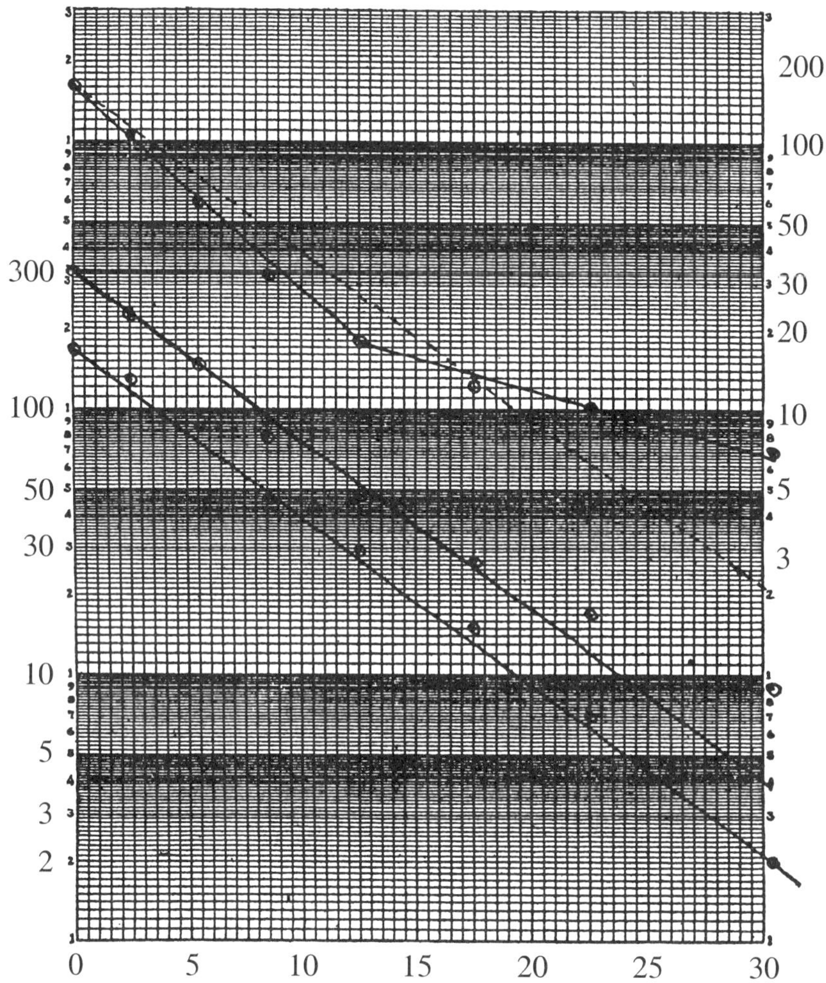

Figure 1.3: Arith-log chart for the first Laplace law using Crum's data. Reprinted with permission from the *Journal of the American Statistical Association*. Copyright 1923 by the American Statistical Association. All rights reserved.

where $\Psi(x_q)$ is a function independent of x. Thus,

$$f_q = e^{\Phi'(x)(x-x_q)-\Phi(x)+\Psi(x_q)}.$$

Setting $\Phi(x) = -\kappa^2 x^2$ and $\Psi(x_q) = -\kappa^2 x_q^2 + \log A$ we obtain

$$f_q = Ae^{-\kappa^2(x-x_q)^2} = Ae^{-\kappa^2 y_q^2},$$

(where y_q is the absolute magnitude of the error in the measurement x_q) the so-called normal law.

Keynes emphasizes that this is only one "amongst a number of possible solutions" but notes that with one additional assumption this is the only law of error leading to the arithmetic mean.

The assumption is that negative and positive errors of the same absolute amount are equally likely.

Indeed in that case f_q will be of the form $Be^{\theta([x-x_q]^2)}$, where $\theta([x-x_q]^2)$ is the value of a certain real function θ evaluated at $(x-x_q)^2$. We have

$$\Phi'(x)(x-x_q) - \Phi(x) + \Psi(x_q) = \theta([x-x_q]^2)$$

or

$$\Phi''(x) = 2\frac{d}{d(x-x_q)^2}\theta([x-x_q]^2)$$

and

$$\frac{d}{d(x-x_q)^2}\theta([x-x_q]^2) = -\kappa^2,$$

where κ is a constant since $\Phi''(x)$ is independent of x_q. Thus,

$$\theta([x-x_q]^2) = -\kappa^2(x-x_q)^2 + \log C$$

and

$$f_q = Ae^{-\kappa^2(x-x_q)^2},$$

with $A = BC$.

2. Next Keynes discusses in detail the case of the law of error if the geometric mean of the measurements leads to the most probable value of the quantity. This yields

$$f_q = A\left(\frac{x}{x_q}\right)^{\kappa x} e^{-\kappa x}.$$

Keynes then compares this with the earlier derivation obtained by D. McAlister in the *Proceedings of the Royal Society*, **29** (1879), p. 365:

$$f_q = Ae^{-\kappa^2 \log^2(x_q/x)},$$

the well-known log-normal law.

He also notes that J.C. Kapteyn in his monograph *Skew Frequency Curves*, Astronomical Laboratory, Groningen (1903), obtained a similar result.

3. Next he discusses the law of errors implied by the harmonic mean leading to

$$f_q = Ae^{-\kappa^2 y_q^2/x_q}.$$

Here positive and negative errors of the same absolute magnitude are not equally likely.

4. Keynes now poses the question:

If the most probable value of the quantity is equal to the median of measurements, what is the law of error?

For this purpose he defines the median of observations and notes its property originally proved by G.T. Fechner (1801–1887),[3] who first introduced median into use: "If x is the median of a

[3] In his book *Kollektivmasslehre*, W. Englemann, Leipzig, 1897.

number of magnitudes, the sum of the absolute differences (i.e., the difference always reckoned positive) between x and each of the magnitudes is a minimum." Now write $|x - x_q| = y_q$. Since $\sum_{i=1}^{m} y_q$ is to be minimum we must have $\sum_{q=1}^{m} \frac{x - x_q}{y_q} = 0$. Whence proceeding as before, we have

$$f_q = Ae^{\int \frac{x - x_q}{y_q} \Phi''(x)dx + \Psi(x_q)}.$$

The simplest case of this is obtained by putting

$$\Phi''(x) = -k^2, \qquad \Psi(x_q) = \frac{x - x_q}{y_q} k^2 x_q,$$

whence

$$f_q = Ae^{-k^2|x - x_q|} = Ae^{-k^2 y_q}.$$

This satisfies the additional condition that positive and negative errors of equal magnitude are equally likely. Thus in this important respect the median is as satisfactory as the arithmetic mean, and the law of error that leads to it is as simple. It also resembles the normal law in that it is a function of the error *only*, not of the magnitude of the measurement as well.

Keynes's (1911) analysis of Laplace's contribution to the first law of error is worth reproducing verbatim.

"The median law of error, $f_q = Ae^{-k^2 y_q}$, where y_q is the absolute amount of the error always reckoned positive, is of some historical interest, because it was the earliest law of error to be formulated. The first attempt to bring the doctrine of averages into definite relation with the theory of probability and with laws of error was published by Laplace in 1774 in a memoir '*Sur la probabilité des causes par les événements*.'[4] This memoir was not subsequently incorporated in his *Théorie Analytique* and does not represent his more mature view. In the *Théorie* he drops altogether the law tentatively adopted in the memoir, and lays down the main lines of the investigation for the next hundred years by the introduction of the *normal* law of error. The popularity of the normal law, with the arithmetic mean and the method of least squares as its corollaries has been very largely due to its overwhelming advantages, in comparison with all other laws of error, for the purposes of mathematical development and manipulation. And in addition to these technical advantages, it is probably applicable as a first approximation to a larger and more manageable group of phenomena than any other single law.[5] So powerful a hold indeed did the normal law obtain on the minds of statisticians that until quite recent times only a few pioneers have seriously considered the possibility of preferring in certain circumstances other means to the arithmetic and other laws of error to the normal. Laplace's earlier memoir fell, therefore, out of remembrance. But it remains interesting, if only for the fact that a law of error there makes its appearance for the first time."

Laplace (1794) sets himself the problem in a somewhat simplified form:

"*Déterminer le milieu que l'on doit prendre entre trois observations données d'un même phénomène*." He begins by assuming a law $y = \phi(x)$ for an error, where y is the probability of an error x; and finally by means of a number of *somewhat arbitrary assumptions* (our emphasis), arrives at the result $\phi(x) = (m/2)e^{-mx}$. If this formula is to follow from his arguments, x must denote the *absolute* error, always taken positive. It is unlikely that Laplace was led to this result by considerations other than those by which he attempts to justify it.

[4] *Mémoires présentés à l'Académie des Sciences Paris*, **vi**, pp. 621–656.

[5] We would add that the Central Limit Theorem should also be credited for this popularity.

"Laplace, however did not notice that his law of error led to the median. For instead of finding the most probable value, which would have led him straight to it, he seeks the "mean of error"—the value, that is to say, which the true value is as likely to fall short of as to exceed. This value is, for the median law, laborious to find and awkward in the result. Laplace works it out correct for the case where the observations are no more than three."

5. Finally Keynes deals with the case where the law of errors leads to a mode without providing an explicit solution and concludes with a discussion of the most general form of the law of errors when it is assumed that positive and negative errors of the same magnitude are equally probable.

   He emphasizes that the most general form leading to the median is

$$f_q = Ae^{\Phi'(x)\frac{x-x_q}{y_q}+\Psi(x_q)},$$

   where f_q is the probability of a measurement x_q given that the true value is x.

Stigler (1986a) provides a somewhat different assessment of Laplace's 1774 memoir. He presents an English translation of the memoir (whose English title is *Probability of the Causes of Events*) and points out that Laplace was just 25 years old when the memoir appeared and that it was his first substantial work in mathematical statistics.

For readers interested in history, it is worthwhile to reproduce Laplace's elegant and ingenious derivation of what is now referred to as the Laplace distribution. We reproduce his illustrative Figure 2 depicting his error distribution (our Figure 1.4). Here V represents the true value of the location parameter (in modern terminology). Denoting by $\phi(x)$ the probability density of the deviation x of an observation from V, in his attempt to determine this function Laplace argues as follows:

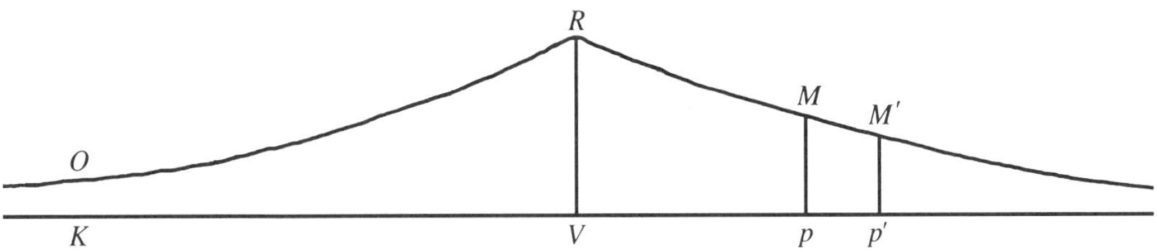

Figure 1.4: An illustration (Figure 2) from Laplace's 1774 memoir.

"But of an infinite number of possible functions, which choice is to be preferred? The following considerations can determine a choice. It is true (Figure 1.4) that if we have no reason to suppose the point p more probable than the point p', we should take $\phi(x)$ to be constant, and the curve ORM' will be a straight line infinitely near the axis Kp. But this supposition must be rejected, because if we suppose there existed a very large number of observations of the phenomenon, it is presumed that they become rarer the farther they are spread from the truth. We can also easily see that this diminution cannot be constant, that it must become less as the observations deviate more from the truth. Thus not only the ordinates of the curve RMM', but also the differences of these ordinates must decrease as they become further from the point V, which in this Figure we always suppose to be the true instant of the phenomenon. Now, as we have no reason to suppose a different law for the ordinates than for their differences,[6] it follows that we must, subject to the rules of probabilities, suppose the ratio of

[6]It is important to note that Laplace is talking here about the difference of the probability density function, not of the observations, i.e., this crucial assumption does not impose that the difference of observations should be distributed in the same way as observations themselves (which is *not* true for the Laplace distribution).

two infinitely small consecutive differences to be equal to that of the corresponding ordinates. We thus will have

$$\frac{d\phi(x+dx)}{d\phi(x)} = \frac{\phi(x+dx)}{\phi(x)}.$$

Therefore

$$\frac{d\phi(x)}{dx} = -m\phi(x),$$

which gives $\phi(x) = Ce^{-mx}$. Thus, this is the value that we should choose for $\phi(x)$. The constant C should be determined from the supposition that the area of the curve ORM equals unity which represents certainty, which gives $C = m/2$. Therefore $\phi(x) = (m/2)e^{-mx}$, e being the number whose hyperbolic logarithm is unity. One can object that this law is repugnant in that if x is supposed extremely large, $\phi(x)$ will not be zero, but to this I reply that while e^{-mx} indeed has a real value of all x, this value is so small for x extremely large that it can be regarded as zero."

Keynes quite justifiably mentions "a number of somewhat arbitrary assumptions" in Laplace's argument. Nevertheless the argument involves several potent ideas. Books by Stigler (1986b) and Hald (1995), and also an article by Eisenhart (1983) contain more rigorous derivations as well as valuable revealing comments.

An interesting "applied" genesis of the Laplace distribution was presented in Mantel and Pasternack (1966) [see also Rohatgi (1984), Example 4, p. 482]. We present it together with a representation of Laplace random variables as the determinant of a random matrix.

Let X_1 and X_2 represent the lifetimes of two identical independent components, an original and its replacement. Suppose that we require the probability that the replacement outlasts the original component. Thus

$$P(X_2 > X_1) = P(X_2 - X_1 > 0) = 1/2.$$

Let us assume that lifetimes are distributed exponentially with common mean λ and compute the density of $Z = X_2 - X_1$. Since Z is a symmetric random variable it is enough to compute the density for $z > 0$. For $z > 0$, the density of the difference of X_2 and X_1 is given by

$$f_Z(z) = \int_0^\infty (\lambda^{-1}e^{-x_1/\lambda})(\lambda^{-1}e^{-(z+x_1)/\lambda})dx_1 = (2\lambda)^{-1}e^{-z/\lambda},$$

and thus for $z \in \mathbb{R}$

$$f_Z(z) = (2\lambda)^{-1}e^{-|z|/\lambda}.$$

We have a verbal proof of this result. Consider two idealized light bulbs in use simultaneously. We are interested in the distribution of the difference in their failure times. Once one bulb fails, the remaining bulb, being as good as new, will have a remaining lifetime given by the standard waiting time distribution (exponential). With probability 1/2, the first failure will correspond either to the first or the second lifetime distribution (exponentials) so that the difference in failure times will be positive or negative with equal probabilities and in each case with absolute value following the standard waiting time distribution, i.e., the *exponential*.

Since a standard exponential random variable multiplied by two has the chi-square distribution with two degrees of freedom, the arguments above show that Z is distributed as half of the difference of two independent chi-square random variables each with two degrees of freedom.

On the other hand, if Z_1, Z_2, Z_3, and Z_4 are independent standard normal random variables, it is easy to see that the distribution of Z is the same as that of $Z_1Z_2 + Z_3Z_4$. Indeed, $U_1 = (X_1 + X_2)/2$, $U_2 = (X_1 - X_2)/2$, $U_3 = (X_3 + X_4)/2$, $U_4 = (X_3 - X_4)/2$ are independent normal variables with variance $1/2$. Thus,

$$X_1X_2 + X_3X_4 = (U_1^2 + U_3^2) - (U_2^2 + U_4^2),$$

which has the same distribution as a difference of two independent χ^2 (chi-square) random variables each with two degrees of freedom.

In general, sums or differences of n normal products — each of two factors— will be distributed like $1/2$ of the differences of two independent χ^2 each with n degrees of freedom, and if n is even this is an *$n/2$-fold convolution of the Laplace distribution*. These sums of n products correspond to the *sample* covariance for bivariate normal samples when the correlation is zero.

To recapitulate, the error distribution, nowadays referred to as the Laplace distribution or the double exponential distribution, originated in Laplace's 1774 memoir. Historically, it was the first continuous distribution of unbounded support. Although since its introduction the distribution has been occasionally recommended as a better fit to certain data, its popularity is unjustifiably by far less than that of its four-years-older "sibling" — Laplace's second law of error — better known in the English language literature as the Gaussian (normal) law.

This monograph is devoted to collecting and presenting properties, generalizations, and applications of the Laplace distribution with a tacit aim of demonstrating that it is a natural and sometimes superior alternative to the normal law. We hope to convince readers that this class of distributions deserves more attention than it has received until very recently.

2

Classical Symmetric Laplace Distribution

In the course of our study of the Laplace distribution and its generalizations we have noticed that quite often in the statistical literature this distribution is used not on its own merits but as a source for counterexamples for other (mainly normal) distributions. It would seem that it has been created solely to provide examples of curiosity, nonregularity, and pathological behavior. In studies with probabilistic content, the distribution serves as a tool for limiting theorems and representations with the emphasis on analyzing its differences from the classical theory based on the "sound" foundations of normality. One gets the impression that the "sharp needle" at the origin of the Laplace distribution where the bulk of the density is concentrated generates a ripple effect that affects the behavior over its whole support including the tails.[7] These observations prompted us to initiate a detailed study of the Laplace distribution on *its own* merits without constant intruding comparisons and analogues.

In Table 2.1 and Figure 2.1, reproduced from Chew (1968), we present definitions and graphs of the six classes of symmetric-about-zero, single-parameter distributions: uniform, triangular, cosine, logistic, Laplace, and normal. Values of the distribution functions are given in Table 2.2. The graphs of their densities for cases of the unit variance convincingly demonstrate the basic features and, in particular, the special position of the Laplace distribution with its towering peak and heavy tails.

Leptokurtic tendencies (see Section 2.1.3.4 for more details) are frequently found among measurements of superior quality and homogeneity. A leptokurtic Laplace curve presents a visible peak: in the vicinity of the center there is a certain excess of (small) elements. As the area under the curve is the same as the area under the normal curve, the peak is counterbalanced by a corresponding diminution of frequencies in the intermediate regions further from the center (tails). Generally, there is an overcompensation so that the leptokurtic curve crosses the normal curve four times, first near the peak and then again at the tails, and tends toward the x-axis by staying slightly above the normal curve.

---

[7] Tails of a random variable X are the probabilities $P(X < -x)$ and $P(X > x)$, $x > 0$. The asymptotic behavior of these functions of x is often referred to as the tail behavior of X, or its distribution.

NAME	DENSITY FUNCTION	DISTRIBUTION FUNCTION	VARIANCE
UNIFORM	$\frac{1}{2a}$, $x \in (-a, a)$ 0, elsewhere	0, $x \le -a$ $\frac{a+x}{2a}$, $x \in (-a, a)$ 1, $x \ge a$	$\frac{a^2}{3}$
TRIANGULAR	$\frac{b+x}{b^2}$, $x \in [-b, 0]$ $\frac{b-x}{b^2}$, $x \in (0, b]$ 0, elsewhere	0, $x < -b$ $\frac{(b+x)^2}{2b^2}$, $x \in [-b, 0]$ $1 - \frac{(b-x)^2}{2b^2}$, $x \in (0, b]$ 1, $x > b$	$\frac{b^2}{6}$
COSINE	$\frac{1+\cos x}{2\pi}$, $x \in [-\pi, \pi]$ 0, elsewhere	0, $x < -\pi$ $\frac{\pi+x+\sin x}{2\pi}$, $x \in [-\pi, \pi]$ 1, $\pi \le x$	$\frac{\pi^2-6}{3}$
LOGISTIC	$\frac{\operatorname{sech}^2(x/d)}{2d}$	$\frac{1}{1+e^{-2x/d}}$	$\frac{(\pi d)^2}{12}$
LAPLACE	$ce^{-2c\|x\|}$	$e^{2cx}/2$, $x < 0$ $1 - e^{-2cx}$, $x \ge 0$	$\frac{1}{2c^2}$
NORMAL	$\frac{1}{\sqrt{2\pi}} e^{-x^2/2}$	$\int_{-\infty}^{x} \frac{1}{\sqrt{2\pi}} e^{-u^2/2} du$	1

Table 2.1: Densities and distribution functions of some symmetrical probability distributions. Reproduced from Chew (1968). Reprinted with permission from *The American Statistician*.

2.1 Definition and basic properties

2.1.1 Density and distribution functions. The *classical Laplace distribution* (also known as the *first law of Laplace*) is a probability distribution on $(-\infty, \infty)$, given by the density function

$$f(x; \theta, s) = \frac{1}{2s} e^{-|x-\theta|/s}, \quad -\infty < x < \infty, \tag{2.1.1}$$

where $\theta \in (-\infty, \infty)$ and $s > 0$ are location and scale parameters, respectively [see, e.g., Ord (1983) and Johnson et al. (1995)]. As discussed in some detail in Chapter 1, it was named after Pierre-Simon Laplace (1749–1827), who in 1774 obtained (2.1.1) as the distribution whose likelihood is maximized when the location parameter is set to the median. As mentioned in Chapter 1 and discussed further in Section 2.2, the Laplace distribution is also known as the law of the difference between two exponential random variables. Consequently, it is also known as *double exponential distribution*,[8] as well as the *two-tailed exponential distribution* [see, e.g., Greenwood et al. (1962)] and the *bilateral exponential law* [see, e.g., Feller (1971)].

---

[8]Note that this name is also used for the extreme value distribution with density $\exp(-\exp(-x))$, $x > 0$, as well as for a distribution from the exponential family studied by Efron (1986). The term *double exponential fitness function* for the probabilities $p = \exp(-\exp(\alpha_0 + \alpha_1 x_1 + \cdots + \alpha_n x_n))$ is common in biostatistic literature [see, e.g., Manly (1976)]. Johnson et al. (1995) recommend calling the extreme value distribution *doubly exponential law*.

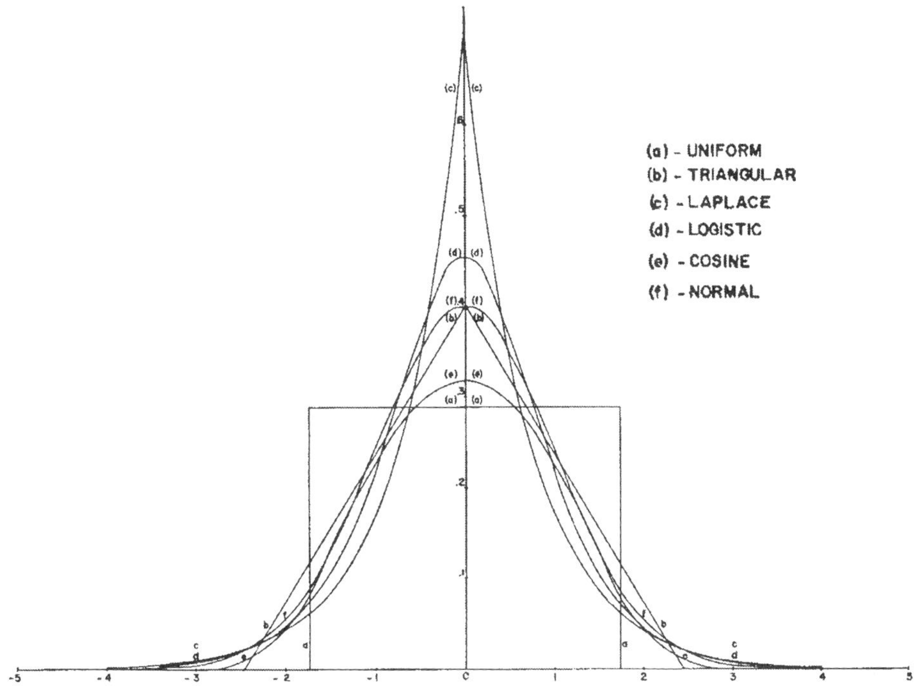

Figure 2.1: Graphs of density functions of several symmetrical populations. Reproduced from Chew (1968). Reprinted with permission from *The American Statistician*.

x	Normal	Logistic	Laplace	Cosine	Triangular
0.0	0.5000	0.5000	0.5000	0.5000	0.5000
0.2	0.5793	0.5897	0.6238	0.5720	0.5785
0.4	0.6554	0.6738	0.7160	0.6422	0.6501
0.6	0.7257	0.7480	0.7860	0.7088	0.7151
0.8	0.7881	0.8102	0.8387	0.7702	0.7734
1.0	0.8413	0.8598	0.8784	0.8252	0.8250
1.2	0.8849	0.8981	0.9084	0.8728	0.8699
1.4	0.9192	0.9269	0.9310	0.9122	0.9082
1.6	0.9452	0.9480	0.9480	0.9436	0.9399
1.8	0.9641	0.9632	0.9608	0.9670	0.9649
2.0	0.9772	0.9741	0.9704	0.9832	0.9832
2.2	0.9861	0.9818	0.9777	0.9931	0.9948
2.4	0.9918	0.9873	0.9832	0.9982	0.9998
2.6	0.9953	0.9911	0.9873	0.9998	1.0000
2.8	0.9974	0.9938	0.9906	1.0000	
3.0	0.9987	0.9957	0.9928		
3.2	0.9993	0.9970	0.9946		
3.4	0.9997	0.9979	0.9959		
3.6	0.9998	0.9985	0.9969		
3.8	0.9999	0.9990	0.9977		
4.0	1.0000	0.9993	0.9983		

Table 2.2: Values of distribution functions of selected distributions. The values of x are in multiples of standard deviation.

It is easy to verify that the variance of (2.1.1) is equal to $2s^2$. Thus the variance of the *standard classical* Laplace distribution, which has the density

$$f(x; 0, 1) = \frac{1}{2}e^{-|x|}, \quad -\infty < x < \infty, \tag{2.1.2}$$

is equal to 2. For various derivations it would seem convenient to consider a reparametrization of Laplace densities

$$g(x; \theta, \sigma) = \frac{1}{\sqrt{2}\sigma}e^{-\sqrt{2}|x-\theta|/\sigma}, \quad -\infty < x < \infty. \tag{2.1.3}$$

In this case the standard Laplace distribution is given by setting $\theta = 0$ and $\sigma = 1$. Here the variance is equal to 1 and the density is of the form

$$g(x; 0, 1) = \frac{1}{\sqrt{2}}e^{-\sqrt{2}|x|}, \quad -\infty < x < \infty. \tag{2.1.4}$$

To distinguish between these two parametrizations we shall refer to the *classical* Laplace $\mathcal{CL}(\theta, s)$ and the *standard* classical Laplace $\mathcal{CL}(0, 1)$ distributions in the cases given by (2.1.1) and (2.1.2), and to Laplace $\mathcal{L}(\theta, \sigma)$ and standard (actually standardized) Laplace $\mathcal{L}(0, 1)$ distributions in the cases represented by (2.1.3) and (2.1.4), respectively. We shall also retain the difference in notation for the scale parameter by reserving s for classical Laplace distributions and σ for those given by (2.1.3). Therefore, reformulating any result from one parametrization to the other is a matter of replacing s by $\sigma/\sqrt{2}$ or σ by $\sqrt{2}s$. In Figure 2.2 we present graphs of the standard classical and the standard Laplace densities.

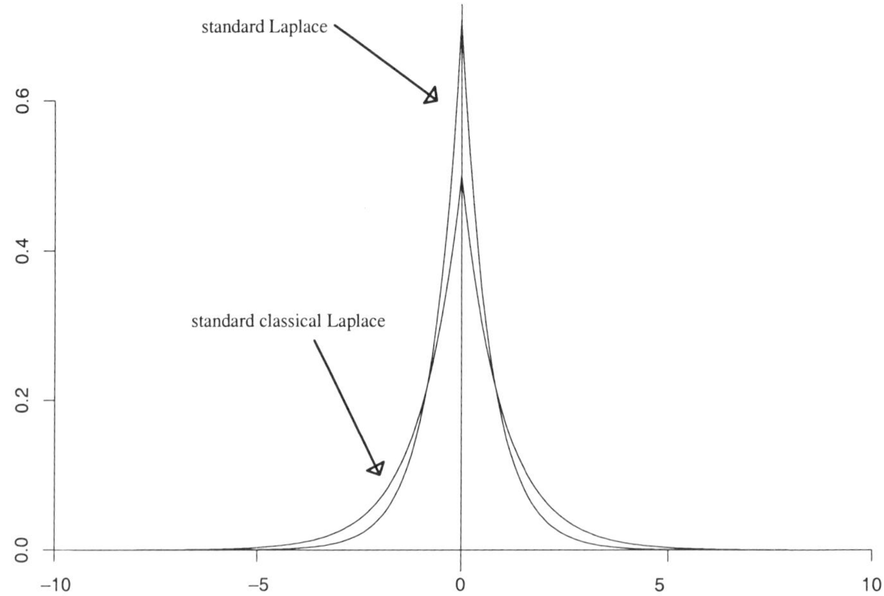

Figure 2.2: Standard classical Laplace [equation (2.1.2)] and standard Laplace [equation (2.1.4)] density functions.

The cumulative distribution function (c.d.f.) corresponding to density (2.1.1) is

$$F(x; \theta, s) = \begin{cases} \frac{1}{2}e^{-|x-\theta|/s} & \text{if } x \le \theta, \\ 1 - \frac{1}{2}e^{-|x-\theta|/s} & \text{if } x \ge \theta. \end{cases} \tag{2.1.5}$$

The distribution is symmetric about θ, i.e., for any real x we have

$$f(\theta - x; \theta, s) = f(\theta + x; \theta, s) \text{ and } F(\theta - x; \theta, s) = 1 - F(\theta + x; \theta, s). \tag{2.1.6}$$

Consequently, the mean, median, and mode of this distribution are all equal to θ.

2.1.2 Characteristic and moment generating functions. The characteristic function (ch.f.) corresponding to the standard classical Laplace $\mathcal{CL}(0, 1)$ random variable (r.v.) X with density (2.1.2) is

$$\psi_X(t) = E[e^{itX}] = \int_{-\infty}^{\infty} e^{itx}\frac{1}{2}e^{-|x|}dx = (1+t^2)^{-1}, \quad -\infty < t < \infty. \tag{2.1.7}$$

For the general classical Laplace r.v. Y with the distribution $\mathcal{CL}(\theta, s)$ we have $Y \stackrel{d}{=} sX + \theta$. Thus

$$\psi_Y(t) = E[e^{it(sX+\theta)}] = e^{it\theta}\psi_X(st) = \frac{e^{it\theta}}{1+s^2t^2}, \quad -\infty < t < \infty. \tag{2.1.8}$$

It is well known but nevertheless a curious fact that the pair of Fourier transforms (2.1.2) and (2.1.7) occur in reverse order for the Cauchy distribution. Namely, the standard Cauchy distribution with density

$$f_c(x) = \frac{1}{\pi(1+x^2)}, \quad -\infty < x < \infty$$

has the characteristic function given by

$$\phi_c(t) = e^{-|t|}, \quad -\infty < t < \infty.$$

The moment generating function of standard classical Laplace r.v. X with density (2.1.2) is

$$M_X(t) = E[e^{tX}] = \int_{-\infty}^{\infty} e^{tx}\frac{1}{2}e^{-|x|}dx = (1-t^2)^{-1}, \quad -1 < t < 1. \tag{2.1.9}$$

For the general classical Laplace r.v. Y with density (2.1.1), we have

$$M_Y(t) = e^{t\theta}M_X(st) = \frac{e^{t\theta}}{1-s^2t^2}, \quad -\frac{1}{s} < t < \frac{1}{s}. \tag{2.1.10}$$

Consequently, the cumulant generating functions, $\log M_Y(t)$ and $\log M_X(t)$, corresponding to (2.1.1) and (2.1.2), are

$$t\theta - \log(1 - s^2t^2) \text{ and } -\log(1-t^2), \tag{2.1.11}$$

respectively.

2.1.3 Moments and related parameters.

2.1.3.1 *Cumulants.* The nth cumulant of a classical Laplace r.v. X, denoted κ_n, is defined as the coefficient of $t^n/n!$ in the Taylor expansion (about $t = 0$) of the cumulant generating function of X. Formulas (2.1.11) for the cumulant generating function generate the cumulants of Laplace distributions in a straightforward manner. Indeed, using the Taylor expansion of $\log(1-z)$ about $z = 0$, we have

$$-\log(1-t^2) = \sum_{k=1}^{\infty}\frac{t^{2k}}{k}.$$

Thus, for the standard classical Laplace r.v. X given by (2.1.2), we have

$$\kappa_n(X) = \begin{cases} 0 & \text{if } n \text{ is odd,} \\ 2(n-1)! & \text{if } n \text{ is even.} \end{cases} \tag{2.1.12}$$

Hence, for a general classical Laplace r.v. Y with $\mathcal{CL}(\theta, s)$ distribution,

$$\kappa_n(Y) = \begin{cases} \theta & \text{if } n = 1, \\ 0 & \text{if } n > 1 \text{ is odd,} \\ 2s^n(n-1)! & \text{if } n \text{ is even,} \end{cases} \tag{2.1.13}$$

since $\kappa_n(Y) = \kappa_n(\theta + sX) = s^n \kappa_n(X)$ for $n \geq 2$.

2.1.3.2 *Moments.* By writing the Taylor expansion of the moment generating function (2.1.10) with $\theta = 0$,

$$M_Y(t) = \sum_{k=0}^{\infty} s^{2k}(2k)! \frac{t^{2k}}{(2k)!},$$

we obtain the nth central moment of general classical Laplace r.v. Y with density (2.1.1):

$$\mu_n(Y) = E(Y-\theta)^n = \begin{cases} 0 & \text{if } n \text{ is odd,} \\ s^n n! & \text{if } n \text{ is even.} \end{cases} \tag{2.1.14}$$

One can obtain the central absolute moment of a classical Laplace distribution by observing that it is equal to the central, raw moment of exponential distribution with parameter $\lambda = 1/s$, or, more directly,

$$\nu_a(Y) = E|Y-\theta|^a = \int_0^\infty x^a \frac{1}{s} e^{-x/s} dx = s^a \Gamma(a+1). \tag{2.1.15}$$

In particular, we have

$$\text{Mean } = \theta, \qquad \text{Variance } = 2s^2, \tag{2.1.16}$$

so that for $\theta \neq 0$, the *coefficient of variation* of Y is

$$\frac{\sqrt{E(Y-EY)^2}}{|EY|} = \frac{\sqrt{2}s}{|\theta|}. \tag{2.1.17}$$

Note that the mean and variance involve different parameters (as is the case of the normal distribution, but unlike the binomial, Poisson and gamma distributions).

The nth moment about zero of the classical Laplace r.v. Y with density (2.1.1) is given by [see, e.g., Farison (1965), Kacki (1965a)]

$$\alpha_n(Y) = EY^n = n! \sum_{j=0}^{n} \frac{1+(-1)^{j+n}}{2j!} \theta^j s^{n-j} = n! \sum_{i=0}^{[[n/2]]} \frac{\theta^{n-2i}}{(n-2i)!} s^{2i}, \tag{2.1.18}$$

where $[[x]]$ denotes the greatest integer less than or equal to x.

2.1.3.3 *Mean deviation.* By (2.1.15), the mean deviation of a classical Laplace r.v. Y with density (2.1.1) is equal to

$$E|Y - E[Y]| = E|Y-\theta| = s. \tag{2.1.19}$$

Furthermore, we have

$$\frac{\text{Mean deviation}}{\text{Standard deviation}} = \frac{s}{\sqrt{2}s} = \frac{1}{\sqrt{2}} \approx 0.707. \tag{2.1.20}$$

Recall that for all normal distributions, the above ratio is given by $\sqrt{2/\pi} \approx 0.798$.

2.1.3.4 *Coefficients of skewness and kurtosis.* For a distribution of an r.v. X with a finite third moment and standard deviation greater than zero, the *coefficient of skewness* is a measure of symmetry defined by

$$\gamma_1 = \frac{E(X - EX)^3}{(E(X - EX)^2)^{3/2}}. \tag{2.1.21}$$

By (2.1.14), the coefficient of skewness of Laplace distribution (2.1.1) is equal to zero (as is the case for any symmetric distribution with a finite third moment).

For an r.v. X with a finite fourth moment, the *excess kurtosis*[9] is defined as

$$\gamma_2 = \frac{E(X - EX)^4}{(Var(X))^2} - 3. \tag{2.1.22}$$

It is a measure of peakedness and of heaviness of the tails (properly adjusted, so that $\gamma_2 = 0$ for a normal distribution) and is independent of the scale. If $\gamma_2 > 0$, the distribution is said to be *leptokurtic*; it is *platykurtic* otherwise. In view of (2.1.14),

$$\gamma_2 = \frac{s^4 4!}{(2s^2)^2} - 3 = 3. \tag{2.1.23}$$

Thus the Laplace distribution is a leptokurtic one, indicating a large degree of peakedness compared to the normal distributions. See Balanda (1987) and Horn (1983) for more details.

2.1.3.5 *Entropy.* Entropy of a classical Laplace variable Y is easy to compute:

$$\begin{aligned} H(Y) = E[-\log f(Y)] &= \int_{-\infty}^{\infty} \left[\log(2s) + \frac{|x-\theta|}{s}\right] \frac{1}{2s} e^{-|x-\theta|/s} dx \\ &= \log(2s) + \nu_1(Y)/s \\ &= \log(2s) + 1. \end{aligned}$$

As will be shown in Section 2.4.5, the Laplace distribution maximizes the entropy within the class of continuous distributions on $\mathbb{R}$ with a given absolute moment [see Kagan et al. (1973)], as well as within the class of *conditionally* Gaussian distributions [see Levin and Tchernitser (1999) or Levin and Albanese (1998)]. These results provide additional arguments for applications of Laplace laws to various practical problems [see Part III].

2.1.3.6 *Quartiles and quantiles.* Because of the availability of an explicit form of the cumulative distribution function, quantiles ξ_q of a classical Laplace distribution can be written explicitly as follows:

$$\xi_q = \begin{cases} \theta + s\log(2q); & q \in (0, 1/2], \\ \theta - s\log[2(1-q)]; & q \in (1/2, 1). \end{cases} \tag{2.1.24}$$

In particular, the first and the third quartiles are given by

$$Q_1 = \xi_{1/4} = \theta - s\log 2, \qquad Q_3 = \xi_{3/4} = \theta + s\log 2.$$

Evidently, the second quartile Q_2 — the median — is equal to θ.

[9] Without centering by 3, it is simply called *kurtosis*.

2.2 Representations and characterizations

In the first part of this section we present various representations of Laplace r.v.'s in terms of other well-known random variables. These representations are also listed in Table 2.3. We shall focus on the standard classical Laplace r.v. X with density (2.1.2) and ch.f. (2.1.7). As already mentioned, for a general Laplace r.v. Y with density (2.1.1) the corresponding representations of Y follow from the relation $Y \stackrel{d}{=} \theta + sX$. When writing equalities in a distribution we shall follow the standard convention that random variables appearing on the same side of the equation are independent.

Characterizations of distributions is a popular and well-developed topic of modern probability theory. This provides additional insight into the structure of distributions, especially those that are, like the Laplace distribution, defined by a simple density and characteristic function. The simplicity of a formula does not always convey obvious features and masks surprises that may be built into a particular distribution. In the case of the Laplace distribution its characterizations unveil quite intriguing properties that one would not suspect from its "modest" density function.

In the second part of this section we describe some characterizations of Laplace distributions, in particular those connected with the *geometric summation*

$$S_p = X_1 + \cdots + X_{\nu_p}, \tag{2.2.1}$$

where ν_p is a geometric random variable with mean $1/p$ and probability function

$$P(\nu_p = k) = (1-p)^{k-1}p, \quad k = 1, 2, 3, \ldots, \tag{2.2.2}$$

while X_i, $i \geq 1$, are i.i.d. r.v.'s independent of ν_p. It turns out that under geometric summation (2.2.1), the Laplace distribution plays a role analogous to that of Gaussian distribution under ordinary summation. As discussed in Kalashnikov (1997), geometric sums (2.2.1) arise naturally in diverse fields in applications such as risk theory, modeling financial asset returns, insurance mathematics and others, and consequently the Laplace distribution is applicable for stochastic modeling.

2.2.1 Mixture of normal distributions. Any Laplace r.v. can be thought of as a Gaussian r.v. with mean zero and *stochastic* variance which has an exponential distribution. More formally, a Laplace r.v. has the same distribution as the product of a normal and an independent exponentially distributed random variable, as sketched in

Proposition 2.2.1 *A standard classical Laplace r.v. X has the representation*

$$X \stackrel{d}{=} \sqrt{2W}Z, \tag{2.2.3}$$

where the random variables W and Z have the standard exponential and normal distributions, respectively.

Proof. Let W be a standard exponential r.v. with density $f_W(w) = e^{-w}$, $w > 0$, and the moment generating function $M_W(t) = E[e^{tW}] = (1-t)^{-1}, t < 1$. Let Z be a standard normal random variable with density $f_Z(z) = \frac{1}{\sqrt{2\pi}}e^{-z^2/2}$, $-\infty < z < \infty$, and the characteristic function $\phi_Z(t) = e^{-t^2/2}$, $-\infty < t < \infty$. The ch.f. of the product $\sqrt{2W}Z$ coincides with the standard classical Laplace ch.f. (2.1.7). Indeed, conditioning on W, we obtain

$$\begin{aligned} E[e^{it\sqrt{2W}Z}] &= E[E[e^{it\sqrt{2W}Z}|W]] = E[\phi_Z(t\sqrt{2W})] \\ &= E[e^{-t^2W}] = M_W(-t^2) = (1+t^2)^{-1}. \end{aligned}$$

The proposition is thus proved. □

Remark 2.2.1 An alternative proof of Proposition 2.2.1 utilizing the densities of W and Z is outlined in Exercise 2.7.10. Relation (2.2.3) written in terms of the densities becomes

$$\frac{1}{2}e^{-|x|} = \int_0^\infty f_Z\left(\frac{x}{\sqrt{2w}}\right)\frac{1}{\sqrt{2w}}f_W(w)dw = \int_0^\infty \frac{1}{2}\frac{1}{\sqrt{\pi w}}e^{-\frac{1}{2}\left(\frac{x^2}{2w}+2w\right)}dw. \qquad (2.2.4)$$

Remark 2.2.2 For a general Laplace r.v. Y with density (2.1.1) we have the representation $Y \stackrel{d}{=} \theta + \sqrt{2}sW^{1/2}Z$.

Remark 2.2.3 Representation (2.2.3) can be written as

$$X \stackrel{d}{=} RZ, \qquad (2.2.5)$$

where Z is as before and the random variable $R = \sqrt{2W}$ has a *Rayleigh distribution* with density $f_R(x) = xe^{-x^2/2}$, $x > 0$.

Remark 2.2.4 Another related representation discussed in Loh (1984) is obtained by denoting $T = 1/\sqrt{W}$. Then

$$X \stackrel{d}{=} \sqrt{2}\frac{Z}{T}. \qquad (2.2.6)$$

Here the r.v. T has a *brittle fracture distribution* with density $f_T(x) = 2x^{-3}e^{1/x^2}$ [such T is used to model breaking stress or strength; see, e.g., Black et al. (1989) or Johnson et al. (1994), p. 694]. A proof of the result is left as an exercise.

2.2.2 Relation to exponential distribution. The ch.f. (2.1.7) of a standard classical Laplace distribution can be factored as follows:

$$\frac{1}{1+t^2} = \frac{1}{1-it}\frac{1}{1+it}. \qquad (2.2.7)$$

Note that the first factor is the ch.f. of a standard exponential r.v. W with density $f_W(w) = e^{-w}$, $w \geq 0$, while the second one is the ch.f. of $-W$. Since for independent random variables the product of ch.f.'s corresponds to their sum, we arrive at a representation of a standard classical Laplace r.v. in terms of two independent exponential random variables. The following proposition is thus valid.

Proposition 2.2.2 *A classical standard Laplace r.v. X admits the representation*

$$X \stackrel{d}{=} W_1 - W_2, \qquad (2.2.8)$$

where W_1 and W_2 are i.i.d. standard exponential random variables.

Remark 2.2.5 For a general Laplace r.v. Y with density (2.1.1), we have

$$Y \stackrel{d}{=} \theta + s(W_1 - W_2).$$

Remark 2.2.6 Denoting $H_i = 2W_i$, $i = 1, 2$, we obtain

$$Y \stackrel{d}{=} \theta + \frac{s}{2}(H_1 - H_2),$$

where H_1 and H_2 are i.i.d. with the χ^2 distribution with two degrees of freedom (having density $f(x) = \frac{1}{2}e^{-x/2}$).

Remark 2.2.7 Note that the following relation for an X distributed according to the standard classical Laplace law follows immediately from (2.2.8):

$$X \stackrel{d}{=} \log \frac{U_1}{U_2},$$

where U_1 and U_2 are independent random variables distributed uniformly on [0, 1] [see, e.g., Lukacs and Laha (1964, p. 61)].

The standard classical Laplace ch.f. (2.1.7) can also be decomposed as follows:

$$\frac{1}{1+t^2} = \frac{1}{2}\frac{1}{1-it} + \frac{1}{2}\frac{1}{1+it}. \tag{2.2.9}$$

The right-hand side of (2.2.9) is the ch.f. of the product IW, where the discrete symmetric variable I takes on values ± 1 with probabilities $1/2$, while W is an independent of I standard exponential (see Exercise 2.7.12). Thus the standard classical Laplace distribution is a simple exponential mixture. This is stated in the following result.

Proposition 2.2.3 *A standard classical Laplace r.v. X admits the representation*

$$X \stackrel{d}{=} IW, \tag{2.2.10}$$

where W is standard exponential while I takes on values ± 1 with probabilities $1/2$.

Remark 2.2.8 For a general Laplace r.v. Y with density (2.1.1), we have

$$Y \stackrel{d}{=} \theta + sIW.$$

Remark 2.2.9 It follows directly from (2.2.10) that if X is a standard classical Laplace r.v., then $|X|$ is a standard exponential r.v. W. Thus, as already noted by Johnson et al. (1995, p. 190), if $X_1, X_2, \dots, X_n$ are i.i.d. standard Laplace r.v.'s, then any statistics depending only on the absolute values $|X_1|, |X_2|, \dots, |X_n|$ can be represented in terms of χ^2 random variables (since as already stated, $2W$ is a χ^2 r.v. with two degrees of freedom).

2.2.3 Relation to the Pareto distribution. A standard exponential r.v. W is related to a Pareto Type I r.v. P with density $f(x) = 1/x^2$, $x \geq 1$, as follows:

$$W \stackrel{d}{=} \log P. \tag{2.2.11}$$

Consequently, representation (2.2.8) can be restated in terms of two independent Pareto random variables.

Proposition 2.2.4 *A standard classical Laplace r.v. X admits the representation*

$$X \stackrel{d}{=} \log \frac{P_1}{P_2}, \tag{2.2.12}$$

where P_1 and P_2 are i.i.d. Pareto Type I *random variables with density* $1/x^2$, $x \geq 1$.

Proof. Note that $W_1 = \log P_1$ has standard exponential distribution with density e^{-x}, $x \geq 0$. The result now follows directly from Proposition 2.2.2. □

Remark 2.2.10 For a general classical Laplace r.v. Y with density (2.1.1) we have

$$Y \stackrel{d}{=} \log \left[e^{\theta} \left(\frac{P_1}{P_2} \right)^s \right].$$

Hence the *log-Laplace random variable* $e^{(Y-\theta)/s}$ has the same distribution as the ratio of two independent Pareto Type I random variables.

2.2.4 Relation to 2×2 **unit normal determinants.** The following connection between Laplace and normal distributions, mentioned in Chapter 1, was established by Nyquist et al. (1954) almost 50 years ago and was the subject of a number of letters to the editor of the *American Statistician* during the last decades.

Proposition 2.2.5 *A standard classical Laplace r.v. X admits the representation*

$$X \stackrel{d}{=} \begin{vmatrix} U_1 & U_2 \\ U_3 & U_4 \end{vmatrix} = U_1U_4 - U_2U_3, \tag{2.2.13}$$

where the U_i*'s are i.i.d. standard normal random variables.*

The proof presented below is based on Proposition 2.2.2 and follows a heuristic derivation due to Mantel and Pasternak (1966). For an alternative formal proof using characteristic functions, see Exercise 2.7.13. For additional comments on this problem, see Nicholson (1958), Mantel (1973), Missiakoulis and Darton (1985), Mantel (1987), and Johnson et al. (1995, p. 191), among others.

Proof. In view of Proposition 2.2.2 and the remark following it, we have $X \stackrel{d}{=} (H_1 - H_2)/2$, where H_1 and H_2 are i.i.d. with the χ^2 distribution with two degrees of freedom. Recall that $H_1 \stackrel{d}{=} (W_1 + W_2)$, where W_1 and W_2 are i.i.d. with the χ^2 distribution with one degree of freedom. (An analogous representation holds for H_2.) Furthermore, $W_1 \stackrel{d}{=} Z_1^2$, where Z_1 is a standard normal variable. Consequently, we have

$$X \stackrel{d}{=} \frac{1}{2}(Z_1^2 + Z_2^2 - Z_3^2 - Z_4^2),$$

where the Z_i's are i.i.d. standard normal variables. Equivalently,

$$X \stackrel{d}{=} \frac{Z_1 - Z_3}{\sqrt{2}}\frac{Z_1 + Z_3}{\sqrt{2}} - \frac{Z_4 - Z_2}{\sqrt{2}}\frac{Z_4 + Z_2}{\sqrt{2}}.$$

Note that the two normal random variables $Z_1 - Z_3$ and $Z_1 + Z_3$ are independent, and so are $Z_4 - Z_2$ and $Z_4 + Z_2$. Thus $U_1 = \frac{Z_1 - Z_3}{\sqrt{2}}$, $U_2 = \frac{Z_4 - Z_2}{\sqrt{2}}$, $U_3 = \frac{Z_4 + Z_2}{\sqrt{2}}$, and $U_4 = \frac{Z_1 + Z_3}{\sqrt{2}}$ are i.i.d. standard normal and (2.2.13) is indeed valid. □

Attempts to generalize this result to determinants of larger size so far have not been successful (see Exercise 2.7.14). All the cited representations are summarized in Table 2.3.

2.2.5 An orthogonal representation. Younes (2000) shows that a classical Laplace r.v. X admits an *orthogonal representation* of the form

$$X = \sum_{n=1}^{\infty} b_n X_n, \tag{2.2.14}$$

where $\{X_n, n \geq 1\}$ is a sequence of uncorrelated random variables (the orthogonality here means uncorrelation). The convergence in (2.2.14) is in the mean square, i.e.,

$$\lim_{n\to\infty} E\left(X - \sum_{k=1}^{n} b_k X_k\right)^2 = 0. \tag{2.2.15}$$

Representation	Variables
$\sqrt{2W} \cdot Z$	Z standard normal r.v. W standard exponential r.v.
$R \cdot Z$	R Rayleigh r.v. (p.d.f. — $f(w) = we^{-w^2/2}$) Z standard normal r.v.
$\sqrt{2}Z/T$	T "brittle fracture" r.v. (p.d.f. — $f(t) = 2t^{-3}e^{1/t^2}$) Z standard normal r.v.
$W_1 - W_2$	W_1, W_2 standard exponential r.v.'s
$(H_1 - H_2)/2$	H_1, H_2 Chi-square r.v.'s with two d.f.
$I \cdot W$	I random sign taking $\pm$ with equal probabilities W standard exponential r.v.
$\log(P_1/P_2)$	P_1, P_2 Pareto Type I r.v.'s (p.d.f. — $f(p) = 1/p^2$, $p > 1$)
$\log(U_1/U_2)$	U_1, U_2 r.v.'s uniformly distributed on [0, 1].
$U_1 \cdot U_4 - U_2 \cdot U_3$	U_1, U_2, U_3, U_4 standard normal r.v.'s
$Y = \sum_{i=1}^n Y_{1i}^{(n)} - Y_{2i}^{(n)}$	Y_{1i}, Y_{2i} gamma distributed r.v.'s with the density given by (2.4.3); see Proposition 2.4.1.

Table 2.3: Summary of the representations of the standard classical Laplace distribution presented in this section. All variables in each representation are mutually independent.

Proposition 2.2.6 *A standard classical Laplace* $\mathcal{CL}(0, 1)$ *r.v.* X *admits the representation* (2.2.14) *with*

$$b_n = \frac{\xi_n}{\sqrt{2}J_0(\xi_n)} \int_0^\infty xe^{-x} J_0(\xi_n e^{-x/2})dx \tag{2.2.16}$$

and

$$X_n = \frac{\sqrt{2}}{\xi_n J_0(\xi_n)} J_0(\xi_n e^{-|X|/2}), \tag{2.2.17}$$

where J_0 *and* J_1 *are the Bessel functions of the first kind of order* 0 *and* 1, *respectively* (*see the appendix*), *and* ξ_n *is the nth root of* J_1.

Proof. See Younes (2000) for a derivation. □

Orthogonal representations play an important role in statistics. For example, they appear in factor analysis, where each of the d observable variables is expressed as the sum of $p < d$ uncorrelated common factors and one unique factor. See, e.g., Younes (2000) for further information on orthogonal representations and their applications in statistics.

2.2.6 Stability with respect to geometric summation. Stability, related to infinite divisibility, is a well-known property of the normal distribution. A formal definition is: If $X, X_1, X_2, \ldots$ are i.i.d. normal, then for every positive integer n, there exist an $a_n > 0$ and a $b_n \in \mathbb{R}$ such that

$$X \stackrel{d}{=} a_n(X_1 + \cdots + X_n) + b_n. \tag{2.2.18}$$

In fact, the normal law is the only nondegenerate one with finite variance having this property.[10] Under geometric summation (2.2.1), the best-known property analogous to (2.2.18) is perhaps the following characterization of the exponential distribution: If $Y, Y_1, Y_2, \ldots$ are positive and nondegenerate i.i.d. random variables with finite variance, then

$$a_p \sum_{i=1}^{\nu_p} Y_i \stackrel{d}{=} Y_1 \quad \text{for all} \quad p \in (0, 1) \tag{2.2.19}$$

if and only if Y_1 has an exponential distribution [see, e.g., Arnold (1973), Kakosyan et al. (1984), Milne and Yeo (1989)]. If instead the Y_i's are symmetric, then (2.2.19) characterizes the class of Laplace distributions. This is not surprising if one notes that—as already mentioned—the Laplace distribution is simply a symmetric extension of the standard exponential distribution.

We shall start the proof with the following lemma.

Lemma 2.2.1 *Let $X_1, X_2, \ldots$ be i.i.d. random variables with ch.f. ψ, and let N be a positive and integer-valued random variable with the generating function defined as $G(z) = E(z^N)$. Then the ch.f. of the r.v. $\sum_{i=1}^N X_i$ is $G(\psi(t))$.*

Proof. Conditioning on N, we obtain directly

$$Ee^{it\sum_{k=1}^N X_i} = \sum_{n=1}^{\infty} \psi^n(t) P(N = n) = E\psi^N(t). \qquad \square$$

Proposition 2.2.7 *Let $Y, Y_1, Y_2, \ldots$ be nondegenerate and symmetric i.i.d. random variables with finite variance $\sigma^2 > 0$, and let ν_p be a geometric random variable with mean $1/p$, independent of the Y_i's. Then the following statements are equivalent:*

(i) *Y is stable with respect to geometric summation, i.e., there exist constants $a_p > 0$ and $b_p \in \mathbb{R}$, such that*

$$a_p \sum_{i=1}^{\nu_p} (Y_i + b_p) \stackrel{d}{=} Y \quad \textit{for all} \quad p \in (0, 1). \tag{2.2.20}$$

(ii) *Y possesses the Laplace distribution with mean zero and variance σ^2.*

Moreover, the constants a_p and b_p must be of the form $a_p = p^{1/2}$, $b_p = 0$.

---

[10] If the finite variance assumption is dropped, then the distributions satisfying (2.2.18) are called *stable* (Paretian stable, α-stable) laws [see, e.g., Zolotarev (1986), Janicki and Weron (1994), Samorodnitsky and Taqqu (1994), and Nikias and Shao (1995)], of which normal distribution is a special case.

Proof. We shall first establish the form of the normalizing constants in (2.2.20). Taking the expected value of both sides of (2.2.20) and exploiting independence, we arrive at

$$0 = E[Y] = E[\nu_p]E[a_p(Y_i + b_p)].$$

Since $E[\nu_p] = 1/p \neq 0$ and $a_p > 0$, in view of the symmetry of Y_i, we have $b_p = -E[Y_i] = 0$. Next we equate the variances of both sides of (2.2.20). Denoting by S_p the left-hand side of (2.2.20), we can write the following well-known decomposition based on conditional variances:

$$\mathrm{Var}[S_p] = \mathrm{Var}[E[S_p|\nu_p]] + E[\mathrm{Var}[S_p|\nu_p]].$$

In this expression the first term is zero, since

$$E[S_p|\nu_p] = \nu_p a_p E[Y_i]$$

and, as shown earlier, $E[Y_i] = 0$. Now $E[\nu_p] = 1/p$, and the second term becomes

$$E[\mathrm{Var}[S_p|\nu_p]] = E[\nu_p a_p^2 \sigma^2] = \left(\frac{a_p}{p^{1/2}}\right)^2 \sigma^2.$$

However, since the variance on the right-hand side of (2.2.20) is σ^2, we have

$$\left(\frac{a_p}{p^{1/2}}\right)^2 = 1,$$

so $a_p = p^{1/2}$.

We now turn to the equivalence between (i) and (ii) with $a_p = p^{1/2}$ and $b_p = 0$. By Lemma 2.2.1, in terms of ch.f.'s relation (2.2.20) is expressed as

$$\frac{p\psi(p^{1/2}t)}{1-(1-p)\psi(p^{1/2}t)} = \psi(t) \quad \text{for all } p \in (0,1) \text{ and all } t \in \mathbb{R}, \tag{2.2.21}$$

where ψ is the ch.f. of Y. [Note that $E(z^{\nu_p}) = pz/(1-(1-p)z)$.] Relation (2.2.21) will often be utilized in what follows. Consequently, we also have for all $t \in \mathbb{R}$,

$$\frac{p\psi(p^{1/2}t)}{1-(1-p)\psi(p^{1/2}t)} \to \psi(t), \quad \text{as } p \to 0. \tag{2.2.22}$$

Since $\psi(p^{1/2}t) \to \psi(0) = 1$, we obtain

$$\frac{p}{1-(1-p)\psi(p^{1/2}t)} \to \psi(t), \quad \text{as } p \to 0, \tag{2.2.23}$$

or, equivalently,

$$\frac{1}{\frac{1}{p}[1-(1-p)\psi(p^{1/2}t)]} \to \psi(t), \quad \text{as } p \to 0 \tag{2.2.24}$$

for all $t \in \mathbb{R}$. Now, since Y possesses the first two moments, its ch.f. can be written as

$$\psi(u) = 1 + iuE[Y] + \frac{(iu)^2}{2}(E[Y^2] + \delta) = 1 - \frac{u^2}{2}(\sigma^2 + \delta), \tag{2.2.25}$$

where $\delta = \delta(u)$ denotes a bounded function of u such that $\lim_{u\to 0}\delta(u) = 0$ [see, e.g., Theorem 8.44 in Breiman (1993)]. Utilizing (2.2.25), we can write the denominator in (2.2.24) as

$$\frac{t^2}{2}(\sigma^2 + \delta) + 1 - \frac{pt^2}{2}(\sigma^2 + \delta), \tag{2.2.26}$$

which converges to $\frac{1}{2}t^2\sigma^2 + 1$ as $p \to 0$. However, $u = p^{1/2}t \to 0$ as $p \to 0$. Consequently,

$$\frac{1}{\frac{1}{2}t^2\sigma^2 + 1} = \psi(t), \tag{2.2.27}$$

so Y has Laplace distribution with mean zero and variance σ^2. We have thus established the implication (i) $\Rightarrow$ (ii). To verify the reverse implication, all that is needed is to verify that the Laplace ch.f. (2.2.27) satisfies (2.2.21). □

Proposition 2.2.7 is perhaps the first theorem in this book that requires somewhat delicate arguments. The result is due to Kakosyan et al. (1984) but the proof presented here differs from the original one.

Remark 2.2.11 If Y_i's are positive r.v.'s but the assumption of finite variance is dropped, (2.2.19) characterizes *Mittag–Leffler distributions* [see, e.g., Gnedenko (1970), Pillai (1990)]. These are distributions of positive r.v.'s with the Laplace transform

$$E[e^{-sX}] = \frac{1}{1 + \sigma^\alpha s^\alpha},$$

where $0 < \alpha \le 1$, and are reduced to an exponential r.v. for $\alpha = 1$.

Remark 2.2.12 If Y_i's are symmetric, but the assumption of finite variance is dropped, (2.2.19) characterizes the *Linnik distributions* [see Lin (1994), Kozubowski (1994b)]. Linnik distributions possess the ch.f.

$$\psi(t) = \frac{1}{1 + \sigma^\alpha |t|^\alpha},$$

where $0 < \alpha \le 2$, and are reduced to Laplace distributions for $\alpha = 2$. We shall study this class in Section 4.3 of Chapter 4.

Remark 2.2.13 If no assumptions on the distribution of the Y_i's are imposed, relation (2.2.19) characterizes the so-called *strictly geometric stable distributions* [see, e.g., Klebanov et al. (1984), Janković (1992), Kozubowski (1994a)]. Further studies dealing with the stability relation (2.2.19) and its generalizations include Janjić (1984), Gnedenko and Janjić (1983), Janković (1993ab), Bunge (1993), Bunge (1996), Baringhaus and Grubel (1997), and Bouzar (1999).

Incidentally, relation (2.2.21) is equivalent to the following relation among random variables:

$$Y \stackrel{d}{=} p^{1/2} I Y_1 + (1 - I)(Y_2 + p^{1/2} Y_3), \tag{2.2.28}$$

where Y, Y_1, Y_2, Y_3 are i.i.d., while I is an indicator (Bernoulli) random variable, independent of Y, Y_1, Y_2, Y_3, with $P(I = 1) = p$ and $P(I = 0) = 1 - p$.

Another relation among random variables that is also equivalent to (2.2.21) is

$$Y \stackrel{d}{=} p^{1/2} Y_1 + (1 - I) Y_2. \tag{2.2.29}$$

The above relation is simply a restatement of representation (2.4.9) for symmetric Laplace r.v.'s with mean zero to be discussed later.

Consequently, we have yet two more characterizations of the Laplace distribution, which can be obtained by computing ch.f.'s of the right-hand sides of (2.2.28) and (2.2.29) and comparing them with the relation (2.2.21).

Proposition 2.2.8 *Let Y, Y_1, Y_2, Y_3 be nondegenerate, symmetric i.i.d. random variables with finite variance $\sigma^2 > 0$. Let I be an indicator random variable with $P(I = 1) = p$ and $P(I = 0) = 1 - p$, independent of Y_1, Y_2, Y_3. Then the following statements are equivalent:*

(i) *Y satisfies relation* (2.2.28) *for all $p \in [0, 1]$.*

(ii) *Y satisfies relation* (2.2.29) *for all $p \in [0, 1]$.*

(iii) *Y has Laplace distribution with mean zero and variance σ^2.*

2.2.7 Distributional limits of geometric sums. An exponential distribution is not only stable with respect to geometric summation, but also appears to be the only possible nondegenerate limiting distribution of normalized geometric sums (2.2.1) with i.i.d. positive terms possessing finite expectations. If X_i's are i.i.d. nonnegative r.v.'s with $\mu = E[X_1] < \infty$, then pS_p, where S_p is given by (2.2.1), converges in distribution (as $p \to 0$) to an exponential r.v. with mean μ. This result is due to Rényi (1956) obtained more than 40 years ago. In Kalashnikov's (1997) opinion, Rényi's theorem may explain the popularity of exponential distribution among researchers in reliability, risk theory, and other fields where geometric sums (2.2.1) frequently arise. The connection between geometric sums, rarefactions of renewal processes, geometric compounding and damage models was emphasized some 20 years later by Galambos and Kotz (1978).

Similarly, Laplace distribution arises as a limit of S_p when X_i's are symmetric with finite variance.

Proposition 2.2.9 *Let S_p be given by* (2.2.1), *where $X_1, X_2, \ldots$ are nondegenerate and symmetric i.i.d. r.v.'s with a finite variance, with ν_p being a geometric r.v. with the mean $1/p$, independent of the X_i's. Then the class of Laplace distributions with zero mean coincides with the class of nondegenerate distributional limits of $a_p S_p$ as $p \to 0$, where $a_p > 0$. Moreover, if* $\mathrm{Var}[X_1] = \sigma^2$ *and*

$$a_p \sum_{i=1}^{\nu_p} X_i \stackrel{d}{\to} Y \quad \text{as } p \to 0, \tag{2.2.30}$$

there exists $\gamma > 0$ such that $a_p = p^{1/2}\gamma + o(p^{1/2})$, and Y has a Laplace distribution with mean zero and variance $\sigma^2\gamma^2$.

Proof. Evidently, if Y has a Laplace distribution, then in view of (2.2.20), the convergence (2.2.30) holds with $X_i \stackrel{d}{=} Y$ and $a_p = p^{1/2}$. It is therefore sufficient to show that if (2.2.30) holds with $\mathrm{Var}[X_1] = \sigma^2$, then for some $\gamma > 0$ the limit must have the Laplace distribution with mean zero and variance $\sigma^2\gamma^2$ where $a_p = p^{1/2}\gamma(1 + o(1))$.

Assume that (2.2.30) holds; the X_i's are symmetric with $\mathrm{Var}[X_1] = \sigma^2$ and Y is nondegenerate. In terms of ch.f.'s, by Lemma 2.2.1, we have

$$\frac{p\phi(a_p t)}{1 - (1-p)\phi(a_p t)} \to \psi(t), \quad \text{as } p \to 0, \text{ for all } t, \tag{2.2.31}$$

where ϕ and ψ are the ch.f.'s of X_1 and Y, respectively. First, note that for all t we must have the convergence

$$\phi(a_p t) \to 1, \quad \text{as } p \to 0. \tag{2.2.32}$$

Indeed, by continuity of ψ and the property $\psi(0) = 1$, we must have $\psi(t) \neq 0$ for all t in an interval $(-\epsilon, \epsilon)$, where $\epsilon > 0$. Then for such a t, the limit in (2.2.31) is nonzero while the limit of the numerator in (2.2.31) is zero. Consequently, the denominator in (2.2.31) ought to converge to zero so that (2.2.32) will hold for such a t. Now take any t in an interval $(-2\epsilon, 2\epsilon)$ and use the inequality

$$0 \leq 1 - \text{Re}\phi(s) \leq 4(1 - \text{Re}\phi(s/2)) \tag{2.2.33}$$

with $s = a_p t$ to conclude that (2.2.32) holds for such a t. Inequality (2.2.33) follows directly from the trigonometric relation

$$1 - \cos 2tx = 2(1 - \cos^2 tx) \leq 4(1 - \cos tx),$$

since Re $\phi(s)$ is the expected value of $\cos tX$. (The last inequality follows directly from $0 \leq (\cos tx - 1)^2$.) This implies that (2.2.32) holds for all t. Next, utilizing (2.2.32), we rewrite (2.2.31) in the form

$$\frac{1}{\frac{1}{p}[1 - (1-p)\phi(a_p t)]} \to \psi(t), \quad \text{as } p \to 0, \tag{2.2.34}$$

for all $t \in \mathbb{R}$. Now, since (2.2.32) holds for all t and ϕ is a ch.f. of a nondegenerate distribution, we must have

$$a_p \to 0, \quad \text{as } p \to 0. \tag{2.2.35}$$

Indeed, if (2.2.35) is not valid, we would have had $a_{p_n} \to c$ for some sequence $p_n \to 0$, where $0 < c \leq \infty$, so as $p \to 0$, we would have had

$$\phi(a_{p_n} t) \to \phi(ct) = 1 \tag{2.2.36}$$

for all t. But (2.2.36) implies that the distribution of X_1 is degenerate, contradicting our assumption. Thus (2.2.35) must be valid.

Now we proceed as in the proof of Proposition 2.2.7 and write the denominator of (2.2.34) in the form

$$\left(\frac{a_p}{p^{1/2}}\right)^2 \frac{t^2}{2}(\sigma^2 + \delta) + 1 - \frac{a_p^2 t^2}{2}(\sigma^2 + \delta), \tag{2.2.37}$$

where again $\delta = \delta(u)$ denotes a bounded function of u such that $\lim_{u \to 0} \delta(u) = 0$. Since as $p \to 0$ the expression (2.2.37) converges to a limit, and moreover, in view of (2.2.35),

$$\frac{t^2}{2}(\sigma^2 + \delta) \to \frac{t^2\sigma^2}{2}; \quad \frac{a_p^2 t^2}{2}(\sigma^2 + \delta) \to 0, \tag{2.2.38}$$

the term $\frac{a_p}{p^{1/2}}$ must converge to some limit $\gamma > 0$ (if the limit were zero, the expression (2.2.37) would converge to 1, implying that $\psi(t) \equiv 1$ and that Y has a degenerate distribution). Consequently, we have verified the convergence in (2.2.34), where the limiting ch.f. is of the form

$$\psi(t) = \frac{1}{1 + \frac{1}{2}\sigma^2\gamma^2 t^2} \tag{2.2.39}$$

and $a_p/p^{1/2} \to \gamma$, so $a_p = p^{1/2}\gamma(1 + o(1))$. This completes the proof. □

Remark 2.2.14 If no assumptions on the distribution of the X_i's are imposed, then, as $p \to 0$, the weak limits of

$$a_p \sum_{i=1}^{\nu_p} (X_i + b_p), \tag{2.2.40}$$

where $a_p > 0$ and $b_p \in \mathbb{R}$, result in *geometric stable* (GS) *laws* [see, e.g., Mittnik and Rachev (1991)].

2.2.8 Stability with respect to the ordinary summation. We saw in Section 2.2.6 that symmetric Laplace distributions are stable with respect to random summation (Proposition 2.2.7). When the summation is "deterministic," the Laplace distribution has the stability property (2.2.18) under a *random* normalization.

Before stating the main result of this subsection, we shall establish some auxiliary properties in which we use the following notation for gamma densities with parameters α and β:

$$f_{\alpha,\beta}(x) = \frac{x^{\alpha-1}e^{-x/\beta}}{\beta^\alpha \Gamma(\alpha)}.$$

A nonrandom sum of the i.i.d. Laplace random variables is no longer a Laplace variable. Instead, the sum admits the representation given below, which is a generalization of the representation (2.2.3) for a single Laplace random variable.

Proposition 2.2.10 *Let $Y_1, Y_2, \ldots$ be i.i.d. $\mathcal{L}(0, 1)$ random variables. Then*

$$Y_1 + \cdots + Y_n \stackrel{d}{=} \sqrt{G_n} Z, \tag{2.2.41}$$

where G_n has a gamma distribution with parameters $\alpha = n$, $\beta = 1$ and Z is a standard normal r.v. independent of G_n.

Proof. Let the Y_i's have the Laplace distribution $\mathcal{L}(0, 1)$, in which case their ch.f. is

$$\psi(t) = \frac{1}{1 + \frac{1}{2}t^2}. \tag{2.2.42}$$

Thus the ch.f. of the sum of n i.i.d. copies of Y_i is

$$\left(\frac{1}{1 + \frac{1}{2}t^2}\right)^n. \tag{2.2.43}$$

Note that the ch.f. of the product $\sqrt{G_n} Z$ of two independent r.v.'s, where Z is standard normal and G_n has a gamma distribution, is of the form

$$\phi(t) = M_{G_n}(t^2/2),$$

where M_{G_n} is the moment generating function of G_n (this relation is evidently true if G_n is replaced by an arbitrary random variable independent of Z). To conclude the proof recall that the moment generating function of a gamma r.v. is of the form

$$M_{G_n}(t) = \left(\frac{1}{1 - t}\right)^n. \tag{2.2.44}$$

□

In what follows, let B_n denote a beta distribution with parameters 1 and n, given by the density

$$f(x) = n(1-x)^{n-1}, \quad 0 < x < 1. \tag{2.2.45}$$

The following result will be needed.

Lemma 2.2.2 *Let B_{n-1} and G_n be independent r.v.'s having the beta distribution with parameters 1 and $n-1$ and the gamma distribution with parameters n and 1, respectively. Let W be a standard exponential variable. Then the representation*

$$W \stackrel{d}{=} G_n B_{n-1}$$

is valid.

Proof. Let $G(\alpha)$ denote the gamma distribution with density

$$f_\alpha(x) = \frac{x^{\alpha-1}e^{-x}}{\Gamma(\alpha)}.$$

If $X_{\alpha_1} \sim G(\alpha_1)$ and $X_{\alpha_2} \sim G(\alpha_2)$ are independent, then it is well known that the two random variables

$$X_{\alpha_1} + X_{\alpha_2} \text{ and } \frac{X_{\alpha_1}}{X_{\alpha_1} + X_{\alpha_2}}$$

are mutually independent, and their distributions are, respectively, $G(\alpha_1 + \alpha_2)$ and a standard beta with parameters α_1 and α_2 [see also pp. 349–350 in Johnson et al. (1994)]. The independence of these two random variables is actually a characterization of the gamma distribution, as established by Lukacs (1955).

Take now $\alpha_1 = 1$ and $\alpha_2 = n-1$ and observe that the standard exponential r.v. X_{α_1} can be expressed as the product of two independent variables,

$$X_{\alpha_1} = (X_{\alpha_1} + X_{\alpha_2}) \frac{X_{\alpha_1}}{X_{\alpha_1} + X_{\alpha_2}},$$

where the first one is a $G(n)$ variable while the second is a beta variable with parameters 1 and $n-1$. □

We now state the main result.

Proposition 2.2.11 *Let $Y, Y_1, Y_2, \ldots$ be i.i.d. random variables with finite variance $\sigma^2 > 0$, and let B_n be an r.v. independent of the Y_i's, with density* (2.2.45). *Then the following statements are equivalent:*

(i) *For all integers n greater than* 1,

$$B_{n-1}^{1/2} \sum_{i=1}^{n} Y_i \stackrel{d}{=} Y. \tag{2.2.46}$$

(ii) *Y has a symmetric Laplace distribution.*

Proof. We shall first deal with the implication (i) $\Rightarrow$ (ii). Taking the expected value on both sides of (2.2.46), we have

$$E[Y] = E[B_{n-1}^{1/2}](E[Y_1] + \cdots + E[Y_n]) = nE[\sqrt{B_{n-1}}]E[Y]. \tag{2.2.47}$$

This implies that $E[Y] = 0$ as $nE[\sqrt{B_{n-1}}] \neq 1$ (for example, $E[\sqrt{B_1}] = 2/3$ since B_1 is uniformly distributed on $[0, 1]$).

Next, write the left-hand side of (2.2.46) in the form $\sqrt{U_n}V_n$, where

$$U_n = nB_{n-1} \text{ and } V_n = \frac{\sum_{i=1}^n Y_i}{n^{1/2}}, \tag{2.2.48}$$

and let $n \to \infty$. Then U_n converges in distribution to a random variable W with the standard exponential distribution. Indeed $P(U_n \leq u) = 1 - (1 - u/n)^{n-1}$, $u \in (0, n)$, which converges to $1 - e^{-u}$, $u \geq 0$. By the Central Limit Theorem, V_n converges to a normal r.v. with mean zero and variance σ^2. Since, by assumption, U_n is independent of V_n, the limit of the product $\sqrt{U_n}V_n$ is the product of the limits, so

$$\sqrt{U_n}V_n \xrightarrow{d} W^{1/2}\sigma Z. \tag{2.2.49}$$

This is, however, a representation of a Laplace r.v. with mean zero and variance σ^2 [see Proposition 2.2.1 and the remarks following it]. To complete the proof of the implication (i) $\Rightarrow$ (ii), observe that Y must have the same distribution as the limit in (2.2.49), since by (i), (2.2.46) holds for all $n > 1$.

We now turn to the proof of the implication (ii) $\Rightarrow$ (i). Multiply both sides of (2.2.41) by $B_{n-1}^{1/2}$ (which is independent of the other r.v.'s) to obtain

$$B_{n-1}^{1/2}(Y_1 + \cdots + Y_n) \stackrel{d}{=} (G_n B_{n-1})^{1/2}\sigma Z. \tag{2.2.50}$$

By Lemma 2.2.2, the product $G_n B_{n-1}$ has the same distribution as a standard exponential r.v. W, so the right-hand side of (2.2.50) has the Laplace distribution (with variance σ^2) by the representation (2.2.41) with $n = 1$. The proof is thus completed. □

Remark 2.2.15 Relation (2.2.46) characterizes the Laplace distribution even if the assumption of finite variance of the Y_i's is dropped. The available proof of this result is highly technical; see Pakes (1992ab).

Remark 2.2.16 Proceeding in the same manner as in the proof of Proposition 2.2.11, one can show that within the class of *positive* r.v.'s the stability relation

$$B_{n-1} \sum_{i=1}^n Y_i \stackrel{d}{=} Y, \quad n \geq 2,$$

characterizes the exponential distributions [see, e.g., Kotz and Steutel (1988), Yeo and Milne (1989), Huang and Chen (1989)]. Similarly, for any $0 < \alpha < 1$, the relation

$$B_{n-1}^{1/\alpha} \sum_{i=1}^n Y_i \stackrel{d}{=} Y, \quad n \geq 2, \tag{2.2.51}$$

characterizes Mittag–Leffler distributions, mentioned earlier, which follows from the results of Pakes (1992ab) and Alamatsaz (1993).

Remark 2.2.17 If the Y_i's are symmetric, then for any $0 < \alpha \leq 2$, relation (2.2.51) characterizes Linnik distributions with index α [see Chapter 3, Section 4.3]. If no assumptions on the distribution of the Y_i's are imposed, then for any $0 < \alpha \leq 2$, relation (2.2.51) characterizes strictly geometric stable distributions, which follows from the results of Pakes (1992ab) and Alamatsaz (1993).

2.2.9 Distributional limits of deterministic sums. One of the basic versions of the central limit theorem (CLT) states that whenever $X_1, X_2, \ldots$ is a sequence of i.i.d. random variables with mean μ and variance $\sigma^2 < \infty$, the sequence of the partial sums,

$$a_n \sum_{i=1}^{n} (X_i - \mu), \tag{2.2.52}$$

where $a_n = n^{-1/2}$, converges in distribution to a normal r.v. with mean zero and variance σ^2. As we saw in Section 2.2.7, the limit may not have a normal distribution if the number of terms in the summation is a *random variable*. Similarly, we may arrive at a nonnormal limit of (2.2.52) if the normalizing sequence a_n is *random*. The following result shows that under beta-distributed a_n's we obtain in the limit a Laplace distribution. We thus have an additional characterization of this class.

Proposition 2.2.12 *Let $X_1, X_2, \ldots$ be nondegenerate i.i.d. r.v.'s with mean μ and finite variance, and for each $n > 1$, let the r.v. B_n be independent of X_i's and have a beta distribution with density* (2.2.45). *Then as $n \to \infty$, the class of nondegenerate distributional limits of* (2.2.52) *with $a_n = B_{n-1}^{1/2}$ coincides with the class of Laplace distributions with zero mean.*

Proof. Evidently, if Y has a Laplace distribution, then in view of (2.2.46), Y is the limit of (2.2.52) with $X_i \stackrel{d}{=} Y$. Thus it is sufficient to show that the sums (2.2.52) with $a_n = B_{n-1}^{1/2}$ converge to a Laplace distribution. To this end, we proceed as in the proof of Proposition 2.2.11, writing (2.2.52) as $U_n V_n$, where

$$U_n = (nB_{n-1})^{1/2} \quad \text{and} \quad V_n = \frac{\sum_{i=1}^{n}(X_i - \mu)}{n^{1/2}}, \tag{2.2.53}$$

and analogously showing that the limit of the product has indeed a Laplace distribution. □

2.3 Functions of Laplace random variables

In this section we discuss distributions of certain standard functions of independent Laplace random variables, including sum, product, and ratio.

2.3.1 The distribution of the sum of independent Laplace variates. Let us first consider two independent classical Laplace random variables X_1 and X_2 with densities

$$f_i(x) = \frac{1}{2s_i} e^{-|x|/s_i}, \quad i = 1, 2, \quad x \in \mathbb{R}. \tag{2.3.1}$$

Our goal is to find the probability distribution of the sum

$$Y = X_1 + X_2. \tag{2.3.2}$$

By symmetry, the difference $X_1 - X_2$ has the same distribution as the sum (2.3.2). Using Proposition 2.2.2 one can write each X_i as the difference of exponential random variables so that the sum of two independent Laplace r.v.'s is a linear combination of four independent standard exponential variables denoted below by Z_i's:

$$Y \stackrel{d}{=} s_1(Z_1 - Z_2) + s_2(Z_3 - Z_4). \tag{2.3.3}$$

(This lack of closure is in contrast with the normal case, where the sum of independent normal variables is normal.) Rearranging the terms, we have

$$Y \stackrel{d}{=} (s_1 Z_1 - s_2 Z_4) - (s_1 Z_2 - s_2 Z_3) = \frac{1}{\sqrt{s_1 s_2}}(W_1 - W_2), \tag{2.3.4}$$

where

$$W_1 = \frac{1}{\kappa} Z_1 - \kappa Z_4 \text{ and } W_2 = \frac{1}{\kappa} Z_2 - \kappa Z_3 \tag{2.3.5}$$

are independent and identically distributed random variables and

$$\kappa = \sqrt{\frac{s_2}{s_1}} \tag{2.3.6}$$

is a positive constant.

We proceed by first finding the distribution of the W_i's and then the distribution of their difference. To accomplish the first step, we use the following result.

Lemma 2.3.1 *Let G_1 and G_2 be i.i.d. random variables with standard gamma distribution given by the density*

$$g(x) = \frac{1}{\Gamma(\nu)} x^{\nu-1} e^{-x}, \quad \nu > 0, \quad x > 0. \tag{2.3.7}$$

Let κ be a positive constant. Then the probability density of the random variable

$$W = \frac{1}{\kappa} G_1 - \kappa G_2 \tag{2.3.8}$$

is

$$h(x) = \frac{1}{\Gamma(\nu)\sqrt{\pi}} \left(\frac{|x|}{\kappa + 1/\kappa} \right)^{\nu - 1/2} e^{\frac{1}{2}(1/\kappa - \kappa)x} K_{\nu-1/2}\left(\frac{1}{2}(1/\kappa + \kappa)|x| \right), \quad x \neq 0, \tag{2.3.9}$$

where K_λ is the modified Bessel function of the third kind with the index λ, given in the appendix.

Remark 2.3.1 The distribution with density (2.3.9) is for obvious reasons known as the Bessel function distribution [see, e.g., Pearson et al. (1929)]. We shall study this class of distributions in Section 4.1 of Chapter 4.

Proof. First, note that the densities of $X_1 = \frac{1}{\kappa} G_1$ and $X_2 = \kappa G_2$ are $\kappa g(\kappa x)$ and $\frac{1}{\kappa} g(\frac{x}{\kappa})$, respectively, where g is the density of G_1 (and G_2) given by (2.3.7). Next, by independence, the joint density of X_1 and X_2 is

$$f(x_1, x_2) = g(\kappa x) g\left(\frac{x}{\kappa}\right) = \frac{1}{[\Gamma(\nu)]^2} (x_1 x_2)^{\nu-1} e^{-\kappa x_1 - \frac{1}{\kappa} x_2}, \quad x_1, x_2 > 0. \tag{2.3.10}$$

Consider a one-to-one transformation $W = X_1 - X_2$, $Z = X_2$. The inverse transformation, $X_1 = W + Z$, $X_2 = Z$, has the Jacobian equal to one, so the joint density of W and Z is

$$p(w,z) = f(w+z,z), \quad z, w+z > 0. \tag{2.3.11}$$

The marginal density of $W = X_1 - X_2$ can be found by integrating the joint density (2.3.11) with respect to z:

$$h(w) = \int_{-\infty}^{\infty} f(w+z,z)dz. \tag{2.3.12}$$

Combining (2.3.10) and (2.3.12), for $w < 0$ we obtain

$$h(w) = \frac{1}{[\Gamma(\nu)]^2} e^{-\kappa w} \int_{-w}^{\infty} z^{\nu-1}(z+w)^{\nu-1} e^{-(\kappa+\frac{1}{\kappa})z} dz. \tag{2.3.13}$$

Now the application of the integration formula (A.0.14) of Bessel functions (see the appendix) with $\mu = \nu$, $u = -w$, and $\beta = \kappa + \kappa^{-1}$ leads to (2.3.9).

Similarly, for $w > 0$, we have

$$h(w) = \frac{1}{[\Gamma(\nu)]^2} e^{-\kappa w} \int_{0}^{\infty} z^{\nu-1}(z+w)^{\nu-1} e^{-(\kappa+\frac{1}{\kappa})z} dz. \tag{2.3.14}$$

The change of variable $x = w + z$ results in

$$h(w) = \frac{1}{[\Gamma(\nu)]^2} e^{\frac{1}{\kappa} w} \int_{w}^{\infty} x^{\nu-1}(x-w)^{\nu-1} e^{-(\kappa+\frac{1}{\kappa})x} dx. \tag{2.3.15}$$

Another application of (A.0.14), this time with $u = w$, produces (2.3.9). The result follows. □

To find the density of the W_i's given by (2.3.5), we apply Lemma 2.3.1 with $\nu = 1$. Here, the Bessel function with index $1/2$ has a closed form given by (A.0.11) in the appendix, and the density of W_1 takes the form

$$\begin{aligned}
h(x) &= \frac{1}{\Gamma(1)\sqrt{\pi}} \left(\frac{|x|}{\kappa + 1/\kappa}\right)^{1/2} e^{\frac{1}{2}(1/\kappa-\kappa)x} K_{1/2}\left(\frac{1}{2}(1/\kappa+\kappa)|x|\right) \\
&= \frac{1}{\sqrt{\pi}} \frac{|x|^{1/2}}{(\kappa+1/\kappa)^{1/2}} e^{\frac{1}{2}(1/\kappa-\kappa)x} \frac{\sqrt{\pi}}{(\kappa+1/\kappa)^{1/2}|x|^{1/2}} e^{-\frac{1}{2}(1/\kappa+\kappa)|x|} \\
&= \frac{1}{\kappa+1/\kappa} e^{\frac{1}{2}(1/\kappa-\kappa)x - \frac{1}{2}(1/\kappa+\kappa)|x|},
\end{aligned}$$

which can be written as

$$h(x) = \frac{1}{1/\kappa + \kappa} \begin{cases} e^{-\kappa|x|}, & \text{for } x \ge 0, \\ e^{-\frac{1}{\kappa}|x|}, & \text{for } x < 0. \end{cases} \tag{2.3.16}$$

Remark 2.3.2 For $\kappa \neq 1$ we obtain an asymmetric Laplace distribution to be studied in detail in Chapter 3.

Next, we shall derive the distribution of the difference $W_1 - W_2$, where the W_i's are i.i.d. variables defined by (2.3.5) with densities given by (2.3.16).

Proposition 2.3.1 *Let W_1 and W_2 be i.i.d. r.v.'s with density* (2.3.16). *Then the density of $V = W_1 - W_2$ is*

$$f_V(x) = \begin{cases} \frac{1}{4}(1+|x|)e^{-|x|}, x \in \mathbb{R}, & \text{for } \kappa = 1, \\ \frac{1}{2}\frac{\kappa}{1-\kappa^4}(e^{-\kappa|x|} - \kappa^2 e^{-\frac{1}{\kappa}|x|}), x \in \mathbb{R}, & \text{for } \kappa \in (0,1)\cup(1,\infty). \end{cases} \tag{2.3.17}$$

Proof. The density of $V = W_1 - W_2$ is related to the common density of W_1 and W_2 as follows:

$$f_V(x) = \int_{-\infty}^{\infty} h(x+y)h(y)dy. \tag{2.3.18}$$

Since the probability density of the difference of two i.i.d. random variables is symmetric, it is sufficient to consider $x > 0$. Splitting the region of integration according to positivity and negativity of the functions $h(x+y)$ and $h(y)$, we obtain $f_V(x) = I_1 + I_2 + I_3$, where

$$I_1 = \int_{-\infty}^{-x} h(x+y)h(y)dy,\ I_2 = \int_{-x}^{0} h(x+y)h(y)dy,\ I_3 = \int_{0}^{\infty} h(x+y)h(y)dy. \tag{2.3.19}$$

We evaluate the above integrals utilizing (2.3.16):

$$I_1 = \left(\frac{1}{1/\kappa+\kappa}\right)^2 \int_{-\infty}^{-x} e^{\frac{1}{\kappa}(x+y)} e^{\frac{1}{\kappa}y} dy = \left(\frac{1}{1/\kappa+\kappa}\right)^2 \frac{\kappa}{2} e^{-\frac{1}{\kappa}x}, \tag{2.3.20}$$

$$\begin{aligned} I_2 &= \left(\frac{1}{1/\kappa+\kappa}\right)^2 \int_{-x}^{0} e^{-\kappa(x+y)} e^{\frac{1}{\kappa}y} dy \\ &= \left(\frac{1}{1/\kappa+\kappa}\right)^2 \begin{cases} \frac{1}{1/\kappa-\kappa}(e^{-\kappa x} - e^{\frac{1}{\kappa}x}), & \text{for } \kappa \neq 1, \\ xe^{-x}, & \text{for } \kappa = 1, \end{cases} \end{aligned} \tag{2.3.21}$$

$$I_3 = \left(\frac{1}{1/\kappa+\kappa}\right)^2 \int_{0}^{\infty} e^{-\kappa(x+y)} e^{-\kappa y} dy = \frac{1}{2\kappa} e^{-\kappa x}. \tag{2.3.22}$$

Combining (2.3.20)–(2.3.22) and simplifying, we obtain the density (2.3.17) of V. □

We now return to the representation (2.3.4) of Y. Using Proposition 2.3.1 along with (2.3.6), we obtain the following density of the sum $X_1 + X_2$ (and the difference $X_1 - X_2$):

$$f_{X_1+X_2}(x) = \begin{cases} \frac{1}{4}s(1+s|x|)e^{-s|x|}, & \text{for } s_1 = s_2 = s, \\ \frac{1}{2}\frac{s_2}{s_1}\frac{1}{1-(s_2/s_1)^2}(s_1 e^{-s_2|x|} - s_2 e^{-s_1|x|}), & \text{for } s_1 \neq s_2. \end{cases} \tag{2.3.23}$$

Remark 2.3.3 Note that the distribution of the sum of two independent Laplace r.v.'s with the same scale parameters is of a different type and much simpler than the one when the scale parameters are different.

Remark 2.3.4 Weida (1935) obtained the distribution of the difference $X_1 - X_2$ by inverting the relevant characteristic function. His derivation, however, seems to be not quite correct.

Next, we consider the case of more than two identically distributed and independent standard classical Laplace r.v.'s with a common density given by (2.3.1) with the scale parameter equal to 1. Recall that the sum of n such variables has a representation in terms of gamma and standard normal random variables (Proposition 2.2.10). Now Lemma 2.3.1 can be used for the derivation of the density of the sum T of these i.i.d. random variables (as well as the density of the corresponding arithmetic mean). Indeed, since for each $i = 1, \dots, n$ we have

$$X_i \stackrel{d}{=} Z_i - Z_i',$$

where Z_i and Z_i' are i.i.d. standard exponential variables (Proposition 2.2.2), it follows that

$$T = n\overline{X}_n = \sum_{i=1}^n X_i \stackrel{d}{=} \sum_{i=1}^n Z_i - \sum_{i=1}^n Z_i' = G_1 - G_2, \tag{2.3.24}$$

where G_1 and G_2 are i.i.d. standard gamma r.v.'s with density (2.3.7) with the shape parameter $\nu = n$. Thus the density of the sum T is given by (2.3.9) with $\nu = n$ and $\kappa = 1$. Since the Bessel function $K_{\nu-1/2}$ admits the closed form (A.0.10) for $\nu = n$, we obtain the following formula for the density of T:

$$f_T(x) = \frac{e^{-|x|}}{(n-1)!2^n} \sum_{j=0}^{n-1} \frac{(n-1+j)!}{(n-1-j)!j!} \frac{|x|^{n-1-j}}{2^j}, \quad x \in \mathbb{R}. \tag{2.3.25}$$

For the arithmetic mean $\overline{X}_n = T/n$, we have the density

$$f_{\overline{X}_n}(x) = nf_T(nx), \quad x \in \mathbb{R}. \tag{2.3.26}$$

In the following result we present a useful representation of T derived in Kou (2000) (see Exercise 2.7.18).

Proposition 2.3.2 *Let $X_1, \dots, X_n$ be i.i.d. standard classical Laplace variables. Then*

$$T = X_1 + \cdots + X_n \stackrel{d}{=} I \cdot \sum_{j=1}^{M_n} Z_j, \tag{2.3.27}$$

where the Z_j's are i.i.d. standard exponential variables, I takes on values ± 1 with probabilities $1/2$, and M_n is an integer-valued r.v. given by the probability function

$$P(M_n = j) = \frac{2^j}{2^{2n-1}} \binom{2n-j-1}{n-1}, \quad j = 1, 2, \dots, n. \tag{2.3.28}$$

[The Z_j's, I, and M_n are mutually independent, and $\binom{0}{0}$ is defined as 1.]

Table 2.4 below contains the densities of X_n for sample sizes $n = 1, 2, 3, 4$, which were worked out in Craig (1932)[11] [see also Edwards (1948)]. Weida (1935) in one of the early papers devoted to the Laplace distribution obtained an expression for the density of $\overline{X}_n$ by inverting the relevant characteristic function. However, his formula is not as simple as ours and involves the derivative of order $n-1$ (with respect to t) of the function $e^{-itnx}(1+it)^n$.

Remark 2.3.5 As noted by Johnson et al. (1995), many authors considered sums or arithmetic means and related statistics under an underlying Laplace model, including Hausdorff (1901), Craig (1932), Weida (1935), and Sassa (1968). In particular, Balakrishnan and Kocherlakota (1986) utilized the density (2.3.26) in studying the effects of nonnormality on $\overline{X}$-charts. They showed that the probabilities α (false alarm) and $1-\beta$ (true alarm) remain almost unchanged when the underlying normal distribution is replaced by the Laplace distribution, and concluded that no modification to the control charts was necessary in this case.

[11] In Craig (1932), the coefficient of $|x|^2$ for $n = 4$ contains a printing error (98 instead of 96).

n	**Density of $\overline{X}_n$**
1	$f(x) = \frac{1}{2}e^{-\lvert x\rvert}, \quad x \in \mathbb{R}$
2	$f(x) = \frac{1}{2}(1+2\lvert x\rvert)e^{-2\lvert x\rvert}, \quad x \in \mathbb{R}$
3	$f(x) = \frac{9}{16}(1+3\lvert x\rvert+3\lvert x\rvert^2)e^{-3\lvert x\rvert}, \quad x \in \mathbb{R}$
4	$f(x) = \frac{1}{24}(15+60\lvert x\rvert+96\lvert x\rvert^2+64\lvert x\rvert^3)e^{-4\lvert x\rvert}, \quad x \in \mathbb{R}$

Table 2.4: Densities of the sample means $\overline{X}_n$ for samples of selected sizes n from a standard classical Laplace distribution with ch.f. $\psi(t) = (1+t^2)^{-1}$.

2.3.2 The distribution of the product of two independent Laplace variates. Consider two independent classical Laplace random variables X_1 and X_2 with densities (2.3.1). We shall find the probability distribution of the random variable

$$Y = X_1 X_2. \tag{2.3.29}$$

Since $X_i \stackrel{d}{=} s_i I_i W_i$, where for $i = 1, 2$, W_i is the standard exponential while I_i is independent of W_i and takes on values ± 1 with probabilities $1/2$ (Proposition 2.2.3), we have

$$Y \stackrel{d}{=} s_1 s_2 (I_1 I_2) W_1 W_2 = s_1 s_2 I W_1 W_2, \tag{2.3.30}$$

where $I = I_1 I_2$ is independent of the W_i's and has the same distribution as each of the I_i's. Consequently, we need to find the distribution of the product of two independent standard exponential random variables. For $x > 0$ we have

$$P(W_1 W_2 \le x) = \int_0^\infty P\left(W_1 \le \frac{x}{z}\right) e^{-z} dz = 1 - \int_0^\infty e^{-(xz^{-1}+z)} dz,$$

as $P(W_1 < u) = 1 - e^{-u}$. We now utilize the definition (A.0.4) of Bessel functions (see the appendix) with $\lambda = -1$ and $u = 2\sqrt{x}$ (noting that $K_\lambda = K_{-\lambda}$) to obtain

$$\int_0^\infty e^{-(xz^{-1}+z)} dz = 2\sqrt{x} K_1(2\sqrt{x}),$$

so the distribution function of $W_1 W_2$ takes the form

$$F_{W_1 W_2}(x) = 1 - 2\sqrt{x} K_1(2\sqrt{x}). \tag{2.3.31}$$

Next, we take the derivative, using the relations (A.0.8) and (A.0.9) of Bessel functions (see the appendix) to obtain an expression for the probability density of $W_1 W_2$:

$$f_{W_1 W_2}(x) = 2K_0(2\sqrt{x}). \tag{2.3.32}$$

Thus, in view of (2.3.30), the density of Y is

$$f_Y(x) = \frac{4}{s_1 s_2} K_0\left(2\sqrt{\frac{|x|}{s_1 s_2}}\right), \quad x \in \mathbb{R}. \tag{2.3.33}$$

It is interesting to compare (2.3.33) with the density of the product of two independent normal variables with means equal to zero and the same variances as those of X_1 and X_2,

$$g(x) = \frac{1}{2\pi s_1 s_2} K_0\left(\frac{|x|}{2 s_1 s_2}\right) \tag{2.3.34}$$

[see, e.g., Craig (1936)]. In both cases the density of the product depends on x through the same Bessel function K_0, and the argument for the Laplace case is essentially the square root of the argument for the normal case (thus in a sense the product retains the original structure of these distributions). Graphs of these two densities are presented in Figure 2.3 (top).

2.3.3 The distribution of the ratio of two independent Laplace variates. Let X_1 and X_2 be two independent classical Laplace random variables with densities (2.3.1). We seek the probability distribution of the random variable

$$Y = \frac{X_1}{X_2}. \tag{2.3.35}$$

Using the representation $X_i \stackrel{d}{=} s_i I_i W_i$ given in Proposition 2.2.3, we have

$$Y \stackrel{d}{=} \frac{s_1}{s_2} \frac{I_1}{I_2} \frac{W_1}{W_2} = \frac{s_1}{s_2} I \frac{W_1}{W_2}, \tag{2.3.36}$$

where $I = I_1/I_2$ takes values ± 1 with equal probabilities and is independent of the standard exponential r.v.'s W_i. We thus must find the distribution of the ratio of two independent standard exponential random variables.

First, we find the distribution function by conditioning. For $x > 0$ we have

$$P\left(\frac{W_1}{W_2} \le x\right) = \int_0^\infty P(W_1 \le xz) e^{-z} dz = 1 - \int_0^\infty e^{-z(x+1)} dz = 1 - \frac{1}{1+x}.$$

Hence the ratio W_1/W_2 has a standard Pareto distribution of the second kind [the so-called Lomax distribution; see, e.g., Johnson et al. (1994), p. 575, or Springer (1979), p. 161] with density

$$f_{W_1/W_2}(x) = \left(\frac{1}{1+x}\right)^2, \quad x \ge 0. \tag{2.3.37}$$

Consequently, the distribution of Y is "double" Pareto with density[12]

$$f_Y(x) = \frac{1}{2} \frac{s_2}{s_1} \left(\frac{1}{1 + (s_2/s_1)|x|}\right)^2, \quad x \in \mathbb{R}. \tag{2.3.38}$$

Note that as in the normal case, where the ratio of two mean zero normal random variables has Cauchy distribution, the distribution with density (2.3.38) has infinite mean and variance. (However,

[12] It should be noted that our result does not fully agree with Weida (1935).

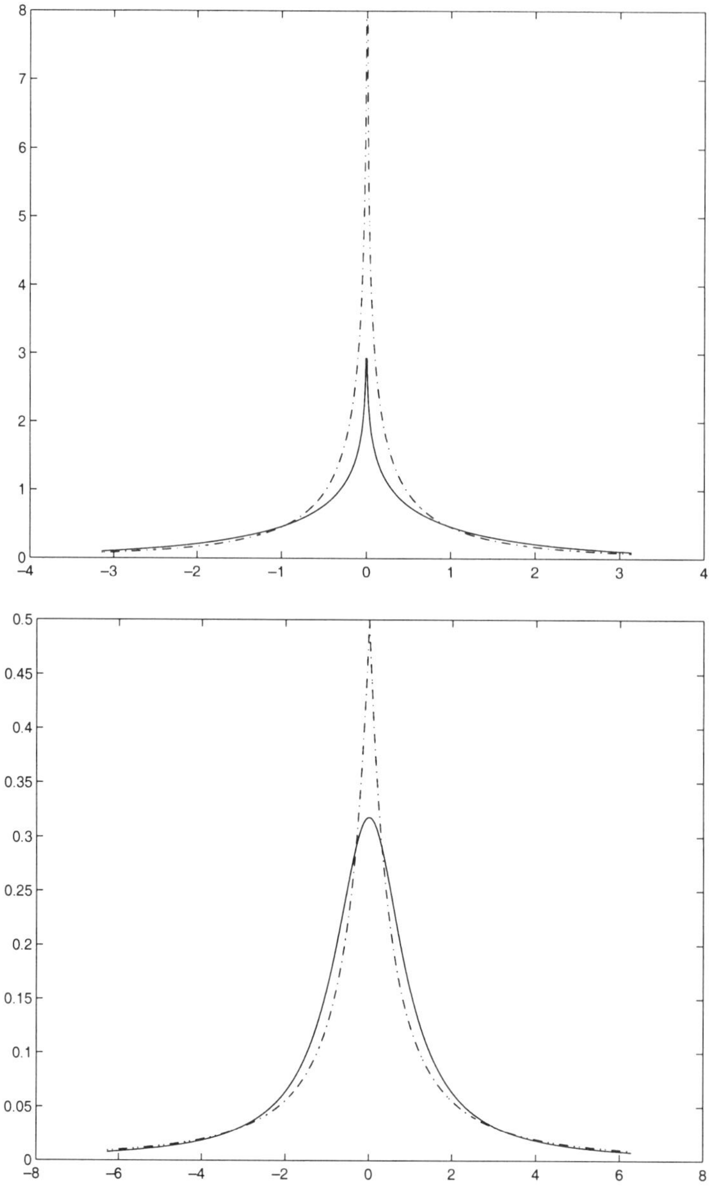

Figure 2.3: Densities of the product (top) and the ratio (bottom) of two i.i.d. standard Laplace random variables (dashed lines) vs. two i.i.d. standard Gaussian random variables (solid lines).

the *fractional moments* $E|Y|^\alpha$ do exist for $0 < \alpha < 1$.) In the i.i.d. case the densities of the ratio of two mean-zero Laplace and two mean-zero normal variables are

$$\frac{1}{2}\left(\frac{1}{1+|x|}\right)^2 \quad \text{and} \quad \frac{1}{\pi}\frac{1}{1+x^2}, \qquad x \in \mathbb{R},$$

respectively. Graphs of these two densities are shown in Figure 2.3 (bottom).

Remark 2.3.6 Note that the same distribution arises under an appropriate randomization of the scale parameter s of the classical Laplace distribution $\mathcal{CL}(0, s)$ (see Exercise 2.7.48).

Table 2.5 below summarizes our results on distributions of common functions of independent Laplace variables.

Function	Distribution
$Y_1 + \cdots + Y_n$	$f(x) = \frac{e^{-\lvert x\rvert}}{(n-1)!2^n} \sum_{j=0}^{n-1} \frac{(n-1+j)!}{(n-1-j)!j!} \frac{\lvert x\rvert^{n-1-j}}{2^j}, \quad x \in \mathbb{R}.$
$X_1 \pm X_2$	$f(x) = \begin{cases} \frac{1}{4}\sqrt{s_1 s_2}(1 + \sqrt{s_1 s_2}\lvert x\rvert)e^{-\sqrt{s_1 s_2}\lvert x\rvert}, & s_1 = s_2, \\ \frac{1}{2}\frac{s_2}{s_1}\frac{1}{1-(s_2/s_1)^2}(s_1 e^{-s_2\lvert x\rvert} - s_2 e^{-s_1\lvert x\rvert}), & s_1 \neq s_2, \end{cases}$
$X_1 \cdot X_2$	$f(x) = \frac{4}{s_1 s_2} K_0\left(2\sqrt{\frac{\lvert x\rvert}{s_1 s_2}}\right), \quad x \in \mathbb{R}.$
X_1/X_2	$f(x) = \frac{1}{2}\frac{s_2}{s_1}\left(\frac{1}{1+(s_2/s_1)\lvert x\rvert}\right)^2, \quad x \in \mathbb{R}.$

Table 2.5: Densities and distributions of sums and products of independent Laplace random variables. Here Y_i, $i = 1, \ldots, n$, are i.i.d. $\mathcal{CL}(0, 1)$ r.v.'s, while X_1 and X_2 are independent $\mathcal{CL}(0, s_1)$ and $\mathcal{CL}(0, s_2)$ r.v.'s.

2.3.4 The t-statistic for a double exponential (Laplace) distribution.

Let $X_1, \ldots, X_n$ be i.i.d. variables with common density f, where $f(x) > 0$ for all x. Define

$$\overline{X}_n = \frac{1}{n}\sum_{i=1}^{n} X_i, \qquad S_n^2 = \sum_{i=1}^{n}(X_i - \overline{X}_n)^2. \tag{2.3.39}$$

The independence of $\overline{X}_n$ and S_n^2 is a unique property of the normal distribution and plays an important role in the derivation of the probability distribution of the t-statistic,

$$T_n = \frac{\overline{X}_n - \theta}{\sigma/\sqrt{n}} \Bigg/ \sqrt{\frac{1}{n-1}S_n^2/\sigma^2} = \frac{\sqrt{n}(\overline{X}_n - \theta)}{S_n/\sqrt{n-1}}, \tag{2.3.40}$$

where θ and σ^2 are the mean and the variance of X_1. In this section, we shall follow Sansing (1976) and discuss the distribution of T_n defined above when the parent population is classical Laplace (with the mean equal to zero). Let

$$f_{\overline{X}_n, S_n^2}(x, y), \qquad -\infty < x < \infty, y > 0, \tag{2.3.41}$$

be the joint density of $\overline{X}_n$ and S_n^2 based on a sample of size $n \geq 2$. Sansing and Owen (1974) derived a recursive relation for $f_{\overline{X}_n, S_n^2}$ presented below.

Lemma 2.3.2 *Let $X_1, X_2, \ldots$ be i.i.d. variables with common density f, where $f(x) > 0$ for all x. Then for any $n \geq 2$, $-\infty < x < \infty$, and $y > 0$, we have*

$$f_{\overline{X}_{n+1}, S_{n+1}^2}(x, y) = \sqrt{\frac{n+1}{n} y} \int_{-1}^{1} w(u) du, \tag{2.3.42}$$

where

$$w(u) = f_{\overline{X}_n, S_n^2}\left(x + \frac{u\sqrt{y}}{\sqrt{n(n+1)}}, y(1-u^2)\right) f\left(x - u\sqrt{\frac{n}{n+1} y}\right). \tag{2.3.43}$$

Proof. Note that

$$\overline{X}_{n+1} = \frac{n}{n+1}\overline{X}_n + \frac{1}{n+1}X_{n+1} \tag{2.3.44}$$

and

$$S_{n+1}^2 = S_n^2 + \frac{n}{n+1}(\overline{X}_n - X_{n+1})^2. \tag{2.3.45}$$

Since X_{n+1} is independent of $\overline{X}_n$ and S_n^2, the joint density of $\overline{X}_n$, S_n^2, and X_{n+1} is

$$f_{\overline{X}_n, S_n^2}(x, y) f(z). \tag{2.3.46}$$

Using the auxiliary variable

$$U = \sqrt{\frac{n}{n+1} \frac{1}{S_n^2}} (\overline{X}_n - X_{n+1}), \tag{2.3.47}$$

we obtain the relation (2.3.42); see Sansing and Owen (1974) for details. □

For $n = 2$, we get directly

$$f_{\overline{X}_2, S_2^2}(x, y) = \sqrt{\frac{2}{y}} f\left(x + \sqrt{\frac{y}{2}}\right) f\left(x - \sqrt{\frac{y}{2}}\right), \quad -\infty < x < \infty, y > 0, \tag{2.3.48}$$

while for $n = 3$ the relation (2.3.42) produces

$$f_{\overline{X}_3, S_3^2}(x, y) = \sqrt{3} \int_{-1}^{1} (1-u^2)^{-1/2} \prod_{i=1}^{3} f(x + \sqrt{y} a_{i3}(u)) du, \tag{2.3.49}$$

where

$$a_{13}(u) = \frac{u}{\sqrt{6}} + \sqrt{\frac{1-u^2}{2}},$$

$$a_{23}(u) = \frac{u}{\sqrt{6}} - \sqrt{\frac{1-u^2}{2}},$$

$$a_{33}(u) = -u\sqrt{\frac{2}{3}},$$

so that

$$\sum_{i=1}^{3} a_{i3}(u) = 0, \qquad \sum_{i=1}^{3} a_{i3}^2(u) = 1.$$

Assume now that f is the density (2.1.1) of the classical Laplace distribution with mean $\theta = 0$ and scale parameter $s > 0$. Then for the random sample of size $n = 2$, the joint density of $\overline{X}_2$ and S_2^2 is

$$f_{\overline{X}_2,S_2^2}(x, y) = \frac{1}{4s^2}\sqrt{\frac{2}{y}} \cdot \begin{cases} e^{-2|x|/s} & \text{if } |x| \geq \sqrt{y/2} \\ e^{-\sqrt{2y}/s} & \text{if } |x| < \sqrt{y/2}. \end{cases} \tag{2.3.50}$$

[Note that here S_2^2 is simply $(X_1 - X_2)^2/2$.] Thus the density of the t-statistic when $n = 2$ is (Exercise 2.7.23)

$$f_{T_2}(t) = \begin{cases} \frac{1}{4} & \text{if } |t| < 1 \\ \frac{1}{4t^2} & \text{if } |t| \geq 1. \end{cases} \tag{2.3.51}$$

For $n = 3$, we obtain from (2.3.49)

$$f_{\overline{X}_3,S_3^2}(x, y) = \frac{\sqrt{3}}{8s^3}\int_{-1}^{1}(1-u^2)^{-1/2}e^{-\frac{\sqrt{y}}{s}\sum_{i=1}^{3}\left|\frac{x}{\sqrt{y}}+a_{i3}(u)\right|}du, \tag{2.3.52}$$

with the functions $a_{i3}(u)$ as before. As noted by Sansing (1976), in the region $|x/\sqrt{y}| \geq \sqrt{2/3}$, we can express (2.3.52) as follows:

$$f_{\overline{X}_3,S_3^2}(x, y) = \frac{\sqrt{3}\pi}{8s^3}e^{-3|x|/s}, \qquad \frac{x}{\sqrt{y}} \geq \sqrt{\frac{2}{3}}. \tag{2.3.53}$$

Further, a similar relation holds for other sample sizes as well [see Sansing (1976)],

$$f_{\overline{X}_n,S_n^2}(x, y) = \frac{\sqrt{n}\pi^{(n-1)/2}\sqrt{y}^{n-3}}{2^n s^n \Gamma\left(\frac{n-1}{2}\right)}e^{-n|x|/s}, \qquad \frac{x}{\sqrt{y}} \geq \sqrt{\frac{n-1}{n}}. \tag{2.3.54}$$

Using (2.3.54), we follow Sansing (1976) to derive the distribution function of the t-statistic (2.3.40) for $t > n - 1$:

$$F_{T_n}(t) = 1 - \frac{\pi^{(n-1)/2}\Gamma(n-1)}{\sqrt{n}2^{n-1}\Gamma\left(\frac{n-1}{2}\right)}\left(\frac{n-1}{n}\right)^{(n-1)/2} t^{-n+1}. \tag{2.3.55}$$

Finally, differentiating (2.3.55), we obtain the p.d.f. of T_n:

$$f_{T_n}(t) = \frac{\pi^{(n-1)/2}\Gamma(n)}{\sqrt{n}2^{n-1}\Gamma\left(\frac{n-1}{2}\right)}\left(\frac{n-1}{n}\right)^{(n-1)/2}|t|^{-n}, \qquad |t| > n - 1. \tag{2.3.56}$$

Remark 2.3.7 Note that the tails of the density (2.3.56) are heavier than those of the corresponding t-distribution with n degrees of freedom (Exercise 2.7.25).

Remark 2.3.8 As noted by Sansing (1976), the evaluation of the joint density of $\overline{X}_n$ and S_n^2 in the region where $|x|/\sqrt{y} < \sqrt{(n-1)/n}$ is quite complicated. For this case Sansing (1976) derived upper and lower bounds for the joint density, leading to the corresponding bounds for the density f_{T_n} of the t-statistic in the region $|t| \leq n - 1$, where the exact formula (2.3.56) is not valid.

Remark 2.3.9 Gallo (1979) considered an analogue of the t-distribution defined as

$$\tilde{T}_n = \frac{U_n - n\theta}{V_n}, \tag{2.3.57}$$

where

$$U_n = \sum_{i=1}^{n} X_i \text{ and } V_n = \sum_{i=1}^{n} |X_i - \theta|. \tag{2.3.58}$$

[and $X_1, \ldots, X_n$ is a random sample from the $\mathcal{CL}(\theta, s)$ distribution]. The joint distribution of U_n and V_n [derived in Gallo (1979)] consists of a continuous part supported by the region

$$I = \{(u, v) : v \geq 0, n\theta < u < n\theta + v\} \tag{2.3.59}$$

and a singular part concentrated on the boundary of I. The corresponding statistic $\tilde{T}_n$ defined in (2.3.57) has support in the interval $[-1, 1]$ (see Exercise 2.7.24). The distribution function of $\tilde{T}_n$ is

$$\tilde{F}_n(x) = \begin{cases} 0 & \text{for } x < -1, \\ \frac{1}{2^n}\left\{1 + \sum_{i=1}^{n-1} a_i \int_0^\infty \Gamma\left(i, \frac{1-x}{1+x}z\right) z^{n-i-1} e^{-z} dz\right\} & \text{for } -1 \leq x < 1, \\ 1 & \text{for } x \geq 1, \end{cases} \tag{2.3.60}$$

where

$$a_i = \binom{n}{i} \frac{1}{\Gamma(i)} \frac{1}{\Gamma(n-i)}$$

and

$$\Gamma(a, y) = \int_y^\infty t^{a-1} e^{-t} dt \tag{2.3.61}$$

is the incomplete gamma function. Note that the distribution of $\tilde{T}_n$ is a mixture of point masses at ± 1 (each with probability $1/2^n$), and a continuous part (occurring with probability $1 - 2/2^n$) with density

$$\tilde{f}_n(x) = \frac{2^n}{2^n - 2} \frac{\Gamma(n)}{2^{2n-1}} \sum_{i=1}^{n-1} a_i (1-t)^{i-1} (1+t)^{n-i-1}, \quad -1 < t < 1 \tag{2.3.62}$$

[see Gallo (1979).][13]

2.4 Further properties

2.4.1 Infinite divisibility. The notion of infinite divisibility plays a fundamental role in the study of central limit theorems and Lévy processes. A probability distribution with ch.f. ψ is infinitely divisible if, for any integer $n \geq 1$, we have $\psi = \phi_n^n$, where ϕ_n is another characteristic function. In other words, an r.v. Y with ch.f. ψ has the representation

$$Y \stackrel{d}{=} \sum_{i=1}^{n} X_i \tag{2.4.1}$$

[13] Note that the c.d.f. and the p.d.f. of $\tilde{T}_n$ derived in Gallo (1979) may contain some typographical errors.

for some i.i.d. random variables X_i. The importance of the class of infinitely divisible distributions follows from the fact that they are limits of the sums of rows of $(X_{n,i})_{n\in\mathbb{N},i=1,\ldots,n}$, and the terms in each row are i.i.d. (Here, $\mathbb{N}$ denotes the set of natural numbers.) Thus, roughly speaking, if we have a large number of independent and similar random effects that add together, the resulting distribution will be approximately infinitely divisible.

According to (2.2.7), the ch.f. (2.1.8) of a classical Laplace distribution $\mathcal{CL}(\theta, s)$ can be factored as follows

$$\frac{e^{i\theta t}}{(1-ist)(1+ist)} = \left[e^{i\theta t/n}\left(\frac{1}{1-ist}\right)^{1/n}\left(\frac{1}{1+ist}\right)^{1/n}\right]^n = \phi_n^n(t). \tag{2.4.2}$$

For each integer $n \geq 1$, the function ϕ_n is the ch.f. of $\theta/n + Y_{1n} - Y_{2n}$, where Y_{1n} and Y_{2n} are i.i.d. with the ch.f. $(1-ist)^{-1/n}$. The latter is the ch.f. of a gamma distribution with density

$$\frac{(1/s)^{1/n}}{\Gamma(1/n)} x^{\frac{1}{n}-1} e^{-x/s}, \quad x \geq 0. \tag{2.4.3}$$

Consequently, Laplace distributions are infinitely divisible,[14] and we state the result formally.

Proposition 2.4.1 *Let Y have a Laplace distribution with ch.f.* (2.1.8). *Then the distribution of Y is infinitely divisible. Furthermore, for every integer $n \geq 1$, representation* (2.4.1) *holds. Each X_i is distributed as $\theta/n + Y_{1n} - Y_{2n}$, where Y_{1n} and Y_{2n} are i.i.d. with gamma density* (2.4.3).

The ch.f. of every infinitely divisible distribution admits a unique canonical *Lévy–Khinchine representation*. Several variations of this representation using different spectral measures are known. Here we consider the representation which states that a ch.f. of an infinitely divisible distribution can be written uniquely in the form

$$\psi(t) = \exp\left(iat - \frac{1}{2}b^2t^2 + \int_{-\infty}^{\infty}(e^{itx} - 1 - it\sin x)d\Lambda(x)\right), \tag{2.4.4}$$

where $-\infty < a < \infty, b \geq 0$, and Λ is a *Lévy measure* on $(-\infty, \infty)$, characterized by the properties: $\Lambda(\{0\}) = 0$ and $\int_{-\infty}^{\infty} \min(1, x^2)d\Lambda(x) < \infty$. Below, we present the Lévy–Khinchine representation of a Laplace distribution. [See Takano (1988) for a detailed treatment of the d-dimensional density $Ce^{-\|\mathbf{x}\|}$, where $\|\mathbf{x}\|$ is the length of the vector $\mathbf{x}$, including the one-dimensional case $d = 1$.]

Proposition 2.4.2 *The ch.f.* (2.1.8) *of a general classical Laplace distribution $\mathcal{CL}(\theta, s)$ admits the Lévy–Khinchine representation* (2.4.4) *with*

$$a = \theta, \quad b = 0, \quad d\Lambda(x) = \frac{1}{|x|}e^{-|x|/s}dx. \tag{2.4.5}$$

Proof. It is sufficient to prove the result for the standard classical Laplace distribution. We need to show that

$$\frac{1}{1+t^2} = e^{2\int_0^\infty (\cos(xt)-1)e^{-x}x^{-1}dx},$$

[14]Dugué (1951) has raised a question of existence of a probability law that is not infinitely divisible but still can be written as a sum of two independent random variables with distributions parametrized by a continuous parameter. Mistakenly, the Laplace distribution was used as an example. As pointed out by Lukacs (1957), the example is not valid as the Laplace distribution *is* infinitely divisible. Lukacs (1957) also constructs another example that answers positively to the question originally raised by Dugué (1951).

or, equivalently,

$$-\log(1+t^2) = 2\int_0^\infty (\cos(xt) - 1)e^{-x}x^{-1}dx. \tag{2.4.6}$$

Since both sides of (2.4.6) have well-defined Taylor series representations about zero, it is enough to demonstrate that the coefficients in these representations coincide.

The left-hand side has the coefficients

$$a_n = \begin{cases} (-1)^{n/2}2(n-1)! & \text{if } n \text{ is even,} \\ 0 & \text{if } n \text{ is odd.} \end{cases}$$

We now compute the coefficients of the right-hand side. Denoting $c(t,x) = \cos(xt) - 1$, we have for $n \geq 1$:

$$\frac{\partial^n c(t,x)}{\partial t^n} = \begin{cases} (-1)^{n/2}x^n \cos(tx) & \text{if } n \text{ is even,} \\ (-1)^{(n-1)/2}x^n \sin(tx) & \text{if } n \text{ is odd.} \end{cases}$$

Consequently, the nth coefficient of the Taylor representation is zero for odd n while for even n it is given by

$$2\int_0^\infty (-1)^{n/2}x^{n-1}e^{-x}dx = 2(-1)^{n/2}(n-1)!.$$

This completes the proof. □

Remark 2.4.1 For comparison, the Lévy–Khinchine representation of the normal distribution with mean μ and variance σ^2 is simply

$$\psi(t) = e^{i\mu t - \frac{\sigma^2 t^2}{2}},$$

and the Lévy measure Λ is zero in this case.

2.4.2 Geometric infinite divisibility. An r.v. Y (and its probability distribution) is said to be *geometric infinitely divisible* if for any $p \in (0,1)$ it satisfies the relation

$$Y \stackrel{d}{=} \sum_{i=1}^{\nu_p} Y_p^{(i)}, \tag{2.4.7}$$

where ν_p is a geometric r.v. with mean $1/p$, the random variables $Y_p^{(i)}$ are i.i.d. for each p, and ν_p and $(Y_p^{(i)})$ are independent [see, e.g., Klebanov et al. (1984)]. It can be shown that geometric infinite divisible laws are the limits of the sums of rows of $(X_{\nu_p,i})_{\nu_p,i=1,\ldots,\nu_p}$, where terms in each row are i.i.d. conditionally on ν_p, and their number ν_p is random, geometrically distributed, and is independent of the $X_{n,i}$. Thus if we have a large random geometrically distributed number of independent and similar random effects (but depending on the number of effects) that add up together, the observed distribution will be approximately geometric infinitely divisible. This property justifies the interest in and importance of this class of distributions for probabilistic model construction and analysis. The following proposition, which is a direct consequence of Proposition 2.2.7, establishes geometric infinite divisibility of Laplace distributions.

Proposition 2.4.3 *Let Y possess a classical Laplace distribution $\mathcal{CL}(0,s)$. Then Y is geometric infinitely divisible and for any $p \in (0,1)$ relation* (2.4.7) *holds with $Y_p^{(i)} \sim \mathcal{CL}(0, s\sqrt{p})$.*

2.4.3 Self-decomposability. A random variable Y (and its probability distribution) is self-decomposable if for each $c \in (0, 1)$ it has the representation

$$Y \stackrel{d}{=} cY + X, \tag{2.4.8}$$

where X and Y are independent (the distribution of X may depend on c). In terms of ch.f.'s this means that the function $\psi(t)/\psi(ct)$, where ψ is the ch.f. of Y, is a ch.f. for each $c \in (0, 1)$. Evidently, normal distributions are self-decomposable as the corresponding ratio is the ch.f. of a normal distribution. Laplace distributions are also self-decomposable, as was shown by Ramachandran (1997). We shall present explicitly the corresponding representation (2.4.8).

Proposition 2.4.4 *Let Y possess a classical Laplace distribution with ch.f.* (2.1.8). *Then Y is self-decomposable and for any $c \in (0, 1)$ we have*

$$Y \stackrel{d}{=} cY + \theta(1-c) + s(\delta_1 W_1 - \delta_2 W_2), \tag{2.4.9}$$

where δ_1 and δ_2 are dependent r.v.'s taking on values of either zero or one with the probabilities

$$P(\delta_1 = 0, \delta_2 = 0) = c^2, \qquad P(\delta_1 = 1, \delta_2 = 1) = 0,$$
$$P(\delta_1 = 1, \delta_2 = 0) = P(\delta_1 = 0, \delta_2 = 1) = \frac{1}{2}(1-c^2).$$

The r.v.'s W_1 and W_2 are standard exponential and Y, W_1, W_2, (δ_1, δ_2) are mutually independent.

Proof. Write $Y = \theta + sX$, where X is the standard classical Laplace variable. Note that the ch.f. of X given by (2.1.7) can be factored as follows:

$$\left(\frac{1}{(1+ict)(1-ict)}\right)\left(c^2 + \frac{1}{2}(1-c^2)\frac{1}{1-it} + \frac{1}{2}(1-c^2)\frac{1}{1+it}\right), \tag{2.4.10}$$

where the first factor is the ch.f. of cX while the second one is the ch.f. of $\delta_1 W_1 - \delta_2 W_2$. Consequently, we obtain the representation

$$X \stackrel{d}{=} cX + \delta_1 W_1 - \delta_2 W_2. \tag{2.4.11}$$

To arrive at (2.4.9), combine (2.4.11) with

$$Y = \theta + sX.$$

□

We summarize stability properties of the Laplace distribution in Table 2.6. In the second part of Section 2.2 and throughout most of Section 2.4, we have studied various distributional relations involving Laplace distributions. In these relations, unlike those presented in the first part of Section 2.2, random variables distributed according to Laplace distributions appear on both sides of distributional equalities. For this reason, we term them stability properties of Laplace distributions. The variables Y and Y_i's are Laplace $\mathcal{CL}(0, s)$. All the variables in (each) representation presented in Table 2.6 are mutually independent.

2.4.4 Complete monotonicity. A function f defined on an interval $I \subset R$ is called completely monotone (respectively, absolutely monotone) if it is infinitely differentiable on I and $(-1)^k f^{(k)}(x) \geq 0$ (respectively, $f^{(k)}(x) \geq 0$) for any $x \in I$ and any $k = 0, 1, 2, \ldots$. Since the derivatives of the Laplace density are straightforward to calculate, it is easy to see that the p.d.f. of the classical Laplace distribution with mean zero is completely monotone on $(0, \infty)$ [and absolutely monotone on $(-\infty, 0)$]. As noted by Dreier (1999), every symmetric density on $(-\infty, \infty)$, which is completely monotone on $(0, \infty)$, is a scale mixture of Laplace distributions.

Stability property	Variables
$Y \stackrel{d}{=} \sqrt{p}\sum_{i=1}^{\nu_p} Y_i = \sum_{i=1}^{\nu_p} Y_p^{(i)}$	ν_p — geometric r.v. with parameter p, $Y_i^{(p)}$ — i.i.d. $\mathcal{CL}(0, \sqrt{p}\cdot s)$ r.v.'s
$Y \stackrel{d}{=} \sqrt{p} I Y_1 + (1-I)(Y_2 + \sqrt{p} Y_3)$	I – 0–1 r.v. with $P(I=1)=p$
$Y \stackrel{d}{=} \sqrt{p} Y_1 + (1-I) Y_2$	I – 0–1 r.v. with $P(I=1)=p$
$Y \stackrel{d}{=} \sqrt{B_{n-1}}(Y_1 + \cdots + Y_n)$	B_{n-1} — beta r.v. with parameters $n-1$ and 1
$Y \stackrel{d}{=} cY + s(\delta_1 W_1 - \delta_2 W_2)$	W_1, W_2 — standard exponential r.v.'s, δ_1, δ_2 — 0–1 r.v.'s given in Proposition 2.4.4

Table 2.6: Summary of stability properties of the classical Laplace distribution. The variables Y and Y_i's are $\mathcal{CL}(0, s)$. All the variables in each representation are mutually independent.

Proposition 2.4.5 *Let f be a symmetric (about zero) probability density on $(-\infty, \infty)$ which is completely monotone on $(0, \infty)$. Then there exists a distribution function G on $(0, \infty)$ such that*

$$f(x) = \int_0^\infty \frac{1}{2} y e^{-y|x|} dG(y), \quad x \neq 0, \tag{2.4.12}$$

while the ch.f. corresponding to f is

$$\psi(t) = \int_0^\infty \frac{1}{1 + t^2/y^2} G(y), \quad -\infty < t < \infty. \tag{2.4.13}$$

Proof. The result follows from the fact that every completely monotone density on $(0, \infty)$ is a scale mixture of exponential densities on $(0, \infty)$; see Steutel (1970). □

Remark 2.4.2 The converse of Proposition 2.4.5 clearly holds as well: every density of the form (2.4.12) with some c.d.f. G on $(0, \infty)$ is a symmetric density on $(-\infty, \infty)$ which is completely monotone on $(0, \infty)$.

Remark 2.4.3 The central moment

$$\mu_{2m} = E[X^{2m}] \tag{2.4.14}$$

of the $\mathcal{CL}(0, s)$ random variable X is equal to $(2m)!s^{2m}$ [cf. (2.1.14)]. Consequently, for every $1 \le l \le r$ we have

$$\left(\frac{\mu_{2l}}{(2l)!}\right)^{\frac{1}{2l}} = \left(\frac{\mu_{2r}}{(2r)!}\right)^{\frac{1}{2r}}, \tag{2.4.15}$$

since each side in (2.4.15) is equal to s. Actually, the Laplace distribution is the only symmetric distribution on $(-\infty, \infty)$ with completely monotone density on $(0, \infty)$ for which the equality in

(2.4.15) holds; for all other symmetric random variables X on $(-\infty, \infty)$ with completely monotone density on $(0, \infty)$ and finite $2m$th moment (2.4.14), we have an inequality

$$\left(\frac{\mu_{2l}}{(2l)!}\right)^{\frac{1}{2l}} \leq \left(\frac{\mu_{2r}}{(2r)!}\right)^{\frac{1}{2r}}, \quad 1 \leq l \leq r \leq m \tag{2.4.16}$$

[see Dreier (1999)].

2.4.5 Maximum entropy property. One of the basic concepts of *information theory* is the notion of *entropy*, which is a measure of uncertainty associated with a probability distribution. The maximum entropy principle states that, of all distributions that satisfy certain constraints, one should select the one with the largest entropy. A maximum entropy distribution is believed not to incorporate any extraneous information other than that specified by the relevant constraints. Thus finding the maximum entropy distribution could be considered as a general inference procedure, and indeed it was proposed initially by Jaynes (1957) in this manner. It has been successfully applied in a great variety of fields including statistical mechanics, statistics, stock market analysis, queuing theory, image analysis, and reliability estimation [see, e.g., Kapur (1993)].

For a one-dimensional r.v. X with density (or probability function) f, the entropy of X is defined by

$$H(X) = E[-\log f(X)]. \tag{2.4.17}$$

It is well known that among all continuous r.v.'s with mean zero and given variance, the Gaussian (normal) distribution provides the largest entropy [see, e.g., Reza (1961)]. Similarly, the Laplace distribution maximizes the entropy among all continuous distributions with given first absolute moment, as noted by Kagan et al. (1973). Both results easily follow from the following proposition, proved in Kagan et al. (1973).

Proposition 2.4.6 (Kagan, Linnik, and Rao) *Let X be a r.v. with density*

$$p(x) > 0 \text{ for } x \in (a, b) \text{ and } p(x) = 0 \text{ otherwise.} \tag{2.4.18}$$

Let $h_1, h_2, \ldots$ be integrable functions on (a, b) satisfying for given constants $g_1, g_2, \ldots$ the conditions

$$\int_a^b h_i p(x) dx = g_i, \quad i = 1, 2, \ldots. \tag{2.4.19}$$

Then the maximum entropy is attained for the distributions with density of the form

$$p(x) = e^{a_0 + a_1 h_1(x) + \cdots} \tag{2.4.20}$$

(and only by them) if there exist constants $a_0, a_1, \ldots$ such that the above density satisfies the conditions (2.4.18) *and* (2.4.19).

How can we deduce the entropy maximization property of the Laplace distribution from the above proposition? Consider continuous random variables with density p satisfying (2.4.18) with $a = -\infty$, $b = \infty$, and such that

$$\int_{-\infty}^{\infty} |x| p(x) dx = c > 0. \tag{2.4.21}$$

Then according to Proposition 2.4.6, the maximum entropy is attained by the density

$$p(x) = e^{a_0 + a_1|x|}, \quad x \in (-\infty, \infty) \tag{2.4.22}$$

for some constants a_0 and a_1. Let us find the constants so that the function (2.4.22) integrates to 1 on $(-\infty, \infty)$ and satisfies the condition (2.4.21). First, note that $a_1 < 0$ to ensure the integrability of p. Then write

$$1 = \int_{-\infty}^{\infty} e^{a_0} e^{a_1|x|} dx = \frac{2e^{a_0}}{|a_1|} \tag{2.4.23}$$

so that

$$e^{a_0} = \frac{|a_1|}{2}. \tag{2.4.24}$$

Finally, by (2.4.21), we have

$$c = \int_{-\infty}^{\infty} |x| \frac{|a_1|}{2} e^{-|a_1 x|} dx = \int_{0}^{\infty} x|a_1| e^{-|a_1|x} dx = \frac{1}{|a_1|}, \tag{2.4.25}$$

so $a_1 = -1/c$ and density (2.4.22) takes the form

$$p(x) = \frac{1}{2c} e^{-|x|/c}, \quad x \in (-\infty, \infty). \tag{2.4.26}$$

The following result summarizes our discussion.

Proposition 2.4.7 *Consider the class $\mathcal{C}$ of all continuous random variables with nonvanishing densities on $(-\infty, \infty)$ such that*

$$E|X| = c > 0 \text{ for } X \in \mathcal{C}. \tag{2.4.27}$$

Then the maximum entropy is attained for the Laplace r.v. X_c with density (2.4.26), *and*

$$\max_{X \in \mathcal{C}} H(X) = H(X_c) = \log(2c) + 1.$$

Remark 2.4.4 If mean deviation about some fixed point θ is prescribed instead of $E|X|$, then the entropy is maximized by the density $\frac{1}{2c} e^{-|x-\theta|/c}$, where $c = E|X - \theta|$ (Exercise 2.7.30).

Remark 2.4.5 If in addition to (2.4.27) we add the condition that $EX = c_1$, where $|c_1| < c$, then the entropy is maximized by the skewed Laplace distribution studied in Chapter 3 (see Proposition 3.4.7). On the other hand, if the mean along with the absolute deviation *about the mean* are prescribed (instead of EX and $E|X|$), then the entropy is maximized by the symmetric Laplace distribution (Exercise 3.6.18).

Recall that the Laplace distribution $\mathcal{L}(0, \sigma)$ (with mean zero and variance σ^2) can be regarded as Gaussian with a stochastic variance $V = \sigma^2 W$, where W has standard exponential distribution (see Proposition 2.2.1). As noted recently by Levin and Tchernitser (1999), among all zero-mean Gaussian r.v.'s with stochastic variance V (independent of the Gaussian term), for any given value of EV, the Laplace distribution maximizes the entropy of V. This follows from the fact that among all distributions with given mean and $(0, \infty)$ support, the maximum entropy corresponds to the exponential distribution [see Gokhale (1975)], which can be established via Proposition 2.4.6. Here is the exact formulation of this result.

Proposition 2.4.8 *Consider the class $\mathcal{M}$ of random variables of the form $\sqrt{D}Z$, where Z and D are independent, Z is standard normal, while D has a continuous distribution on $(0, \infty)$ with mean σ^2. Then the maximum entropy,*

$$\max_{Y \stackrel{d}{=} \sqrt{D}Z \in \mathcal{M}} H(Y) = \log(\sqrt{2}\sigma) + 1,$$

is attained for the Laplace r.v. $Y \stackrel{d}{=} \sigma\sqrt{W}Z$, where W is standard exponential.

2.5 Order statistics

In this section we shall discuss order statistics of random variables having a Laplace distribution.

Let the measurements obtained from a sample of size n be represented by random variables $X_1, \ldots, X_n$. The X_i's are mutually independent and each one has the same cumulative distribution function (and probability density function, if it exists).

We now introduce n new random variables

$$X_{1:n}, X_{2:n}, \ldots, X_{n:n},$$

which are the original random variables arranged in ascending order of magnitude so that

$$X_{1:n} \le X_{2:n} \le \cdots \le X_{n:n}.$$

The random variables $X_{r:n}$, where $1 \le r \le n$, are called order statistics (as distinguished from rank order statistics equal to $1, 2, 3, \ldots, n$ for $X_{1:n}, X_{2:n}, X_{3:n}, \ldots, X_{n:n}$, which are occasionally also referred to as order statistics).

In particular, $X_{1:n}$ is the minimum of the X_i's, and $X_{n:n}$ is the maximum. Another common order statistic is $X_{k+1:2k+1}$, which coincides with the sample median when the sample size is odd ($n = 2k + 1$). For the last 50 years, order statistics have been playing an increasingly important role in statistical inference and have appeared in many areas of statistical theory and practice. We shall encounter them in later chapters as well.

2.5.1 Distribution of a single order statistic. Given the parent distribution of X_1 (or, equivalently, any one of $X_i, i = 1, \ldots, n$), it is an elementary exercise in probability theory to find the distribution of any order statistic. For instance, if F denotes the c.d.f. of X_1, then the c.d.f. of $X_{n:n}$ is obtained as follows:

$$F_{n:n}(x) - P(X_{n:n} \le x) = P\,(\text{all } X_i \le x) = [F(x)]^n.$$

Similarly, for a general order statistic, we have

$$F_{r:n}(x) = P(X_{r:n} \le x) = P\left(\sum_{i=1}^{n} I_i \ge r\right),$$

where I_i's are i.i.d. indicator r.v.'s defined as

$$I_i = \begin{cases} 1 & \text{if } X_i \le x, \\ 0 & \text{if } X_i > x. \end{cases}$$

The sum $\sum_{i=1}^{n} I_i$ is a binomial r.v. with probability of success $p = P(X_i \le x) = F(x)$ so that

$$F_{r:n}(x) = \sum_{i=r}^{n} \binom{n}{i} [F(x)]^i [1 - F(x)]^{n-i}. \tag{2.5.1}$$

In the continuous case, the corresponding p.d.f. (obtained by differentiation) is

$$f_{r:n}(x) = r\binom{n}{r}[F(x)]^{r-1}[1-F(x)]^{n-r}f(x), \tag{2.5.2}$$

where f is the density corresponding to F. We shall now assume that $X_1, \ldots, X_n$ are i.i.d. from the classical Laplace distribution $\mathcal{CL}(\theta, s)$. Denote the c.d.f. and p.d.f. of the rth order statistics by $F_{r:n}(\cdot; \theta, s)$ and $f_{r:n}(\cdot; \theta, s)$, respectively. For the standard distribution $\mathcal{CL}(0, 1)$, we shall omit the parameters and simply write $F_{r:n}(\cdot)$ and $f_{r:n}(\cdot)$. Below we shall derive the distributions of order statistics connected with the standard classical Laplace distribution. To obtain the corresponding distribution in the case of a general Laplace distribution, use the relations

$$F_{r:n}(x; \theta, s) = F_{r:n}\left(\frac{x-\theta}{s}\right) \quad \text{and} \quad f_{r:n}(x; \theta, s) = \frac{1}{s} f_{r:n}\left(\frac{x-\theta}{s}\right).$$

The following result is obtained by direct application of formulas (2.5.1)–(2.5.2).

Proposition 2.5.1 *Let $X_{r:n}$ be the rth order statistic connected with a sample of size n from the standard classical Laplace distribution $\mathcal{CL}(0, 1)$. Then the c.d.f. and p.d.f. of $X_{r:n}$ are*

$$F_{r:n}(x) = \left(\frac{1}{2}\right)^n \sum_{i=r}^{n} \binom{n}{i} \begin{cases} e^{ix}(2-e^x)^{n-i} & \text{if } x \le 0, \\ e^{-(n-i)x}(2-e^{-x})^i & \text{if } x \ge 0 \end{cases} \tag{2.5.3}$$

and

$$f_{r:n}(x) = r\left(\frac{1}{2}\right)^n \binom{n}{r} \begin{cases} e^{rx}(2-e^x)^{n-r} & \text{if } x \le 0, \\ e^{-(n-r+1)x}(2-e^{-x})^{r-1} & \text{if } x \ge 0, \end{cases} \tag{2.5.4}$$

respectively.

Remark 2.5.1 For the classical Laplace $\mathcal{CL}(\theta, s)$ distribution, we have the density

$$f_{r:n}(x; \theta, s) = \frac{r}{s}\left(\frac{1}{2}\right)^n \binom{n}{r} \cdot \begin{cases} e^{r(x-\theta)/s}(2-e^{(x-\theta)/s})^{n-r} & \text{if } x \le \theta, \\ e^{(n-r+1)(\theta-x)/s}(2-e^{(\theta-x)/s})^{r-1} & \text{if } x \ge \theta. \end{cases} \tag{2.5.5}$$

In particular, we have the following special cases.

2.5.1.1 *The minimum.* The first order statistic connected with a sample of size n from the $\mathcal{CL}(\theta, s)$ distribution has the following c.d.f. and p.d.f.:

$$F_{1:n}(x; \theta, s) = \begin{cases} 1-(1-\frac{1}{2}e^{(x-\theta)/s})^n & \text{if } x \le \theta, \\ 1-\left(\frac{1}{2}\right)^n e^{n(\theta-x)/s} & \text{if } x \ge \theta \end{cases} \tag{2.5.6}$$

and

$$f_{1:n}(x; \theta, s) = \frac{n}{2^n s} \begin{cases} e^{(x-\theta)/s}(2-e^{(x-\theta)/s})^{n-1} & \text{if } x \le \theta, \\ e^{n(\theta-x)/s} & \text{if } x \ge \theta. \end{cases} \tag{2.5.7}$$

2.5.1.2 *The maximum.* The nth order statistic connected with a sample of size n from the $\mathcal{CL}(\theta, s)$ distribution has the following c.d.f. and p.d.f.:

$$F_{n:n}(x; \theta, s) = \left(\frac{1}{2}\right)^n \begin{cases} e^{n(x-\theta)/s} & \text{if } x \le \theta, \\ (2 - e^{(\theta-x)/s})^n & \text{if } x \ge \theta \end{cases} \tag{2.5.8}$$

and

$$f_{n:n}(x; \theta, s) = \frac{n}{2^n s} \begin{cases} e^{n(x-\theta)/s} & \text{if } x \le \theta, \\ e^{(\theta-x)/s}(2 - e^{(\theta-x)/s})^{n-1} & \text{if } x \ge \theta. \end{cases} \tag{2.5.9}$$

The symmetry in the expressions for $f_{1:n}$ and $f_{n:n}$ results from the relation

$$X_{1:n} \stackrel{d}{=} 2\theta - X_{n:n}.$$

2.5.1.3 *The median.* Let $n = 2k + 1$, $k = 0, 1, 2, \ldots$, and let $X_{k+1:n}$ be the sample median $\tilde{X}$ of $X_1, X_2, \ldots, X_n$. Then the p.d.f. of $X_{k+1:n}$ is

$$f_{k+1:n}(x) = \frac{n!}{(k!)^2} \left(\frac{1}{2}\right)^{2k+1} \frac{1}{s} e^{-(k+1)|x-\theta|/s}(2 - e^{-|x-\theta|/s})^k, \tag{2.5.10}$$

and the distribution is symmetric about θ. This distribution was derived in Fisher (1934); see also Karst and Polowy (1963).

2.5.2 Joint distributions of order statistics. Proceeding as in Section 2.5.1, we can find the joint distributions of two or more order statistics. Consider a random sample $X_1, \ldots, X_n$ from a continuous distribution with the c.d.f. F and p.d.f. f. Let

$$1 \le n_1 < n_2 < \cdots < n_k \le n,$$

where $1 \le k \le n$. Then the joint p.d.f. of $X_{n_1:n}, X_{n_2:n}, \ldots, X_{n_k:n}$ is nonzero at $\mathbf{x} = (x_1, \ldots, x_k)'$ only if $x_1 \le x_2 \le \cdots \le x_k$, in which case it is equal to

$$f_{n_1,\ldots,n_k:n}(\mathbf{x}) = n! \left[\prod_{j=1}^{k} f(x_j)\right] \prod_{j=0}^{k} \frac{[F(x_{j+1}) - F(x_j)]^{n_{j+1}-n_j-1}}{(n_{j+1} - n_j - 1)!}, \tag{2.5.11}$$

with $x_0 = -\infty$, $x_{k+1} = +\infty$, $n_0 = 0$, and $n_{k+1} = n+1$ [see, e.g., David (1981, p. 10)]. In particular, the joint distribution of two order statistics, $X_{r:n}$ and $X_{r':n}$, where $1 \le r < r' \le n$, has density

$$f_{r,r':n}(x, y) = C(n, r, r')F^{r-1}(x)f(x)[F(y) - F(x)]^{r'-r-1}f(y)[1 - F(y)]^{n-r'} \tag{2.5.12}$$

for $x \le y$ [and $f_{r,r':n}(x, y) = 0$ for $x > y$], where

$$C(n, r, r') = \frac{n!}{(r-1)!(r'-r-1)!(n-r')!}. \tag{2.5.13}$$

An application of the above to order statistics associated with the Laplace distribution leads immediately to the following result.

Proposition 2.5.2 *Let $X_1, \ldots, X_n$ be a random sample from standard classical Laplace distribution $\mathcal{CL}(0, 1)$. Then for any $1 \le r < r' \le n$, the joint distribution of $X_{r:n}$ and $X_{r';n}$ has density*

$$f_{r,r':n}(x, y) = \left(\frac{1}{2}\right)^n C(n, r, r')u(x, y), \tag{2.5.14}$$

where the constant $C(n, r, r')$ *is given by* (2.5.13) *and*

$$u(x, y) = \begin{cases} e^{rx+y}[e^y - e^x]^{r'-r-1}[2 - e^y]^{n-r'} & \text{if } x \le y \le 0, \\ e^{rx-(n-r'+1)y}[2 - e^{-y} - e^x]^{r'-r-1} & \text{if } x \le 0 \le y, \\ e^{-(x+(n-r'+1)y)}[e^{-x} - e^{-y}]^{r'-r-1}[2 - e^{-x}]^{r-1} & \text{if } 0 \le x \le y, \\ 0 & \text{if } x > y. \end{cases} \tag{2.5.15}$$

Remark 2.5.2 The joint distribution of the minimum and maximum is thus given by

$$f(x, y) = \frac{n(n-1)}{2^n} \begin{cases} e^{x+y}(e^y - e^x)^{n-2} & \text{if } x \le y \le 0, \\ e^{x-y}(2 - e^{-y} - e^x)^{n-2} & \text{if } x \le 0 \le y, \\ e^{-x-y}(e^{-x} - e^{-y})^{n-2} & \text{if } 0 \le x \le y, \\ 0 & \text{if } x > y. \end{cases} \tag{2.5.16}$$

Remark 2.5.3 When the sample is drawn from a general $\mathcal{CL}(\theta, s)$ distribution, the joint density of $X_{r:n}$ and $X_{r';n}$ is

$$f_{r,r':n}(x, y; \theta, s) = \frac{1}{s^2} f_{r,r':n}((x-\theta)/s, (y-\theta)/s), \tag{2.5.17}$$

with $f_{r,r':n}(x, y)$ given by (2.5.14) and (2.5.15).

The joint distributions of order statistics play an important role in statistical applications. Many common statistics utilized in statistical inference are functions of order statistics, and we can obtain their distributions via (2.5.11) coupled with standard transformation methods. Here we present several examples of such derivations for the Laplace distribution.

2.5.2.1 *Range, midrange, sample median.* The three commonly used statistics that are functions of just two order statistics are

$$\begin{aligned} R &= X_{n:n} - X_{1:n} \text{ — the } \textit{range} \text{ of } X_i\text{'s} \\ MR &= \frac{X_{n:n} + X_{1:n}}{2} \text{ — the } \textit{midrange} \text{ of } X_i\text{'s} \\ \tilde{X} &= \frac{X_{k:2k} + X_{k+1:2k}}{2} \text{ — the } \textit{sample median} \text{ when } n = 2k \text{ is even.} \end{aligned}$$

In the next proposition, we derive the distribution of R [see, e.g., Edwards (1948)].

Proposition 2.5.3 *Let* $X_1, \ldots, X_n$, $n \ge 1$, *be a random sample from the standard classical Laplace distribution* $\mathcal{CL}(0, 1)$. *Then the range* R *has the density function*

$$f_R(z) = \frac{n-1}{2^{n-1}} e^{-z}\left[(1 - e^{-z})^{n-2} + \frac{n}{2} I_n(z)\right], \quad z > 0,$$

where $I_n(z) = \int_{-z}^{0} (2 - e^{-x-z} - e^x)^{n-2} dx$ *can be computed from the following recurrent relations:*

$$\begin{aligned} I_2 &= z, \quad A_2 = 1 - e^{-z}, \quad B_2 = e^z - 1, \\ I_n &= 2I_{n-1} - A_{n-1} - e^{-z} B_{n-1}, \\ A_n &= \frac{2}{n-1}\left[(n-2)(A_{n-1} - e^{-z} I_{n-1}) + \frac{1}{2}(1 - e^{-z})^{n-1}\right], \\ B_n &= \frac{2}{n-1}\left[(n-2)(B_{n-1} - I_{n-1}) - \frac{1}{2}(1 - e^{-z})^{n-2}(1 - e^z)\right], \end{aligned}$$

where $A_n = \int_{-z}^{0} e^x (2 - e^{-x-z} - e^x)^{n-2} dx$, $B_n = \int_{-z}^{0} e^{-x}(2 - e^{-x-z} - e^x)^{n-2} dx$.

Proof. Let $f(x, y)$ denote the density of $(X_{1:n}, X_{n:n})$ given by (2.5.16). The density of R can be written as the following sum of three integrals:

$$f_R(z) = \int_{-\infty}^{-z} f(x, z+x)dx + \int_{-z}^{0} f(x, x+z)dx + \int_{0}^{\infty} f(x, z+x)dx.$$

The first and the third integral are equal to each other and equal to

$$\frac{n-1}{2^n}(1-e^{-z})^{n-2}e^{-z},$$

while the middle integral is equal to

$$\frac{n(n-1)}{2^n}e^{-z}I_n(z).$$

Thus it remains to prove the recurrent relations.

First, note that $I_2(z) = \int_{-z}^{0} 1dx = z$, $A_2(z) = \int_{-z}^{0} e^x dx = 1-e^{-z}$, and $B_2(z) = \int_{-z}^{0} e^{-x}dx = e^z - 1$. Next we have

$$\begin{aligned} I_{n+1}(z) &= \int_{-z}^{0} (2-e^{-x-z}-e^x)^{n-1}dx \\ &= 2I_n(z) - A_n(z) - e^{-z}B_n(z), \\ A_{n+1}(z) &= \int_{-z}^{0} e^x(2-e^{-x-z}-e^x)^{n-1}dx \\ &= 2A_n(z) - e^{-z}I_n(z) - \int_{-z}^{0} e^{2x}(2-e^{-x-z}-e^x)^{n-2}dx, \\ B_{n+1}(z) &= \int_{-z}^{0} e^{-x}(2-e^{-x-z}-e^x)^{n-1}dx \\ &= 2B_n(z) - I_n(z) - e^{-z}\int_{-z}^{0} e^{-2x}(2-e^{-x-z}-e^x)^{n-2}dx. \end{aligned}$$

Integration by parts of $A_{n+1}(z)$ and $B_{n+1}(z)$ leads to

$$\begin{aligned} A_{n+1}(z) &= (1-e^{-z})^n - (n-1)e^{-z}I_n(z) \\ &\quad + (n-1)\int_{-z}^{0} e^{2x}(2-e^{-x-z}-e^x)^{n-2}dx, \\ B_{n+1}(z) &= -(1-e^{-z})^{n-1}(1-e^z) - (n-1)I_n(z) \\ &\quad + (n-1)e^{-z}\int_{-z}^{0} e^{-2x}(2-e^{-x-z}-e^x)^{n-2}dx, \end{aligned}$$

and after some elementary algebra we arrive at the recurrent relations stated in the theorem. □

It is interesting to see how the distribution of R differs from the case when the sample is from a Gaussian population. Unfortunately, for the latter case, to the best of our knowledge, the exact distributions can be computed explicitly only for special cases. McKay and Pearson (1933) studied the case $n = 3$, obtaining the density

$$\tilde{f}_R(z) = \frac{6}{\sqrt{\pi}}e^{-z^2/4}\Phi(z/\sqrt{6}), \quad z > 0,$$

where Φ is the c.d.f. of the standard normal distribution. Larger values of n would require numerical computations of certain integrals. [The elaborated computations for $n = 2$ to 20 of the cumulative distribution functions in the pre-computer era are given in Pearson and Hartley (1942).] The case of a Laplace population is thus computationally easier since our recursive formulas allow for explicit form of the densities for an arbitrary n.

The Laplace variable $\mathcal{L}(0, 1) = \mathcal{CL}(0, \sqrt{2}/2)$ has the mean equal to zero and variance equal to one, so it is appropriate for comparisons. For this random variable the density of the range for sample size equal to three is given by

$$f_R(z) = e^{-\sqrt{2}z}\left(3z - \sqrt{2} + \sqrt{2}e^{-\sqrt{2}z}\right), \quad z > 0.$$

The graphs of these two densities are presented in Figure 2.4. The heavier tails of the Laplace distribution are evident.

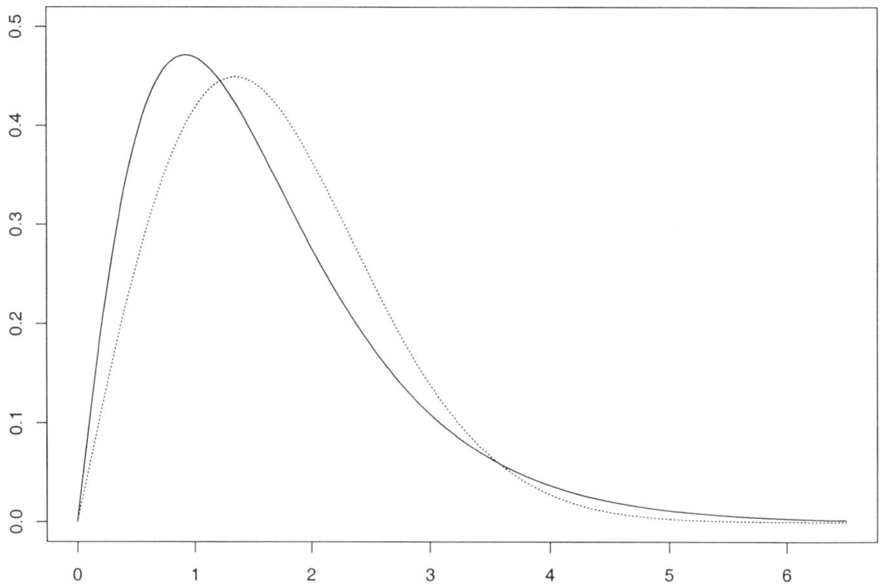

Figure 2.4: The comparison of the p.d.f. of the range for sample size $n = 3$: normal (dotted line) vs. Laplace (solid line) cases.

Consider now another function of the maximal and minimal order statistics—the midrange MR. Using a similar technique, we obtain the following result.

Proposition 2.5.4 *Let $X_1, \ldots, X_n$, $n \geq 1$, be a random sample from the standard classical Laplace distribution $\mathcal{CL}(0, 1)$. Then the midrange MR has the density $f_{MR}(z) = 2h(2z)$, where h, the density of $X_{1:n} + X_{n:n}$, is given by*

$$h(z) = \frac{e^{-|z|}}{(1 + e^{-|z|})^2}\left[1 - \frac{n-1}{2^n}(1 - e^{-|z|})^{n-1}\left(\frac{n+1}{n-1} + e^{-|z|}\right)\right] + \frac{n(n-1)}{2^n}e^{-|z|}J_n(|z|). \tag{2.5.18}$$

Here

$$J_n(z) = \int_0^{z/2}(e^{-x} - e^{-z+x})^{n-2}dx, \quad z > 0,$$

and it can be computed from the following recurrent relations:

$$J_2 = z/2, \quad J_3 = 1 - 2e^{-z/2} + e^{-z},$$
$$J_n = \frac{2}{n-2}\left(\frac{1}{2}(1+e^{-z})(1-e^{-z})^{n-3} - 2(n-3)e^{-z}J_{n-2}\right). \tag{2.5.19}$$

Proof. Since the distribution of MR is symmetric around zero it is sufficient to compute the density $f_{MR}(z)$ for the positive z. As in the previous proof, let $f(x, y)$ given by (2.5.16) be the joint density of $X_{1:n}$ and $X_{n:n}$ (the minimal and maximal order statistics). Then the density h of the sum $X_{1:n} + X_{n:n}$ is

$$h(z) = \int_{-\infty}^{\infty} f(x, z-x)dx = \frac{n(n-1)}{2^n}e^{-z} \times$$
$$\times \left[\int_{-\infty}^{0} e^{2x}(2 - e^{x-z} - e^x)^{n-2}dx + \int_0^{z/2}(e^{-x} - e^{-z+x})^{n-2}dx\right].$$

The first integral can be computed directly by substitution and it leads directly to (2.5.18).

The recursive relation (2.5.19) for computing the second integral can be obtained as follows. First,

$$J_n = \int_0^{z/2} e^{-x}(e^{-x} - e^{-z+x})^{n-3}dx - e^{-z}\int_0^{z/2} e^x(e^{-x} - e^{-z+x})^{n-3}dx.$$

For the two integrals in the above equation, denoted as I_1 and I_2, respectively, we have

$$I_1 = \int_0^{z/2} e^{-2x}(e^{-x} - e^{-z+x})^{n-4}dx - e^{-z}J_{n-2}, \tag{2.5.20}$$

$$I_2 = -e^{-z}\int_0^{z/2} e^{2x}(e^{-x} - e^{-z+x})^{n-4}dx + J_{n-2}. \tag{2.5.21}$$

To compute $\int_0^{z/2} e^{-2x}(e^{-x} - e^{-z+x})^{n-4}dx$ and $\int_0^{z/2} e^{2x}(e^{-x} - e^{-z+x})^{n-4}dx$, let us apply the integration by parts technique to the integrals on the left-hand side of (2.5.20) and (2.5.21). We get

$$\begin{aligned}
&\int_0^{z/2} e^{-x}(e^{-x} - e^{-z+x})^{n-3}dx \\
&\quad = (1 - e^{-z})^{n-3} + \int_0^{z/2} e^{-x}(n-3)(e^{-x} - e^{-z+x})^{n-4}(-e^{-x} - e^{-z+x})dx \\
&\quad = (1 - e^{-z})^{n-3} - \frac{n-3}{e^z}\left\{J_{n-2} + e^z\int_0^{z/2} e^{-2x}(e^{-x} - e^{-z+x})^{n-4}dx\right\},
\end{aligned}$$

$$\begin{aligned}
&\int_0^{z/2} e^{x}(e^{-x} - e^{-z+x})^{n-3}dx \\
&\quad = -(1 - e^{-z})^{n-3} - \int_0^{z/2} e^{x}(n-3)(e^{-x} - e^{-z+x})^{n-4}(-e^{-x} - e^{-z+x})dx \\
&\quad = -(1 - e^{-z})^{n-3} + (n-3)\left\{J_{n-2} + e^{-z}\int_0^{z/2} e^{2x}(e^{-x} - e^{-z+x})^{n-4}dx\right\}.
\end{aligned}$$

Thus

$$\int_0^{z/2} e^{-2x}(e^{-x} - e^{-z+x})^{n-4}dx = \frac{(1-e^{-z})^{n-3}}{n-2} - \frac{n-4}{n-2}e^{-z}J_{n-2}$$

and

$$e^{-z}\int_0^{z/2} e^{2x}(e^{-x}-e^{-z+x})^{n-4}dx = \frac{(1-e^{-z})^{n-3}}{n-2} - \frac{n-4}{n-2}J_{n-2}.$$

Substituting these integrals into (2.5.20) and (2.5.21) leads to recursive formula (2.5.19). □

It is well known [see, e.g., Gumbel (1944)] that the distribution of the midrange converges (when appropriately normalized) to the logistic distribution given by the density

$$f(z) = \frac{e^{-|z|}}{(1+e^{-|z|})^2}.$$

This limiting density is the first factor in the expression (2.5.18) for the density h of the sum of the two extremal order statistics. Clearly, no normalization (scaling) is required for the sum $X_{1:n} + X_{n:n}$ to converge to this logistic variable as n increases to infinity. Consequently, we see that for the Laplace distribution a simple multiplication of the midrange by 2 is required to achieve the limiting standard logistic distribution.

The distribution of $\tilde{X}$ for $n = 2k+1$ was given in (2.5.10). In our next result, we present the density of $\tilde{X}$ in case of an even sample size, as derived by Asrabadi (1985), omitting the details of its technical derivation.

Proposition 2.5.5 *The distribution of the sample median $\tilde{X}$ for $n = 2k$ is given by the density*

$$f_{\tilde{X}}(z) = \frac{n!}{2^k[(k-1)!]^2}\sum_{i=0}^{k-2}\frac{(-1)^i\binom{k-1}{i}}{2^i(k-1-i)}e^{-(k+1+i)|z|}(1-e^{-(k-1-i)|z|})$$
$$- \frac{(-1)^k}{2^{k-1}}|z|e^{-2k|z|} + \frac{1}{k2^k}e^{-2k|z|}.$$

2.5.3 Moments of order statistics. The computation of central moments of order statistics connected with a general classical Laplace distribution is straightforward. Using the explicit density (2.5.5) of the rth order statistic $X_{r:n}$, we obtain

$$E[(X_{r:n}-\theta)^k] = s^k\frac{n!\Gamma(k+1)}{(r-1)!(n-r)!} \times \left\{(-1)^k\sum_{j=0}^{n-r}a_j + \sum_{j=0}^{r-1}b_j\right\}, \tag{2.5.22}$$

where

$$a_j = (-1)^j\frac{(n-r)!}{j!(n-r-j)!}2^{-(r+j+1)}(r+j)^{-(k+1)} \tag{2.5.23}$$

and

$$b_j = (-1)^j\frac{(r-1)!}{j!(r-1-j)!}2^{-(n-r+2+j)}(n-r+1+j)^{-(k+1)}. \tag{2.5.24}$$

In particular, for odd n, the mean of the sample median $X_{(n+1)/2:n}$ is equal to θ; the variance of the sample median is

$$E[(X_{(n+1)/2:n}-\theta)^2] = \frac{4s^2n!}{[(n-1)/2]!}\sum_{j=0}^{(n-1)/2}c_j, \tag{2.5.25}$$

where

$$c_j = (-1)^j \left[j! \left(\frac{n-1}{2} - j \right) ! 2^{j+(n+1)/2} \left(\frac{n+1}{2} + j \right)^3 \right]^{-1}. \quad (2.5.26)$$

When the sample size $n = 2k$ is even, the mean of the sample median is still equal to θ; the variance of the sample median was derived in Asrabadi (1985). Its value for the standard classical Laplace distribution is

$$\frac{n!}{[(k-1)!]^2} 2^{2-k} \left(\sum_{j=0}^{k-2} d_j + 2^{-k-3} k^{-4} \{(-1)^{k-1} + 1\} \right), \quad (2.5.27)$$

where

$$d_j = \frac{(k-1)!}{j!(k-1-j)!} (-2)^{-j} (k-1-j)^{-1} \{(k+1+j)^{-3} - (2k)^{-3}\}. \quad (2.5.28)$$

Govindarajulu (1966) obtained expressions for the means, variances and covariances of order statistics from the standard classical Laplace distribution in terms of those from the standard exponential distribution. His method applies to a general distribution that is symmetric about the origin [see also Balakrishnan et al. (1993)]. Let $X_{1:n}, \ldots, X_{n:n}$ denote the order statistics corresponding to a random sample of size n from a symmetric distribution with c.d.f. F_X, and let $Y_{1:n}, \ldots, Y_{n:n}$ be the order statistics obtained from a similar sample from the corresponding folded distribution with c.d.f. $F_Y(y) = 2F_X(y) - 1$, $y \geq 0$ (so that $Y \stackrel{d}{=} |X|$). Then we have the relations

$$\begin{aligned} E[X_{r:n}^k] = \frac{1}{2^n} \Bigg\{ & \sum_{i=0}^{r-1} \binom{n}{i} E[Y_{r-i:n-i}^k] \\ & + (-1)^k \sum_{l=r}^{n} \binom{n}{i} E[Y_{i-r+1:i}^k] \Bigg\}, \quad 1 \leq r \leq n, \end{aligned} \quad (2.5.29)$$

and for $1 \leq r < s \leq n$,

$$\begin{aligned} E[X_{r:n} X_{s:n}] = \frac{1}{2^n} \Bigg\{ & \sum_{i=0}^{r-1} \binom{n}{i} E[Y_{r-i:n-i} Y_{s-i:n-i}] \\ & - \sum_{i=r}^{s-1} \binom{n}{i} E[Y_{i-r+1:i}] E[Y_{s-i:n-i}] \\ & + \sum_{i=s}^{n} \binom{n}{i} E[Y_{i-s+1:i} Y_{i-r+1:i}] \Bigg\} \end{aligned} \quad (2.5.30)$$

[see Govindarajulu (1963)]. Recalling that if X is a standard classical Laplace variable, then $Y = |X|$ is a standard exponential variable, Govindarajulu (1966) used well-known explicit expressions of the means of exponential order statistics in (2.5.29)–(2.5.30) to obtain the following moments of order statistics connected with the $\mathcal{CL}(0, 1)$ distribution:

$$E[X_{r:n}] = \frac{1}{2^n}\left\{\sum_{i=0}^{r-1}\binom{n}{i}S_1(r-i, n-i) - \sum_{i=r}^{n}\binom{n}{i}S_1(i-r+1, i)\right\}, \quad 1 \le r \le n, \tag{2.5.31}$$

$$E[X_{r:n}^2] = \frac{1}{2^n}\left\{\sum_{i=0}^{r-1}\binom{n}{i}S_2(r-i, n-i) + \sum_{i=r}^{n}\binom{n}{i}S_1(i-r+1, i)\right\}, \quad 1 \le r \le n, \tag{2.5.32}$$

and for $1 \le r < s \le n$,

$$E[X_{r:n}X_{s:n}] = \frac{1}{2^n}\left\{\sum_{i=0}^{r-1}\binom{n}{i}S_3(r-i, s-i, n-i) - \sum_{i=r}^{s-1}\binom{n}{i}S_1(i-r+1, i)S_1(s-i, n-i) + \sum_{i=s}^{n}\binom{n}{i}S_3(i-s+1, i-r+1, i)\right\}. \tag{2.5.33}$$

Here, for $1 \le r \le n$,

$$S_1(r, n) = \sum_{i=n-r+1}^{n}\frac{1}{i}, \quad S_2(r, n) = \sum_{i=n-r+1}^{n}\frac{1}{i^2} + [S_1(r, n)]^2, \tag{2.5.34}$$

and for $1 \le r < s \le n$,

$$S_3(r, s, n) = \sum_{i=n-r+1}^{n}\frac{1}{i^2} + S_1(r, n)\cdot S_1(s, n). \tag{2.5.35}$$

Utilizing the relations (2.5.31)–(2.5.33), Govindarajulu (1966) calculated the means, variances, and covariances of order statistics connected with the standard classical Laplace distribution for sample sizes up to 20.

Remark 2.5.4 Balakrishnan (1988) extended the relations (2.5.29) and (2.5.30) to the case of a single-scale outlier model (when the random sample consists of $n-1$ i.i.d. symmetric variables and one symmetric scale outlier). Balakrishnan and Ambagaspitiya (1988) used this extension in studying robustness of various linear estimators of the location and scale parameters of the classical Laplace distribution. The results have also been extended by Balakrishnan (1989) to the case of independent but not necessarily identically distributed observations from the Laplace distribution.

Remark 2.5.5 Akahira and Takeuchi (1990) studied the loss of information associated with the order statistics and estimators related to the Laplace distribution.

Remark 2.5.6 Lien et al. (1992) derived moments of order statistics and related best linear unbiased estimators of the location and scale parameters connected with the standard *doubly truncated* Laplace distribution with density

$$f(x) = \frac{1}{2(1-P-Q)} e^{-|x|}, \quad \log(2Q) \le x \le -\log(2P), \tag{2.5.36}$$

where P and Q represent the proportions of truncation on the left and right of the standard classical Laplace density. Khan and Khan (1987) obtained recurrence relations for the moments of order statistics connected with the doubly truncated Laplace distribution (2.5.36).

2.5.4 Representation of order statistics via sums of exponentials. In many considerations we found it useful to represent order statistics in the form of sums of independent exponential random variables [see, for example, Subsection 2.6.1.1].

It follows from (2.2.10) that a vector $(X_1, \ldots, X_n)$ of i.i.d. standard classical Laplace random variable has a distributional representation of the form

$$(X_1, \ldots, X_n) \stackrel{d}{=} (\delta_1 W_1, \ldots, \delta_n W_n), \tag{2.5.37}$$

where $(\delta_1, \ldots, \delta_2)$ are i.i.d. Rademacher r.v.'s (random signs taking pluses and minuses with equal probabilities) and $(W_1, \ldots, W_n)$ are i.i.d. standard exponential variables independent of the δ_i's.

Let B_n be a Bernoulli random variable counting the number of pluses among the δ_i's. The number of minuses is denoted by $\bar{B}_n = n - B_n$.

Proposition 2.5.6 *Let $(X_1, \ldots, X_n)$ be a vector of i.i.d. $\mathcal{CL}(0,1)$ random variables, and let B_n be a Bernoulli random variable with $p = 1/2$ independent of two independent sequences $(\bar{W}_i)_{i=1}^{\infty}$, $(W_i)_{i=1}^{\infty}$ of i.i.d. standard exponential random variables.*

Then the order statistics of $(X_1, \ldots, X_n)$ have the following distributional representations:

$$\begin{aligned}(X_{1:n}, \ldots, X_{n:n}) &\stackrel{d}{=} (-\bar{W}_{\bar{B}_n:\bar{B}_n}, \ldots, -\bar{W}_{1:\bar{B}_n}, W_{1:B_n}, \ldots, W_{B_n:B_n}) \\ &\stackrel{d}{=} \left(-\left(\sum_{l=1}^{i} \frac{\bar{W}_l}{n-l+1}\right)_{i=1}^{\bar{B}_n}, \left(\sum_{l=1}^{i} \frac{W_l}{n-l+1}\right)_{i=1}^{B_n} \right).\end{aligned}$$

Proof. It is enough to notice that conditionally on the δ_i's, $\{W_i, \delta_i = 1\}$ are independent of $\{W_i, \delta_i = -1\}$. Thus we can represent $\{W_i, \delta_i = 1\}$ by $\{\bar{W}_i, i - 1, \ldots, \bar{B}_n\}$, and $\{W_i, \delta_i = 1\}$ by $\{W_i, i = 1, \ldots, B_n\}$. The first representation then follows by appropriate ordering of these two sequences.

The second representation follows from the well-known representation of the exponential order statistics:

$$(W_{i:n})_{i=1}^{n} \stackrel{d}{=} \left(\sum_{l=1}^{i} \frac{W_l}{n-l+1}\right)_{i=1}^{n}. \tag{2.5.38}$$

[See, e.g., Balakrishnan and Cohen (1991), p. 34.] □

Remark 2.5.7 Consider $n = 2k+1$. Let $K_n = \max(B_n, \bar{B}_n)$ and $\delta_n = \operatorname{sign}(B_n - k - 1/2)$, where B_n is as in the above representation. We then have the following representations for the median:

$$X_{k+1:n} \stackrel{d}{=} \delta_n \sum_{l=1}^{K_n - k} \frac{W_l}{K_n - l + 1} \stackrel{d}{=} \delta_n \sum_{l=k+1}^{K_n} \frac{W_l}{l}.$$

Here K_n and δ_n are dependent but jointly independent of the W_i's.

2.6 Statistical inference

In this rather lengthy section we discuss basic statistical theory and methodology for the Laplace distribution. We warn readers that some of the proofs presented herein may be tough going but in our opinion quite a rewarding experience. When collecting material for this section we were pleasantly surprised by the abundance of available results scattered in the literature. Before proceeding with results on estimation and testing, let us make some remarks concerning the classical Laplace location-scale family of distributions with the density

$$f(x;\theta,s) = \frac{1}{s} f\left(\frac{x-\theta}{s}\right), \qquad -\infty < \theta < \infty, \quad 0 < s < \infty, \quad -\infty < x < \infty, \tag{2.6.1}$$

where f is the standard classical Laplace density (2.1.2). We start with an observation that our class is not a member of the exponential family of distributions, i.e., the density (2.6.1) cannot be written as

$$a(\theta,s)b(x)e^{\sum_{i=1}^{k} c_i(\theta,s)d_i(x)}, \qquad -\infty < \theta < \infty, \quad 0 < s < \infty, \quad -\infty < x < \infty, \tag{2.6.2}$$

where $a(\theta,s)$ and $c_i(\theta,s)$, $1 \le i \le k$, are some functions of the vector parameter (θ,s) and $b(x)$ and $d_i(x)$, $1 \le i \le k$, are some functions of x. Consequently, many standard results that are valid for exponential families of distributions are not available for the Laplace distribution.

Let $X_1, \dots, X_n$ be i.i.d. each with density (2.6.1). If the density was of the form (2.6.2), then the data could be reduced to the set of k sufficient statistics $(T_1, \dots, T_k)$, where

$$T_i = T_i(X_1, \dots, X_n) = \sum_{j=1}^{n} d_i(X_j). \tag{2.6.3}$$

Since we are not dealing with exponential family, this is not the case. Clearly, the set of all order statistics

$$T = (X_{1:n}, \dots, X_{n:n}) \tag{2.6.4}$$

is sufficient, as it is for any i.i.d. observations. Moreover, greater reduction of the data is not possible here, since the statistic T is also *minimal sufficient* [see, e.g., Lehmann and Casella (1998)].

Proposition 2.6.1 *Let $\mathcal{P}$ be the family of densities* (2.6.1), *and let the variables $X_1, \dots, X_n$ be i.i.d. each with density $f(\cdot;\theta,s) \in \mathcal{P}$. Then the statistic T given by* (2.6.4) *is minimal sufficient for $\mathcal{P}$.*

The proof of Proposition 2.6.1 hinges on the following lemma, presented in Lehmann and Casella (1998).

Lemma 2.6.1 *If $\mathcal{P}$ is a family of distributions with common support and $\mathcal{P}_0 \subset \mathcal{P}$, and if T is minimal sufficient for $\mathcal{P}_0$ and sufficient for $\mathcal{P}$, it is minimal sufficient for $\mathcal{P}$.*

Proof. To establish Proposition 2.6.1, note that the statistic T is sufficient for $\mathcal{P}$ by the Factorization Criterion [see, e.g., Lehmann and Casella (1998), Theorem 6.5]. It remains to show that T is also minimal sufficient. Let $\mathcal{P}_0$ be the subset of $\mathcal{P}$ of these densities (2.6.1) where $s = 1$. In view of Lemma 2.6.1, it is enough to show that T is minimal sufficient for $\mathcal{P}_0$. Consider a subset $\mathcal{P}_1$ of $\mathcal{P}_0$ consisting of densities with a rational value of θ. Since the family $\mathcal{P}_1$ is countable, the set of statistics of the form

$$S_j(X_1, \dots, X_n) = \frac{\prod_{i=1}^{n} f(X_i;\theta_j,s)}{\prod_{i=1}^{n} f(X_i;0,s)}, \tag{2.6.5}$$

where θ_j is the jth rational number different from zero (since there are countably many rational numbers, they can be enumerated), is minimal sufficient for $\mathcal{P}_1$ [see Lehmann and Casella (1998), Theorem 6.12]. Since for the Laplace distribution

$$S_j(X_1, \ldots, X_n) = e^{-\sum_{i=1}^n |X_i - \theta_j| + \sum_{i=1}^n |X_i|}, \tag{2.6.6}$$

it is clear that the set of statistics (2.6.6) is equivalent to the set of order statistics, that is,

$$S_j(X_1, \ldots, X_n) = S_j(Y_1, \ldots, Y_n), \quad j = 1, 2, \ldots, \tag{2.6.7}$$

if and only if $(X_1, \ldots X_n)$ and $(Y_1, \ldots, Y_n)$ have the same order statistics. Thus the set of order statistics T is minimal sufficient for $\mathcal{P}_1$, and also for $\mathcal{P}_0$ via another application of Lemma 2.6.1. □

We now turn to a study of the amount of Fisher information contained in a random sample from the distribution with density (2.6.1). For the location-scale family with density (2.6.1), the entries of the Fisher information matrix,

$$I(\theta, s) = \begin{bmatrix} I_{11} & I_{12} \\ I_{21} & I_{22} \end{bmatrix}, \tag{2.6.8}$$

are given by

$$I_{11} = \frac{1}{s^2} \int \left(\frac{f'(y)}{f(y)} \right)^2 f(y) dy, \tag{2.6.9}$$

$$I_{22} = \frac{1}{s^2} \int \left(\frac{y f'(y)}{f(y)} + 1 \right)^2 f(y) dy, \tag{2.6.10}$$

and

$$I_{12} = I_{21} = \frac{1}{s^2} \int y \left(\frac{f'(y)}{f(y)} \right)^2 f(y) dy \tag{2.6.11}$$

[see, e.g., Lehmann and Casella (1998)]. After routine calculations (see Exercise 2.7.31), we obtain

$$I(\theta, s) = \begin{bmatrix} 1/s^2 & 0 \\ 0 & 1/s^2 \end{bmatrix}. \tag{2.6.12}$$

It is worth noting that $\int [f'(y)/f(y)]^2 f(y) dy$ is 1 for both Laplace and normal densities but has a different value for other symmetric distributions such as logistic and Cauchy.

Remark 2.6.1 Note that the Laplace density does not satisfy the standard differentiability assumptions required for the computation of the Fisher information matrix, since f is not differentiable at zero. However, the relations (2.6.9)–(2.6.11) are valid under a weaker assumption that f is absolutely continuous, which is the case for the Laplace density [see, e.g., Huber (1981), Section 4.4].

2.6.1 Point estimation. We start with the problem of estimating the parameters of the Laplace distribution. Since the theory of estimation for the classical Laplace distribution is well developed, we shall stick to the $\mathcal{CL}(\theta, s)$ parametrization. We shall assume that $X_1, \ldots, X_n$ are n mutually independent random variables with probability density function (2.1.1), while $x_1, \ldots, x_n$ are their particular realizations.

2.6.1.1 *Maximum likelihood estimation.* The likelihood function based on a sample of size n from the classical Laplace distribution with scale s and location θ is

$$f_n(x_1, \dots, x_n; \theta, s) = \prod_{i=1}^{n} f(x_i; \theta, s) = \left(\frac{1}{2s}\right)^n e^{-\frac{1}{s}\sum_{i=1}^{n}|x_i-\theta|}. \tag{2.6.13}$$

Let us consider three cases, two where one of the parameters is known and one where both parameters are unknown.

Case 1: *The value of s is known.* Clearly, to find the maximum value of f_n with respect to θ, is the same as to minimize the expression

$$\frac{1}{n}\sum_{i=1}^{n}|x_i - \theta| \tag{2.6.14}$$

with respect to θ. Note that (2.6.14) is the expected value $E|Y - \theta|$, where Y is a discrete random variable taking each of the values $x_1, \dots, x_n$ with probability $1/n$. Consequently, the value of θ that minimizes (2.6.14) is the median of Y, which here coincides with the sample median of the observations $x_1, \dots, x_n$ [see Hombas (1986)]. Norton (1984) established this result by using calculus (see Exercise 2.7.34).

Thus for n odd, the maximum likelihood estimator (MLE) of θ, denoted $\hat{\theta}_n$, is uniquely defined as the middle observation $X_{(n+1)/2:n}$. For n even, $\hat{\theta}_n$ can be chosen as any value between the two middle observations. For convenience, in this case the *canonical median*, which is the arithmetic mean of the two middle values, is usually used in practice.

Proposition 2.6.2 *Let $X_1, \dots, X_n$ be i.i.d. with the $\mathcal{CL}(\theta, s)$ distribution* (2.1.1), *where s is known and $\theta \in \mathbb{R}$ is unknown. Then the MLE of θ,*

$$\hat{\theta}_n = \begin{cases} X_{k+1:n}, & \text{for } n = 2k+1, \\ \frac{1}{2}\{X_{k:n} + X_{k+1:n}\}, & \text{for } n = 2k, \end{cases} \tag{2.6.15}$$

where $X_{r:n}$ denotes the rth order statistic, is

(i) *unbiased;*

(ii) *consistent;*

(iii) *asymptotically normal; i.e., $\sqrt{n}(\hat{\theta}_n - \theta)$ converges in distribution to a normal distribution with mean zero and variance s^2.*

Proof. The result can be established by using the explicit form of the density and moments of sample median, derived in Section 2.5.

(i) Using the formulas for the moments of order statistics (see Section 2.5), we find that the mean of the sample median defined by (2.6.15) is equal to θ.

(ii) The consistency of $\hat{\theta}_n$ follows from part (i) and the fact the variance of $\hat{\theta}_n$ converges to zero as $n \to \infty$ (Exercise 2.7.39).

(iii) The standard regularity conditions usually stated in theorems on asymptotic normality of MLE's do not hold for Laplace distribution. To establish the asymptotic normality, use Theorem 3.2, Chapter 5 of Lehman (1983), which asserts that the sequence $\sqrt{n}(\hat{\theta}_n - \theta)$ converges to the normal distribution with mean zero and variance $1/[4f^2(0)]$, where f is the p.d.f. of X_1. □

Remark 2.6.2 The median may not be the best estimator to use for the $\mathcal{CL}(\theta, s)$ distribution, since there are other unbiased estimators of θ with smaller variances. For example, Rosenberger and Gasko (1983) found that the variances of both the *midmean*[15] and the *broadened median*[16] are less than that of the median. However, the median has a desirable property of robustness (as do most other trimmed means) as it performs well (in terms of efficiency) if the assumed model departs from the Laplace distribution; Rosenberger and Gasko (1983) recommend the median as an estimator of location based on samples of size $n \leq 6$ from a symmetric, possibly heavy tailed distribution.

Keynes (1911) conjectured that the property that the sample median is a MLE of the location parameter is a characterization of the Laplace distribution. This indeed is the case, as shown in Kagan et al. (1973) for the case of $n = 4$ and under the assumption that the density function of the considered distribution is lower semicontinuous. Recall that normal distribution admits a similar characterization, where the MLE of the shift parameter is the sample mean for sample sizes $n = 2, 3$ [see, Teicher (1961)]. It is interesting to note that the result for Laplace fails for sample sizes $n = 2, 3$ [see Rao and Ghosh (1971) and Exercise 2.7.35].

This characterization problem of the Laplace distribution has been thoroughly studied in Findeisen (1982), who showed that the following conditions imply that f is a Laplace density (with the mode at zero), where $X_1, \ldots, X_n$ are i.i.d. with density $f(x - \theta)$, $-\infty < x, \theta < \infty$.

(i) For all n, *every* median of the random sample of size n is the MLE of θ.

(ii) There is at least one *even* n, such that *every* median of the random sample of size n is the MLE of θ.

(iii) There are infinitely many n's such that for every random sample of size n *at least one* median is the MLE of θ.

(iv) For sufficiently large n, the *canonical median* given by (2.6.15) is always a MLE of θ.

In addition, Findeisen (1982) demonstrated that conditions (v) and (vi) given below are not sufficient to conclude that f is a Laplace density (see Exercises 2.7.36 and 2.7.37).

(v) There exists at least one n such that *every* median of a random sample of size n is the MLE of θ.

(vi) There exists an even n such that the two particular medians, $X_{n/2:n}$ and $X_{n/2+1:n}$, are the MLE's of θ.

Buczolich and Székely (1989) improved these results by showing that the characterization of the Laplace distribution of Kagan et al. (1973) holds for an arbitrary even sample size $n \geq 4$ and without any regularity conditions on the density, and by replacing "every median" with "some median" in condition (ii) of Findeisen (1982) given above. Thus we have the following characterization of the Laplace distribution.

Proposition 2.6.3 *Let* $\{F(x - \theta), \theta \in \mathbb{R}\}$ *be a family of absolutely continuous distribution functions on* $\mathbb{R}$ *depending on a shift parameter* θ. *If the canonical sample median given by* (2.6.15) *is the MLE*

[15]The midmean is the average of the central half of the order statistics (the 25% trimmed mean).

[16]For n odd, the broadened median is the average of the three middle order statistics for $5 \leq n \leq 12$ and the five middle order statistics for $n \geq 13$. For n even, it is a weighted average of the four middle order statistics for $1 \leq n \leq 12$ with weights 1/6, 1/3, 1/3, and 1/6, while for $n \geq 13$ it is a weighted average of six middle order statistics with weights 1/10, 1/5, 1/5, 1/5, 1/5, and 1/10 [see, e.g., Rosenberger and Gasko (1983)].

of θ for some even sample size $n \geq 4$, then F must be a Laplace distribution function so that

$$F'(x) = f(x) = \frac{1}{2a} e^{-a|x|}, \quad x \neq 0.$$

We refer the reader to Buczolich and Székely (1989) for a fairly advanced proof of the result.

Remark 2.6.3 More generally, if for some $i \in \{1, 2, \ldots, n-1\}$ a linear combination of two consecutive order statistics of the form

$$W = a_i X_{i:n} + a_{i+1} X_{i+1:n} \tag{2.6.16}$$

is the MLE of θ, where $n \geq 3$ and

$$a_i + a_{i+1} = 1, \quad a_i, a_{i+1} > 0, \tag{2.6.17}$$

then F must be a skewed Laplace distribution function corresponding to the density

$$f(x) = \begin{cases} ce^{-b_1|x|} & \text{if } x \leq 0, \\ ce^{-b_2|x|} & \text{if } x \geq 0, \end{cases} \tag{2.6.18}$$

where b_1 is some positive constant, $b_2 = \frac{i}{n-i} b_1$, and c is chosen so that the density (2.6.18) integrates to 1 [Buczolich and Székely (1989)]. In particular, Proposition 2.6.3 still holds if the canonical sample median is replaced by an arbitrary median [of the form (2.6.16) with $i = n/2$].

Remark 2.6.4 We see that the MLE of the location parameter when sampling from a Laplace distribution is the sample median (the empirical 0.5-quantile). A question arises as to whether there are any distributions for which the MLE's of the location parameters are given by other empirical quantiles. It turns out that this is generally true for skewed Laplace distributions (2.6.18) (see Section 3.5 of Chapter 3). One family of skewed Laplace distributions is given by the p.d.f.

$$f(x) = \alpha(1-\alpha) \begin{cases} e^{-(1-\alpha)|x-\theta|}, & \text{for } x < \theta, \\ e^{-\alpha|x-\theta|}, & \text{for } x \geq \theta, \end{cases} \tag{2.6.19}$$

where $\theta \in (-\infty, \infty)$ and $\alpha \in (0, 1)$ [see Poiraud-Casanova and Thomas-Agnan (2000)]. Here, given i.i.d. observations from the density (2.6.19) (with a given value of α), the MLE of θ is the empirical α-quantile (see Exercise 2.7.38). For $\alpha = 1/2$, the density (2.6.19) reduces to a symmetric Laplace density and the MLE of θ is the empirical 0.5-quantile (the median).

The two-tailed power distribution with the c.d.f.

$$F(x) = \begin{cases} x^n/\theta^{n-1} & \text{for } 0 \leq x \leq \theta \\ 1-(1-x)^n/(1-\theta)^{n-1} & \text{for } \theta \leq x \leq 1 \end{cases}$$

and density

$$f(x) = \begin{cases} nx^{n-1}/\theta^{n-1} & \text{for } 0 \leq x \leq \theta \\ n(1-x)^{n-1}/(1-\theta)^{n-1} & \text{for } \theta \leq x \leq 1 \end{cases}$$

has a similar property: the MLE of the parameter θ (which is not actually a location parameter as described in (2.6.1)) is given by an order statistic. (This distribution serves as an alternative to beta distributions. For $n = 2$, we have the triangular distribution.)

Remark 2.6.5 Marshall and Olkin (1993) extended the above maximum likelihood characterization of Laplace distribution to the multivariate case. They showed that if $\mathbf{x}_1, \mathbf{x}_2, \mathbf{x}_3, \mathbf{x}_4$ is a random sample of size $n = 4$ from a location family $\{F(\mathbf{x} - \theta), \theta \in \mathbb{R}^d\}$ of distributions in $\mathbb{R}^d$, where $f = F'$ is lower semicontinuous at $\mathbf{x} = \mathbf{0}$, and the vector of sample medians is a MLE of θ, then f must be the product of univariate Laplace densities.

Case 2: *The value of θ is known.* Here the likelihood function is maximized by the sample first absolute moment.

Proposition 2.6.4 *Let $X_1, \ldots, X_n$ be i.i.d. with the $\mathcal{CL}(\theta, s)$ distribution* (2.1.1), *where θ is known and $s > 0$ is unknown. Then the MLE of s,*

$$\hat{s}_n = \frac{1}{n} \sum_{i=1}^{n} |X_i - \theta|, \tag{2.6.20}$$

is

(i) *unbiased;*

(ii) *strongly consistent;*

(iii) *asymptotically normal; i.e., $\sqrt{n}(\hat{s}_n - s)$ converges in distribution to a normal distribution with mean zero and variance s^2;*

(iv) *efficient.*

Proof. To establish (2.6.20), write the log-likelihood,

$$\log f_n(x_1, \ldots, x_n; \theta, s) = -n \log 2 - n \log s - \frac{1}{s} \sum_{i=1}^{n} |x_i - \theta|, \tag{2.6.21}$$

and note that its derivative with respect to s,

$$\frac{1}{s}\left(\frac{1}{s}\sum_{i=1}^{n} |X_i - \theta| - n\right),$$

is decreasing for $s < \hat{s}_n$ and increasing for $s > \hat{s}_n$.

(i) The unbiasedness of $\hat{s}_n$ follows from the representation

$$|X_i - \theta| \stackrel{d}{=} sW, \tag{2.6.22}$$

where W is standard exponential with mean and variance equal to one.

(ii) The strong consistency of $\hat{s}_n$ follows from the Strong Law of Large Numbers, since the random variables (2.6.22) are i.i.d. with mean s.

(iii) The asymptotic normality follows from the classical version of the Central Limit Theorem, as the random variables (2.6.22) are i.i.d. with mean and standard deviation both equal to s.

(iv) The efficiency of $\hat{s}_n$ follows from the fact that the variance of $\hat{s}_n$ coincides with the Cramér–Rao lower bound (for the variance of any unbiased estimator of s). Indeed, the Cramér–Rao lower bound is $[nI(s)]^{-1}$, where

$$I(s) = -E\left(\frac{\partial^2}{\partial s^2} \log f(x; \theta, s)\right) \tag{2.6.23}$$

is the Fisher information in one observation from $f(x;\theta,s)$. The second derivative of $\log f(x;\theta,s)$ with respect to s is

$$\frac{\partial^2}{\partial s^2}\log f(x;\theta,s) = \frac{1}{s^2} - \frac{2|x-\theta|}{s^3}, \tag{2.6.24}$$

so $I(s) = 1/s^2$ and the Cramér–Rao lower bound is s^2/n. □

Note that since $\hat{s}_n$ is unbiased and efficient, it is a uniformly minimum variance unbiased estimator (UMVU) of s.

We have shown that for a scale parameter family of Laplace distributions, a MLE of the scale parameter is the first absolute moment given by (2.6.20). Is the converse true? Recall that for the corresponding scale parameter family of normal distributions, a MLE of the scale parameter is $\sqrt{\frac{1}{n}\sum_{i=1}^n X_i^2}$, which actually is a characterization of a normal distribution [see Teicher (1961)]. For the Laplace distribution, such characterization holds as well.

Proposition 2.6.5 *Let $\{F(x/s), s > 0\}$ be a family of absolutely continuous distributions on $\mathbb{R}$, depending on a scale parameter s. Suppose that the density $f(x) = F'(x)$ satisfies the following conditions:*

(i) *f is continuous on $(-\infty, \infty)$;*

(ii)

$$\lim_{y\to 0}\frac{f(\lambda y)}{f(y)} = 1 \quad \textit{for all } \lambda > 0. \tag{2.6.25}$$

If for all sample sizes n, a MLE of s is given by $\frac{1}{n}\sum_{i=1}^n |X_i|$, then F is Laplace and $f(x) = \frac{1}{2}e^{-|x|}$.

Proof. Suppose that $\hat{s}_n = \frac{1}{n}\sum_{i=1}^n |X_i|$ is a MLE of s for all sample sizes n. Then $\hat{s}_n$ maximizes the likelihood function, so we have the inequality

$$\left(\frac{1}{\hat{s}_n}\right)^n \prod_{i=1}^n f\left(\frac{x_i}{\hat{s}_n}\right) \geq \left(\frac{1}{s}\right)^n \prod_{i=1}^n f\left(\frac{x_i}{s}\right) \tag{2.6.26}$$

for all $s > 0$ and $x_i \in \mathbb{R}$, $i = 1, \ldots, n$. Let $y_i = x_i/\hat{s}_n$ and $\lambda = \hat{s}_n/s$. Then we can write (2.6.26) as

$$\prod_{i=1}^n f(y_i) \geq \lambda^n \prod_{i=1}^n f(\lambda y_i), \tag{2.6.27}$$

where $\lambda > 0$ and $y_1, \ldots, y_n$ satisfy the condition

$$\sum_{i=1}^n |y_i| = n. \tag{2.6.28}$$

Consider the function f for $x > 0$. With positive y_i's satisfying (2.6.28) and arbitrary $\lambda > 0$, the condition (2.6.26) leads to an exponential function,

$$f(x) = c_1 e^{-x}, \quad x > 0 \tag{2.6.29}$$

[see Teicher (1961, Theorem 2)]. Similarly, for $x > 0$, denote $g(x) = f(-x)$ and write (2.6.27) as

$$\prod_{i=1}^{n} g(y_i) \geq \lambda^n \prod_{i=1}^{n} g(\lambda y_i), \tag{2.6.30}$$

where $\lambda > 0$, and $y_i > 0$ satisfy $\sum_{i=1}^{n} y_i = n$. Proceeding as above, we arrive at the conclusion that

$$f(x) = g(-x) = c_2 e^x, \quad x < 0. \tag{2.6.31}$$

Since f is a probability density on $(-\infty, \infty)$, we must have $c_1 + c_2 = 1$. To conclude the proof, note that only the choice $c_1 = c_2 = \frac{1}{2}$ leads to a MLE given by the sample first absolute moment. □

Remark 2.6.6 Cifarelli and Regazzini (1976) considered the problem of characterization of probability distributions for which the mean absolute deviation (2.6.20) is an unbiased and efficient estimator of the scale parameter. Suppose that $X_1, \ldots, X_n$ are i.i.d. with density

$$g(x) = \frac{1}{s} f\left(\frac{x}{s}\right), \tag{2.6.32}$$

where f is positive for all real x and $s > 0$, continuous at $x = 0$, and satisfies some technical conditions. Cifarelli and Regazzini (1976) showed that if the statistic (2.6.20) (with $\theta = 0$) is unbiased and efficient for the scale parameter s of (2.6.32), then f is the standard classical Laplace distribution. Cifarelli and Regazzini (1976) also obtained a generalization, showing that if for some $\gamma > 0$ the statistic

$$\hat{s}_{n,\gamma} = \frac{1}{n} \sum_{i=1}^{n} |X_i|^\gamma \tag{2.6.33}$$

is an unbiased and efficient estimator for the parameter s^γ [under the model (2.6.32)], then g must be the *exponential power density*

$$g(x) = \frac{\gamma^{1-\gamma^{-1}}}{2s\Gamma(\gamma^{-1})} e^{-(\gamma s^\gamma)^{-1}|x|^\gamma},$$

which we shall (briefly) consider in Section 4.4.2 of Chapter 4.

Case 3: *Both s and θ are unknown.* Similarly as above, here the MLE of θ is the sample median $\hat{\theta}_n$ given by (2.6.15), while the MLE of the scale parameter s is equal to the mean absolute deviation

$$\hat{s}_n = \frac{1}{n} \sum_{j=1}^{n} |X_j - \hat{\theta}_n|. \tag{2.6.34}$$

We shall demonstrate that these estimators are consistent and asymptotically normal. To prove these results one could use the general theory of maximum likelihood estimation and its asymptotics. Instead we have decided to give more explicit derivations using the specific structure of maximum likelihood estimators for Laplace distributions. We restrict ourselves to the case of an odd sample size, i.e., $n = 2k + 1$. The case of an even sample size can be derived in an analogous way with some minor adjustments for the different form of the median. Thus we shall assume that $n = 2k + 1$.

Let us start with an interesting representation of the median and the mean absolute deviation for Laplace distributions. First, note the following general relations for the mean absolute deviation:

$$\begin{aligned}
\frac{1}{n}\sum_{i=1}^{n}|X_i - X_{k+1:n}| &= \frac{1}{n}\sum_{i=1}^{n}|X_{i:n} - X_{k+1:n}| \\
&= \frac{1}{n}\left[\sum_{i=1}^{k}(X_{k+1:n} - X_{i:n}) + \sum_{i=k+2}^{n}(X_{i:n} - X_{k+1:n})\right] \\
&= \frac{1}{n}\left[\sum_{i=k+2}^{n} X_{i:n} - \sum_{i=1}^{k} X_{i:n}\right]. \qquad (2.6.35)
\end{aligned}$$

Now let us consider X_i's being i.i.d. from the standard classical Laplace distribution. We use the representation of their order statistics given in Proposition 2.5.6 to obtain the following result.

Proposition 2.6.6 *Let $(X_1, \ldots, X_n)$ be a vector of i.i.d. $\mathcal{CL}(0, 1)$ random variables, $n = 2k + 1$, and let B_n be a Bernoulli random variable with $p = 1/2$ independent of two independent sequences $(\bar{W}_i)_{i=1}^{\infty}$, $(W_i)_{i=1}^{\infty}$ of i.i.d. standard exponential random variables. Define $\bar{B}_n = n - B_n$, $K_n = \max(B_n, \bar{B}_n)$, $\bar{K}_n = n - K_n$, and $\delta_n = \operatorname{sign}(B_n - k - 1/2)$.*

Then we have the following three joint representations of $\hat{\theta}_n$ and $\hat{s}_n$:

$$\begin{aligned}
\hat{\theta}_n &\stackrel{d}{=} \delta_n W_{K_n-k:K_n} \\
&\stackrel{d}{=} \delta_n \sum_{l=1}^{K_n-k} \frac{W_l}{K_n - l + 1} \\
&\stackrel{d}{=} \delta_n \sum_{l=k+1}^{K_n} \frac{W_l}{l}, \\
\hat{s}_n &\stackrel{d}{=} \frac{1}{n}\left(\sum_{i=1}^{\bar{K}_n} \bar{W}_{i:\bar{K}_n} + \sum_{i=K_n-k+1}^{K_n} W_{i:K_n} - \sum_{i=1}^{K_n-k-1} W_{i:K_n}\right) \\
&\stackrel{d}{=} \frac{1}{n}\left(\sum_{l=1}^{\bar{K}_n} \bar{W}_l + \sum_{l=K_n-k+1}^{K_n} W_l + \frac{k}{k+1} W_{K_n-k} + \sum_{l=1}^{K_n-k-1} \frac{2k - K_n + l}{K_n - l + 1} W_l\right) \\
&\stackrel{d}{=} \frac{1}{n}\left(\sum_{l=1}^{\bar{K}_n} \bar{W}_l + \sum_{l=1}^{k} W_l + \frac{k}{k+1} W_{k+1} + \sum_{l=k+2}^{K_n} \left(\frac{2k+1}{l} - 1\right) W_l\right).
\end{aligned}$$

Here and below, if the upper limit of summation is smaller than the lower limit, then the sum is assumed to be zero.

Proof. The representation for the median was explained in Remark 2.5.7. For the mean absolute deviation, let us consider two cases.

First, let $B_n \geq k + 1$, i.e., $K_n = B_n$. We have

$$\sum_{i=k+2}^{n} X_{i:n} \stackrel{d}{=} \sum_{i=B_n-k+1}^{B_n} W_{i:B_n}$$

and

$$\sum_{i=1}^{k} X_{i:n} \stackrel{d}{=} -\sum_{i=1}^{\bar{B}_n} \bar{W}_{i:\bar{B}_n} + \sum_{i=1}^{B_n-k-1} W_{i:B_n}.$$

Thus in this case the first representation for $\hat{s}_n$ follows from the relation (2.6.35).

The second case of $B_n \le k$, i.e., $K_n = \bar{B}_n$, can be treated similarly. We obtain

$$\hat{s}_n \stackrel{d}{=} \frac{1}{n}\left(\sum_{i=1}^{B_n} W_{i:B_n} + \sum_{i=\bar{B}_n-k+1}^{\bar{B}_n} \bar{W}_{i:\bar{B}_n} - \sum_{i=1}^{\bar{B}_n-k-1} \bar{W}_{i:\bar{B}_n}\right).$$

The first representation of $\hat{s}_n$ follows from the fact that B_n is independent of the W_i's and $\bar{W}_i$'s, which allows for the replacement of W_i's by $\bar{W}_i$'s (and vice versa) in the last equation.

To prove the second representation, we apply the representation of order statistics of exponential random variables given in (2.5.38). Let us consider only the case of $B_n \ge k+1$, the other case being symmetric. Since the representation for the median was discussed in Remark 2.5.7, here we consider the mean absolute deviation.

We have (for $B_n \ge k+1$)

$$\hat{s}_n \stackrel{d}{=} \frac{1}{n}\left(\sum_{i=B_n-k+1}^{B_n} W_{i:B_n} - \sum_{i=1}^{B_n-k-1} W_{i:B_n} + \sum_{i=1}^{\bar{B}_n} \bar{W}_{i:\bar{B}_n}\right). \tag{2.6.36}$$

By representation (2.5.38), the distribution of the first term in the above equation is the same as that of

$$\begin{aligned}
\sum_{i=B_n-k+1}^{B_n} &\left(\sum_{l=1}^{B_n-k} \frac{W_l}{B_n-l+1} + \sum_{l=B_n-k+1}^{i} \frac{W_l}{B_n-l+1}\right) \\
&= k\sum_{l=1}^{B_n-k} \frac{W_l}{B_n-l+1} + \sum_{l=B_n-k+1}^{B_n} \sum_{i=l}^{B_n} \frac{W_l}{B_n-l+1} \\
&= k\sum_{l=1}^{B_n-k} \frac{W_l}{B_n-l+1} + \sum_{l=B_n-k+1}^{B_n} W_l.
\end{aligned}$$

The second and the third terms in (2.6.36) can be written as follows:

$$\begin{aligned}
\sum_{i=1}^{B_n-k-1} \sum_{l=1}^{i} \frac{W_l}{B_n-l+1} &= \sum_{l=1}^{B_n-k-1} \sum_{i=l}^{B_n-k-1} \frac{W_l}{B_n-l+1} \\
&= \sum_{l=1}^{B_n-k-1} \frac{B_n-k-l}{B_n-l+1} W_l, \\
\sum_{i=1}^{\bar{B}_n} \sum_{l=1}^{i} \frac{\bar{W}_l}{\bar{B}_n-l+1} &= \sum_{l=1}^{\bar{B}_n} \sum_{i=l}^{\bar{B}_n} \frac{\bar{W}_l}{\bar{B}_n-l+1} \\
&= \sum_{l=1}^{\bar{B}_n} \bar{W}_l.
\end{aligned}$$

Combining these three distributional relations results in the second representation of the mean absolute deviation.

Finally, the third representation is obtained by replacing the sequence $(W_1, \dots, W_{B_n})$ by $(W_{B_n}, \dots, W_1)$ and $(\bar{W}_1, \dots, \bar{W}_{\bar{B}_n})$ by $(\bar{W}_{\bar{B}_n}, \dots, \bar{W}_1)$. □

Now we prove the main theorem about consistency and asymptotic efficiency of $\hat{\theta}_n$ and $\hat{s}_n$ as estimators of θ and s. The proof is rather involved. We hope that our readers will communicate to us a simplified proof. Note, however, that consistency, asymptotic normality, and efficiency of MLE's for various distributions is a challenging problem, and a number of prominent mathematical statisticians struggled with it over the last 30 years.

Theorem 2.6.1 *Let $(X_i)_{i=1}^{\infty}$ be a sequence of i.i.d. random variables having $\mathcal{CL}(\theta, s)$ distribution. Then the pair of maximum likelihood estimators $(\hat{\theta}_n, \hat{s}_n)$ of (θ, s) is consistent, asymptotically normal and efficient. The asymptotic covariance matrix has the form*

$$\mathbf{\Sigma} = \begin{bmatrix} s^2 & 0 \\ 0 & s^2 \end{bmatrix}.$$

(See also Fisher's information matrix at the beginning of this section.)

Proof. It is sufficient to assume that $\theta = 0$ and $s = 1$ and show that

$$\sqrt{n}(\hat{\theta}_n - E\hat{\theta}_n, \hat{s}_n - E\hat{s}_n)$$

converges in distribution to the standard bivariate normal distribution while $\sqrt{n}E(\hat{\theta}_n)$ and $\sqrt{n}[E(\hat{s}_n) - 1]$ converge to zero.

We shall use the representation of the estimators given in Proposition 2.6.6. By the Central Limit Theorem and Skorohod's representation theorem we can assume that

$$(B_n - n/2)/\sqrt{n/4}$$

converges almost surely to a standard normal random variable Z which is independent of the W_i's and $\bar{W}_i$'s.

Let us first consider the median $\hat{\theta}_n$. By Proposition 2.6.6, we need to find the limiting distribution of the variable A_n equal to the middle expression in the following inequalities multiplied by δ_n:

$$\sqrt{n}\frac{1}{k+1}\sum_{l=k+1}^{K_n} W_l \le \sqrt{n}\sum_{l=k+1}^{K_n}\frac{W_l}{l} \le \sqrt{n}\frac{1}{K_n}\sum_{l=k+1}^{K_n} W_l. \tag{2.6.37}$$

Consider the right-hand side expression, say R_n, and take its characteristic function with respect to the conditional distribution given B_n:

$$\begin{aligned}
\phi_{R_n}(t|B_n) &= E\left(\exp\left(it\sqrt{n}\frac{1}{K_n}\delta_n\sum_{l=k+1}^{K_n} W_l\right)\middle| B_n\right) \\
&= \frac{1}{(1 - it\delta_n\sqrt{n}/K_n)^{K_n-k}} \\
&= \left(\frac{1}{(1 - it\delta_n\sqrt{n}/K_n)^{K_n/(it\delta_n\sqrt{n})}}\right)^{it\delta_n\sqrt{n}(K_n-k)/K_n}.
\end{aligned}$$

Note that $i\delta_n\sqrt{n}/K_n$ converges in the absolute value to zero, and $\delta_n\sqrt{n}(K_n - k)/K_n$ converges by the assumption to Z a.e. Consequently, the considered characteristic function converges (a.e. with respect to K_n) to e^{itZ}. Thus the conditional distribution of the right-hand side of (2.6.37), R_n, converges to a degenerated distribution at Z. Thus the convergence is in probability. Exactly the same arguments can be repeated for the left-hand side of (2.6.37). This implies that A_n, conditionally on B_n, converges in probability to Z. To obtain the unconditional limiting distribution of A_n, note that

$$\phi_{A_n}(t) = E(\phi_{A_n}(t|B_n)).$$

Since $\phi_{A_n}(t|B_n)$ is bounded and convergent almost everywhere, it follows from the Dominated Convergence Theorem that $\phi_{A_n}(t)$ converges to $E(e^{itZ}) = e^{-t^2/2}$.

Now we consider the mean absolute deviation. We again consider the distribution of $\hat{s}_n$ conditionally on B_n. Set

$$C_n = \sqrt{n}(\hat{s}_n - E(\hat{s}_n|B_n))$$

and note the representation

$$C_n \stackrel{d}{=} \frac{\sum_{l=1}^{\bar{K}_n}(\bar{W}_l - E(\bar{W}_l))}{\sqrt{n}} + \frac{\sum_{l=1}^{k}(W_l - E(W_l))}{\sqrt{n}}$$
$$+ \frac{1}{\sqrt{n}}\frac{k}{k+1}(W_{k+1} - E(W_{k+1})) + \sqrt{n}\sum_{l=k+2}^{K_n}\left(\frac{1}{l} - \frac{1}{n}\right)(W_l - E(W_l)).$$

Note that the four terms in the above representation are mutually independent. Also, the first two terms are independent of the median. It follows from the Central Limit Theorem that each of the first two terms is convergent in distribution to the standard normal distribution multiplied by $\sqrt{2}/2$ (we need also to invoke the Law of Large Numbers to show that $\bar{K}_n/n$ converges almost surely to $1/2$). Thus their sum is convergent to the standard normal distribution. Clearly,

$$\frac{1}{\sqrt{n}}\frac{k}{k+1}(W_{k+1} - E(W_{k+1}))$$

converges to zero.

It remains to consider the distributional limit of the last term,

$$\sqrt{n}\sum_{l=k+2}^{K_n}\left(\frac{1}{l} - \frac{1}{n}\right)(W_l - E(W_l)).$$

Note the following inequalities ($E(W_1) = 1$):

$$\sqrt{n}(K_n - k - 2)\left(\frac{1}{K_n} - \frac{1}{n}\right) \le \sqrt{n}\sum_{l=k+2}^{K_n}\left(\frac{1}{l} - \frac{1}{n}\right)E(W_l)$$
$$\le \sqrt{n}(K_n - k - 2)\left(\frac{1}{k+2} - \frac{1}{n}\right)$$

and

$$\sqrt{n}\sum_{l=k+2}^{K_n} W_l\left(\frac{1}{K_n} - \frac{1}{n}\right) \le \sqrt{n}\sum_{l=k+2}^{K_n}\left(\frac{1}{l} - \frac{1}{n}\right)W_l \le \sqrt{n}\sum_{l=k+2}^{K_n} W_l\left(\frac{1}{k+2} - \frac{1}{n}\right).$$

Since K_n/n converges in probability to $1/2$, and $(K_n - k - 1/2)/\sqrt{n}$ converges almost surely to $|Z|/2$, we conclude that

$$\sqrt{n}(K_n - k - 2)\left(\frac{1}{K_n} - \frac{1}{n}\right) \text{ and } \sqrt{n}(K_n - k - 2)\left(\frac{1}{k+2} - \frac{1}{n}\right)$$

converge in probability to $|Z|/2$ (conditionally on B_n). Observe that

$$\sqrt{n}\sum_{l=k+2}^{K_n} W_l\left(\frac{1}{K_n} - \frac{1}{n}\right) \text{ and } \sqrt{n}\sum_{l=k+2}^{K_n} W_l\left(\frac{1}{k+2} - \frac{1}{n}\right)$$

have the same limit (conditionally on B_n) since

$$\frac{1/K_n - 1/n}{1/(k+2) - 1/n}$$

converges in probability to one. In addition,

$$\sqrt{n}\sum_{l=k+2}^{K_n} W_l\left(\frac{1}{k+2} - \frac{1}{n}\right) = \frac{k}{k+2}\sum_{l=k+2}^{K_n}\frac{W_l}{\sqrt{n}}.$$

The characteristic function (conditionally on B_n) of $\sum_{l=k+2}^{K_n} W_l/\sqrt{n}$ is convergent to $e^{it|Z|/2}$. This shows that, in probability (conditionally on B_n),

$$\lim_{n\to\infty}\sqrt{n}\sum_{l=k+2}^{K_n}\left(\frac{1}{l} - \frac{1}{n}\right)(W_l - E(W_l)) = \frac{|Z|}{2} - \frac{|Z|}{2} = 0.$$

Consequently, $\hat{s}_n$ converges to the standard normal distribution and asymptotically is independent of $\hat{\theta}_n$ (the only terms in the representation of $\hat{s}_n$ that are dependent on $\hat{\theta}_n$ are convergent in probability to zero).

To conclude the proof, we need to show that

$$\lim_{n\to\infty}\sqrt{n}[E(\hat{s}_n) - 1] = 0.$$

We have

$$\begin{aligned}
E(\hat{s}_n) &= \frac{E(K_n)}{n} + \frac{k}{n} + \frac{1}{n}\frac{k}{k+1} + E\left(\sum_{l=k+2}^{K_n}\frac{1}{l}\right) - \frac{E(K_n) - k - 1}{n} \\
&= \frac{1}{2} + \frac{1}{2 + 1/k} + \frac{1}{n}\frac{k}{k+1} + E\left(\sum_{l=k+2}^{K_n}\frac{1}{l}\right) - \frac{n/2 - k - 1}{n}.
\end{aligned}$$

We see that all but the first two terms converge to zero at the rate $o(n^{-1/2})$ and the first two terms converge to one at the same rate. This concludes the proof. □

Remark 2.6.7 Harter et al. (1979) discuss adaptive MLE's of the location and scale parameters (θ and s, respectively) of a symmetric population, where a sample is first classified as having come from uniform, normal, or Laplace distribution, and then the MLE's of θ and s, appropriate for the chosen population, are computed. See Harter et al. (1979) and references therein for further information, including the classification criteria.

2.6.1.2 *Maximum likelihood estimation under censoring.* Let $X_1, \ldots, X_n$ be an i.i.d. sample from the classical Laplace distribution with density $f(\cdot; \theta, s)$ given by (2.1.1) and distribution function $F(\cdot; \theta, s)$ given by (2.1.5). When the smallest r and the largest r observations are censored we obtain a Type II (symmetrically) censored sample:

$$X_{r+1:n} \leq \cdots \leq X_{n-r:n}. \tag{2.6.38}$$

If $x_{r+1:n} \leq \cdots \leq x_{n-r:n}$ is a particular realization of (2.6.38), then the likelihood function is

$$L(\theta, s) = \frac{n!}{(r!)^2}\{F(x_{r+1:n}; \theta, s)[1 - F(x_{n-r:n}; \theta, s)]\}^r \prod_{i=r+1}^{n-r} f(x_{i:n}; \theta, s). \tag{2.6.39}$$

Utilizing (2.1.1) and (2.1.5) we obtain

$$L(\theta, s) = \frac{n!}{2^n (r!)^2 s^{2n-2}} \tag{2.6.40}$$

$$\times \begin{cases} \frac{e^{-(x_{n-r:n}-\theta)/s}(2-e^{-(x_{r+1:n}-\theta)/s})}{\exp\{\sum_{i=r+1}^{n-r}(x_{i:n}-\theta)/s\}}, & \theta < x_{r+1:n}, \\ \exp\left\{\frac{-r}{s}(x_{n-r:n} - x_{r+1:n}) - \sum_{i=r+1}^{n-r}\left|\frac{x_{i:n}-\theta}{s}\right|\right\}, & \theta \in [x_{r+1:n}, x_{n-r:n}], \\ \frac{e^{(x_{r+1:n}-\theta)/s}(2-e^{(x_{n-r:n}-\theta)/s})}{\exp\{\sum_{i=r+1}^{n-r}(\theta-x_{i:n})/s\}}, & \theta > x_{n-r:n}. \end{cases}$$

We now fix $s > 0$ and maximize the function L with respect to θ. The likelihood function is monotonically increasing in θ on $(-\infty, x_{r+1:n})$ clearly and monotonically decreasing in θ on $(x_{n-r:n}, \infty)$, so that the maximum values of L must occur for some θ in $[x_{r+1:n}, x_{n-r:n}]$; see Exercise 2.7.44. But on the latter interval, the function L is maximized if the sum

$$\sum_{i=r+1}^{n-r}\left|\frac{x_{i:n} - \theta}{s}\right|$$

is minimal, so the MLE of θ is sample median of the censored sample (which is the same as that of the original sample). Substituting the sample median $\hat{\theta}_n$ given by (2.6.15) into the likelihood function (2.6.40) results in the following function of s to be maximized,

$$g(s) = L(\hat{\theta}_n, s) = \frac{n!}{2^n (r!)^2 s^{n-2r}} e^{-C/s}, \tag{2.6.41}$$

where

$$C = r(x_{n-r:n} - x_{r+1:n}) + \sum_{i=r+1}^{n-r} |x_{i:n} - \hat{\theta}_n| > 0. \tag{2.6.42}$$

Since the function g is maximized at $s = C/(n-2r)$ (Exercise 2.7.44), we obtain the following MLE of s [see Balakrishnan and Cutler (1994)]:

$$\hat{s}_n = \frac{1}{n-2r}\left\{\sum_{i=[[(n+1)/2]]+1}^{n-r} x_{i:n} - \sum_{i=r+1}^{[[n/2]]} x_{i:n} + r(x_{n-r:n} - x_{r+1:n})\right\}. \tag{2.6.43}$$

Remark 2.6.8 Balakrishnan and Cutler (1994) derived the bias and efficiencies of the above estimators (compared to the BLUE's discussed below); see also Childs and Balakrishnan (1997a) for the derivation of the mean square error of these estimators. Balakrishnan and Cutler (1994) obtained similar explicit estimators of θ and s under Type II right-censoring, while Childs and Balakrishnan (1997b) extended the results to a general Type II censored samples.

2.6.1.3 *Maximum likelihood estimation of monotone location parameters.* Let for each $i = 1, 2, \ldots, k$, $f(x; \theta_i)$ be the density (2.1.1) of the classical Laplace $\mathcal{CL}(\theta_i, s)$ distribution with the location parameter θ_i and the scale parameter $s = 1$. Assume that n_i items,

$$X_{i1}, X_{i2}, \ldots, X_{in_i}, \tag{2.6.44}$$

are chosen from the distribution with density $f(x; \theta_i)$, and that the resulting k samples are independent. Our goal is to find estimates $\hat{\theta}_1, \hat{\theta}_2, \ldots, \hat{\theta}_k$ of $\theta_1, \theta_2, \ldots, \theta_k$ such that

$$\hat{\theta}_1 \geq \hat{\theta}_2 \geq \cdots \geq \hat{\theta}_k. \tag{2.6.45}$$

Brunk (1955) considered problems of this type when $f(x; \theta)$ is a member of an exponential family of distributions (that includes the normal distribution with either unknown mean or unknown standard deviation but does not include the Laplace distribution), while Robertson and Waltman (1968) developed a procedure for finding restricted estimates (2.6.45) for a class of distributions containing the classical Laplace law. More information on the early history of such problems is given in Brunk (1965).

A procedure for obtaining restricted maximum likelihood estimates developed by Robertson and Waltman (1968) assumes that the family of functions $\{f(x; \theta), \theta \in \Theta\}$, where Θ is a connected set of real numbers, satisfies the following four conditions:

(A1) $f(x; \theta)$ has support S which is the same for all $\theta \in \Theta$.

(A2) For each $x \in S$ the function $f(x; \theta)$ is continuous in θ.

(A3) If $x_1, \ldots, x_n \in S$, then the likelihood function

$$L(\theta; x_1, \ldots, x_n) = \prod_{i=1}^{n} f(x_i; \theta) \tag{2.6.46}$$

is unimodal with mode M (not necessary unique).

(A4) If $x_1, \ldots, x_n \in S$ and $y_1, \ldots, y_m \in S$, and M_x , M_y are the modes of the likelihood functions $L(\theta; x_1, \ldots, x_n)$ and $L(\theta; y_1, \ldots, y_m)$, respectively, then M_{xy} is between M_x and M_y, where M_{xy} is the mode of $L(\theta; x_1, \ldots, x_n, y_1, \ldots, y_m)$.

Conditions A3 and A4 do not assume that the mode is unique [similar earlier results by van Eeden (1957) did assume the uniqueness of the mode], although condition A4 requires the existence of a certain rule by which the mode is to be selected.

In this setting, let M_i be the mode of the likelihood function of the ith sample (2.6.44), and for $1 \leq R \leq S \leq k$, let $M(R, S)$ denote the mode of the likelihood function

$$\prod_{i=R}^{S} \prod_{j=1}^{n_i} f(x_{ij}; \theta) \tag{2.6.47}$$

of the combined observations of the Rth through Sth samples. The objective is to find a point $\hat{\theta}_1, \hat{\theta}_2, \ldots, \hat{\theta}_k$ in the set

$$S_k = \{(\alpha_1, \ldots, \alpha_k) : \alpha_i \in \Theta, \alpha_1 \geq \alpha_2 \geq \cdots \geq \alpha_k\} \tag{2.6.48}$$

for which the likelihood function

$$L(\alpha_1, \ldots, \alpha_k) = \prod_{i=1}^{k} \prod_{j=1}^{n_i} f(x_{ij}; \alpha_i) \tag{2.6.49}$$

is maximized. The main result of Robertson and Waltman (1968) asserts that under conditions A1–A4 there exists a point in S_k maximizing the likelihood function (2.6.49) and that it admits the representation

$$\hat{\theta}_j = \min_{1\le R\le j}\max_{R\le S\le k} M(R,S) = \max_{j\le S\le k}\min_{1\le R\le S} M(R,S). \tag{2.6.50}$$

In addition, if $\theta_1 \ge \theta_2 \ge \cdots \ge \theta_k$ and if

$$\lim_{m\to\infty}\sum_{i=1}^{k}|M_i-\theta_i| = 0, \tag{2.6.51}$$

then with probability one

$$\lim_{m\to\infty}\sum_{i=1}^{k}|\hat{\theta}_i-\theta_i| = 0, \tag{2.6.52}$$

where $m = \min(n_1, \dots, n_k)$ [see Robertson and Waltman (1968)].

Evidently, the family of Laplace densities with location $\theta \in \Theta = (-\infty, \infty)$ and a given scale parameter s (for convenience assumed to be one) satisfies conditions A1–A3 above. Here the mode of the likelihood function (the MLE of θ) is the sample median. Further, if in case of an even sample size the median is chosen as in (2.6.15) to be the average of the two middle values, then condition A4 is satisfied as well (see Exercise 2.7.40). Consequently, we have the following result [see Robertson and Waltman (1968)].

Proposition 2.6.7 *Assume that we have k independent random samples, where the ith sample, given in* (2.6.44), *is from the classical Laplace distribution with the location parameter θ_i and the scale parameter $s = 1$. Then $\hat{\theta}_1 \ge \hat{\theta}_2 \ge \cdots \ge \hat{\theta}_k$, where $\hat{\theta}_j$ is given by* (2.6.50), *is the MLE of $\theta_1, \theta_2, \dots, \theta_k$ subject to the condition* (2.6.45).

Further, as noted by Robertson and Waltman (1968), the sample median of the ith sample, M_i, converges almost surely to θ_i by the Glivenko–Cantelli Theorem, so by (2.6.51) we have the almost sure convergence (2.6.52) of the restricted MLE's.

2.6.1.4 *The method of moments.* Let $X_1, \dots, X_n$ be a random sample from the classical Laplace distribution with density (2.1.1). As in the case of MLE's, we shall consider three cases, two when one of the parameters is known, and one when both are unknown.

Case 1: The value of s is known. Since the mean of the $\mathcal{CL}(\theta, s)$ random variable is equal to θ, the method of moments estimator (MME) of θ is the sample mean

$$\tilde{\theta}_n = \frac{1}{n}\sum_{i=1}^{n} X_i. \tag{2.6.53}$$

Clearly, the estimator (2.6.53) is unbiased for θ. Further, by the Strong Law of Large Numbers and the Central Limit Theorem, it is consistent and asymptotically normal.

Proposition 2.6.8 *Let $X_1, \dots, X_n$ be i.i.d. with the $\mathcal{CL}(\theta, s)$ distribution* (2.1.1), *where s is known and $\theta \in \mathbb{R}$ is unknown. Then the MME of θ given by* (2.6.53) *is*

(i) *unbiased;*

(ii) *strongly consistent;*

(iii) *asymptotically normal; i.e.,* $\sqrt{n}(\tilde{\theta}_n - \theta)$ *converges in distribution to a normal distribution with mean zero and variance* $2s^2$.

Note that the asymptotic variance of the MME of θ is twice as large as that of the MLE of θ, so that for the Laplace distribution the *asymptotic relative efficiency* (ARE) of the sample median $\hat{\theta}_n$ relative to the sample mean $\tilde{\theta}_n$ is

$$ARE(\hat{\theta}_n) = \frac{2s^2}{s^2} = 2.$$

For any finite sample size n, the variance of the MME is

$$\text{Var}(\tilde{\theta}_n) = \frac{\text{Var}(X_1)}{n} = \frac{2s^2}{n}, \tag{2.6.54}$$

while the variance of the MLE (the canonical median) is given in Section 2.5 [see also the relations (2.7.24)–(2.7.25), Exercise 2.7.39]. Table 2.7 contains the variances of $\hat{\theta}_n$ and $\tilde{\theta}_n$ for sample sizes $n = 1(1)7$. We see that

$$\text{Var}(\hat{\theta}_n) \leq \text{Var}(\tilde{\theta}_n) \tag{2.6.55}$$

when the sample size n is between 3 and 7, the difference being rather substantial. Chu and Hotelling (1955) established the relation

$$B_k\left(1 - \frac{1}{2k+2}\right)^{3/2} \leq \frac{\text{Var}(\hat{\theta}_{2k+1})}{1/(2k+1)} \leq 1.51 B_k\left(1 + \frac{1}{2k}\right)^{3/2}, \quad k \geq 1, \tag{2.6.56}$$

where

$$B_k = \frac{(2k+1)!}{(k!)^2}\left(\frac{1}{2}\right)^{2k+1}\sqrt{\frac{2\pi}{2k+1}}, \tag{2.6.57}$$

and concluded that if $n = 2k + 1 \geq 7$, then the relation (2.6.55) holds as well (Exercise 2.7.42).

n	1	2	3	4	5	6	7
$\text{Var}(\tilde{\theta}_n)$	2	1	0.667	0.500	0.400	0.333	0.286
$\text{Var}(\hat{\theta}_n)$	2	1	0.639	0.406	0.351	0.261	0.236

Table 2.7: The variances of $\tilde{\theta}_n$ (the sample mean) and $\hat{\theta}_n$ (the sample median) for samples of size n from the standard classical Laplace distribution.

Case 2: The value of θ is known. Since the r.v. $X_i - \theta$ has the $\mathcal{CL}(0, s)$ distribution, without loss of generality we shall assume that $\theta = 0$. By the moment relation (2.1.14), we have $EX_i^2 = 2s^2$, so the MME of s is

$$\tilde{s}_n = \sqrt{\frac{1}{2n}\sum_{i=1}^{n} X_i^2}. \tag{2.6.58}$$

The following result summarizes the asymptotic properties of $\tilde{s}_n$.

Proposition 2.6.9 *Let $X_1, \ldots, X_n$ be i.i.d. with the $\mathcal{CL}(0, s)$ distribution. Then the MME of s given by* (2.6.58) *is*

(i) *strongly consistent;*

(ii) *asymptotically normal; i.e., $\sqrt{n}(\tilde{s}_n - s)$ converges in distribution to a normal distribution with mean zero and variance* $1.25s^2$.

Proof. To establish (i), note that by the Strong Law of Large Numbers,

$$\frac{1}{n}\sum_{i=1}^{n} X_i^2 \stackrel{a.s.}{\to} E[X_i^2] = 2s^2. \tag{2.6.59}$$

Thus

$$\tilde{s}_n = g\left(\frac{1}{n}\sum_{i=1}^{n} X_i^2\right) \stackrel{a.s.}{\to} g(2s^2) = s, \tag{2.6.60}$$

where

$$g(x) = \sqrt{x/2}. \tag{2.6.61}$$

Similarly, part (ii) can be established via the Central Limit Theorem. Since $X_i^2, i = 1, 2, \ldots$ are i.i.d. with

$$E[X_1^2] = 2s^2 \text{ and } \operatorname{Var}[X_i^2] = E[X_i]^4 - (E[X_i^2])^2 = 20s^4$$

[see the moment formula (2.1.14)], the sequence

$$n^{1/2}\left(\frac{1}{n}\sum_{i=1}^{n} X_i^2 - 2s^2\right) \tag{2.6.62}$$

converges in distribution to a normal distribution with mean zero and variance $20s^4$. Thus by standard arguments of the large sample theory [see, e.g., Rao (1965)], the sequence

$$n^{1/2}\left[g\left(\frac{1}{n}\sum_{i=1}^{n} X_i^2\right) - g(2s^2)\right] = n^{1/2}(\tilde{s}_n - s) \tag{2.6.63}$$

converges in distribution to a normal distribution with mean zero and variance

$$[g'(2s^2)]^2(20s^4) = \frac{5}{4}s^2. \tag{2.6.64}$$

□

Remark 2.6.9 Note that the asymptotic variance of the $\tilde{s}_n$ is larger than that of the MLE $\hat{s}_n$. The relation between the variances for a finite sample size n is investigated in Exercise 2.7.43.

Case 3: *Both s and θ are unknown.* Let

$$\widehat{m}_{1n} = \frac{1}{n}\sum_{i=1}^{n} X_i \text{ and } \widehat{m}_{2n} = \frac{1}{n}\sum_{i=1}^{n} X_i^2 \tag{2.6.65}$$

be the first and second sample moments for the random sample $X_1, \ldots, X_n$ from the $\mathcal{CL}(\theta, s)$ distribution. Since the first two moments of X_1 are

$$E[X_1] = \theta, \qquad E[X_1^2] = \theta^2 + 2s^2 \tag{2.6.66}$$

[see (2.1.18)], solving equations (2.6.66) for θ and s in terms of the first two moments and substituting the sample moments (2.6.65), we arrive at the following MME's of θ and s:

$$\tilde{\theta}_n = \widehat{m}_{1n} = \frac{1}{n}\sum_{i=1}^{n} X_i, \quad \tilde{s}_n = \sqrt{\frac{\widehat{m}_{2n} - \widehat{m}_{1n}^2}{2}} = \sqrt{\frac{1}{2n}\sum_{i=1}^{n}(X_i - \overline{X}_n)^2}. \tag{2.6.67}$$

As before, the consistency and asymptotic normality of the estimators (2.6.67) follow from standard arguments of the large sample theory [see, e.g., Rao (1965)].

Proposition 2.6.10 *Let $X_1, \ldots, X_n$ be i.i.d. from the $\mathcal{CL}(\theta, s)$ distribution, where $\theta \in \mathbb{R}$ and $s > 0$. Let*

$$\tilde{\xi}_n = \begin{bmatrix} \tilde{\theta}_n \\ \tilde{s}_n \end{bmatrix}, \tag{2.6.68}$$

where $\tilde{\theta}_n$ and $\tilde{s}_n$ are given by (2.6.67), *be the MME of the vector parameter*

$$\xi = \begin{bmatrix} \theta \\ s \end{bmatrix}. \tag{2.6.69}$$

Then the estimator $\tilde{\xi}_n$ is

(i) *strongly consistent;*

(ii) *asymptotically normal; i.e., the sequence $\sqrt{n}(\tilde{\xi}_n - \xi)$ converges in distribution to a bivariate normal distribution with the (vector) mean zero and the covariance matrix*

$$\mathbf{\Sigma}_{MME} = \begin{bmatrix} 2s^2 & 0 \\ 0 & \frac{5}{4}s^2 \end{bmatrix}. \tag{2.6.70}$$

Proof. Consider an auxiliary sequence of i.i.d. bivariate random vectors

$$\mathbf{Y}_i = \begin{bmatrix} X_i \\ X_i^2 \end{bmatrix}, i = 1, 2, \ldots. \tag{2.6.71}$$

The vector mean and the covariance matrix of $\mathbf{Y}_i$ are as follows:

$$\mathbf{m_Y} = \begin{bmatrix} \theta \\ \theta^2 + 2s^2 \end{bmatrix}, \quad \mathbf{\Sigma_Y} = \begin{bmatrix} 2s^2 & 4\theta s^2 \\ 4\theta s^2 & 8\theta^2 s^2 + 20s^4 \end{bmatrix}. \tag{2.6.72}$$

[We have used the moment formulas (2.1.18).] Clearly, the Strong Law of Large Numbers and Central Limit Theorem apply to the sequence $(\mathbf{Y}_i)$, so

$$\lim_{n\to\infty} \frac{1}{n}\sum_{i=1}^{n} \mathbf{Y}_i \stackrel{\text{a.s.}}{=} \mathbf{m_Y} \tag{2.6.73}$$

and

$$\lim_{n\to\infty} \sqrt{n}\left(\frac{1}{n}\sum_{i=1}^{n}\mathbf{Y}_i - \mathbf{m_Y}\right) \stackrel{\text{d}}{=} N_2(\mathbf{0}, \mathbf{\Sigma_Y}). \tag{2.6.74}$$

(The notation $N_d(\mathbf{m}, \mathbf{\Sigma})$ denotes the d-dimensional normal distribution with mean vector $\mathbf{m}$ and the covariance matrix $\mathbf{\Sigma}$.) Observe that the estimator (2.6.68) can be expressed in terms of the $\mathbf{Y}_i$'s as

$$\tilde{\xi}_n = g\left(\frac{1}{n}\sum_{i=1}^{n}\mathbf{Y}_i\right), \tag{2.6.75}$$

where

$$g(x_1, x_2) = \left(x_1, \sqrt{\frac{x_2 - x_1^2}{2}}\right). \tag{2.6.76}$$

To prove the strong consistency, use (2.6.73) together with the continuity of g to conclude that

$$\lim_{n\to\infty} g\left(\frac{1}{n}\sum_{i=1}^{n}\mathbf{Y}_i\right) = \lim_{n\to\infty}\tilde{\xi}_n \stackrel{\text{a.s.}}{=} g(\mathbf{m_Y}) = \xi. \tag{2.6.77}$$

Similarly, we establish the asymptotic normality of $\tilde{\xi}_n$ by standard result from the large sample theory [see, e.g., Rao (1965)]. Since the function g has a nonsingular matrix of partial derivatives at the point $\mathbf{m_Y}$,

$$\mathbf{D} = \left[\left.\frac{\partial g_i}{\partial x_j}\right|_{\mathbf{x}=\mathbf{m_Y}}\right] = \frac{1}{s}\begin{bmatrix} s & 0 \\ -\theta/s & 1/4 \end{bmatrix}, \tag{2.6.78}$$

the convergence (2.6.74) produces

$$\lim_{n\to\infty}\sqrt{n}\left[g\left(\frac{1}{n}\sum_{i=1}^{n}\mathbf{Y}_i\right) - g(\mathbf{m_Y})\right] \stackrel{\text{d}}{=} N_2(\mathbf{0}, \mathbf{D\Sigma_Y D'}), \tag{2.6.79}$$

or

$$\lim_{n\to\infty}\sqrt{n}\left[\tilde{\xi}_n - \xi\right] \stackrel{\text{d}}{=} N_2(\mathbf{0}, \mathbf{\Sigma}_{MME}), \tag{2.6.80}$$

since

$$g\left(\frac{1}{n}\sum_{i=1}^{n}\mathbf{Y}_i\right) = \tilde{\xi}_n, \quad g(\mathbf{m_Y}) = \xi, \quad \text{and } \mathbf{D\Sigma_Y D'} = \mathbf{\Sigma}_{MME}.$$

□

Remark 2.6.10 For $0 < p < 1$, the function

$$f(x) = p\frac{1}{2s_1}e^{-|x-\theta_1|/s_1} + (1-p)\frac{1}{2s_2}e^{-|x-\theta_2|/s_2}, \quad -\infty < x < \infty, \tag{2.6.81}$$

is the density of the mixture of two Laplace distributions $\mathcal{CL}(\theta_1, s_1)$ and $\mathcal{CL}(\theta_2, s_2)$. Such distributions may no longer be unimodal [see Exercise 2.7.46]. The method of moments estimation of the parameters of (2.6.81) is considered in Kacki (1965b), Krysicki (1966ab), and Kacki and Krysicki (1967).

2.6.1.5 *Linear estimation.* In this section we consider the so-called L-estimators of the parameters θ and s of the classical Laplace distribution, which are linear combinations of order statistics.

Best linear unbiased estimation. Let $X_1, \ldots, X_n$ be a random sample from the $\mathcal{CL}(\theta, s)$ distribution, and let

$$X_{k+1:n} \le \cdots \le X_{n-m:n} \tag{2.6.82}$$

be the corresponding Type II censored sample. For $i = 1, \ldots, n$, let

$$\mu_i = E\left[\frac{X_{i:n} - \theta}{s}\right], \quad \sigma_{ii} = \mathrm{Var}\left[\frac{X_{i:n} - \theta}{s}\right], \quad \sigma_{ij} = \mathrm{Cov}\left[\frac{X_{i:n} - \theta}{s}, \frac{X_{j:n} - \theta}{s}\right] \tag{2.6.83}$$

be the means, variances, and covariances of the order statistics from the standard classical Laplace distribution, with values given in (2.5.31), (2.5.32), and (2.5.33), respectively. Then the *best linear unbiased estimators* (BLUE's—unbiased estimators of minimum variance in the class of linear unbiased estimators) of θ and s based on (2.6.82) are [see, e.g., Sarhan (1954, 1955), Govindarajulu (1966), David (1981), Balakrishnan and Cohen (1991)]

$$\theta_n^* = \frac{\mathbf{m}'\boldsymbol{\Sigma}^{-1}\mathbf{m}\mathbf{1}'\boldsymbol{\Sigma}^{-1} - \mathbf{m}'\boldsymbol{\Sigma}^{-1}\mathbf{1}\mathbf{m}'\boldsymbol{\Sigma}^{-1}}{(\mathbf{m}'\boldsymbol{\Sigma}^{-1}\mathbf{m})(\mathbf{1}'\boldsymbol{\Sigma}^{-1}\mathbf{1}) - (\mathbf{m}'\boldsymbol{\Sigma}^{-1}\mathbf{1})^2} \cdot \mathbf{X} = \sum_{i=k+1}^{n-m} a_i X_{i:n} \tag{2.6.84}$$

and

$$s_n^* = \frac{\mathbf{1}'\boldsymbol{\Sigma}^{-1}\mathbf{1}\mathbf{m}'\boldsymbol{\Sigma}^{-1} - \mathbf{1}'\boldsymbol{\Sigma}^{-1}\mathbf{m}\mathbf{1}'\boldsymbol{\Sigma}^{-1}}{(\mathbf{m}'\boldsymbol{\Sigma}^{-1}\mathbf{m})(\mathbf{1}'\boldsymbol{\Sigma}^{-1}\mathbf{1}) - (\mathbf{m}'\boldsymbol{\Sigma}^{-1}\mathbf{1})^2} \cdot \mathbf{X} = \sum_{i=k+1}^{n-m} b_i X_{i:n}, \tag{2.6.85}$$

where

$$\begin{aligned} \mathbf{X} &= (X_{k+1:n}, \ldots, X_{n-m:n})' \\ \mathbf{m} &= (\mu_{k+1}, \ldots, \mu_{n-m})' \\ \mathbf{1} &= (1, \ldots, 1)' \end{aligned} \tag{2.6.86}$$

are $n-k-m$-dimensional vectors and

$$\boldsymbol{\Sigma} = [\sigma_{ij}]_{i,j=k+1\ldots n-m} \tag{2.6.87}$$

is an $n-k-m \times n-k-m$ covariance matrix. The variances and covariances of the estimators (2.6.84) and (2.6.85) are

$$\mathrm{Var}(\theta_n^*) = s^2 \frac{\mathbf{m}'\boldsymbol{\Sigma}^{-1}\mathbf{m}}{(\mathbf{m}'\boldsymbol{\Sigma}^{-1}\mathbf{m})(\mathbf{1}'\boldsymbol{\Sigma}^{-1}\mathbf{1}) - (\mathbf{m}'\boldsymbol{\Sigma}^{-1}\mathbf{1})^2}, \tag{2.6.88}$$

$$\mathrm{Var}(s_n^*) = s^2 \frac{\mathbf{1}'\boldsymbol{\Sigma}^{-1}\mathbf{1}}{(\mathbf{m}'\boldsymbol{\Sigma}^{-1}\mathbf{m})(\mathbf{1}'\boldsymbol{\Sigma}^{-1}\mathbf{1}) - (\mathbf{m}'\boldsymbol{\Sigma}^{-1}\mathbf{1})^2}, \tag{2.6.89}$$

$$\mathrm{Cov}(\theta_n^*, s_n^*) = -s^2 \frac{\mathbf{m}'\boldsymbol{\Sigma}^{-1}\mathbf{1}}{(\mathbf{m}'\boldsymbol{\Sigma}^{-1}\mathbf{m})(\mathbf{1}'\boldsymbol{\Sigma}^{-1}\mathbf{1}) - (\mathbf{m}'\boldsymbol{\Sigma}^{-1}\mathbf{1})^2}. \tag{2.6.90}$$

Note that under symmetric censoring ($k = m$) the covariance (2.6.90) is equal to 0 (since in this case $\mathbf{m}'\boldsymbol{\Sigma}^{-1}\mathbf{1} = 0$), the coefficients of $X_{i:n}$ and $X_{n-i+1:n}$ in θ_n^* in (2.6.84) are equal and in s_n^* in (2.6.85) are equal in absolute value and opposite in sign.

The coefficients a_i and b_i in (2.6.84) and (2.6.85) were tabulated by Sarhan (1954, 1955) for sample sizes up to 5 and by Govindarajulu (1966) for sample sizes up to 20 (and all choices of symmetric censoring). Balakrishnan, Chandramouleeswaran, and Ambagaspitiya (1994) give tables of a_i and b_i for the case of Type II right censored samples of sizes up to 20 [with $k = 0$ and $m = 0(1)(n-2)$]. In Table 2.8 one can find the coefficients a_i and b_i of θ_n^* and s_n^* based on complete samples for sample sizes $n = 2(1)10$ [calculated by Govindarajulu (1966)].

n		$X_{n:n}$	$X_{n-1:n}$	$X_{n-2:n}$	$X_{n-3:n}$	$X_{n-4:n}$	Variances
2	θ_n^*	0.5000					1.000
	s_n^*	0.6667					0.7778
3	θ_n^*	0.1481	0.7037				0.5895
	s_n^*	0.4444	0.0000				0.4321
4	θ_n^*	0.0473	0.4527				0.4155
	s_n^*	0.3077	0.2145				0.2986
5	θ_n^*	0.0166	0.2213	0.5241			0.3169
	s_n^*	0.2331	0.2264	0.0000			0.2290
6	θ_n^*	0.0063	0.1006	0.3931			0.2548
	s_n^*	0.1876	0.1943	0.1132			0.1858
7	θ_n^*	0.0025	0.0455	0.2386	0.4267		0.2122
	s_n^*	0.1572	0.1631	0.1439	0.0000		0.1565
8	θ_n^*	0.0010	0.0208	0.1316	0.3465		0.1814
	s_n^*	0.1355	0.1391	0.1391	0.0718		0.1351
9	θ_n^*	0.0004	0.0097	0.0698	0.2374	0.3654	0.1581
	s_n^*	0.1191	0.1211	0.1251	0.1013	0.0000	0.1190
10	θ_n^*	0.0002	0.0046	0.0364	0.1478	0.3110	0.1399
	s_n^*	0.1063	0.1074	0.1110	0.1061	0.0504	0.1062

Table 2.8: Coefficients of the BLUE's of the parameters θ and s of the classical Laplace distribution. The last column gives the values of $\mathrm{Var}(\theta_n^*)/s^2$ and $\mathrm{Var}(s_n^*)/s^2$.

By definition, the variance of the BLUE of θ is smaller than that of MLE (the sample median) and MME (the mean) as these are also linear combinations of order statistics and unbiased for θ. Sarhan (1954) compared the efficiencies[17] of the latter two estimators, as well as of the midrange $(X_{1:n}+X_{n:n})/2$ (which is also unbiased), relative to the BLUE of θ. The efficiencies are presented in Table 2.9, and also graphically in Figure 2.5 [taken from Sarhan (1954)]. As noted by Sarhan (1954), the MLE (the median) is more efficient than the MME (the mean) and the midrange (and less efficient than the BLUE).

Remark 2.6.11 Chan and Chan (1969) derived the BLUE's of θ and s based on k selected order statistics (k-optimum BLUE's) connected with a random sample of size n from the classical Laplace distribution $\mathcal{CL}(\theta, s)$. In Chan and Chan (1969), the authors provided tables containing the optimum ranks, the coefficients, biases, variances, and efficiencies (relative to the corresponding BLUE's based on all order statistics for complete samples) of the k-optimum BLUE's for $k = 1, 2, 3, 4$ and $n = k(1)20$.

[17] The efficiency of an estimator $\hat{\theta}_1$ relative to another estimator $\hat{\theta}_2$ is the ratio $\mathrm{Var}(\hat{\theta}_2)/\mathrm{Var}(\hat{\theta}_1)$ expressed as a percentage.

Sample Size n Estimator	2	3	4	5
Mean	100.00	88.43	82.80	79.21
Midrange	100.00	67.90	49.65	38.29
Median	100.00	92.27	98.90	90.23

Table 2.9: Efficiencies of various estimators of the location parameter θ of the classical Laplace distribution, relative to the BLUE of θ.

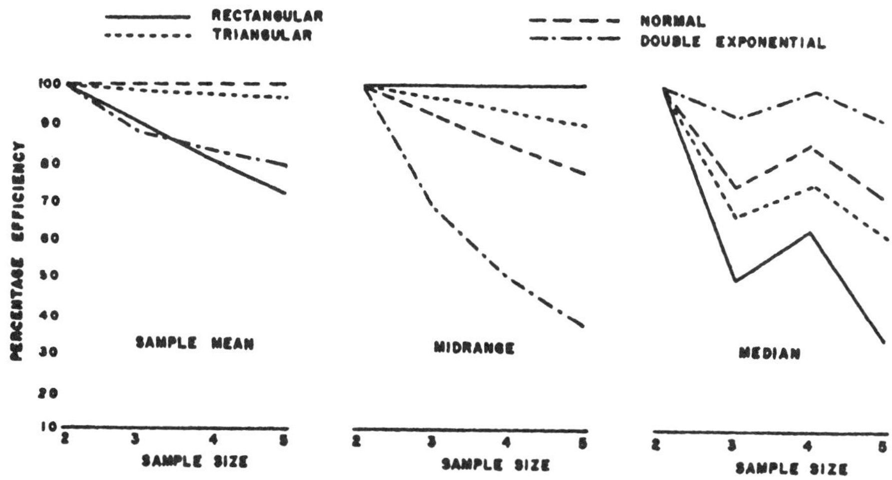

Figure 2.5: Percentage efficiencies of the three estimators of the location parameter θ: the sample mean, the midrange, and the median, relative to the BLUE of θ, in different populations [Republished with permission of Institute of Mathematical Statistics, from A.E. Sarhan, *Annals of Mathematical Statistics*, **25**, copyright 1954.

Remark 2.6.12 Rao et al. (1991) derived an optimum linear (in absolute values of order statistics) unbiased estimator of the scale parameter s in complete and censored samples. The estimator reduces to the sample mean absolute deviation (the MLE of s when θ is known) for complete samples and is generally more efficient than the BLUE of s.

Remark 2.6.13 Ahsanullah and Rahim (1973) noted some practical situations where a number of observations somewhere in the middle of an ordered sample may be missing [see, e.g., Sarhan and Greenberg (1967)]. For a given sample size n, $1 \leq R_1 < R_2 \leq n$, and $k = k_1 + k_2$, where $k_1 < R_1$ and $k_2 < n - (R_2 - 1)$, Ahsanullah and Rahim (1973) determined the optimum ranks

$$1 \leq n_1^0 < n_2^0 < \cdots < n_{k_1}^0 \leq R_1 \text{ and } R_2 \leq n_{k_1+1}^0 < \cdots < n_{k_1+k_2}^0 \leq n$$

and derived the BLUE's of θ and s based on the order statistics

$$X_{n_1^0:n}, X_{n_2^0:n}, \ldots, X_{n_{k_1}^0:n}, X_{n_{k_1+1}^0:n}, \ldots, X_{n_{k_1+k_2}^0:n},$$

observing that the efficiency of their estimates (relative to the BLUE's based on a complete sample) was quite high.

Remark 2.6.14 Let $X_{1:n}, \ldots, X_{n:n}$ be the order statistics corresponding to a random sample of size $n = 2k+1$ from the classical Laplace distribution with an unknown θ and the scale parameter $s = 1$. Akahira (1986) showed that variance of the linear estimator

$$\hat{\theta}_{AK} = \frac{1}{2}(X_{k+1-r\sqrt{k}:n} + X_{k+1+r\sqrt{k}:n}) \tag{2.6.91}$$

with the optimal choice of $r = 0.48$ is asymptotically smaller than that of the MLE of θ (the sample median $\hat{\theta}_n$):

$$\mathrm{Var}(\hat{\theta}_n) = \frac{1}{n}\left\{1 + \frac{1.13}{\sqrt{k}} + o\left(\frac{1}{n}\right)\right\} \tag{2.6.92}$$

while

$$\mathrm{Var}(\hat{\theta}_{AK}) = \frac{1}{n}\left\{1 + \frac{0.90}{\sqrt{k}} + o\left(\frac{1}{n}\right)\right\}. \tag{2.6.93}$$

Generalizing, Sugiura and Naing (1989) showed that an appropriate linear estimator of θ of the form

$$\hat{\theta}_{SN,m} = \sum_{i=1}^{m} a_i[X_{k+1-r_i\sqrt{k}:n} + X_{k+1+r_i\sqrt{k}:n}] + bX_{k+1:n}, \tag{2.6.94}$$

where $0 < r_m < \cdots < r_2 < r_1$ (and with $r_i\sqrt{k}$ assumed to be an integer), has smaller asymptotic variance than the estimator $\hat{\theta}_{AK}$ defined in (2.6.91), as the constant 0.90 in (2.6.93) is reduced to $\sqrt{2/\pi} \approx 0.80$ [see also Akahira (1987, 1990) and Akahira and Takeuchi (1993)]. Sugiura and Naing (1989) observed that the variance of their estimator admits the same asymptotic expansion [given by (2.6.93) with 0.90 replaced by $\sqrt{2/\pi}$] as Bayes risk with respect to a prior having finite interval support (and satisfying some technical conditions) derived by Joshi (1984).

Remark 2.6.15 Let

$$X_{1:n} \le \cdots \le X_{n-s:n} \tag{2.6.95}$$

be a Type II right censored sample associated with a random sample of size n from the $\mathcal{CL}(\theta, s)$ distribution. Balakrishnan and Chandramouleeswaran (1994b) utilized the pivotal variables

$$Q_1 = \frac{X_{n-s+1:n} - X_{n-s}}{s_n^*} \quad \text{and} \quad Q_2 = \frac{X_{n:n} - X_{n-s}}{s_n^*} \tag{2.6.96}$$

in predicting $X_{n-s+1:n}$ and $X_{n:n}$ (the percentage points of Q_1 and Q_2 were determined by Monte-Carlo simulations). The quantity s_n^* in (2.6.96) denotes the BLUE of the scale parameter s based on the censored sample (2.6.95). In addition, these authors derived *prediction intervals* for extreme order statistics $Y_{1:m}$ and $Y_{m:m}$ connected with a *future* sample of size m from the Laplace distribution. The prediction intervals utilize the (simulated) percentage points of the pivotal quantities

$$Q_3 = \frac{Y_{1:m} - \theta_n^*}{s_n^*} \quad \text{and} \quad Q_4 = \frac{Y_{m:m} - \theta_n^*}{s_n^*}, \tag{2.6.97}$$

where θ_n^* and s_n^* are the BLUE's of θ and s, respectively, based on the censored sample (2.6.95). Ling (1977) and Ling and Lim (1978) approached these prediction problems from the Bayesian perspective.

Simplified linear estimation. Let

$$W_i = X_{n-i+1:n} - X_{i:n} \tag{2.6.98}$$

and

$$V_i = \frac{1}{2}(X_{n-i+1:n} + X_{i:n}) \tag{2.6.99}$$

be the ith quasi-range and the ith quasi-midrange, respectively, connected with the random sample $X_1, \ldots, X_n$ from the classical Laplace distribution $\mathcal{CL}(\theta, s)$. Raghunandanan and Srinivasan (1971) considered *simplified linear estimators* of θ and s based on V_i and linear combinations of W_i's for complete as well as symmetrically censored samples. Similar estimators for the parameters of a normal distribution were obtained in Dixon (1957, 1960).

When k largest and k smallest observations are censored, where $k \geq 0$, the simplified estimator of θ is that V_i (with $i \geq k+1$) that has the smallest variance. Under the same censoring, the simplified estimator of s, denoted by $\hat{s}_{k,n}$, is the estimator with minimum variance among estimators of the form

$$C \sum_{i=k+1}^{[[n/2]]} c_i W_i, \tag{2.6.100}$$

where the W_i's are given by (2.6.98), the c_i's take the values of 0 or 1, and C is a normalizing constant that makes the estimator (2.6.100) unbiased. Table 2.10 contains the values of the index i corresponding to the simplified estimator of θ, V_i, based on complete samples with $n = 3(1)20$. The relative efficiency of this estimator relative to the BLUE of θ is also included in Table 2.10 (note that when $n = 3$ and 5 the estimator coincides with the MLE of θ, the sample median).

Table 2.11 contains the values of the simplified estimator $\hat{s}_{k,n}$ of the form (2.6.100), along with its efficiency relative to the BLUE s_n^* of s, defined as

$$\text{Eff}(\hat{s}_{k,n}) = \text{Var}(s_n^*)/\text{Var}(\hat{s}_{k,n}) \times 100\%.$$

More extensive tables can be found in Raghunandanan and Srinivasan (1971).

Remark 2.6.16 Iliescu and Vodă (1973) considered asymptotically unbiased estimators of s of the form

$$\alpha(n) \sum_{i=1}^{[[n/2]]} W_i, \tag{2.6.101}$$

which have the same structure as the simplified estimator (2.6.100) of the scale parameter.

Asymptotic best linear unbiased estimation. Cheng (1978) remarked that for a large sample size n, the BLUE's of θ and s are too tedious to calculate. Consequently, using the theory of *asymptotically best linear unbiased estimates* (ABLUE) developed by Ogawa (1951), he derived a method for an optimal selection of the order statistics from complete as well as singly or doubly censored large samples to estimate parameters of the Laplace distribution. The method utilizes the *sample quantiles*

$$X_{[[n\lambda_1]]+1:n} < \cdots < X_{[[n\lambda_k]]+1:n}, \tag{2.6.102}$$

where the real numbers

$$0 = \lambda_0 < \lambda_1 < \cdots < \lambda_k < \lambda_{k+1} = 1 \tag{2.6.103}$$

n	i	$\mathrm{Var}(V_i)/s^2$	$Eff(V_i)$
3	2	0.638890	92.3
4	2	0.420135	98.9
5	3	0.351180	90.2
6	3	0.260905	97.7
7	3	0.225805	94.0
8	4	0.187310	96.8
9	4	0.164795	95.9
10	5	0.145225	96.3
11	5	0.129605	96.7
12	6	0.118125	96.0
13	6	0.106670	97.0
14	7	0.099285	95.9
15	7	0.090540	97.2
16	7	0.085190	96.0
17	8	0.078575	97.2
18	8	0.074175	96.7
19	9	0.069350	97.3
20	9	0.065670	97.0

Table 2.10: Simplified linear estimator of θ, V_i, its variance, and its (percent) efficiency relative to the BLUE of θ, based on a complete random sample of size n from the classical Laplace distribution $\mathcal{CL}(\theta, s)$.

n	k	$\hat{s}_{k,n}$	$\mathrm{Var}(\hat{s}_{k,n})/s^2$	$\mathrm{Eff}(\hat{s}_{k,n})$
4	0	$0.289157(W_1 + W_2)$	0.300624	99.3
5	0	$0.231325(W_1 + W_2)$	0.229000	100.0
6	0	$0.183486(W_1 + W_2 + W_3)$	0.186515	99.6
6	1	$0.666667W_2$	0.304009	98.5
7	0	$0.157274(W_1 + W_2 + W_3)$	0.156500	100.0
7	1	$0.390721(W_2 + W_3)$	0.234731	97.5
8	0	$0.134254(W_1 + W_2 + W_3 + W_4)$	0.135438	99.7
8	1	$0.324571(W_2 + W_3)$	0.188570	98.4
8	2	$0.967133W_3$	0.303726	99.4
9	0	$0.119337(W_1 + W_2 + W_3 + W_4)$	0.119000	100.0
9	1	$0.282882(W_2 + W_3)$	0.158812	98.3
9	2	$0.790855W_3$	0.233068	98.5
10	0	$0.108696(W_1 + W_2 + W_3 + W_4)$	0.106392	99.8
10	1	$0.238741(W_2 + W_3 + W_5)$	0.137784	98.0
10	2	$0.681084W_3$	0.190810	97.3
10	3	$1.267536W_4$	0.305295	99.7

Table 2.11: Simplified linear estimator of s, $\hat{s}_{k,n}$, its variance, and its (percent) efficiency relative to the BLUE of s, based on a random sample of size n from the classical Laplace distribution $\mathcal{CL}(\theta, s)$, where k observations are censored from each end.

are called the *spacings* and the u_i's defined by

$$\lambda_i = \int_{-\infty}^{u_i} f(x)dx = F(u_i) \tag{2.6.104}$$

are the population quantiles of the standard classical Laplace distribution with density f and distribution function F. Under this, the ABLUE of θ (when s is known) is

$$\theta_n^{**} = \sum_{i=1}^{k} a_i X_{[[n\lambda_i]]+1:n} - \frac{K_3}{K_2}s, \tag{2.6.105}$$

the ABLUE of s (when θ is known) is

$$s_n^{**} = \sum_{i=1}^{k} b_i X_{[[n\lambda_i]]+1:n} - \frac{K_3}{K_2}\theta \tag{2.6.106}$$

and their asymptotic variances are

$$\text{Var}_{ASY}(\theta_n^{**}) = \frac{s^2}{nK_1}, \qquad \text{Var}_{ASY}(s_n^{**}) = \frac{s^2}{nK_2}, \tag{2.6.107}$$

where

$$\begin{aligned} K_1 &= \textstyle\sum_{i=1}^{k+1} \frac{(f_i - f_{i-1})^2}{\lambda_i - \lambda_{i-1}} \\ K_2 &= \textstyle\sum_{i=1}^{k+1} \frac{(f_i u_i - f_{i-1}u_{i-1})^2}{\lambda_i - \lambda_{i-1}} \\ K_3 &= \textstyle\sum_{i=1}^{k+1} \frac{(f_i - f_{i-1})(f_i u_i - f_{i-1}u_{i-1})}{\lambda_i - \lambda_{i-1}} \end{aligned} \tag{2.6.108}$$

and

$$\begin{aligned} a_i &= \tfrac{f_i}{K_1}\left(\tfrac{f_i - f_{i-1}}{\lambda_i - \lambda_{i-1}} - \tfrac{f_{i+1} - f_i}{\lambda_{i+1} - \lambda_i}\right), \\ b_i &= \tfrac{f_i}{K_2}\left(\tfrac{f_i u_i - f_{i-1}u_{i-1}}{\lambda_i - \lambda_{i-1}} - \tfrac{f_{i+1}u_{i+1} - f_i u_i}{\lambda_{i+1} - \lambda_i}\right), \\ f_i &= f(u_i), \quad i = 1, 2, \ldots, k, \quad f_0 = f_{k+1} = f_0 u_0 = f_{k+1}u_{k+1} = 0. \end{aligned} \tag{2.6.109}$$

The asymptotic efficiencies (ARE) of θ_n^{**} and s_n^{**} relative to the Cramér–Rao lower bound are

$$ARE(\theta_n^{**}) = K_1, \qquad ARE(s_n^{**}) = K_2. \tag{2.6.110}$$

The estimates based on the *optimal* spacings (2.6.103) are those that maximize the ARE's (2.6.110) and are referred to as the $\{\lambda_i\}$-ABLUE [see Chan (1970)].

As shown in Cheng (1978) the coefficients a_i in (2.6.105) for the $\{\lambda_i\}$-ABLUE of θ are zero except for the coefficient of 1 corresponding to a single-point spacing $\{1/2\}$.

Proposition 2.6.11 *Let $X_1, \ldots, X_n$ be a random sample of size n from the classical Laplace distribution $\mathcal{CL}(\theta, s)$ with known value of s. The optimum spacing for the $\{\lambda_i\}$-ABLUE of θ, θ_n^{**}, is a single-point spacing $\{1/2\}$, which is independent of the number of order statistics k. The ARE of θ_n^{**} is* 1.

Thus in large samples, we can uniquely estimate the location parameter θ of the $\mathcal{CL}(\theta, s)$ distribution (with known value of s) by θ_n^{**}, from either a full sample or a censored one, as long as the middle observation is not missing.

The estimation of the parameter s is more complicated. Here, maximizing K_2 (the ARE of s_n^{**}) with respect to the spacings (2.6.103) leads to a system of equations [see Cheng (1978)]:

$$\left(\frac{f_{i+1}u_{i+1} - f_i u_i}{\lambda_{i+1} - \lambda_i} + \frac{f_i u_i - f_{i-1}u_{i-1}}{\lambda_i - \lambda_{i-1}}\right) f(u_i) - 2\frac{d(f_i u_i)}{du_i} = 0.$$

Cheng (1978) noted that in this case the optimal spacings may not be unique, and they may be symmetric about the point $1/2$ only when the number k is even. We refer the reader to Cheng (1978) for further information and an extensive set of tables containing the optimal spacings $\{\lambda_i\}$ and the corresponding coefficients b_i for the $\{\lambda_i\}$-ABLUE of s given by (2.6.106), as well as the asymptotic efficiencies of s_n^{**} relative to the Cramér–Rao lower bound.

Ali et al. (1981) derived estimators for the ξ-quantiles, x_ξ, of the classical Laplace $\mathcal{CL}(\theta, s)$ distribution. Their estimators are

$$\tilde{x}_\xi = a_l X_{l:n} + a_m X_{m:n}, \quad 1 \le l \le m \le n,$$

where the ranks l, m and the coefficients a_l, a_m are chosen so that $\tilde{x}_\xi$ is asymptotically best (minimum variance) linear unbiased estimator (ABLUE) of x_ξ. The procedure does not involve estimation of the location and scale parameters and does not require the use of tables, since the estimator admits the following explicit form:

$$\tilde{x}_\xi = \begin{cases} 0.255 X_{[[0.30506\xi n]]+1:n} + 0.745 X_{[[1.50134\xi n]]+1:n} \\ \qquad \text{for } 0.0352 \le \xi \le 0.3330 \\ -\frac{z_\xi}{1.59362} X_{[[0.10159n]]+1:n} + \left(1 + \frac{z_\xi}{1.59362}\right) X_{[[n/2]]+1:n} \\ \qquad \text{for } \xi < 0.0352 \text{ and } 0.3330 < \xi < 0.5 \\ X_{[[n/2]]+1:n} \\ \qquad \text{for } \xi = 0.5 \\ \left(1 - \frac{z_\xi}{1.59362}\right) X_{[[n/2]]+1:n} + \frac{z_\xi}{1.59362} X_{[[0.89841n]]+1:n} \\ \qquad \text{for } 0.5 < \xi < 0.6670 \text{ and } \xi > 0.9648 \\ 0.745 X_{[[(1.50134\xi - 0.50134)n]]+1:n} + 0.255 X_{[[(0.30536\xi + 0.69494)n]]+1:n} \\ \qquad \text{for } 0.6670 \le \xi \le 0.9648, \end{cases} \tag{2.6.111}$$

where z_ξ is the ξ-quantile of the standard classical Laplace distribution. They compared the asymptotic variance of their estimator with that of the standard quantile estimator $X_{[[n\xi]]+1:n}$, concluding that x_ξ^* performs much better. Table 2.12 contains the asymptotic relative efficiencies (ARE) of x_ξ^* relative to $X_{[[n\xi]]+1:n}$, computed by Ali et al. (1981). See Saleh et al. (1983) for further discussion on quantile estimation for double exponential distribution, and Umbach et al. (1984) for applications of ABLUE's based on optimal spacings in testing hypothesis.

2.6.2 Interval estimation. We shall now discuss confidence intervals for parameters of the classical Laplace distribution. Let $X_1, \dots, X_n$ be a random sample from the $\mathcal{CL}(\theta, s)$ distribution. If the scale parameter s is known, then a confidence interval for θ may be constructed utilizing the distribution of the sample median given in (2.5.10) and Proposition 2.5.5. If the location parameter θ is known, then since the r.v.'s $|X_i - \theta|/s$ are i.i.d. standard exponential (see Proposition 2.2.3), the MLE of

ξ	0.1	0.2	0.3339	0.4	0.5
ARE	122	128	191	147	100

Table 2.12: Asymptotic relative (percent) efficiencies (ARE) for x_ξ^* relative to $X_{[[n\xi]]+1:n}$ for the Laplace distribution.

s given by (2.6.20) is distributed as $(2n)^{-1}sV$, where V has a χ^2 distribution with $2n$ degrees of freedom. Consequently, the $100(1-\alpha)\%$ confidence interval for s is given by

$$\left(2\sum_{j=1}^{n}\frac{|X_j-\theta|}{\chi^2_{2n,1-\alpha/2}},\, 2\sum_{j=1}^{n}\frac{|X_j-\theta|}{\chi^2_{2n,\alpha/2}}\right), \tag{2.6.112}$$

where $\chi^2_{2n,p}$ denotes the pth quantile of the χ^2 distribution with $2n$ degrees of freedom. If both θ and s are unknown, confidence intervals for θ and s can be obtained via the distributions of the pivotal quantities

$$V_n = \frac{1}{s}\sum_{j=1}^{n}|X_j-\hat{\theta}_n| \text{ and } W_n = \frac{\hat{\theta}_n-\theta}{\sum_{j=1}^{n}|X_j-\hat{\theta}_n|}, \tag{2.6.113}$$

where $\hat{\theta}_n$ is the MLE of θ given by (2.6.15), as V_n and W_n are distributed independently of the parameters [see Bain and Engelhardt (1973)]. The distributions of V_n and W_n can be derived exactly for small values of n, but calculations become quite tedious as the value of n increases [cf. Bain and Engelhardt (1973)]. For $n=3$, we have

$$V_3 \stackrel{d}{=} Y_{3:3}-Y_{1:3} \text{ and } W_3 \stackrel{d}{=} \frac{Y_{2:3}}{Y_{3:3}-Y_{1:3}}, \tag{2.6.114}$$

where $Y_{1:3}\le Y_{2:3}\le Y_{3:3}$ are the order statistics connected with a random sample of size three from the standard classical Laplace distribution. Since V_3 coincides with the range, its p.d.f. follows from Proposition 2.5.3 in Section 2.5,

$$f_{V_3}(x) = e^{-x}(e^{-x}+1.5x-1), \quad x>0. \tag{2.6.115}$$

The p.d.f. of W_3 can be derived from the joint p.d.f. of the order statistics given in (2.5.11),

$$f_{W_3}(x) = \begin{cases} \frac{9}{2}|x|(1-9|x|^2)^{-2} & \text{if } |x|>1 \\ \frac{3}{8}\left(\frac{8}{(1+|x|)^3}-\frac{3}{(1+|x|)^2}-\frac{1}{(1+3|x|)^2}\right) & \text{otherwise} \end{cases} \tag{2.6.116}$$

[see Bain and Engelhardt (1973)]. For $n\ge 3$ one can use either asymptotic distributions of V_n and W_n [see, e.g., Bain and Engelhardt (1973)] or Monte-Carlo approximations to derive the confidence intervals. Using the latter approach, one would first approximate the value $w_{\alpha/2}$ such that

$$P(W_n > w_{\alpha/2}) = \frac{\alpha}{2} \tag{2.6.117}$$

from the empirical distribution of W_n obtained by Monte-Carlo simulations. Then an approximate $(1-\alpha)100\%$ confidence interval for θ is

$$\left(\hat{\theta}_n - w_{\alpha/2}\sum_{j=1}^{n}|X_j-\hat{\theta}_n|,\, \hat{\theta}_n + w_{\alpha/2}\sum_{j=1}^{n}|X_j-\hat{\theta}_n|\right). \tag{2.6.118}$$

Similarly, an approximate $(1-\alpha)100\%$ confidence interval for s would be

$$\left(\frac{\sum_{j=1}^{n}|X_j-\hat{\theta}_n|}{v_{1-\alpha/2}}, \frac{\sum_{j=1}^{n}|X_j-\hat{\theta}_n|}{v_{\alpha/2}}\right), \tag{2.6.119}$$

where v_β denotes an estimate of the βth quantile obtained by Monte-Carlo simulations. More details can be found in Bain and Engelhardt (1973).

Remark 2.6.17 Balakrishnan, Chandramouleeswaran, and Ambagaspitiya (1994) studied the inference on θ when s is assumed either known or unknown, and on s when θ is unknown, for complete as well as Type II censored samples, through the three pivotal quantities

$$\frac{\theta_n^*-\theta}{s\sqrt{V_1}}, \quad \frac{\theta_n^*-\theta}{s_n^*\sqrt{V_1}}, \quad \frac{s_n^*/s-1}{\sqrt{V_2}}, \tag{2.6.120}$$

where θ_n^* and s_n^* are the BLUE's of θ and s and s^2V_1 and s^2V_2 are the variances of θ_n^* and s_n^*. See Balakrishnan, Chandramouleeswaran, and Ambagaspitiya (1994) for the percentage points of the pivotal quantities (2.6.120) and also Balakrishnan, Chandramouleeswaran, and Govindarajulu (1994) for further results on the approximations of the distributions of (2.6.120) and their accuracy.

2.6.2.1 *Confidence bands for the Laplace distribution function.* Let $F(\cdot;\theta,s)$ be the c.d.f. of the classical Laplace distribution given by (2.1.5). Srinivasan and Wharton (1982) constructed one-sided and two-sided confidence bands on $F(\cdot;\theta,s)$ using the Kolmogorov–Smirnov-type statistics

$$L_n = \sup_{-\infty<x<\infty} |F(x;\theta,s)-F(x;\theta_n^*,s_n^*)| \tag{2.6.121}$$

and

$$L_n^+ = \sup_{x\geq 0}\{F(x;\theta,s)-F(x;\theta_n^*,s_n^*)\}, \tag{2.6.122}$$

where θ_n^* and s_n^* are the BLUE's of θ and s. For any $0<\alpha<1$, let the αth quantile of L_n be l_α (so that $P(L_n\leq l_\alpha)=\alpha$). Then a two-sided $\alpha 100\%$ confidence band for $F(\cdot;\theta,s)$ is given by

$$\left(\max\{F(x;\theta_n^*,s_n^*)-l_\alpha,0\}, \min\{F(x;\theta_n^*,s_n^*)+l_\alpha,1\}\right), \tag{2.6.123}$$

with a similar one-sided confidence band based on L_n^+. Tables 2.13 and 2.14 below present simulated percentage points of L_n and L_n^+ for n up to 20, derived by Srinivasan and Wharton (1982). For larger values on n, Srinivasan and Wharton (1982) recommended certain large-sample approximations for the percentage points of L_n and L_n^+. For example, the quantiles of L_n may be approximated through the limiting distribution of $\sqrt{n}L_n$, which is the same as that of the random variable $\sup|X_0(y)|$, where $X_0(y)$ is a Gaussian process with the representation

$$X_0(y)=\frac{1}{2}e^{-|y|}(U+Vy), \quad -\infty<y<\infty. \tag{2.6.124}$$

In (2.6.124), the variables U and V are i.i.d. standard normal. We refer the reader to Srinivasan and Wharton (1982) for more technical details regarding this problem.

$n \setminus \alpha$	0.80	0.85	0.90	0.95	0.99
5	0.31	0.35	0.39	0.45	0.56
6	0.29	0.32	0.35	0.41	0.52
7	0.26	0.29	0.33	0.38	0.48
8	0.25	0.27	0.31	0.36	0.46
9	0.23	0.26	0.29	0.34	0.44
10	0.22	0.24	0.27	0.32	0.41
11	0.21	0.23	0.26	0.31	0.39
12	0.20	0.22	0.25	0.30	0.38
13	0.19	0.22	0.24	0.28	0.36
14	0.18	0.21	0.23	0.27	0.34
15	0.18	0.20	0.22	0.26	0.33
16	0.17	0.19	0.22	0.25	0.32
17	0.16	0.18	0.21	0.24	0.31
18	0.16	0.18	0.20	0.24	0.31
19	0.16	0.18	0.20	0.23	0.31
20	0.15	0.17	0.19	0.23	0.29

Table 2.13: Simulated percentage points l_α of the statistic L_n.

$n \setminus \alpha$	0.80	0.85	0.90	0.95	0.99
5	0.23	0.27	0.31	0.38	0.51
6	0.21	0.24	0.29	0.35	0.47
7	0.19	0.22	0.26	0.32	0.44
8	0.18	0.21	0.25	0.31	0.42
9	0.16	0.19	0.23	0.38	0.39
10	0.16	0.18	0.22	0.27	0.38
11	0.15	0.17	0.21	0.26	0.36
12	0.14	0.17	0.20	0.25	0.34
13	0.13	0.16	0.19	0.24	0.34
14	0.13	0.15	0.18	0.23	0.32
15	0.12	0.14	0.18	0.22	0.30
16	0.12	0.14	0.17	0.21	0.29
17	0.12	0.14	0.17	0.21	0.28
18	0.12	0.14	0.17	0.21	0.28
19	0.11	0.13	0.16	0.20	0.27
20	0.11	0.13	0.15	0.19	0.26

Table 2.14: Simulated percentage points l_α^+ of the statistic L_n^+.

2.6.2.2 *Conditional inference.* The confidence intervals discussed in Section 2.6.2 are based on the MLE's $\hat{\theta}_n$ and $\hat{s}_n$ of the parameters θ and s of the classical Laplace distribution $\mathcal{CL}(\theta, s)$. As noted by Kappenman (1975), these estimators are not sufficient statistics so that inference about θ and s based on these statistics leads to some loss of information contained in the random sample. It is generally accepted that the lost information may be recovered (on the average) by conditioning on the ancillary statistics, which was first suggested by Fisher (1934) [see also remarks by Edwards (1974)].

Kappenman (1975) followed the conditional approach and obtained conditional confidence intervals for the Laplace parameters, based on the conditional distributions of the pivotal quantities (2.6.113) given the ancillary statistics. Here we shall first examine the loss of information associated with the median and related estimators in the Laplace case, and then discuss the conditional inference.

Loss of information. The loss of information associated with the median when estimating the location parameter of the classical Laplace distribution was discussed by Fisher (1922, 1925, 1934). We consider the location family given by the density

$$f(x;\theta) = f(x-\theta) = \frac{1}{2}e^{-|x-\theta|}, \quad -\infty < x, \theta < \infty, \tag{2.6.125}$$

where f is the standard classical Laplace density. Let $X_1, \ldots, X_n$ be a random sample of size $n = 2k+1$ from the distribution given by the density (2.6.125). Then by (2.6.12), the Fisher information supplied by the sample is $n = 2k+1$. On the other hand, when we use the MLE for estimating the location parameter θ, which by Proposition 2.6.2 is the sample median $\hat{\theta}_n = X_{k+1:n}$, we are replacing $n = 2k+1$ observations from the distribution (2.6.125) by a *single* observation from the distribution with the density $f_{k+1:n}(x)$ of the median given by (2.5.10). Since the latter distribution is also a location family,

$$f_{k+1:n}(x) = g(x-\theta), \quad -\infty < x, \theta < \infty, \tag{2.6.126}$$

where

$$g(x) = \frac{(2k+1)!}{(k!)^2}\left(\frac{1}{2}\right)^{2k+1} e^{-(k+1)|x|}(2-e^{-|x|})^k, \quad -\infty < x < \infty, \tag{2.6.127}$$

is an absolutely continuous density function, the Fisher information contained in the median is

$$I(\theta) = \int_{-\infty}^{\infty}\left(\frac{g'(y)}{g(y)}\right)^2 g(y)dy, \tag{2.6.128}$$

with g given by (2.6.127) [see Huber (1981), Lehmann and Casella (1998), and also Exercise 2.7.31]. After a lengthy calculation we obtain (Exercise 2.7.32)

$$I(\theta) = \begin{cases} 12[\log 2 - 0.5] & \text{if } k = 1 \\ \frac{(k+1)(2k+1)}{k-1}\left(1 - \frac{(2k)!}{(k!)^2}\left(\frac{1}{2}\right)^{2k-1}\right) & \text{if } k > 1 \end{cases} \tag{2.6.129}$$

[see Fisher (1934)]. As noted by Fisher (1934), although the median is asymptotically efficient (the ratio of $2k+1$ to $I(\theta)$ given by (2.6.129) tends to 1 as $k \to \infty$), the amount lost,

$$2k+1-I(\theta) = \frac{2(2k+1)}{k-1}\left\{(k+1)\frac{(2k)!}{(k!)^2}\left(\frac{1}{2}\right)^{2k} - 1\right\}, \quad k > 1, \tag{2.6.130}$$

increases to infinity. As $k \to \infty$, we obtain an asymptotic approximation of the loss,

$$2k+1-I(\theta) \approx 4(\sqrt{k/\pi}-4), \quad k \to \infty, \tag{2.6.131}$$

using Stirling's Formula (Exercise 2.7.32). Fisher (1934) noted that with the sample size $n = 2k+1 = 629$, this loss is about 36.

More generally, we can calculate the loss of information associated with the statistic

$$T_l = (X_{k-l+1:n}, \ldots, X_{k+l+1:n}), \tag{2.6.132}$$

which is the set of the central $2l+1$ order statistics obtained from a sample of size $n = 2k+1$ from the Laplace distribution (2.6.125). It is well known [see, e.g., Fisher (1925), Rao (1961)] that the loss of information associated with an arbitrary statistic T obtained from a sample of size n from the population with density $f(\cdot;\theta)$ is

$$E_\theta \left\{ \mathrm{Var}_\theta \left(\sum_{i=1}^{n} \frac{\partial}{\partial\theta} \log f(X_i;\theta) | T \right) \right\}, \tag{2.6.133}$$

where $\mathrm{Var}_\theta(\cdot|T)$ is the conditional variance given T and E_θ is an unconditional expectation. In case of the Laplace distribution (2.6.125), we have

$$\frac{\partial}{\partial\theta} \log f(X_i;\theta) = \mathrm{sign}(X_i - \theta), \tag{2.6.134}$$

and the conditional variance takes the form

$$\mathrm{Var}\left(\sum_{i=1}^{2k+1} \mathrm{sign}(X_i - \theta) | T_l \right) = (k-l)(V_1 + V_2), \tag{2.6.135}$$

where

$$V_1 = \begin{cases} 0 & \text{for } X_{k-l+1:n} \le \theta \\ (2u-1)/u^2 & \text{for } X_{k-l+1:n} > \theta, \end{cases} \tag{2.6.136}$$

$$V_2 = \begin{cases} 0 & \text{for } X_{k+l+1:n} \ge \theta \\ (2v-1)/v^2 & \text{for } X_{k+l+1:n} < \theta, \end{cases} \tag{2.6.137}$$

and

$$u = F(X_{k-l+1:n}), \qquad v = 1 - F(X_{k+l+1:n}), \tag{2.6.138}$$

with F being the distribution function of the standard classical Laplace distribution [see Akahira and Takeuchi (1990) for details]. Hence the loss of information associated with T_l is

$$\begin{aligned} L_l &= (k-l)(E(V_1) + E(V_2)) \\ &= \frac{2(2k+1)!}{(k-l-1)!(k+l)!} \int_{1/2}^{1} \frac{2u-1}{u^2} u^{k-l}(1-u)^{k+l} du, \end{aligned} \tag{2.6.139}$$

since both V_1 and V_2 have support on $[1/2, 1]$ ($F(x) > 1/2$ if $x > \theta$) and

$$E(V_1) = E(V_2) = \frac{(2k+1)!}{(k-l)!(k+l)!} \int_{1/2}^{1} \frac{2u-1}{u^2} u^{k-l}(1-u)^{k+l} du. \tag{2.6.140}$$

Relating the integral in (2.6.139) to an incomplete beta function, Akahira and Takeuchi (1990) obtained the following result for the loss of information.

Proposition 2.6.12 *For each integer $0 \le l \le k-2$, the loss of information L_l associated with the statistic T_l given by* (2.6.132) *is*

$$\frac{2^{2k} L_l}{2(2k+1)} = \frac{(2k)!}{(k!)^2} - \frac{(l+1)2^{2k}}{k-l-1} + \sum_{j=0}^{l} \frac{2(l-j+1)}{k-l-1} \frac{(2k)!}{(k-l)!(k+l)!}. \tag{2.6.141}$$

Note that for $l = 0$, in which case T_l is the median $X_{k+1:n}$, the relation (2.6.141) reduces to (2.6.130). Asymptotically, for fixed l and large k, the loss of information (2.6.141) is given by

$$L_l = \frac{4k}{\sqrt{\pi}}(1 + o(1)) - 4(l+1) + O\left(\frac{l^2}{\sqrt{k}}\right) \tag{2.6.142}$$

and coincides with (2.6.131) for $l = 0$ [see Akahira and Takeuchi (1990)]. We refer interested readers to Akahira (1987, 1990), Akahira and Takeuchi (1990, 1993), and Takeuchi and Akahira (1976) for more information on loss of information and second-order asymptotic results for order statistics and related estimators of the location parameter in the case of the Laplace distribution.

Conditional confidence intervals. Let $X_1, \dots, X_n$ be i.i.d. random variables with the common classical Laplace distribution with density (2.1.1), and let $X_{1:n} \le \cdots \le X_{n:n}$ be the corresponding order statistics. Define the statistic

$$\mathbf{a} = (a_1, \dots, a_n)', \tag{2.6.143}$$

where

$$a_i = \frac{X_{i:n} - \hat{\theta}_n}{\hat{s}_n}, \qquad i = 1, \dots, n, \tag{2.6.144}$$

and $\hat{\theta}_n$ and $\hat{s}_n$ are the MLE's of the location and scale parameters given by (2.6.15) and (2.6.34), respectively. Note that for $n = 2m+1$ we have $a_{m+1} = 0$, while for $n = 2m$ we have $a_m = -a_{m+1}$. In addition,

$$\sum_{i=1}^{n} |a_i| = 0, \tag{2.6.145}$$

so that only $n-2$ of the components of $\mathbf{a}$ are independent. Further, since the pivotal quantities

$$U_n = \frac{\hat{\theta}_n - \theta}{\hat{s}_n} \quad \text{and} \quad V_n = \frac{\hat{s}_n}{s} \tag{2.6.146}$$

have distributions that do not depend on the parameters θ and s [see Antle and Bain (1969)], it follows that $\mathbf{a}$ is an ancillary statistic for θ and s [cf. Kappenman (1975)]. The joint conditional density function of $\hat{\theta}_n$ and $\hat{s}_n$, given the value of the ancillary statistics $\mathbf{a}$, is proportional to

$$\frac{1}{s^2}\left(\frac{\hat{s}_n}{s}\right)^{n-2} \exp\left\{-\frac{\hat{s}_n}{s}\sum_{i=1}^{n}\left|\frac{\hat{\theta}_n - \theta}{\hat{s}_n} + a_i\right|\right\}. \tag{2.6.147}$$

Note that the Jacobian of $(\hat{s}_n, \hat{\theta}_n)$ as a function of U_n and V_n is $s^2 V_n$ so the conditional joint density of U_n and V_n, given the value of the ancillary statistic $\mathbf{a}$, is equal to

$$p_{U_n, V_n}(u, v|\mathbf{a}) = K v^{n-1} e^{-v \sum_{i=1}^{n} |u + a_i|}. \tag{2.6.148}$$

The normalizing constant in (2.6.148) is equal to

$$K = \frac{1}{2\Gamma(n-1)}[B_n(\mathbf{a})c(\hat{\theta}_n)]^{n-1}, \tag{2.6.149}$$

where

$$c(t) = \sum_{i=1}^{n} |a_i - t| = \begin{cases} \sum_{i=1}^{n} a_i - nt & \text{for } t \le a_1 \\ (2i-n)t + \sum_{j=i+1}^{n} a_j - \sum_{j=1}^{i} a_j & \text{for } a_i \le t \le a_{i+1} \\ nt - \sum_{i=1}^{n} a_i & \text{for } t \ge a_n \end{cases} \tag{2.6.150}$$

and $B_n(\mathbf{a})$ is equal to

$$\left\{ \sum_{i=1}^{n} \frac{[c(\hat{\theta}_n)/c(a_i)]^{n-1}}{(2i-n)(n+2-2i)} \right\}^{-1/(n-1)} \tag{2.6.151}$$

if n is odd and to

$$\left\{ \frac{(n-1)(a_{n/2+1} - a_{n/2})}{2c(\hat{\theta}_n)} + \frac{1}{2} \sum_{\substack{i=1 \\ i \ne n/2, n/2+1}}^{n} \frac{[c(\hat{\theta}_n)/c(a_i)]^{n-1}}{(2i-n)(n+2-2i)} \right\}^{-1/(n-1)} \tag{2.6.152}$$

if n is even [see Kappenmann (1975) and Uthoff (1973)]. Utilizing (2.6.148), one can now derive the marginal conditional density of U_n,

$$p_{U_n}(u|\mathbf{a}) = K\Gamma(n) \left\{ \sum_{i=1}^{n} |u + a_i| \right\}^{-n}, \tag{2.6.153}$$

and use it to produce the conditional $100(1-\alpha)\%$ confidence interval for θ,

$$(\hat{\theta}_n - u_2 \hat{s}_n, \hat{\theta}_n - u_1 \hat{s}_n), \tag{2.6.154}$$

where the constants u_1 and u_2 satisfy the conditions

$$P(U_n \le u_1|\mathbf{a}) = P(U_n \ge u_2|\mathbf{a}) = \alpha/2. \tag{2.6.155}$$

Similarly, we can derive the marginal conditional density of V_n and consequently obtain the expression

$$\begin{aligned} K \Bigg\{ & \frac{\gamma(n-1; v_2 c(a_1)) - \gamma(n-1; v_1 c(a_1))}{n(c(a_1))^{n-1}} \\ & + \sum_{i=1}^{n-1} \frac{\gamma(n-1; v_2 c(a_i)) - \gamma(n-1; v_1 c(a_i))}{(2i-n)(c(a_i))^{n-1}} \\ & - \sum_{i=1}^{n-1} \frac{\gamma(n-1; v_2 c(a_{i+1})) - \gamma(n-1; v_1 c(a_{i+1}))}{(2i-n)(c(a_{i+1}))^{n-1}} \\ & + \frac{\gamma(n-1; v_2 c(a_n)) - \gamma(n-1; v_1 c(a_n))}{n(c(a_n))^{n-1}} \Bigg\} \end{aligned} \tag{2.6.156}$$

for the probability

$$P(v_1 < V_n < v_2|\mathbf{a}) = P\left(\frac{\hat{s}_n}{v_2} < s < \frac{\hat{s}_n}{v_1} |\mathbf{a} \right), \tag{2.6.157}$$

where

$$\gamma(n-1; x) = \int_0^x e^{-t} t^{n-2} dt, \quad 0 < x < \infty, \tag{2.6.158}$$

is the incomplete gamma function; see Kappenman (1975). Thus the conditional $100(1-\alpha)\%$ confidence interval for s is

$$(\hat{s}_n/v_2, \hat{s}_n/v_1), \tag{2.6.159}$$

where the constants v_1 and v_2 are chosen so that the conditional probability (2.6.157) given by (2.6.156) is equal to $1-\alpha$.

Grice et al. (1978) compared the conditional confidence intervals for θ given by (2.6.154) with the unconditional ones given by (2.6.118) in terms of their expected lengths. Using Monte-Carlo techniques they concluded that the conditional approach yields slightly narrower intervals on average, and that the two methods are essentially in agreement for large sample sizes. Table 2.15, taken from Grice et al. (1978), contains the expected lengths of the conditional and unconditional confidence intervals for selected sample sizes.

$1-\alpha$	0.90	0.90	0.95	0.95	0.98	0.98
n	Cond.	Uncond.	Cond.	Uncond.	Cond.	Uncond.
3	3.352	3.641	4.740	4.975	7.495	7.649
5	2.113	2.273	2.575	2.912	3.542	3.787
9	1.375	1.498	1.698	1.949	2.119	2.316
15	0.997	1.061	1.214	1.326	1.484	1.525
33	0.631	0.682	0.761	0.830	0.917	0.942

Table 2.15: Expected lengths of conditional and unconditional $100(1-\alpha)\%$ confidence intervals for θ based on random samples with selected size n from the $\mathcal{CL}(\theta, 1)$ distribution.

Remark 2.6.18 Conditional inference for the Laplace distribution under Type II right-censoring is discussed in Childs and Balakrishnan (1996).

2.6.3 Tolerance intervals. Let $X_1, \ldots, X_n$ be a random sample of size n from a distribution with density f, and let

$$U = U(X_1, \ldots, X_n) \text{ and } L = L(X_1, \ldots, X_n)$$

be two statistics such that

$$P\left(\int_L^\infty f(x)dx \geq \beta\right) = \gamma \tag{2.6.160}$$

and

$$P\left(\int_{-\infty}^U f(x)dx \geq \beta\right) = \gamma. \tag{2.6.161}$$

Then L and U are said to be *lower* and *upper* (β, γ) tolerance limits, while the intervals (L, ∞) and $(-\infty, U)$ are, respectively, lower and upper γ probability *tolerance intervals* for proportion β

(β-content tolerance intervals at level γ). Similarly, for $L < U$, the interval (L, U) is a *two-sided* γ probability tolerance interval for proportion β (β-content tolerance interval at level γ) if

$$P\left(\int_L^U f(x)dx \geq \beta\right) = \gamma. \tag{2.6.162}$$

We shall discuss tolerance intervals when the random sample is from the two-parameter classical Laplace distribution with density (2.1.1). Let us first consider the lower tolerance interval of the form

$$(L, \infty) = (\hat{\theta}_n - b\hat{s}_n, \infty), \tag{2.6.163}$$

where $\hat{\theta}_n$ and $\hat{s}_n$ are the MLE's of the parameters θ and s given by (2.6.15) and (2.6.34), respectively. Thus the problem is to determine the *tolerance factor* b in (2.6.163). Upon substituting the Laplace density (2.1.1) and L given by (2.6.163) into (2.6.160) and changing the variable $u = (x - \theta)/s$, we obtain the following equation for b:

$$P\left(\int_{\frac{\hat{\theta}_n - \theta}{s} - b\frac{\hat{s}_n}{s}}^{\infty} \frac{1}{2}e^{-|u|}du \geq \beta\right) = \gamma. \tag{2.6.164}$$

Restricting β to $\beta \geq 1/2$ (in practice, the proportion β is close to one) we can write equivalently

$$P\left(\frac{\hat{\theta}_n - \theta}{s} - b\frac{\hat{s}_n}{s} \leq k_\beta\right) = \gamma, \tag{2.6.165}$$

where

$$k_\beta = \log[2(1 - \beta)] \leq 0. \tag{2.6.166}$$

Bain and Engelhardt (1973) expressed (2.6.165) as

$$P\left(U_n\left(\frac{b}{n}\right) \leq k_\beta\right) = \gamma, \tag{2.6.167}$$

where

$$U_n(c) = \frac{\hat{\theta}_n - \theta}{s} - cn\frac{\hat{s}_n}{s}, \tag{2.6.168}$$

and used the approximation

$$P\left(U_n\left(\frac{b}{n}\right) \leq k_\beta\right) \approx \Phi\left(\frac{\sqrt{n}(k_\beta - b)}{\sqrt{1 + b^2}}\right) \tag{2.6.169}$$

to obtain an approximate value of the tolerance factor:

$$b \approx \frac{1}{n - z_\gamma^2}\left\{-nk_\beta + z_\gamma\sqrt{n(1 + k_\beta^2) - z_\gamma^2}\right\}. \tag{2.6.170}$$

[Φ and z_γ are the standard normal c.d.f. and γth quantile, respectively.] Note that by symmetry, the interval

$$(-\infty, U) = (-\infty, \hat{\theta}_n + b\hat{s}_n), \tag{2.6.171}$$

with b as in (2.6.170), is an approximate upper γ probability tolerance interval.

Kappenman (1977) derived *conditional tolerance intervals* following the conditional approach presented in Section 2.6.2.2. Here the interval of the form (2.6.163) is a lower γ probability conditional tolerance interval for proportion β if

$$P\left(\int_{\hat{\theta}_n - b\hat{s}_n}^{\infty} f(x;\theta,s)dx \geq \beta \,|\mathbf{a}\right) = \gamma, \tag{2.6.172}$$

where $f(x;\theta,s)$ is the Laplace p.d.f. (2.1.1) and $\mathbf{a}$ is the vector of ancillary statistics given by (2.6.143) and (2.6.144) in Section 2.6.2.2. (The upper and the two-sided conditional tolerance intervals are defined similarly.) Using the conditional joint distribution of $(\hat{\theta}_n - \theta)/s$ and $\hat{s}_n/s$, Kappenman (1977) obtained the following value for the tolerance factor b:

$$\begin{aligned} b = {} & -a_h - \frac{c(a_h)}{n-2h} + \frac{1}{n-2h} \\ & \times \left\{ e^{k_\beta(n-2h)} \left[(c(a_h))^{1-n} + \frac{p(n-2h)}{K\Gamma(n-1)} \right] \right\}^{-1/(n-1)}, \end{aligned} \tag{2.6.173}$$

where k_β is given by (2.6.166), $\mathbf{a}$ is a s before, $c(t)$ is given by (2.6.150), K is the normalizing constant (2.6.149), h is the largest integer ($h \geq 2$) such that

$$\begin{aligned} Q(h) = {} & K\Gamma(n-1) \left\{ \frac{1}{n(c(a_1))^{n-1}} + \sum_{i=1}^{h-1} \frac{1}{n-2i} \right. \\ & \times \left. \left[\frac{1}{(c(a_{i+1}))^{n-1}} - \frac{1}{(c(a_i))^{n-1}} \right] \right\} \leq 1-\gamma, \end{aligned} \tag{2.6.174}$$

and $p = 1 - \gamma - Q(h)$. To actually calculate b, one must first find h, usually by setting $h = 2, 3, \ldots$ in (2.6.174).

By symmetry, the upper γ probability conditional tolerance interval for proportion β is

$$(-\infty, \hat{\theta}_n - b\hat{s}_n), \tag{2.6.175}$$

where b is obtained from (2.6.173) and (2.6.174) with k_β replaced by $-k_\beta$ and with p equal to $\gamma - Q(h)$, where h now is the largest integer ($h \geq 2$) such that $Q(h) < \gamma$.

Shyu and Owen (1986a) remarked that the approximate tolerance intervals (2.6.163), which are based on the approximation (2.6.170), can miss the exact values significantly in some applications, while the conditional tolerance factors (2.6.173) are not easy to compute even for small sample sizes. They proposed a method based on Monte-Carlo simulations sketched below, leading to useful tables for the tolerance factor b. Denoting

$$W_n = \frac{(\hat{\theta}_n - \theta)/s - k_\beta}{\hat{s}_n/s}, \tag{2.6.176}$$

we see that the relation (2.6.165) is equivalent to

$$P(W_n \leq b) = \gamma. \tag{2.6.177}$$

Since the distribution of $(\hat{\theta}_n - \theta)/s$ and $\hat{s}_n/s$ is independent of the parameters θ and s [see Antle and Bain (1969)], the same property is shared by the statistic W_n defined in (2.6.176). Consequently, the tolerance factor b can be determined from the relation (2.6.177) for any given values of β, γ, and n.

For $n = 2$, the p.d.f. of W_2 takes the following form for $x \neq 0$:

$$g(x) = \begin{cases} \frac{1}{4}\left\{u(x)e^{\frac{2k_\beta}{x-1}} + [1-u(x)]\,e^{\frac{2k_\beta}{x+1}} + \frac{e^{2k_\beta}}{x^2}\right\} & \text{for } x > 1, \\ \frac{1}{4}\exp\left\{[1-u(x)]\,e^{\frac{2k_\beta}{x+1}} + \frac{1}{x^2}e^{2k_\beta}\right\} & \text{for } -1 < x \le 1, \\ \frac{1}{4}\frac{1}{x^2}e^{2k_\beta} & \text{for } x \le -1, \end{cases} \tag{2.6.178}$$

where

$$u(x) = \frac{1}{x^2} + \frac{2k_\beta}{x}$$

(see Exercise 2.7.41).

Thus the exact value of b can be obtained by solving (2.6.177) (numerically, since the relevant distribution function does not admit a closed form). Shyu and Owen (1986a) provide a table for the resulting values of b for $n = 2$ and

$$\beta = 0.750, 0.900, 0.950, 0.990, 0.995, 0.999,$$
$$\gamma = 0.500, 0.750, 0.900, 0.950, 0.975, 0.990, 0.995.$$

They also note that when $n > 2$ the exact distribution of W_n is difficult to obtain and hence they derive approximations based on simulations. The values of the tolerance factor b for sample sizes $n = 3(1)11, 50, 100$ and the same values of β and γ as those for $n = 2$ can be found in Shyu and Owen (1986a).

Similarly, Shyu and Owen (1986b) developed analogous procedures for obtaining the two-sided tolerance intervals of the form

$$(L, U) = (\hat{\theta}_n - b\hat{s}_n, \hat{\theta}_n + b\hat{s}_n), \tag{2.6.179}$$

where $\hat{\theta}_n$ and $\hat{s}_n$ are as before, and they presented useful tables for the tolerance factor b, for the same values of n, β, and γ as those used in Shyu and Owen (1986a) for the one-sided tolerance limits.

In Shyu and Owen (1987), the authors consider β-expectation tolerance intervals of the form (2.6.179) defined by the condition

$$E\left[\int_L^U f(x;\theta,s)dx\right] = \beta, \tag{2.6.180}$$

where $f(\cdot;\theta,s)$ is the double exponential density (2.1.1). Shyu and Owen (1987) note that (2.6.180) is equivalent to

$$P(-b < Y_n < b) = \beta, \tag{2.6.181}$$

where

$$Y_n = \frac{X - \hat{\theta}_n}{\hat{s}_n}, \tag{2.6.182}$$

the variable X has a standard classical Laplace distribution, and $\hat{\theta}_n$ and $\hat{s}_n$ are as before, and are independent from X. Subsequently, by simulations, they developed useful tables for the tolerance factor b, with the same values of n, β, and γ as those used in Shyu and Owen (1986ab).

Remark 2.6.19 Balakrishnan and Chandramouleeswaran (1994a) developed upper and lower tolerance intervals based on Type II censored samples from the Laplace distribution. Their intervals are of the form

$$(-\infty, U) = (-\infty, \theta_n^* + bs_n^*) \text{ and } (L, \infty) = (\theta_n^* - bs_n^*, \infty),$$

where θ_n^* and s_n^* are the BLUE's of θ and s. They developed tables of the tolerance factor b for sample size $n = 5(1)10, 12, 15, 20$, right-censoring level $s = 0(1)[[n/2]]$, and

$$\beta = 0.500(0.025)0.975,$$
$$\gamma = 0.750, 0.850, 0.900.0.950, 0.980, 0.990, 0.995.$$

In addition, Balakrishnan and Chandramouleeswaran (1994a) proposed an estimator of the reliability

$$R_X(t) = P(X > t) = 1 - F(t; \theta, s) \tag{2.6.183}$$

of the $\mathcal{CL}(\theta, s)$ r.v. X at time t of the form

$$R_X^*(t) = \begin{cases} 1 - \frac{1}{2}e^{(t-\theta_n^*)/s_n^*} & \text{for } t \le \theta_n^* \\ \frac{1}{2}e^{-(t-\theta_n^*)/s_n^*} & \text{for } t \ge \theta_n^*, \end{cases} \tag{2.6.184}$$

and they described how to use their tables of the tolerance factor b to obtain confidence intervals for the reliability (2.6.183).

2.6.4 Testing hypothesis.

2.6.4.1 *Testing the normal versus the Laplace.* Let $X_1, \ldots, X_n$ be i.i.d. with the common density

$$\frac{1}{\sigma} f\left(\frac{x-\theta}{\sigma}\right), \tag{2.6.185}$$

where the function f is symmetric about zero, and consider the problem of testing

$$H_0 : f = f_0 \quad \text{against} \quad H_1 : f = f_1, \tag{2.6.186}$$

where f_0 and f_1 are the standard normal and the standard Laplace densities, respectively. Let us derive the likelihood ratio test for this problem.

Writing the density (2.6.185) in the form

$$f(x; \theta, \sigma, \alpha) = \frac{c_\alpha}{\sigma} e^{b_\alpha \left|\frac{x-\theta}{\sigma}\right|^\alpha} \tag{2.6.187}$$

and choosing the parameter space to be

$$\Omega = \{(\theta, \sigma, \alpha) : \theta \in \mathbb{R}, 0 < \sigma, \alpha = 1, 2\} = \Omega_0 \cup \Omega_1, \tag{2.6.188}$$

we are testing whether the vector parameter belongs to

$$\Omega_0 = \{(\theta, \sigma, \alpha) : \theta \in \mathbb{R}, 0 < \sigma, \alpha = 2\}$$

(the normal distribution) or to

$$\Omega_1 = \{(\theta, \sigma, \alpha) : \theta \in \mathbb{R}, 0 < \sigma, \alpha = 1\}$$

(the Laplace distribution). The likelihood ratio criterion rejects H_0 if the ratio

$$\frac{\sup_{(\theta,\sigma,\alpha)\in\Omega_0}\prod_{i=1}^n f(x_i;\theta,\sigma,\alpha)}{\sup_{(\theta,\sigma,\alpha)\in\Omega}\prod_{i=1}^n f(x_i;\theta,\sigma,\alpha)} \tag{2.6.189}$$

is less than some constant c. Clearly, on Ω_0 the supremum is attained by the MLE's of the mean and the standard deviation under the normal model:

$$\hat{\theta}_n^N = \frac{1}{n}\sum_{i=1}^n x_i = \bar{x}_n, \tag{2.6.190}$$

$$\hat{\sigma}_n^N = \sqrt{\frac{1}{n}\sum_{i=1}^n (x_i - \bar{x}_n)^2}. \tag{2.6.191}$$

Similarly, the supremum of the joint density over the set Ω_1 is attained when the parameters are the MLE's under the Laplace model:

$$\hat{\theta}_n^L = \tilde{x}_n \text{ (the sample median)}, \tag{2.6.192}$$

$$\hat{\sigma}_n^L = \frac{\sqrt{2}}{n}\sum_{i=1}^n |x_i - \tilde{x}_n|. \tag{2.6.193}$$

Thus the likelihood ratio (2.6.189) becomes

$$\frac{\prod_{i=1}^n f(x_i;\hat{\theta}_n^N,\hat{\sigma}_n^N,2)}{\max\left\{\prod_{i=1}^n f(x_i;\hat{\theta}_n^N,\hat{\sigma}_n^N,2), \prod_{i=1}^n f(x_i;\hat{\theta}_n^L,\hat{\sigma}_n^L,1)\right\}}. \tag{2.6.194}$$

The substitution of the density (2.6.187) (where $c_2 = 1/\sqrt{2\pi}$, $b_2 = 1/2$ for the normal and $c_1 = 1/\sqrt{2}$, $b_1 = \sqrt{2}$ for the Laplace) and the statistics (2.6.190), (2.6.191), (2.6.192), and (2.6.193) into (2.6.194) results in the following expression for the likelihood ratio:

$$\left(\max\left\{1, \left(\frac{\pi n}{2e}\frac{\sqrt{\sum(x_i-\bar{x}_n)^2}}{\sum|x_i-\tilde{x}_n|}\right)\right\}\right)^{-1}. \tag{2.6.195}$$

Thus the likelihood ratio test rejects H_0 if

$$V_n = \frac{\frac{1}{n}\sum|x_i-\tilde{x}_n|}{\sqrt{\frac{1}{n-1}\sum(x_i-\bar{x}_n)^2}} < C, \tag{2.6.196}$$

where C is chosen to produce the required size of the test.

Remark 2.6.20 A similar test when testing for normality, based on the ratio

$$\frac{\sum|x_i-\bar{x}_n|}{\sqrt{\sum(x_i-\bar{x}_n)^2}}, \tag{2.6.197}$$

was proposed by Geary (1935) and investigated by Pearson (1935). Note that here we use the sample mean when calculating the mean deviation.

The test (2.6.196) is not a uniformly most powerful (UMP) test [unless $n = 1$; see Rohatgi (1984)]. However, as shown by Uthoff (1973), there exists a most powerful scale and location invariant test for (2.6.185), which is asymptotically equivalent to but different from the likelihood ratio test (2.6.196). This test rejects H_0 if

$$B_n V_n < k,$$

where V_n is given in (2.6.196) and B_n is a certain function of the order statistics [see Uthoff (1973) for details]. On the other hand, in case θ is known (for convenience set to zero) the likelihood ratio and the most powerful scale and location invariant test are both equivalent to rejecting H_0 when

$$\frac{\sum |x_i|}{\sqrt{\sum x_i^2}} < C \tag{2.6.198}$$

[see Hogg (1972)].

The approximate critical region of the test (2.6.196) may be based on the asymptotic distribution of the test statistic in (2.6.196). It was shown in Uthoff (1973) that if the underlying probability distribution is symmetric and absolutely continuous with a finite fourth moment and with a density f continuous in the neighborhood of the median, then the statistic V_n (as well as $B_n V_n$) is asymptotically normal with the mean $\nu_1 \nu_2^{-1/2}$ and the variance

$$\frac{1}{n}[1 - \nu_1 \nu_3 \nu_2^{-2} + 4^{-1} \nu_1^2 \nu_2^{-1} (\nu_4 \nu_2^{-2} - 1)], \tag{2.6.199}$$

where $\nu_i = E|X - m|^i$ and m is the median of f. Thus under H_0, where the distribution is normal, the distribution of V_n is approximately normal with the mean of 0.798 and the variance of $0.045/n$ [Uthoff (1973)].

2.6.4.2 *Goodness-of-fit tests.* In this section we follow Yen and Moore (1988) and discuss two nonparametric goodness-of-fit tests for the Laplace distribution. The tests are used to determine whether for a given random sample $X_1, \ldots, X_n$, the underlying probability distribution is a $\mathcal{CL}(\theta, s)$ distribution (with some unknown values of the parameters).

Anderson–Darling test. The test statistic for the (modified) Anderson–Darling (AD) test is

$$A_n^2 = -n - \frac{1}{n}\sum_{j=1}^{n}(2j-1)[\log F(X_{j:n}; \theta, s) + \log F(X_{n-j+1:n}; \theta, s)], \tag{2.6.200}$$

where $F(\cdot; \theta, s)$ is the classical Laplace distribution function (2.1.5) and $X_{j:n}$ is the jth order statistic connected with the given random sample [see Yen and Moore (1988)]. The values of the parameters θ and s are usually not known and must be estimated before the test statistic (2.6.200) can be computed. Yen and Moore (1988) obtained the critical values for the test by Monte-Carlo simulations. For each $n = 5(5)50$, a random sample of size n was generated from Laplace distribution and the MLE's (2.6.15) and (2.6.34) of the parameters were substituted into (2.6.200) to obtain a value of the test statistic. The procedure was repeated 5000 times producing an empirical distribution of the test statistic (2.6.200), from which sample quantiles approximating the critical values were obtained. Table 2.16, taken from Yen and Moore (1988), contains the critical values of the test statistic (2.6.200) for selected sample sizes and significance levels α.

The Cramér–von Mises test. The test statistic for the (modified) Cramér–von Mises (CvM) test is

$$W_n^2 = \frac{1}{12n} + \sum_{j=1}^{n}\left[F(X_{j:n}; \theta, s) - \frac{2j-1}{2n}\right]^2, \tag{2.6.201}$$

$n \setminus \alpha$	0.20	0.15	0.10	0.05	0.01
5	0.607	0.682	0.789	0.948	1.256
10	0.558	0.618	0.707	0.854	1.224
15	0.611	0.686	0.801	0.989	1.409
20	0.592	0.658	0.758	0.919	1.264
25	0.622	0.691	0.793	0.999	1.435
30	0.599	0.667	0.773	0.949	1.416
35	0.628	0.698	0.800	0.975	1.457
40	0.639	0.706	0.817	1.012	1.461
45	0.619	0.692	0.807	0.980	1.441
50	0.607	0.673	0.783	0.967	1.393

Table 2.16: Critical values for the modified Anderson–Darling test for the Laplace distribution, for selected values of the sample size n and significance level α.

where $F(\cdot; \theta, s)$ and $X_{j:n}$ are as before [see Yen and Moore (1988)]. As in the former test, the values of the parameters θ and s must be estimated before the test statistic (2.6.201) can be computed. Yen and Moore (1988) obtained the critical values for the test by Monte-Carlo simulations similar to those for the AD test. Table 2.17, taken from Yen and Moore (1988), contains the critical values of the test statistic (2.6.201) for selected sample sizes and significance levels α.

$n \setminus \alpha$	0.20	0.15	0.10	0.05	0.01
5	0.080	0.090	0.105	0.131	0.193
10	0.076	0.084	0.096	0.116	0.172
15	0.085	0.096	0.112	0.142	0.205
20	0.082	0.092	0.104	0.128	0.186
25	0.088	0.100	0.114	0.145	0.220
30	0.084	0.095	0.109	0.137	0.207
35	0.089	0.101	0.116	0.146	0.213
40	0.092	0.104	0.121	0.148	0.222
45	0.088	0.099	0.116	0.145	0.215
50	0.085	0.096	0.113	0.142	0.212

Table 2.17: Critical values for the modified Cramér–von Mises test for the Laplace distribution, for selected values of the sample size n and significance level α.

Yen and Moore (1988) tabulated the power of the two (level $\alpha = 0.01$ and $\alpha = 0.05$) tests discussed above under six different alternative hypotheses with normal, Weibull, uniform, Cauchy, gamma, and exponential distributions. The power function of the AD test was higher than that for the CvM test under the uniform, Cauchy, gamma, and exponential alternatives across all sample sizes and significance levels considered. Under the normal and Weibull alternatives, the power functions were comparable.

2.6.4.3 *Neyman–Pearson test for location.* In this section we shall consider two simple hypotheses about the location of the Laplace distribution when the scale is known. Namely, let $X_1, \dots, X_n$ be an i.i.d. sample from the Laplace distribution $\mathcal{CL}(\theta, s)$. We want to test

$$H_0 : \theta = \theta_1 \quad \text{against} \quad H_1 : \theta = \theta_2,$$

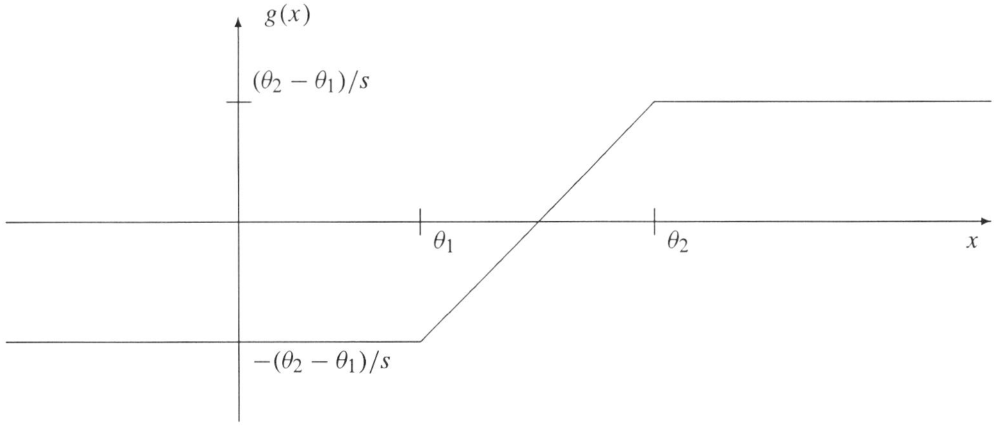

Figure 2.6: Function $g(x)$ used in the Neyman–Pearson test.

where θ_1 and θ_2 are some known prescribed numbers.

It follows from the Neyman–Pearson Lemma that the optimal test (i.e., the most powerful test) of the significance level α rejects H_0 if

$$\frac{\prod_{i=1}^n f(X_i; \theta_1, s)}{\prod_{i=1}^n f(X_i; \theta_2, s)} < k_\alpha,$$

where k_α satisfies the equation

$$P\left(\frac{\prod_{i=1}^n f(X_i; \theta_1, s)}{\prod_{i=1}^n f(X_i; \theta_2, s)} < k_\alpha \middle| \theta = \theta_1\right) = \alpha, \tag{2.6.202}$$

where $f(x; \theta, s)$ is the density function of $\mathcal{CL}(\theta, s)$.

We shall consider the case $\theta_2 > \theta_1$, since otherwise we would rewrite the sample as $(-X_1, \dots, -X_n)$ replacing θ_1 and θ_2 by $-\theta_1$ and $-\theta_2$, respectively. Substituting the density $f(x; \theta, s)$ into (2.6.202) it is easy to observe that the above testing procedure is equivalent to rejecting H_0 provided that

$$\sum_{i=1}^n g(X_i) > t_\alpha,$$

where

$$g(x) = \begin{cases} -(\theta_2 - \theta_1)/s & \text{for} \quad x < \theta_1, \\ 2x/s - (\theta_2 + \theta_1)/s & \text{for} \quad \theta_1 \le x \le \theta_2, \\ (\theta_2 - \theta_1)/s & \text{for} \quad x > \theta_2. \end{cases} \tag{2.6.203}$$

The graph of the function g is sketched in Figure 2.6.

To determine the value of t_α, we are required to solve the equation

$$P\left(\sum_{i=1}^n g(X_i) > t_\alpha \middle| \theta = \theta_1\right) = \alpha.$$

This requires knowledge of the distribution of the test statistic $\sum_{i=1}^n g(X_i)$ under the H_0 hypothesis. This distribution is given in Marks et al. (1978). We now present this result and its proof.

Theorem 2.6.2 *Let $X_1, \ldots, X_n$ be a random sample from the $\mathcal{CL}(\theta, s)$ distribution. Then under the null hypothesis $\theta = \theta_1$, the distribution of*

$$T_n = \sum_{i=1}^{n} g(X_i),$$

where $g(x)$ is defined in 2.6.203 and $\theta_2 > \theta_1$, is given by the following c.d.f.:

$$\begin{aligned} F_n^{(0)}(x) = \frac{1}{2^n} \Bigg\{ & \sum_{k=1}^{n} \sum_{l=0}^{n-k} \sum_{r=0}^{k} \binom{n}{k} \binom{n-k}{l} (-1)^r \binom{k}{r} e^{-(r+l)(\theta_2-\theta_1)/s} \\ & \times \left[1 - e^{-v(x)/2} \cdot e_{k-1}\left(v(x)/2\right) \right] u\left(v(x)\right) \\ & + \sum_{m=0}^{n} \binom{n}{m} e^{-m(\theta_2-\theta_1)/s} u\left(x + (n-2m)(\theta_2-\theta_1)/s\right) \Bigg\}, \end{aligned}$$

where $v(x) = x + (n - 2l - 2r)(\theta_2 - \theta_1)/s$, $e_k(\cdot)$ is the incomplete exponential function, i.e.,

$$e_k(z) = \sum_{i=0}^{k} \frac{z^i}{i!},$$

and

$$u(x) = \begin{cases} 0 & \textit{for} \quad z < 0 \\ 1 & \textit{for} \quad z \geq 0. \end{cases}$$

The expected value and the variance of T_n under H_0 are

$$\mathrm{E}^{(0)}(T_n) = n\left(1 - e^{-(\theta_2-\theta_1)/s} - (\theta_2 - \theta_1)/s\right)$$

and

$$\mathrm{Var}^{(0)}(T_n) = n\left(3 - 2e^{-(\theta_2-\theta_1)/s} - e^{-2(\theta_2-\theta_1)/s} - \frac{4(\theta_2-\theta_1)}{s} e^{-(\theta_2-\theta_1)/s}\right).$$

If $\theta = \theta_2$ (H_1 hypothesis), the distribution of T_n is given by the c.d.f.

$$F_n^{(1)}(x) = 1 - F_n^{(0)}(-x),$$

and in this case the expected value and the variance are given by

$$\mathrm{E}^{(1)}[T_n] = -\mathrm{E}^{(0)}[T_n], \qquad \mathrm{Var}^{(1)}[T_n] = \mathrm{Var}^{(0)}[T_n].$$

The statistic T_n is asymptotically normal, i.e.,

$$\lim_{n\to\infty} \frac{T_n - \mathrm{E}[T_n]}{\sqrt{\mathrm{Var}[T_n]}} \stackrel{d}{=} N(0, 1).$$

Proof. Consider first the distribution of T_n under H_0. Since $g(X_i)$ is a truncated Laplace random variable, its distribution is given by

$$F(x) = \begin{cases} 0 & \text{for} \quad x < -(\theta_2 - \theta_1)/s, \\ F(x; (\theta_1 - \theta_2)/s, 2) & \text{for} \quad -(\theta_2 - \theta_1)/s \leq x \leq (\theta_2 - \theta_1)/s, \\ 1 & \text{for} \quad x > (\theta_2 - \theta_1)/s, \end{cases}$$

where $F(x;\theta,s)$ is the c.d.f. of the $\mathcal{CL}(\theta,s)$ distribution.

Straightforward calculations yield the following characteristic function for this truncated distribution

$$\phi(t) = e^{-\frac{\theta_2-\theta_1}{2s}}\left\{\cosh\left[\left(\frac{1}{2}-it\right)\frac{\theta_2-\theta_1}{s}\right]+\frac{\sinh\left[(\frac{1}{2}-it)(\theta_2-\theta_1)/s\right]}{1-2it}\right\}.$$

Consequently, the characteristic function of T_n, $\phi^{(0)}(t)$ becomes

$$e^{-n\frac{\theta_2-\theta_1}{2s}}\left\{\cosh\left[\left(\frac{1}{2}-it\right)\frac{\theta_2-\theta_1}{s}\right]+\frac{\sinh\left[(\frac{1}{2}-it)(\theta_2-\theta_1)/s\right]}{1-2it}\right\}^n.$$

Expressing the hyperbolic sine and cosine in terms of complex exponentials, and using the binomial expansion of the nth power of the sum, we obtain (after rather tedious but straightforward simplifications)

$$\begin{aligned}\phi^{(0)}(t) = \frac{1}{2^n}\Bigg\{&\sum_{k=1}^{n}\sum_{l=0}^{n-k}\sum_{r=0}^{k}\binom{n}{k}\binom{n-k}{l}(-1)^r\binom{k}{r}\\ &e^{-(r+l)(\theta_2-\theta_1)/s}\cdot\frac{e^{-it(n-2r-2l)(\theta_2-\theta_1)/s}}{(1-2it)^k}\\ &+\sum_{m=0}^{n}\binom{n}{m}e^{-m(\theta_2-\theta_1)/s}e^{-it(n-2m)(\theta_2-\theta_1)/s}\Bigg\}.\end{aligned}$$

Note that

$$\psi_1(t) = \frac{e^{-it(n-2r-2l)(\theta_2-\theta_1)/s}}{(1-2it)^k}$$

and

$$\psi_2(t) = e^{-it(n-2m)(\theta_2-\theta_1)/s}$$

are, respectively, the characteristic functions of the χ^2 r.v. with $2k$ degrees of freedom (shifted by $(2r+2l-n)(\theta_2-\theta_1)/s$ to the right) and the constant random variable equal to $(2m-n)(\theta_2-\theta_1)/s$. The final formula for the c.d.f. $F_n^{(0)}$ follows from the forms of the c.d.f.'s for these two distributions. The formulas for the expected value and variance can be obtained easily by integration of the truncated Laplace random variable $g(X)$.

The corresponding formulas under H_1 follow from the symmetry of Laplace distribution. First, note the relation

$$g(x) = -g(-(x-\theta_2)+\theta_1).$$

Thus

$$\begin{aligned}P(T_n \le x\,|\,\theta=\theta_2) &= P\left(\sum_{i=1}^{n} g(-(X_i-\theta_2)+\theta_1) \ge -x\,\middle|\,\theta=\theta_2\right)\\ &= P\left(\sum_{i=1}^{n} g(X_i) \ge -x\,\middle|\,\theta=\theta_1\right)\\ &= 1-P(T_n \le -x\,|\,\theta=\theta_1).\end{aligned}$$

The second-to-last equality above follows from the fact that if X has the $\mathcal{CL}(\theta_2, s)$ distribution, then $Y = -(X - \theta_2) + \theta_1$ has the $\mathcal{CL}(\theta_1, s)$ distribution.

The asymptotic normality is a direct consequence of the Central Limit Theorem. □

The importance of the explicit formula for the test statistic in the above problem is due to the fact that the asymptotic Gaussian approximation is not usually very accurate for small and moderate sample sizes. For example, it was shown in Dadi and Marks (1987) that for samples size in the range from 5 to 50 the Gaussian approximation can be quite conservative, some yielding the t_α-value substantially larger than its exact value (see the Dadi and Marks paper for numerical results).

2.6.4.4 *Asymptotic optimality of the Kolmogorov–Smirnov test.* The asymptotic optimality of the Kolmogorov goodness-of-fit test for the location Laplace family was studied in Nikitin (1995), who derived the following characterization of the Laplace distribution: The Kolmogorov goodness-of-fit test is locally asymptotic optimal in the Bahadur sense if and only if the underlying family of distributions are symmetric Laplace laws. To state this result more precisely, let us recall some basic notions from the theory of asymptotic efficiency for statistical tests.

Let us consider a location family given by the densities f_θ, $\theta \in \mathbb{R}$, and let $F(x;\theta)$ be the corresponding cumulative distribution functions. Let $K(\theta, \theta_0)$ be the information number, i.e., $K(\theta, \theta_0) = E_{\theta_0} \log(f_\theta / f_{\theta_0})$. The Smirnov one-sided statistics are defined as

$$D_n^\pm = \sup_{x \in \mathbb{R}} \pm[F_n(x) - F(x;0)],$$

and the Kolmogorov statistic is

$$D_n = \sup_{x \in \mathbb{R}} |F_n(x) - F(x;0)|.$$

The statistics $D_n^\pm$ (or D_n) are locally optimal in the Bahadur sense if and only if

$$\lim_{n \to \infty} \frac{1}{n} \log P_{\theta,n} = -K(\theta, 0),$$

where $P_{\theta,n}$ is the observed P-value based on $D_n^\pm$ (or D_n) under the assumption that the sample is obtained from the distribution given by f_θ.

Let $\mathcal{G}$ be the class of absolutely continuous densities on the real line such that for $g \in \mathcal{G}$, we have

$$0 < \lim_{\theta \to 0} \left\{ \theta^{-2} \int \log\left(\frac{g(x+\theta)}{g(x)} \right) g(x+\theta) dx \right\} = \frac{1}{2} \int \frac{[g'(x)]^2}{g(x)} dx < \infty.$$

The following theorem was proved in Nikitin (1995, Theorem 6.3.1).

Theorem 2.6.3 *Consider a location testing problem with $f_\theta = g(x+\theta)$. Then the sequences of statistics D_n and D_n^+ are locally asymptotically optimal in the Bahadur sense within the class $\mathcal{G}$ only for the Laplace distribution, i.e., for $g(x) = 1/2e^{-|x|}$. The sequence of statistics D_n^- is never optimal in the Bahadur sense in the class $\mathcal{G}$.*

2.6.4.5 *Comparison of nonparametric tests of location.* Ramsey (1971) examines eight nonparametric tests of location in a small sample setting and investigates power functions for samples drawn from Laplace distribution. His main conclusion is that the Mood median test, which is the asymptotically most powerful (AMP) rank test, performs poorly for the alternatives that are not close to the null hypothesis.

Consider a rank sum statistic. Let $X_1, X_2, \ldots, X_m$ and let $Y_1, \ldots, Y_n$ be independent random samples from populations $F(x)$ and $G(y)$, respectively. We test $H_0 : G(x) \equiv F(x)$ versus the location shift alternative $H_A : G(x) = F(x - \theta)$ for some $\theta > 0$. Let δ_i $(i = 1, 2, \ldots, N = m + n)$ be a zero/one random variable indicating whether the ith smallest value in the *combined* sample is a Y.

A rank sum statistic is a linear combination

$$T_N = \sum_{i=1}^{N} a_{N,i}\delta_i,$$

where the $a_{N,i}$ $(i = 1, \ldots, N)$ are the so-called ith "scores."

When $F(x)$ is known to belong to a family (for example normal) that admits a UMP test, the choice of a test is clear and unique. When nothing is known about $F(\cdot)$ except for the information provided by the samples, one should select a nonparametric procedure with good efficiency in a wide class of distributional families.

For an intermediate situation when partial knowledge about $F(x)$ is available, Ramsey (1971) proposes using the Laplace distribution for the null hypothesis. It is not quite clear why this is an appropriate assumption—presumably the idea is that the data is long tailed—however, the behavior of eight standard nonparametric tests of location under the assumption that the null distribution is Laplace is, of course, of interest on its own.

The eight nonparametric tests for the Laplace distribution investigated in Ramsey (1971) and in Conover et al. (1978) (a follow-up to the first paper) are as follows:

1. The *locally most powerful* (LMP) rank test. Under the Laplace distribution the LMP scores are

   $$a_{N,i} = 2P(Z_N \le i - 1) - 1,$$

   where Z_N is binomial variable with the parameters N and $p = 1/2$.

2. The *Mood* median test (M), where

   $$a_{N,i} = \text{sign}(2i - N - 1)$$

   and the statistic is an AMP rank test [see, e.g., Hájek (1969)].

3. The normal scores (F) test, where $a_{N,i}$ is the expected value of the ith order statistic in a random sample of N observations from the standard normal distribution.

4. The Wilcoxon test (W) [see Wilcoxon (1945)], where $a_{N,i} = i$, the rank itself. Note that indicators of Y-ranks (rather than X-ranks) form the test statistic, so the alternative hypothesis is favored by the large values of the test statistic.

5. The van der Waerden (V) test [see van der Waerden (1952)] uses quantiles of the standard normal distribution as scores.

6. The Tukey quick test (T) counts the number of Y's exceeding the largest X and the number of X's that are less than the smallest Y. (If in the combined sample the largest and smallest observations come from the same sample, then $T = 0$.)

7. The Neave–Tukey quick test (N) statistic, which maximizes the Tukey statistic over subsamples in which one observation is omitted.

8. The Kolmogorov–Smirnov test (K): If F_m and G_n are the *sample* c.d.f.'s of X's and Y's, respectively, then

   $$KS = \sup_x |F_m(x) - G_n(x)|.$$

Ramsey chooses sample sizes $n = m = 5$ and calculates the power functions of each test as a function of the location shift θ. The power functions are of the form

$$p(\theta) = 1 + \frac{1}{a_{00}} \sum_{i=1}^{5} e^{-i\theta} \sum_{j=0}^{i} a_{j,i}\theta^j.$$

The results are presented in Figure 2.7. Here the power of the LMP test (p_{LMP}) is used as a standard and thus the LMP test is represented by the zero line. For other tests, the diagrams show the differences

$$p_{\cdot}(\theta) - p_{LMP}(\theta),$$

where $p_{\cdot}(\theta)$ is the power function of another test.

It is quite surprising that the Mood median test (which is AMP) performs poorly except for a small local region in which it is an approximation to the LMP test. Note also that the F, W, and V tests behave almost as well as the LMP tests.

This example with Laplace distribution shows that sometimes with an unfamiliar distributional family the cost of deriving the LMP test may not be justified and serves a warning to those who "purchase a shred of optimality (i.e., the use of asymptotically most powerful test) at the expense of a large sample assumption" [Ramsey (1971)].

2.7 Exercises

In this section we present some 60 exercises of various degrees of difficulty related to the material discussed in Chapter 2. We urge our readers to at least skim this section since it contains information that will enhance their understanding of the properties of the classical symmetric Laplace distribution.

Exercise 2.7.1 Show that the nth moment about zero of the classical Laplace r.v. Y with density (2.1.1) is given by (2.1.18). Compare with the corresponding result for a normal r.v. with mean θ and variance σ^2.

Exercise 2.7.2 Show that the density function $f(x; \theta, s)$ given by (2.1.1) has derivatives of any order, except at $x = \theta$, where there is a cusp. Demonstrate the following explicit form of these derivatives:

$$\frac{d}{dx^n} f(x; \theta, s) = \begin{cases} (-1)^n \frac{1}{2} \frac{1}{s^{n+1}} e^{-|x-\theta|/s} & \text{if } x > \theta, \\ \frac{1}{2} \frac{1}{s^{n+1}} e^{-|x-\theta|/s} & \text{if } x < \theta. \end{cases} \tag{2.7.1}$$

Exercise 2.7.3 The *Gini mean difference* for the distribution of a r.v. X is defined as

$$\gamma(X) = E|X_1 - X_2|,$$

where X_1, X_2 are i.i.d. copies of X. Show that if $X \sim \mathcal{CL}(\theta, s)$, then $\gamma(X) = \frac{3}{2s}$

Exercise 2.7.4 Let X be a classical Laplace r.v. with density $f(x) = f(x; \theta, s)$ as in (2.1.1).

(a) Show that for $\theta < 0$ the *geometric mean* of X, defined as

$$\lambda = \exp\left[\int_0^\infty \log x f(x) dx\right],$$

is

$$\lambda = \exp\left\{-\frac{1}{2}(\gamma - \log s)e^{\theta/s}\right\},$$

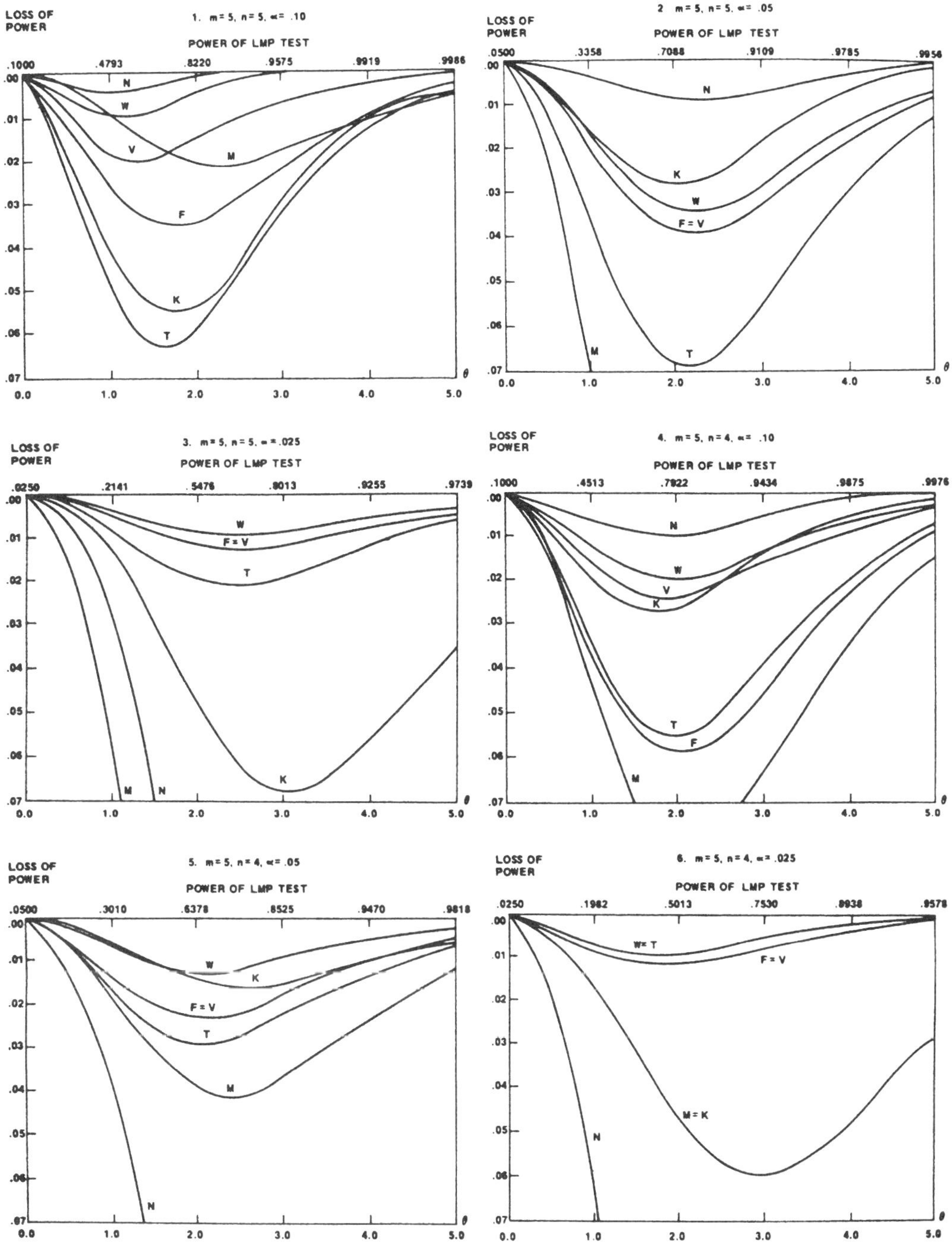

Figure 2.7: Power functions of eight nonparametric tests of location in the Laplace family for various values of significance level α (0.1 — top left, middle right; 0.05 — top right, bottom left; 0.025 — middle left, bottom right) and sample sizes $m = 5$ and $n = 5$ (first three graphs) and $n = 4$ (last three graphs). Reproduced from Conover et al. (1978). Reprinted with permission from the *Journal of the American Statistical Association*. Copyright 1978 by the American Statistical Association. All rights reserved.

where

$$\gamma = -\int_0^\infty e^{-y} \log y \, dy \approx 0.5772156\ldots$$

is Euler's constant [Christensen (2000)]. What is the value of λ when $\theta > 0$?

(b) Calculate the *harmonic mean* of X, defined as

$$\eta = \left[\int_{-\infty}^{\infty} \frac{1}{x} f(x) dx \right]^{-1},$$

where the integral is understood in the Cauchy's principal value sense.

Exercise 2.7.5 Let Y have a classical Laplace $\mathcal{CL}(0, s)$ distribution with density $f(x) = f(x; 0, s)$ given by (2.1.1).

(a) Verify that

$$\int_{-\infty}^{\infty} \frac{\log f(x)}{1 + x^2} dx = -\infty \tag{2.7.2}$$

and

$$-xf'(x)/f(x) \text{ is increasing without bound as } x \to \infty. \tag{2.7.3}$$

Recall that for a real r.v. Y whose c.d.f. is absolutely continuous with density f the conditions (2.7.2) (the so-called *Krein condition*) and (2.7.3) (the so-called *Lin condition*) are sufficient for the moments

$$\alpha_n = E[Y^n] = \int_{-\infty}^{\infty} x^n f(x) dx \tag{2.7.4}$$

to determine the distribution of Y uniquely [see Krein (1944) and Stoyanov (2000)]. Thus the $\mathcal{CL}(0, s)$ distribution is uniquely determined by the sequence $\{\alpha_n\}$ of its moments.

(b) Another sufficient condition for the moments (2.7.4) to determine the distribution uniquely is the so-called *Carleman condition*:

$$\sum_{n=1}^{\infty} \alpha_{2n}^{-\frac{1}{2n}} = \infty \tag{2.7.5}$$

[see, e.g., Harris (1966)]. Does the Laplace distribution $\mathcal{CL}(0, s)$ satisfy the Carleman condition?

(c) Is the general classical Laplace distribution $\mathcal{CL}(\theta, s)$ determined uniquely by the sequence $\{\alpha_n\}$ of its moments?

Exercise 2.7.6 Let X be a random variable with coefficients of skewness and kurtosis γ_1 and γ_2, respectively. The quantities

$$\gamma_3 = \frac{E[(X - EX)^5]}{[E(X - EX)]^{5/3}} - 10\gamma_1$$

and

$$\gamma_4 = \frac{E[(X - EX)^6]}{[E(X - EX)]^3} - 15\gamma_2 - 10\gamma_1^2 - 15$$

may be viewed as generalizations of γ_1 and γ_2. Compute these quantities for the standard Laplace and the standard normal distributions.

Exercise 2.7.7 In this exercise you will study the effect of rounding on the mean and the variance of the Laplace distribution. If the values of a continuous r.v. X are rounded into intervals of width ω, where the center of the interval containing zero is $a\omega$, then the values of the resulting discrete r.v. $\tilde{X}$ are

$$a\omega, a\omega \pm \omega, a\omega \pm 2\omega, \dots .$$

Moreover,

$$\tilde{X} = a\omega + n\omega, \quad n = 0, \pm 1, \dots$$

whenever

$$a\omega + n\omega - \omega/2 \le x < a\omega + n\omega + \omega/2.$$

(a) Let X have the $\mathcal{CL}(0, s)$ distribution, so that $E[X] = 0$ and $\mathrm{Var}[X] = \sigma^2 = 2s^2$. Show that the probability function of the r.v. $\tilde{X}$ admits the following explicit form:

$$P(\tilde{X} = a\omega + n\omega) = \begin{cases} \frac{1}{2}\left(e^{\omega/2s} - e^{-\omega/2s}\right) e^{a\omega/s + \omega n/s} & \text{for } n \le -1, \\ 1 - \frac{1}{2} e^{-\omega/2s}\left(e^{a\omega/s} + e^{-a\omega/s}\right) & \text{for } n = 0, \\ \frac{1}{2}\left(e^{\omega/2s} - e^{-\omega/2s}\right) e^{-a\omega/s - \omega n/s} & \text{for } n \ge 1. \end{cases} \tag{2.7.6}$$

(b) Derive closed form expressions for the mean and the variance of the r.v. $\tilde{X}$ given by (2.7.6). Discuss the effects of rounding on the mean and variance. You may want to follow Tricker (1984), writing $\omega = r\sigma$ and considering the behavior of the bias

$$\frac{E[\tilde{X}] - E[X]}{\omega}$$

and the ratio

$$V = \frac{\mathrm{Var}[\tilde{X}]}{\mathrm{Var}[X]}$$

for various values of a and r.

(c) Repeat the above for the normal distribution with mean zero and variance σ^2. Does the probability function of $\tilde{X}$ admit an explicit form in this case? What about $E[X]$ and $\mathrm{Var}[\ddot{X}]$? In which case is the effect of rounding more severe?

Exercise 2.7.8 Let F and G be the d.f.'s of two continuous distributions symmetric about θ_F and θ_G, respectively. We say that F is *lighter tailed* than G, denoted by

$$F <_s G,$$

if the function $G^{-1}[F(x)]$ is convex for $x > \theta_F$ [see van Zwet (1964)].

(a) Show that the s-ordering defined above is location and scale invariant.

(b) Assume that $\theta_F = \theta_G = 0$ and show that if $F <_s G$, then $G(x) \le F(x)$ for $x > 0$. Thus G has *more probability in the tail* than F does [Hettmansperger and Keenan (1975)].

(c) Show that

$$\text{uniform} <_s \text{normal} <_s \text{logistic} <_s \text{Laplace}.$$

(d)* Further, show that although we have *Logistic* $<_s$ *Cauchy*, the Laplace and the Cauchy distributions are not comparable with respect to the $<_s$ ordering [see Latta (1979) and Balanda

(1987)]. In practice, the uniform is usually referred to as *light tailed*, the normal and logistic as *medium tailed*, while the Laplace and the Cauchy as *heavy tailed*, so in a sense, the s-ordering corresponds to a common perception of tail heaviness. See, e.g., Hettmansperger and Keenan (1975) for more information on ordering of distributions by tail heaviness.

Exercise 2.7.9 Let X have the standard classical Laplace distribution with density $p(x) = \frac{1}{2}e^{-|x|}$ $(-\infty < x < \infty)$. Show that the ordinate $p(X)$, considered as a random variable [the so-called *vertical density function*; see Troutt (1991)] has uniform distribution on $(0, 1/2)$. Note that the same is true for the ordinate $p(X)$ when X has the standard exponential density $p(x) = e^{-x}$ $(x > 0)$ (in which case we obtain the standard uniform distribution). Investigate the corresponding case of the standard normal distribution: derive the density of the ordinate $p(X)$ when X is standard normal with p.d.f. $p(x) = \frac{1}{\sqrt{2\pi}}e^{-x^2/2}$ $(-\infty < x < \infty)$, which is not uniform!

Exercise 2.7.10 Let W be a standard exponential r.v. with the density $f_W(w) = e^{-w}$, $w \geq 0$, and let Z be a standard normal random variable, independent of W, with density

$$f(x) = \frac{1}{\sqrt{2\pi}}e^{-x^2/2}, \quad -\infty < x < \infty.$$

Show that the density of the product $X = \sqrt{2W}Z$ is given by the right-hand side of relation (2.2.4).

Hint: Consider the transformation $Y_1 = W$, $Y_2 = \sqrt{2W}Z$ and derive the joint density of Y_1 and Y_2. Then integrate the joint density with respect to y_1 to obtain the marginal density of Y_2.

Exercise 2.7.11 Let W have a standard exponential distribution with density $f_W(w) = e^{-w}$, $w \geq 0$. Show that the random variable $T = 1/\sqrt{W}$ has the density $f_T(x) = 2x^{-3}e^{1/x^2}$, $x > 0$.

Exercise 2.7.12 Let W have a standard exponential distribution with density $f_W(w) = e^{-w}$, $w \geq 0$. Let I be r.v. taking on values ± 1 with probabilities $1/2$ and independent of W. Show that the ch.f. of IW is given by the right-hand side of (2.2.9).

Exercise 2.7.13 Let U_1, U_2, U_3, U_4 be i.i.d. standard normal random variables. By computing relevant characteristic functions, show that the r.v. $X = U_1U_4 - U_2U_3$ has the standard Laplace distribution.

Hint: First, show that the ch.f. of X is $(E[e^{itU_1U_4}])^2$ and compute this expectation by conditioning on U_4.

Exercise 2.7.14 Explain why a three-dimensional extension of (2.2.13) given by a 3×3 matrix does not result in a Laplace distribution or its modifications. Investigate an n-dimensional extension.

Exercise 2.7.15 Let δ_1 and δ_2 be r.v.'s taking values of either zero or one with probabilities given in Proposition 2.4.4. Let W_1, W_2 be i.i.d. standard exponential r.v.'s, independent of (δ_1, δ_2). Let X have a standard Laplace distribution with ch.f. (2.1.7).

(a) Show that the ch.f. of cX, where $c \in (0, 1)$, is given by the first factor of (2.4.10).

(b) Show that the ch.f. of $\delta_1 W_1 - \delta_2 W_2$ is given by the second factor of (2.4.10).

(c) Show that the product (2.4.10) is equal to the ch.f. of X.

Exercise 2.7.16 Show that if X_1 and X_2 are i.i.d. $\mathcal{CL}(0, s)$ random variables, then $Y = |X_1/X_2|$ has F-distribution with $\nu_1 = 2$ and $\nu_2 = 2$ degrees of freedom.

Exercise 2.7.17 Show that if Z_i, $i = 1, 2, \ldots, 6$, are i.i.d. standard normal r.v.'s, then

$$Y = |Z_1|\sqrt{Z_2^2 + Z_3^2} - |Z_4|\sqrt{Z_5^2 + Z_6^2}$$

has the standard classical Laplace distribution.

Exercise 2.7.18 Let $X_1, \ldots, X_n$ be i.i.d. standard classical Laplace r.v.'s. Show that the sum $T = \sum_{j=1}^{n} X_j$ admits the random sum representation (2.3.27) of Proposition 2.3.2.

Hint: Write the ch.f. $\phi(t)$ of the right-hand side of (2.3.27) by conditioning on I and M_n to obtain

$$\phi(t) = \frac{1}{2}\sum_{j=1}^{n}\left[\left(\frac{1}{1-it}\right)^j + \left(\frac{1}{1+it}\right)^j\right]\frac{2^j}{2^{2n-1}}\binom{2n-j-1}{n-1}. \tag{2.7.7}$$

Then show that (2.7.7) coincides with $[1 + t^2]^{-n}$, which is the ch.f. of T.

Exercise 2.7.19 Let X_1 and X_2 be i.i.d. random variables with density $f(x) = px^{p-1}$, $p > 0$, $x \in (0, 1)$ [the standard power function distribution with parameter p; see, e.g., Johnson et al. (1994, p. 607)]. Show that the r.v.

$$Y = p \log \frac{X_1}{X_2}$$

has the standard classical Laplace distribution.

Hint: Relate X_1 to the standard Pareto Type I r.v. with p.d.f. $1/x^2$, $x > 1$, and use Proposition 2.2.4.

Exercise 2.7.20 Recall that the standard classical Laplace r.v. X has the same distribution as the difference of two i.i.d. standard exponential variables (see Proposition 2.2.2). Investigate whether there are any other i.i.d. r.v.'s V_1 and V_2 such that

$$X \stackrel{d}{=} V_1 - V_2. \tag{2.7.8}$$

Proceed by writing the relation (2.7.8) in terms of ch.f.'s,

$$\frac{1}{1+t^2} = \psi_{V_1}(t)\psi_{V_1}(-t), \tag{2.7.9}$$

where ψ_{V_1} is the ch.f. of V_1, and note that the ch.f.

$$\psi_{V_1}(t) = (1 - it)^{-\alpha}(1 + it)^{\alpha - 1}, \quad 0 \le \alpha \le 1,$$

is a solution of (2.7.9). What is the corresponding r.v. V_1? Are there any other solutions to (2.7.9)? [See Problem 64–13, *SIAM Review*, **8**(1), (1966), pp. 108–110].

Exercise 2.7.21 Let $X_1, X_2, \ldots$ be i.i.d. random variables with finite mean μ, and let N be a positive and integer-valued random variable with finite mean $E[N]$. Show that if N and X_i's are independent, then the mean of the random sum $\sum_{i=1}^{N} X_i$ is equal to the product $\mu E[N]$.

Exercise 2.7.22 Define $f_n(t) = [\phi(\sqrt{t})]^{1/n}$ for $t > 0$ and $n = 1, 2, \ldots$, where ϕ is a real-valued characteristic function. If the function f_n is completely monotone on $(0, \infty)$ for each n (that is, $(-1)^k f_n^{(k)}(t) \ge 0$ for $t > 0$, $k = 0, 1, \ldots$), then the ch.f. ϕ is infinitely divisible [Kelker (1971)]. Apply the above result to the ch.f. of the standard classical Laplace distribution to establish its infinite divisibility.

Exercise 2.7.23 Suppose that X_1 and X_2 are i.i.d. classical Laplace r.v.'s with p.d.f. (2.1.1), where $\theta = 0$ and $s > 0$. Let

$$\overline{X}_2 = \frac{1}{2}(X_1 + X_2) \quad \text{and} \quad S_2^2 = (X_1 - \overline{X}_2)^2 + (X_2 - \overline{X}_2)^2.$$

Show that the p.d.f. of the t-statistic (2.3.40) with $n = 2$ is given by (2.3.51).

Exercise 2.7.24 Let $X_1, \ldots, X_n$ be a random sample from the classical Laplace distribution $\mathcal{CL}(\theta, s)$.

(a) Show that the distribution of the t-type statistic $\tilde{T}_n$ given by (2.3.57) is concentrated on the interval $[-1, 1]$ and does not depend on the parameters θ and s.

(b) Show that the distribution function of the statistic $\tilde{T}_n$ is given by (2.3.60).

(c) Investigate the distribution of another analog of the t-distribution, the statistic

$$\frac{\sum_{i=1}^n (X_i - \theta)}{\sum_{i=1}^n |X_i - \hat{\theta}_n|},$$

where $\hat{\theta}_n$ is the sample median of the X_i's.

Exercise 2.7.25 Let

$$g_n(t) = \frac{\Gamma\left(\frac{n+1}{2}\right)}{\sqrt{n\pi}\,\Gamma\left(\frac{n}{2}\right)} \left(1 + \frac{t^2}{n}\right)^{-(n+1)/2}, \quad -\infty < t < \infty,$$

be the density of the t-distribution with n degrees of freedom, and let f_{T_n} be the density (2.3.56) of the t-statistic (2.3.40) based on a random sample of size n from the classical Laplace distribution with density (2.1.1) with $\theta = 0$. Investigate the behavior of the ratio

$$\gamma_n(t) = g_n(t)/f_{T_n}(t)$$

as $t \to \infty$. Specifically, show that $\gamma_n(t)$ is monotonically increasing to infinity for $t \in (t_0, \infty)$ for some $t_0 > 0$. Conclude that the tails of density f_{T_n} are heavier than those of the student t-density g_n. What are the implications when one uses the critical points of the t-distribution when calculating the Type I error probabilities, the power function, or the confidence levels connected with samples from the Laplace distribution?

Exercise 2.7.26 Compare products and ratios of two independent Laplace random variables with products and ratios of two independent normal random variables.

Exercise 2.7.27 Let X_1, X_2, X_3, X_4 be independent standard classical Laplace random variables. Find the p.d.f.'s of their following functions:

$$\frac{X_3}{\sqrt{(X_1^2 + X_2^2)/2}}, \qquad \frac{2X_3^2}{X_1^2 + X_2^2}, \qquad \frac{X_1^2 + X_2^2}{X_3^2 + X_4^2}.$$

Exercise 2.7.28 If X_1 has density

$$f_1(x) = \begin{cases} \frac{1}{2a}; & -a < x < a, \\ 0; & \text{otherwise,} \end{cases}$$

if X_2 has density

$$f_2(x_2) = \frac{1}{\sqrt{2\pi}\sigma} e^{-\frac{1}{2}x_2^2/\sigma^2}, \quad -\infty < x_2 < \infty,$$

and X_1 and X_2 are independent, then $Y = X_1X_2$ has density

$$h(y) = \frac{1}{2\sqrt{2\pi}a\sigma}E_1\left(\frac{y^2}{2a^2\sigma^2}\right), \quad -\infty < y < \infty,$$

where

$$E_1(x) = \int_x^\infty \frac{e^{-t}}{t}dt, \quad x > 0,$$

is the exponential integral. What is the corresponding result when X_2 is replaced by a Laplace r.v. with mean zero and scale parameter σ?

Exercise 2.7.29 Let B_n have beta distribution with parameters 1 and n, with density given by (2.2.45). Show that as $n \to \infty$, the sequence nB_{n-1} converges in distribution to a standard exponential random variable.

Exercise 2.7.30 Show that if in Proposition 2.4.7 the condition (2.4.27) is replaced by

$$E|X - \theta| = c > 0 \text{ for } X \in \mathcal{C},$$

then the maximum entropy is attained by the classical Laplace distribution with density $f(x) = \frac{1}{2c}e^{-|x-\theta|/c}$ [Kapur (1993)].

Exercise 2.7.31 (a) Consider a location family with density

$$f(x-\theta), \quad -\infty < x, \theta < \infty, \tag{2.7.10}$$

where f is the standard classical Laplace density $f(x) = \frac{1}{2}e^{-|x|}$. Show that the Fisher information $I(\theta)$, given by

$$I(\theta) = \int_{-\infty}^{\infty} \frac{[f'(y)]^2}{f(y)}dy, \tag{2.7.11}$$

is equal to one. Compare it with the corresponding values of $I(\theta)$ when f is the standard normal, standard logistic, and standard Cauchy density.

(b) Now consider a location-scale family with density (2.6.1). Using the relations (2.6.9)–(2.6.11), show that the Fisher information matrix is given by (2.6.12).

(c) Show that for a location-scale family of $\mathcal{L}(\theta, \sigma)$ distributions given by the density (2.1.3), the Fisher information matrix is

$$\begin{bmatrix} 2/\sigma^2 & 0 \\ 0 & 1/\sigma^2 \end{bmatrix}.$$

(d) What is the corresponding Fisher information matrix when f in (2.6.1) is the standard normal density $f(x) = \frac{1}{\sqrt{2\pi}}e^{-x^2/2}$?

Exercise 2.7.32 Let $X_1, \ldots, X_n$ be a random sample of size $n = 2k+1$ from the classical Laplace location family with density (2.6.125), and let $\hat{\theta}_n = X_{k+1:n}$ be the sample median with the density given by (2.6.126)–(2.6.127).

(a) Following (2.6.128), show that the Fisher information about θ contained in $\hat{\theta}_n$ is given by (2.6.129).

(b) Show that the amount of Fisher information lost when using $\hat{\theta}_n$ is given by (2.6.130).

(c) Show that the loss (2.6.130) converges to infinity as $k \to \infty$.

(d) Establish the asymptotic relation (2.6.131).

Exercise 2.7.33 Given a random sample $X_1, \ldots, X_n$ (from a continuous distribution with density f and distribution function F) and a score function $J(u)$, $0 < u < 1$, (corresponding to a one-sample linear rank test of symmetry), the R-estimator of the location parameter θ is defined as the solution of

$$\sum_{i=1}^{n} \text{sign}(X_i - \theta) J^{+}\left(\frac{R(|X_i - \theta|)}{n+1}\right) = 0, \tag{2.7.12}$$

where

$$J^{+}(u) = J(1/2 + u/2)$$

and $R(w)$ is the rank of w [see, e.g., Hall and Joiner (1983)]. Under some regularity conditions, the efficient score function, corresponding to the asymptotically most powerful rank test [see, e.g., Hájek (1969)] is

$$J(u) = \frac{-f'(F^{-1}(u))}{f'(F^{-1}(u))}. \tag{2.7.13}$$

(a) Show that if the sample is from the $\mathcal{CL}(\theta, 1)$ distribution, then the efficient score function (2.7.13) is

$$J(u) = \text{sign}(u - 1/2) \tag{2.7.14}$$

(so that the corresponding asymptotically most powerful rank test is the sign test).

(b) Show that if the score function is given by (2.7.14), then the R-estimator of location given by (2.7.12) is the sample median.

(c) What is the most efficient score function (and the corresponding asymptotically most powerful rank test) if the underlying distribution is normal?

(d) What is the most efficient score function (and the corresponding asymptotically most powerful rank test) under the underlying logistic distribution?

Exercise 2.7.34 Let $X_1, \ldots, X_n$ be a random sample from the density

$$f(x; \theta) = \frac{1}{2} e^{-|x-\theta|}, \quad -\infty < x < \infty, \quad -\infty < \theta < \infty. \tag{2.7.15}$$

Use calculus to show that the MLE of θ is the sample median. Proceed by writing the log-likelihood as

$$\psi(\theta) = -n \log 2 - \sum_{i=1}^{n} \{(x_i - \theta)^2\}^{1/2}, \tag{2.7.16}$$

and then by taking the derivative with respect to θ to find the intervals where ψ is increasing and decreasing.

Exercise 2.7.35 The following example was derived in Rao and Ghosh (1971). Consider the location family $\{f(x - \theta), \theta \in \mathbb{R}\}$, where

$$f(y) = \begin{cases} e^{k_1 - \alpha_1 |y|}, & \text{for } 0 \le |y| \le c_1, \\ e^{k_2 - \alpha_2 |y|}, & \text{for } c_1 \le |y| < \infty, \end{cases} \tag{2.7.17}$$

where, for continuity, we have $(\alpha_2 - \alpha_1)c_1 + k_1 = k_2$, and

$$0 < \alpha_1 < \alpha_2 < 2\alpha_1. \tag{2.7.18}$$

The constants $k_1, k_2, \alpha_1, \alpha_2$, and c_1 are such that f is a valid probability density on $(-\infty, \infty)$.

(a) Show that if g is convex and symmetric function on $\mathbb{R}$, then for any $x_1, x_2 \in \mathbb{R}$, the function

$$h(\theta) = g(x_1 - \theta) + g(x_2 - \theta) \tag{2.7.19}$$

is minimized for $\theta^* = \frac{x_1+x_2}{2}$.

(b) Let X_1, X_2 be a random sample of size $n = 2$ from density $f(x - \theta)$. Apply part (a) to the function

$$g(x) = -\log f(x) \tag{2.7.20}$$

to show that the likelihood function is maximized when θ is set to the sample median.

(c) Let $y_1 < y_0 < y_2$ be a random sample of size $n = 3$ form $f(x - \theta)$. Assuming that $y_0 = 0$, write the negative log-likelihood function and show that it is convex. Further, show that for θ near zero the negative of the log-likelihood function is minimized by $\theta = 0$ (sample median). Argue that the global minimum exists, and is also attained at $\theta = 0$. Thus we have a non-Laplace distribution, such that the MLE of the location parameter for sample sizes $n = 2, 3$ is sample median.

Exercise 2.7.36 Let f be the skewed Laplace density given by (2.6.18).

(a) Show that the function f is a probability density on $(-\infty, \infty)$ if $c = (1/b_1 + 1/b_2)^{-1}$.

(b) Let n be odd, and let $X_1, \dots, X_n$ be a random sample of size n from the distribution with density $f(x - \theta)$, where f is the density (2.6.18) with the constants b_1 and b_2 such that

$$b_1\frac{n-1}{n+1} \le b_2 \quad \text{and} \quad b_2\frac{n-1}{n+1} \le b_1. \tag{2.7.21}$$

Show that every median of $X_1, \dots, X_n$ is the MLE of θ.

(c) Let n be odd and let $b_1 > 0$ and $b_2 = \frac{n+2}{n}b_1$. Show that the above b_1 and b_2 satisfy the conditions (2.7.21).

(d) In view of the above results, show that the condition (v) preceding Proposition 2.6.3 (see Section 2.6.1.1) is not enough to conclude that the population is Laplace [Findeisen (1982)].

Exercise 2.7.37 Consider the function

$$f(x) = c(2 + |x|)^{-1}e^{-|x|}, \quad -\infty < x < \infty. \tag{2.7.22}$$

(a) Argue that f with an appropriate $c > 0$ is a probability density function on $(-\infty, \infty)$.

(b) Show that for every $-\infty < x, y, \theta < \infty$ we have

$$\log f(x - \theta) + \log f(y - \theta) \le \log f(0) + \log f(y - x). \tag{2.7.23}$$

(c) Using part (b), show that if X_1 and X_2 are i.i.d. with density $f(x - \theta)$, where f is given by (2.7.22), then both X_1 and X_2 are the MLE's of θ.

(d) In view of the above results, show that the condition (vi) preceding Proposition 2.6.3 (see Section 2.6.1.1) is not sufficient to conclude that the population is Laplace [Findeisen (1982)].

Exercise 2.7.38 Let $X_1, \dots, X_n$ be a random sample from the density (2.6.19) with a given value of α and an unknown value of θ. Show that the MLE of θ is the empirical α-quantile of the sample (defined to be a number $\hat{\xi}_\alpha$ such that at least $\alpha \times 100\%$ of the observations are less than or equal to $\hat{\xi}_\alpha$, and at least $(1 - \alpha) \times 100\%$ of the observations are greater or equal to $\hat{\xi}_\alpha$).

Exercise 2.7.39 Let $X_{1:n}, \dots, X_{n:n}$ denote order statistics from a standard classical Laplace distribution $\mathcal{CL}(0, 1)$. Then the variance of the sample median (2.6.15) is given by

$$\sigma_n^2 = \begin{cases} \frac{n!}{(k!)^2} 2^{1-k} \sum_{i=0}^{k} \frac{k!}{i!(k-i)!} (-2)^{-i} (k+1+i)^{-3}, & \text{for } n = 2k+1, \\ \frac{n!}{[(k-1)!]^2} 2^{2-k} \left(\sum_{i=0}^{k-2} a(i,k) + \frac{3(-1)^{k-1}+1}{2^{k+3} k^4} \right), & \text{for } n = 2k \end{cases} \tag{2.7.24}$$

(in case $n = 2$ the sum $\sum_{i=0}^{-1}$ should be set to zero), where

$$a(i,k) = \frac{(k-1)!}{i!(k-1-i)!} (-2)^{-i} (k-1-i)^{-1} \{(k+1+i)^{-3} - (2k)^{-3}\}. \tag{2.7.25}$$

Show that $\sigma_n^2 \to 0$ as $n \to \infty$.

Exercise 2.7.40 Let M_x and M_y be the sample medians (2.6.15) of $x_1, \dots, x_n$ and $y_1, \dots, y_m$, respectively, and let M_{xy} be the sample median of $x_1, \dots, x_n, y_1, \dots, y_m$. Show that M_{xy} is between M_x and M_y.

Exercise 2.7.41 Let $X_1, \dots, X_n$ be a random sample from the standard classical Laplace distribution, and let

$$W_n = \frac{\hat{\theta}_n - \log[2(1-\beta)]}{\hat{s}_n},$$

where $0.5 < \beta < 1$ and the statistics $\hat{\theta}_n$ and $\hat{s}_n$ are, respectively, the (canonical) sample median (2.6.15) and the sample mean absolute deviation (2.6.34) (the MLE's of the Laplace parameters). Show that if $n = 2$, then the p.d.f. of W_2 is given by (2.6.178) with $k_\beta = \log[2(1-\beta)]$ [Shyu and Owen (1986a)].

Exercise 2.7.42 Let $X_1, \dots, X_n$ be i.i.d. from the $\mathcal{CL}(\theta, s)$ distribution, and let $\hat{\theta}_n$ and $\tilde{\theta}_n$ be the MLE and MME of θ given by (2.6.15) and (2.6.53), respectively.

(a) Show that if $s = 1$ and $n = 2k+1$, then for any integer $k \geq 3$ the right-hand side of (2.6.56) satisfies the relation

$$(1.51) \frac{(2k+1)!}{(k!)^2} \left(\frac{1}{2}\right)^{2k+1} \sqrt{\frac{2\pi}{2k+1}} \left(1 + \frac{1}{2k}\right)^{3/2} \leq 2. \tag{2.7.26}$$

Conclude that for $n = 2k+1 \geq 7$ the variance of $\hat{\theta}_n$ is less than the variance of $\tilde{\theta}_n$ (which is $2/n$).

(b) Investigate the corresponding case when the sample size is even.

Exercise 2.7.43 Let $X_1, \dots, X_n$ be i.i.d. from the $\mathcal{CL}(0, s)$ distribution, and let $\hat{s}_n$ and $\tilde{s}_n$ be the MLE and MME of s given by (2.6.20) and (2.6.58), respectively.

(a) We saw in Proposition 2.6.4 that $\hat{s}_n$ is unbiased for s. Investigate whether this property is shared by $\tilde{s}_n$.

(b) Asymptotically, the variance of $\hat{s}_n$ is smaller than that of $\tilde{s}_n$. For $n \geq 1$, derive the variances of $\hat{s}_n$ and $\tilde{s}_n$ and examine which one is larger.

Exercise 2.7.44 Consider a Type II censored sample (2.6.38) from the classical Laplace distribution and the corresponding likelihood function (2.6.40).

(a) Show that the likelihood function is continuous in θ for any fixed $s > 0$.

(b) Show that for any fixed $s > 0$ the likelihood function is monotonically increasing in θ for $\theta \in (-\infty, x_{r+1:n})$ and monotonically decreasing in θ for $\theta \in (x_{n-r:n}, \infty)$.

(c) Show that for $\theta \in [x_{r+1:n}, x_{n-r:n}]$ and for any fixed $s > 0$ the likelihood function is maximized by sample median of $x_{r+1:n}, \ldots, x_{n-r:n}$.

(d) Show that the MLE of θ is the sample median.

(e) Show that when we substitute the sample median $\hat{\theta}_n$ into the likelihood function (2.6.40) we obtain the function g given by (2.6.41)–(2.6.42).

(f) Show that the function g is maximized by $s = C/(n-2r)$ and deduce that the MLE of s is given by (2.6.43).

(g) Investigate the case of Type II right censored samples and general Type II censored samples.

Exercise 2.7.45 Let $X_1, \ldots, X_n$ be i.i.d. with the $\mathcal{CL}(0, s)$ distribution.

(a) Show that

$$\delta_1 = \frac{1}{2n}\sum_{i=1}^{n} X_i^2 \tag{2.7.27}$$

is an unbiased and consistent but not efficient estimator of the parameter s^2.

(b) Show that under the loss function of the form

$$L(\delta, s^2) = f(s^2)(\delta - s^2)^2, \tag{2.7.28}$$

where f is an arbitrary positive function, the risk of the estimators of the form

$$\delta_\alpha = \frac{\alpha}{n}\sum_{i=1}^{n} X_i^2 \tag{2.7.29}$$

is minimized for

$$\alpha^* = \frac{n}{2(5+n)} \tag{2.7.30}$$

[Jakuszenkow (1978)]. Is the resulting estimator consistent for s^2? Compare the variances of δ_1 and δ_{α^*}.

Exercise 2.7.46 Consider the mixture of two Laplace distributions with density (2.6.81). Show the following.

(a) If $\theta_1 = \theta_2 = 0$, then for any $0 < p < 1$, the distribution is unimodal with the mode at θ.

(b) If $\theta_1 < \theta_2$ and

$$\frac{s_1^2}{s_1^2 + s_2^2 e^{(\theta_2-\theta_1)/s_2}} < p < \frac{s_1^2}{s_1^2 + s_2^2 e^{(\theta_1-\theta_2)/s_1}},$$

then the distribution is bimodal with the modes at θ_1 and θ_2.

(c) If $\theta_1 < \theta_2$ and

$$0 < p < \frac{s_1^2}{s_1^2 + s_2^2 e^{(\theta_2-\theta_1)/s_2}},$$

then the distribution is unimodal with the mode at θ_2.

(d) If $\theta_1 < \theta_2$ and

$$\frac{s_1^2}{s_1^2 + s_2^2 e^{(\theta_1-\theta_2)/s_1}} < p < 1,$$

then the distribution is unimodal with the mode at θ_1 [Kacki and Krysicki (1967)].

Exercise 2.7.47 Let Y have a classical Laplace distribution with density (2.1.1) so that

$$Y \stackrel{d}{=} \theta + sX, \tag{2.7.31}$$

where X has the $\mathcal{CL}(0, 1)$ distribution. Then the mixture on θ of the distribution of Y is the type I compound Laplace distribution with parameters μ, σ, and s, if θ in (2.7.31) has the normal distribution with mean μ and variance σ^2 [see, e.g., Johnson et al. (1995)]. Show that the p.d.f. of this distribution is

$$f(x) = C\left\{\Phi\left(\frac{x-\mu}{\sigma} - \frac{\sigma}{s}\right)e^{-(x-\mu)/s} + \Phi\left(-\frac{x-\mu}{\sigma} - \frac{\sigma}{s}\right)e^{(x-\mu)/s}\right\},$$

where Φ is the c.d.f. of the standard normal distribution,

$$C = \frac{1}{2s}e^{\frac{1}{2}\left(\frac{\sigma}{s}\right)^2},$$

and $-\infty < x < \infty$, $-\infty < \mu < \infty$, $\sigma > 0$, and $s > 0$.

Exercise 2.7.48 Let Y have a classical Laplace distribution with density (2.1.1) and representation (2.7.31). The mixture on $1/s$ of the distribution of Y is the type II compound Laplace distribution with parameters θ, α, and β if $1/s$ in (2.7.31) has the $\Gamma(\alpha, \beta)$ distribution with density

$$f_{\alpha,\beta}(x) = \frac{x^{\alpha-1}e^{-x/\beta}}{\beta^\alpha\Gamma(\alpha)}, \quad \alpha > 0, \beta > 0, x > 0,$$

[see, e.g., Johnson et al. (1995)].

(a) Show that the p.d.f. and the c.d.f. of this distribution are

$$f(x) = \frac{1}{2}\alpha\beta[1 + |x-\theta|\beta]^{-(\alpha+1)}, \quad \alpha > 0, \beta > 0, -\infty < x < \infty, \tag{2.7.32}$$

and

$$F(x) = \begin{cases} \frac{1}{2}[1 + |x-\theta|\beta]^{-\alpha}, & \text{for } x < \theta, \\ 1 - \frac{1}{2}[1 + |x-\theta|\beta]^{-\alpha}, & \text{for } x \geq \theta, \end{cases}$$

respectively. Note that for $\theta = 0$, $\alpha = 1$, and $\beta = s_2/s_1$, the density (2.7.32) coincides with that of the ratio of two independent, mean zero, classical Laplace r.v.'s with scale parameters $s_1 > 0$ and $s_2 > 0$, respectively (see Section 2.3.3).

(b) Further, show that as $\alpha \to \infty$ and $\beta \to 0$ with $\alpha\beta = s > 0$, then $f(x)$ in (2.7.32) converges to the classical Laplace density (2.1.1). The relation between Laplace distributions and distributions with densities given by (2.7.32) is analogous to that between normal and Pearson Type VII distributions [see, e.g., Johnson et al. (1995)].

Exercise 2.7.49 Let Y have the type II compound Laplace distribution with density (2.7.32).

(a) Show that for $\alpha > 1$ the mean of Y is equal to θ, and for $\alpha > 2$ the variance of Y is

$$\sigma^2 = \frac{2\beta^2}{(\alpha-1)(\alpha-2)}, \quad \alpha > 2.$$

Note that the distribution is symmetric about θ, so that (for $\alpha > 1$) we have median = mean = mode.

(b) More generally, show that the moments of order α or greater do not exist, and for $0 < r < \alpha$ we have

$$E(X-\theta)^r = \begin{cases} \alpha\beta^r \sum_{j=0}^{r}(-1)^j \frac{r!}{j!(r-j)!}(\alpha+j-r)^{-1}, & \text{for } r \text{ even}, \\ 0, & \text{for } r \text{ odd}. \end{cases}$$

(c) Show that the mean deviation is $\beta/(\alpha-1)$ (for $\alpha > 1$). Derive an expression for the Mean deviation/Standard deviation and compare this with the corresponding value for the Laplace distribution.

(d) Show that the coefficient of kurtosis, defined in (2.1.22), is given by

$$\gamma_2 = \frac{6(\alpha-1)(\alpha-2)}{(\alpha-3)(\alpha-4)} - 3, \quad \alpha > 4.$$

What is the range of γ_2? How does γ_2 above compare with the corresponding value for the Laplace distribution? Is the type II compound Laplace distribution leptokurtic ($\gamma_2 > 0$) or platykurtic ($\gamma_2 < 0$)?

Exercise 2.7.50 Let $Y_1, \ldots, Y_n$ be i.i.d. normal variables with mean μ and variance σ^2. Assume that the variance is a constant, and the mean is a random variable with the Laplace $\mathcal{L}(\theta, \eta)$ prior distribution (so that μ has the mean and variance equal to θ and η^2, respectively). Let Y be the corresponding sample mean and let f be the marginal density of Y.

(a) Show that

$$f(x) = \frac{1}{\eta} e^{(\sigma/\eta)^2/n} \{F(z) + F(-z)\}, \tag{2.7.33}$$

where

$$z = \frac{\sqrt{n}}{\sigma}(y-\theta), \quad F(z) = e^{b^* z}\Phi(-z-b^*), \quad b^* = \frac{\sigma}{\eta}\sqrt{\frac{2}{n}},$$

and Φ is the c.d.f. of the standard normal distribution.

(b) Determine the posterior p.d.f. of μ given $Y = y$. Show that the posterior mean and variance arc

$$E(\mu|Y-y) = w(z)(y + \frac{\sigma}{\sqrt{n}}b^*) + (1-w(z))\left(y - \frac{\sigma}{\sqrt{n}}b^*\right)$$

and

$$\mathrm{Var}(\mu|Y=y) = \frac{\sigma^2}{n} - \frac{4\sigma^4}{n^2\eta^2}H(z),$$

respectively, where

$$w(z) = \frac{F(z)}{F(z)+F(-z)},$$
$$H(z) = \frac{[F(z)+F(-z)]g(z) - 2F(z)F(-z)}{[F(z)+F(-z)]^2},$$

and

$$g(z) = e^{-b^* z}\phi(-z-b^*).$$

(The function ϕ above denotes the standard normal p.d.f.)

(c) Investigate the dependence of the posterior mean and variance on y. How do they change as y varies from $-\infty$ to ∞? Does the posterior variance attain a minimum value for some y? Is the posterior distribution symmetric or a skewed one? What happens to the posterior distribution as $y \to \infty$? [Mitchell (1994)].

Exercise 2.7.51 Let X have a normal distribution with variance equal to 1 and with a random mean μ having the Laplace distribution $\mathcal{CL}(0, \eta)$ (the Laplace prior).

(a) Using the previous exercise, show that

$$E(\mu|X) = X - h(X)\eta,$$

where

$$h(x) = \frac{1 - e^{2cx}\psi(x)}{1 + e^{2cx}\psi(x)},$$

$$\psi(x) = \frac{\Phi(-x-c)}{\Phi(x-c)},$$

and Φ is the standard normal distribution function.

(b) Show that h is a monotonically increasing and odd function from $(-\infty, \infty)$ onto $(-1, 1)$ with $h(0) = 0$.

(c) A prior for μ is said to be neutral if the median of μ is 0 and the median of μ^2 is 1. Show that the above Laplace prior is neutral for $\eta = \log 2$.

(d) Show that the risk of $\hat{\mu}(X)$, defined as

$$E[(\hat{\mu}(X) - \mu)^2|\mu],$$

is a bounded function of μ.

Magnus (2000) refers to $\hat{\mu}(X) = X - h(X)\log 2$ as the *neutral Laplace estimator* of the mean μ. Further properties of $\hat{\mu}(X)$ can be found in the above paper.

Exercise 2.7.52 Let X_1, X_2, X_3 be i.i.d. logistic random variables with the distribution function

$$F(x) = (1 + e^{-x})^{-1}, \quad -\infty < x < \infty, \tag{2.7.34}$$

and let Y be a standard classical Laplace variable with p.d.f. (2.1.2).

(a) Show that

$$X_{2:3} + Y \stackrel{d}{=} X_1, \tag{2.7.35}$$

where $X_{2:3}$ is the second order statistic (the sample median) of the X_i's. The above result involving the Laplace distribution is actually a characterization of the logistic distribution [see George and Mudholkar (1981)]. If $Y \sim \mathcal{CL}(0, 1)$ and the relation (2.7.35) holds, then under some technical conditions on the distribution of X_1, the c.d.f. of X_1 is given by (2.7.34). George and Mudholkar (1981) provide an interesting interpretation of (2.7.35) utilizing the decomposition of the Laplace r.v. into a difference of two i.i.d. exponential variables W_1 and W_2: if adding and subtracting W_1 and W_2 to and from the median $X_{2:3}$ produces the distribution of X_1, then X_1 must have a logistic distribution.

(b) Under the above conditions, establish the relation

$$\frac{X_{1:3} + X_{3:3}}{2} + Y \stackrel{d}{=} X_1. \tag{2.7.36}$$

Deduce from (2.7.35) and (2.7.36) that for a random sample of size $n = 3$ from the standard logistic distribution the sample median has the same distribution as the midrange [George and Rousseau (1987)]. Investigate whether this property is actually a characterization of the logistic distribution.

(c) Generalize part (a) by showing that if $X_1, X_2, \ldots$ are i.i.d. with c.d.f. (2.7.34) and $Y_1, Y_2, \ldots$ are i.i.d. Laplace $\mathcal{CL}(0, 1)$ random variables, then

$$X_{k+1:2k+1} + \sum_{j=1}^{k} \frac{Y_j}{j} \stackrel{d}{=} X_1, \quad k \geq 1 \tag{2.7.37}$$

[George and Rousseau (1987)].

(d) Generalize part (b) by showing that under the conditions of part (c) we have

$$\frac{X_{1:2k+1} + X_{2k+1:2k+1}}{2} + \sum_{j=1}^{k} \frac{Y_j}{2j-1} \stackrel{d}{=} X_1, \quad k \geq 1. \tag{2.7.38}$$

Further, show that when the midrange is based on an even number of i.i.d. logistic random variables, then

$$\frac{X_{1:2k} + X_{2k:2k}}{2} + \frac{1}{2}\sum_{j=1}^{k-1} \frac{Y_j}{j} \stackrel{d}{=} \frac{X_1 + X_2}{2}, \quad k \geq 1 \tag{2.7.39}$$

[George and Rousseau (1987)].

Exercise 2.7.53 Let X_1 and X_2 be i.i.d. standard normal random variables, and let W be an exponential random variable with mean two and independent of X_1 and X_2. Then by Proposition 2.2.1, the r.v. $Y = \sqrt{W} X_2$ has the standard classical Laplace distribution $\mathcal{CL}(0, 1)$.

(a) Show that for any positive constants σ and η, the density of the r.v.

$$\sigma X_1 + \eta Y = \sigma X_1 + \eta \sqrt{W} X_2 \tag{2.7.40}$$

(which is the sum of zero mean and independent normal and Laplace variables) is given by

$$g(x) = \frac{1}{\eta} e^{\sigma^2/(2\eta^2)} \left[\frac{1}{2} e^{-x/\eta} \Phi\left(\frac{\eta x - \sigma^2}{\eta\sigma} \right) + \frac{1}{2} e^{x/\eta} \Phi\left(-\frac{\eta x + \sigma^2}{\eta\sigma} \right) \right],$$

where Φ is the distribution function of X_1 [Kou (2000)].

(b) Show that if (2.7.40) is divided by $\sqrt{\sigma^2 + \eta^2 W}$, then the resulting r.v.,

$$U_1 = \frac{\sigma X_1 + \eta\sqrt{W} X_2}{\sqrt{\sigma^2 + \eta^2 W}},$$

has the standard normal distribution. Further, show that this result remains valid for an arbitrary positive r.v. W [Sarabia (1993)].

(c) Generalize by showing that if X_1, X_2, and X_3 are i.i.d. standard normal r.v.'s and V is an arbitrary r.v., then the r.v.

$$U_2 = \frac{X_1 + V X_2 + V^2 X_3}{\sqrt{1 + V^2 + V^4}}$$

is standard normal [Sarabia (1993)]. Investigate an extension with more than three normal variables.

Exercise 2.7.54 Extend parts (b) and (c) of Exercise 2.7.53 by showing that if X_1, X_2, and X_3 are i.i.d. symmetric stable r.v.'s with ch.f. $\phi(t) = e^{-|t|^\alpha}$, where $0 < \alpha \le 2$, then the r.v.'s

$$U_{1,\alpha} = \frac{X_1 + VX_2}{(1 + V^\alpha)^{1/\alpha}}$$

and

$$U_{2,\alpha} = \frac{X_1 + VX_2 + V^\alpha X_3}{(1 + V^\alpha + V^{\alpha^2})^{1/\alpha}},$$

where V is an arbitrary nonnegative r.v. independent of the X_i's, have the same distribution as X_1 [Sarabia (1994)]. Investigate an extension where the number of X_i's is more than three.

Exercise 2.7.55 Let $Y_1, Y_2, \ldots$ be an i.i.d. sequence of $\mathcal{CL}(0, 1)$ random variables.

(a) Show that the r.v.

$$X = \sum_{j=1}^{\infty} \frac{Y_j}{j}$$

has the standard logistic distribution with c.d.f. (2.7.34) and ch.f.

$$\varphi_X(t) = t\pi \operatorname{cosech} \pi t$$

[see Pakes (1997) for further discussion and generalizations].

(b) Using the above representation deduce that the logistic distribution is infinitely divisible. *Hint*: Note the following infinite product representation of the hyperbolic cosecant function:

$$\operatorname{cosech}(z) = \frac{1}{z} \prod_{j=1}^{\infty} \left(1 + \frac{z^2}{j^2\pi^2}\right)^{-1}$$

[see, e.g., Abramowitz and Stegun (1965)].

(c) Using parts (a) and (d) of Exercise 2.7.52, deduce the limiting distribution of the logistic midrange $(X_{1:2k} + X_{2k:2k})/2$ as $k \to \infty$.

Exercise 2.7.56 Let $X_{1:n} \le \cdots \le X_{n:n}$ be the order statistics connected with a random sample from a uniform distribution on the interval $(-1, 1)$.

(a) Derive the joint distribution of the statistics

$$U_n = \frac{X_{n:n} - X_{1:n}}{2} \quad \text{and} \quad V_n = \frac{X_{n:n} + X_{1:n}}{2}.$$

(b) Show that the marginal p.d.f. of V_n is

$$g_n(x) = \frac{n}{2}(1 - |x|)^{n-1}, \quad |x| \le 1, \tag{2.7.41}$$

and that the variance of V_n is

$$\sigma_n^2 = \frac{2}{(n+1)(n+2)} \tag{2.7.42}$$

[see Neyman and Pearson (1928); Carlton (1946)].

(c) Show that as $n \to \infty$, the p.d.f. of the standardized variable $W_n = V_n/\sigma_n$, which is given by

$$\frac{1}{s_n} g_n\left(\frac{x}{s_n}\right) \tag{2.7.43}$$

with

$$s_n = \frac{1}{\sigma_n} = \sqrt{\frac{(n+1)(n+2)}{2}}, \tag{2.7.44}$$

converges to the standard Laplace density (2.1.4).

(d) Note that in part (c), the limit

$$\lim_{n\to\infty} \frac{s_n}{n} \tag{2.7.45}$$

is equal to $s = 1/\sqrt{2}$. Generalize part (c) by showing that if for a positive sequence $\{s_n\}$ the limit (2.7.45) is equal to s, where $0 < s < \infty$, then the p.d.f.'s (2.7.43) converge to the Laplace distribution with mean zero and scale parameter s with density (2.1.1) [Dreier (1999)]. What happens if the limit (2.7.45) is equal to zero? What if it is equal to ∞?

(e) Now let the sample be from the uniform distribution on the interval $(0, a)$ with some $a > 0$. By considering an appropriate linear transformation, derive the p.d.f. of V_n, show that V_n is unbiased for the population mean $a/2$, and find the variance of V_n. Further, show that the standardized random variable

$$W_n = \frac{V_n - E(V_n)}{\sqrt{\mathrm{Var}(V_n)}}$$

still converges in distribution to the standard Laplace distribution with density (2.1.4).

(f) Under the conditions of part (e), show that standardized sample mean,

$$Z_n = \frac{\overline{X}_n - E(\overline{X}_n)}{\sqrt{\mathrm{Var}(\overline{X}_n)}},$$

converges in distribution to the standard normal distribution. In view of these results, discuss the use of W_n and $\overline{X}_n$ as estimates of the mean of the uniform distribution on the interval $(0, a)$ with some $a > 0$ [Biswas and Sehgal (1991)].

Exercise 2.7.57 Let g_n be the density (2.7.41). Show that for every $x > 0$ there exists an $n_0 \in N$, such that

$$\left|\frac{1}{2} y e^{-yx} - \frac{y}{n} g_n\left(\frac{xy}{n}\right)\right| \le \frac{1}{2nx}$$

for all $n \ge n_0$ and all $y \ge 0$. Conclude that the convergence to the Laplace density,

$$\lim_{n\to\infty} \frac{y}{n} g_n\left(\frac{xy}{n}\right) = \frac{y}{2} e^{-y|x|}, \quad -\infty < x < \infty,$$

is uniform in y for every $x \ne 0$ [Dreier (1999)].

Exercise 2.7.58 Navarro and Ruiz (2000) define a discrete Laplace distribution by the probability function

$$f(k) = c(s)e^{|k-\theta|/s}, \quad k = 0, \pm 1, \pm 2, \ldots, \tag{2.7.46}$$

where θ is an integer, s is a positive real number, and $c(s)$ is a norming constant (the authors also mention a possible extension where θ is a real number and the support of the distribution is a countable set of real numbers).

(a) Show that in order for the function (2.7.46) to be a genuine probability function we must have

$$c(s) = \frac{1-e^{-1/s}}{1+e^{-1/s}}. \tag{2.7.47}$$

(b) Show that a r.v. Y with the probability function (2.7.46) admits the representation

$$Y \stackrel{d}{=} \theta + X_1 - X_2, \tag{2.7.48}$$

where X_1 and X_2 are i.i.d. geometric variables given by the probability function

$$P(X_1 = k) = (1-p)^k p, \quad k = 0, 1, 2, \ldots \tag{2.7.49}$$

with

$$p = 1 - e^{-1/s}. \tag{2.7.50}$$

(c) Show that if a geometric distribution (2.7.49) with p as in (2.7.50) is extended symmetrically to the set of negative integers, then we obtain the distribution (2.7.46) with $\theta = 0$. Thus, analogous to the Laplace case, we might call this distribution a *double geometric distribution*.

Exercise 2.7.59 If F is a distribution function with the corresponding cumulants κ_i, then the *Edgeworth expansion* of F is given by

$$\begin{aligned} F(x) = \Phi(x) &- \frac{\kappa_3}{6}(x^2-1)\phi(x) - \frac{\kappa_4}{24}(x^3-3x)\phi(x) \\ &- \frac{\kappa_3^2}{72}(x^5 - 10x^3 + 15x)\phi(x) + \cdots, \end{aligned}$$

where Φ and ϕ are the c.d.f. and the p.d.f. of the standard normal distribution [see, e.g., Kotz and Johnson (1982)].

(a) Let $X_1, \ldots, X_n$ be i.i.d. from the $\mathcal{CL}(\theta, s)$ distribution, and consider the standardized sample mean

$$T_n = \frac{1}{\sqrt{2}sn}\sum_{j=1}^{n}(X_j - \theta).$$

Show that the jth cumulant of T_n is given by

$$n^{1-j/2}(\sqrt{2}s)^{-j}\kappa_j,$$

where the κ_j is the jth cumulant of $X_1 - \theta$.

(b) Using the expression (2.1.13) for the cumulants of the Laplace distribution, derive the following (Edgeworth) approximation of the c.d.f. of T_n:

$$F_n(x) = \Phi(x) - \frac{1}{8n}\phi(x)(x^3 - 3x) + O(n^{-2})$$

[Pace and Salvan (1997)].

Exercise 2.7.60 Let $X_1, X_2, \ldots$ be i.i.d. standard Laplace $\mathcal{L}(0, 1)$ random variables. Then the sequence $\{X_n, n \geq 1\}$ obeys the *law of the iterated logarithm*,

$$\limsup_{n\to\infty} \frac{\sum_{k=1}^{n} X_k}{\sqrt{2n\log(\log n)}} = 1 \ \ a.s., \tag{2.7.51}$$

since (2.7.51) holds for any i.i.d. sequence of standardized random variables [see, e.g., Breiman (1993), Theorem 13.25]. Generalize (2.7.51) by showing that for any $\alpha \geq 0$ the sequence $\{X_n, n \geq 1\}$ satisfies

$$\limsup_{n\to\infty} \frac{\sum_{k=1}^{n} k^\alpha X_k}{n^\alpha\sqrt{2n\log(\log n)}} = \frac{1}{\sqrt{2\alpha+1}} \ \ a.s. \tag{2.7.52}$$

[Tomkins (1972)].

Hint: Denote $c_n = n^{-1/4}$ and show that for large n the double inequality

$$e^{t^2(1-c_n|t|)/(2n)} \leq E[e^{tX_k/\sqrt{n}}] \leq e^{t^2(1+c_n|t|/2)/(2n)} \tag{2.7.53}$$

holds for each positive integer $k \leq n$ and any t such that $|t| \leq 1/c_n$. Then use the fact that the condition (2.7.53) is sufficient for (2.7.52) [Tomkins (1972)].

Exercise 2.7.61 A random variable X on $[0, \infty)$ with the Laplace transform $\eta(s) = Ee^{-sX}$ is called a *generalized gamma convolution* (GGC) if

$$\eta(s) = \exp\left\{-as - \int_0^\infty \log\left(1 + \frac{2}{w}\right) dU(w)\right\}, \quad a \geq 0, \mathrm{Re}(s) \geq 0, \tag{2.7.54}$$

where U is a nonnegative measure on $(0, \infty)$ such that

$$\int_0^1 |\log w| dU(w) < \infty \ \text{ and } \ \int_1^\infty \frac{1}{w} dU(w) < \infty$$

[see, e.g., Bondesson (1992)].

(a) Show that the standard exponential distribution belongs to the class of GGC laws and the measure U is a unit mass at $u = 1$. Consequently, symmetric Laplace distributions, as well as their asymmetric and multivariate generalizations studied in this book, are mean-variance mixtures of normal laws by generalized gamma convolutions.

(b) Similarly, show that every gamma distributions is a GGC. What is the measure U in this case?

3

Asymmetric Laplace Distributions

Chapter 3 is devoted to asymmetric Laplace distributions — a skewed family of distributions that in our opinion is the most appropriate skewed generalization of the classical Laplace law. In the last several decades, various forms of skewed Laplace distributions have sporadically appeared in the literature. One of the earliest is due to McGill (1962), who considers distributions with p.d.f.

$$f(x) = \begin{cases} \frac{\phi_1}{2} e^{-\phi_1 |x-\theta|}, & x \le \theta, \\ \frac{\phi_2}{2} e^{-\phi_2 |x-\theta|}, & x > \theta, \end{cases} \tag{3.0.1}$$

while Holla and Bhattacharya (1968) study the distribution with p.d.f.

$$f(x) = \begin{cases} p\phi e^{-\phi |x-\theta|}, & x \le \theta, \\ (1-p)\phi e^{-\phi |x-\theta|}, & \theta < x, \end{cases} \tag{3.0.2}$$

where $0 < p < 1$. Lingappaiah (1988) derived some properties of (3.0.1), terming the distribution *two-piece double exponential*. Poiraud-Casanova and Thomas-Agnan (2000) exploited a skewed Laplace distribution with p.d.f.

$$f(x) = \alpha(1-\alpha) \begin{cases} e^{-(1-\alpha)|x-\theta|}, & \text{for } x < \theta, \\ e^{-\alpha |x-\theta|}, & \text{for } x \ge \theta, \end{cases} \tag{3.0.3}$$

where $\theta \in (-\infty, \infty)$ and $\alpha \in (0, 1)$, to show the equivalence of certain quantile estimators.

Azzalini (1985) noted that if X and Y are symmetric (about zero) and independent r.v.'s with densities f_X, f_Y and distribution functions F_X, F_Y, respectively, then for any λ,

$$\frac{1}{2} = P(X - \lambda Y < 0) = \int_{-\infty}^{\infty} f_Y(y) F_X(\lambda y) dy. \tag{3.0.4}$$

Consequently, the function

$$g(y) = 2 f_Y(y) F_X(\lambda y) \tag{3.0.5}$$

is a p.d.f. for any λ. If we take X and Y to be i.i.d. standard normal variables, then (3.0.5) gives the density of the skew-normal distribution, extensively studied since its introduction in O'Hagan and Leonhard (1976) mainly by Azzalini and associates [see Azzalini (1985, 1986), Henze (1986), Liseo (1990), Azzalini and Dalla Valle (1996), Azzalini and Capitanio (1999)]. Similarly, if X and Y are i.i.d. standard Laplace r.v.'s, utilizing (3.0.5) with $\lambda > 0$, we obtain a skewed Laplace distribution with density

$$g(x) = \begin{cases} \frac{1}{2}e^{(1+\lambda)x}, & -\infty < x \le 0, \\ e^{-x} - \frac{1}{2}e^{-(1+\lambda)x}, & 0 < x < \infty, \end{cases} \tag{3.0.6}$$

studied by Balakrishnan and Ambagaspitiya (1994) in an unpublished technical report.

Another manner of introducing skewness into a symmetric distribution has been proposed by Fernández and Steel (1998) [see also Fernández et al. (1995)]. Here the idea is to convert a symmetric p.d.f. into a skewed one by postulating inverse scale factors in the positive and negative orthants. Thus a symmetric density f generates the following class of skewed distributions, indexed by $\kappa > 0$,

$$f(x|\kappa) = \frac{2\kappa}{1+\kappa^2}\begin{cases} f(\kappa x), & x \ge 0, \\ f(\kappa^{-1}x), & x < 0. \end{cases} \tag{3.0.7}$$

When f is the standard classical Laplace density (2.1.2), then (3.0.7), with the addition of a location and scale parameters, leads to a three-parameter family with density

$$p(x) = \frac{1}{\sigma}\frac{\kappa}{1+\kappa^2}\begin{cases} \exp\left(-\frac{\kappa}{\sigma}(x-\theta)\right), & \text{for } x \ge \theta, \\ \exp\left(\frac{1}{\sigma\kappa}(x-\theta)\right), & \text{for } x < \theta, \end{cases} \tag{3.0.8}$$

introduced by Hinkley and Revankar (1977). These distributions, termed *asymmetric Laplace* (AL) laws by Kozubowski and Podgórski (2000), show promise in financial modeling (see Part III of the monograph devoted to applications and references therein). It is our opinion that members of this particular class deserve to be called *the* asymmetric Laplace (AL) distributions. There are at least three reasons why these laws warrant special treatment.

First, these distributions *arise naturally as limiting distributions* in a random summation scheme. Recall that symmetric Laplace laws are the only possible limiting distributions for (normalized) sums of i.i.d. symmetric random variables with a finite variance, when the number of terms in the summation has a geometric distribution with the mean converging to infinity (see Proposition 2.2.9). Similarly, if the assumption of symmetry of the summands is omitted, we obtain AL laws as the limiting distributions (see Proposition 3.4.4).

Second, the *AL laws extend naturally all the basic properties of symmetric Laplace distributions*.

- *Mixtures of normal distributions*. A classical symmetric Laplace r.v. may be viewed as a normal r.v. with mean zero and a stochastic variance (see Proposition 2.2.1). Analogously, an AL r.v. has a similar interpretation, where the mean of the normal distribution is now stochastic (see Proposition 3.2.1). This fact is of particular importance for application in finance where stochastic variance models are being used [see, e.g., Madan, et al. (1998), Levin and Tchernitser (1999)].

- *Stability with respect to geometric summation*. A symmetric Laplace r.v. Y has the same distribution as a (appropriately scaled) sum of a geometric number of i.i.d. copies of Y (see Proposition 2.2.7). More generally, we obtain a similar characterization of an AL r.v., when the equality of distributions is replaced by the weak convergence (see Proposition 3.4.5).

- *Distributions with maximal entropy*. As we saw in Proposition 2.4.7, among all continuous distributions on $(-\infty, \infty)$ with a given first absolute moment, the one with a maximal entropy is provided by a symmetric Laplace distribution. As we shall show in the present chapter, under an additional restriction on the value of the mean, the entropy is maximized by an AL law.

- *Convolution of exponential distributions.* A classical Laplace r.v. can be represented as a difference of two i.i.d. exponential random variables (see Proposition 2.2.2). If the two exponential r.v.'s are independent but no longer identically distributed, their difference has an AL law (see Proposition 3.2.2).

Finally, it is the properties and features of AL distributions that are similar in nature to these features of the normal distribution that make them particularly attractive in applications.

- *Infinite divisibility.* Variables appearing in many applications in various sciences can often be represented as sums of a large number of tiny variables, often independent and identically distributed. This is a practical interpretation of the notion of infinite divisibility Thus, when dealing with such a phenomenon, a "proper" model ought to be infinitely divisible. It is well known that all normal distributions are infinitely divisible, and so are the AL laws.

- *Limiting laws.* The normal distribution arises as a limit of a deterministic sum of i.i.d. random variables with a finite variance, where the number of terms in the summation tends to infinity. Consequently, if a variable of interest can be viewed as a result of a large number of independent increments (with a finite variance), then its distribution may be approximated by the normal law. Similarly, a *random* sum of i.i.d. random variables with finite variance converges to an AL r.v. when the *average* number of terms in the summation tends to infinity. Thus in practice we could use an AL approximation for a variable resulting from a random number (a geometric variable with a large mean) of independent innovations (with a finite variance).

- *Maximum entropy property.* The principle of maximum entropy, which states that out of all the distributions satisfying a given set of constraints one should choose the one with the largest entropy, is considered as general inference procedure and has been applied successfully in a wide variety of fields, including statistical mechanics, statistics, economics, queuing theory, and image analysis [see, e.g., Kapur (1993)]. Thus distributions maximizing the entropy under suitable constraints provide useful models in applications. It is well known that among all continuous distributions on $(-\infty, \infty)$ with a given mean and variance, the Gaussian (normal) distribution provides the largest entropy. Analogously, the entropy is maximized by the AL distribution, when the mean and the first absolute moment are specified (Proposition 3.4.7).

- *Finiteness of moments.* It is often argued that most variables appearing in applications should have finite moments of all orders (or at least the mean and the variance). This holds for the normal as well as for the AL laws.

- *Symmetry.* Probability distributions of variables arising in the real-world are often symmetric. The normal distribution is symmetric, and as such, it is often used as a model in practice. An AL distribution can also be symmetric (in which case it reduces to the classical Laplace distribution), but the AL model actually provides more flexibility, allowing for asymmetry.

- *Simplicity.* The distributions applied in practice ought to be handled easily. It is highly advantageous if their densities, distribution functions, and other characteristics allow for straightforward calculations, and estimation procedures should also be preferably implemented with ease. Ideally, the c.d.f. and the p.d.f. should have closed form expressions, which would substantially facilitate the derivation and implementation of estimation and simulation procedures. This is indeed the case with the normal distribution, although the distribution function here lacks an explicit form and requires a numerical approximation. We shall see that the corresponding formulas and procedures for the AL laws are at least as simple if not simpler than their normal counterparts.

- *Extensions.* An appropriate model should allow for various extensions, particularly to the multivariate setting. This is the case with both the normal and the AL laws. The multivariate extensions of a univariate AL law is quite natural (and are discussed in Part II of this text).

3.1 Definition and basic properties

A formal definition of the class of asymmetric Laplace distributions is as follows.

Definition 3.1.1 *A random variable Y is said to have an asymmetric Laplace (AL) distribution if there exist parameters $\theta \in \mathbb{R}$, $\mu \in \mathbb{R}$ and $\sigma \geq 0$ such that the characteristic function of Y has the form*

$$\psi(t) = \frac{e^{i\theta t}}{1 + \frac{1}{2}\sigma^2 t^2 - i\mu t}. \tag{3.1.1}$$

We denote the distribution of Y by $\mathcal{AL}(\theta, \mu, \sigma)$ and write $Y \sim \mathcal{AL}(\theta, \mu, \sigma)$.

Remark 3.1.1 Asymmetric Laplace laws with $\theta = 0$ constitute a subclass of GS distributions defined in Subsection 4.4.4. Namely,

$$\mathcal{AL}(0, \mu, \sigma) = GS_2(\sigma/\sqrt{2}, \beta, \mu), \quad \beta = \text{sign}(\mu), \tag{3.1.2}$$

where $GS_\alpha(\sigma, \beta, \mu)$ denotes the distribution given by ch.f. (4.4.7) (see Exercise 3.6.15).

3.1.1 An alternative parametrization and special cases. While the distribution is properly defined for every $\theta \in \mathbb{R}$, $\mu \in \mathbb{R}$, and $\sigma \geq 0$, we note specifically the following special cases:

- If $\theta = \mu = \sigma = 0$, then $\psi(t) = 1$ for every $t \in \mathbb{R}$ and the distribution is degenerate at 0.
- For $\theta = \sigma = 0$ and $\mu \neq 0$, we have an exponential r.v. with mean μ [concentrated on $(0, \infty)$ for $\mu > 0$ and on $(-\infty, 0)$ for $\mu < 0$].
- For $\mu = 0$ and $\sigma \neq 0$, we have a symmetric Laplace distribution with mean θ and variance σ^2.

The ch.f. (3.1.1) with $\sigma > 0$ can be expressed in the following manner:

$$\psi(t) = e^{i\theta t}\left(\frac{1}{1 + i\frac{\sigma\kappa}{\sqrt{2}}t}\right)\left(\frac{1}{1 - i\frac{\sigma}{\sqrt{2}\kappa}t}\right) = \frac{e^{i\theta t}}{1 + \frac{1}{2}\sigma^2 t^2 - i\frac{\sigma}{\sqrt{2}}\left(\frac{1}{\kappa} - \kappa\right)t}, \tag{3.1.3}$$

where the additional parameter $\kappa > 0$ is related to μ and σ is as follows:

$$\kappa = \frac{\sqrt{2}\sigma}{\mu + \sqrt{2\sigma^2 + \mu^2}} = \frac{\sqrt{2\sigma^2 + \mu^2} - \mu}{\sqrt{2}\sigma}, \tag{3.1.4}$$

while

$$\mu = \frac{\sigma}{\sqrt{2}}\left(\frac{1}{\kappa} - \kappa\right). \tag{3.1.5}$$

Note that for each fixed $\sigma > 0$, expression (3.1.4), considered as a function of μ and written $\kappa = \kappa(\mu)$, is decreasing on $(-\infty, \infty)$ with $\kappa(0) = 1$ and

$$\lim_{\mu \to -\infty} \kappa(\mu) = \infty, \qquad \lim_{\mu \to \infty} \kappa(\mu) = 0. \tag{3.1.6}$$

We use the abbreviation AL to denote all distributions with ch.f. given either by (3.1.1) or by (3.1.3), including those with $\mu = 0$ (symmetric ones) and $\sigma = 0$.

We find it convenient to express certain properties of the asymmetric Laplace distributions in the (θ, κ, σ) parametrization, using the notation $\mathcal{AL}^*(\theta, \kappa, \sigma)$ for the distribution given by (3.1.3). The parameter κ is scale invariant, so the random variables Y and cY have the same κ parameter whenever Y is $\mathcal{AL}^*(\theta, \sigma, \kappa)$ distributed and $c > 0$. Note also that in the (θ, σ, κ) parametrization, σ is a bona fide scale parameter.

The following relations are often used in what follows:

$$\frac{1}{\kappa} - \kappa = \frac{\sqrt{2}\mu}{\sigma}, \quad \frac{1}{\kappa} + \kappa = \sqrt{4 + \frac{2\mu^2}{\sigma^2}}, \quad \frac{1}{\kappa^2} + \kappa^2 = 2\left(\frac{\mu^2}{\sigma^2} + 1\right). \tag{3.1.7}$$

The result below follows easily from the form of the AL characteristic function.

Proposition 3.1.1 *Let $X \sim \mathcal{AL}^*(\theta, \kappa, \sigma)$ and let c be a nonzero real constant. Then*

(i) $c + X \sim \mathcal{AL}^*(c + \theta, \kappa, \sigma)$.

(ii) $cX \sim \mathcal{AL}^*(c\theta, \kappa_c, |c|\sigma)$, *where* $\kappa_c = \kappa^{sign(c)}$.

Remark 3.1.2 Note that in particular, if $X \sim \mathcal{AL}^*(\theta, \kappa, \sigma)$, then $-X \sim \mathcal{AL}^*(-\theta, 1/\kappa, \sigma)$.

3.1.2 Standardization. Since θ is simply a location parameter, we shall often assume $\theta = 0$. To simplify the notation in this case, we write $\mathcal{AL}(\mu, \sigma)$ and $\mathcal{AL}^*(\kappa, \sigma)$ for the distributions $\mathcal{AL}(0, \mu, \sigma)$ and $\mathcal{AL}^*(0, \kappa, \sigma)$, respectively. Further, for $\theta = 0$ and $\sigma = 1$ we say that the distribution is *standard*, and write $\mathcal{AL}(\mu)$ and $\mathcal{AL}^*(\kappa)$, respectively [for the distributions $\mathcal{AL}(0, \mu, 1)$ and $\mathcal{AL}^*(0, \kappa, 1)$].

Tables 3.1 and 3.2 below contain summary of our notation and the special cases in the two parametrizations.

3.1.3 Densities and their properties. Using the factorization (3.1.3), we can represent an asymmetric Laplace r.v. Y as follows:

$$Y \stackrel{d}{=} \theta + Y_1 - Y_2, \tag{3.1.8}$$

where the two variables on the right-hand side are independent and exponentially distributed with means $\sigma/(\sqrt{2}\kappa)$ and $\sigma\kappa/\sqrt{2}$, respectively. Equivalently, we have

$$Y \stackrel{d}{=} \theta + \frac{\sigma}{\sqrt{2}}\left(\frac{1}{\kappa}W_1 - \kappa W_2\right), \tag{3.1.9}$$

where W_1 and W_2 are two i.i.d. standard exponential random variables. This representation leads to explicit formulas for the corresponding density and distribution function [cf. formula (2.3.16) and the computations preceding it].

Proposition 3.1.2 *Let $f_{\theta,\kappa,\sigma}$ and $F_{\theta,\kappa,\sigma}$ denote the p.d.f. and c.d.f. of an $\mathcal{AL}^*(\theta, \kappa, \sigma)$ distribution, respectively. Then*

$$f_{\theta,\kappa,\sigma}(x) = \frac{\sqrt{2}}{\sigma}\frac{\kappa}{1+\kappa^2}\begin{cases} \exp\left(-\frac{\sqrt{2}\kappa}{\sigma}|x-\theta|\right), & \text{if } x \ge \theta \\ \exp\left(-\frac{\sqrt{2}}{\sigma\kappa}|x-\theta|\right), & \text{if } x < \theta, \end{cases} \tag{3.1.10}$$

and

Case	Name	Notation	Char. funct.
$\theta \in \mathbb{R}$ $\sigma \geq 0$ $\mu \in \mathbb{R}$	Asymm. Laplace	$\mathcal{AL}(\theta, \mu, \sigma)$	$\frac{e^{i\theta t}}{1+\frac{1}{2}\sigma^2 t^2 - i\mu t}$
$\theta = 0$ $\sigma \geq 0$ $\mu \in \mathbb{R}$	Asymm. Laplace	$\mathcal{AL}(0, \mu, \sigma)$, $\mathcal{AL}(\mu, \sigma)$	$\frac{1}{1+\frac{1}{2}\sigma^2 t^2 - i\mu t}$
$\theta \in \mathbb{R}$ $\sigma \geq 0$ $\mu = 0$	Symm. Laplace	$\mathcal{AL}(\theta, 0, \sigma)$, $\mathcal{L}(\theta, \sigma)$	$\frac{e^{i\theta t}}{1+\frac{1}{2}\sigma^2 t^2}$
$\theta = 0$ $\sigma = 1$ $\mu \in \mathbb{R}$	Standard AL	$\mathcal{AL}(0, \mu, 1)$, $\mathcal{AL}(\mu)$	$\frac{1}{1+\frac{1}{2}t^2 - i\mu t}$
$\theta = 0$ $\sigma = 0$ $\mu \neq 0$	Exponential	$\mathcal{AL}(0, \mu, 0)$, $\mathcal{E}(\mu)$	$\frac{1}{1-i\mu t}$
$\theta \in \mathbb{R}$ $\sigma = 0$ $\mu = 0$	Degenerated		$e^{i\theta t}$

Table 3.1: Special cases and notation for an asymmetric Laplace distribution in the $\mathcal{AL}(\theta, \mu, \sigma)$ parametrization.

$$F_{\theta,\kappa,\sigma}(x) = \begin{cases} 1 - \frac{1}{1+\kappa^2} \exp\left(-\frac{\sqrt{2}\kappa}{\sigma}|x-\theta|\right), & \text{if } x \geq \theta \\ \frac{\kappa^2}{1+\kappa^2} \exp\left(-\frac{\sqrt{2}}{\sigma\kappa}|x-\theta|\right), & \text{if } x < \theta. \end{cases} \tag{3.1.11}$$

Figure 3.1 shows AL densities for various values of the parameters.

Remark 3.1.3 Note that for $\kappa = 1$ we obtain the p.d.f. and the c.d.f. of the symmetric Laplace distribution.

Remark 3.1.4 To obtain expressions of the AL p.d.f. and c.d.f. in the $\mathcal{AL}(\theta, \mu, \sigma)$ parametrization, substitute in (3.1.10)–(3.1.11) the expression for κ given by (3.1.4).

Remark 3.1.5 If Y is an AL random variable given by (3.1.10)–(3.1.11), then

$$P(Y \leq \theta) = F_{\theta,\kappa,\sigma}(\theta) = \frac{\kappa^2}{1+\kappa^2} = q_\kappa \tag{3.1.12}$$

and

$$P(Y > \theta) = 1 - F_{\theta,\kappa,\sigma}(\theta) = \frac{1}{1+\kappa^2} = p_\kappa. \tag{3.1.13}$$

Case	Name	Notation	Char. funct.
$\theta \in \mathbb{R}$ $\sigma \geq 0$ $\kappa > 0$	Asymm. Laplace	$\mathcal{A}L^*(\theta, \kappa, \sigma)$	$\frac{e^{i\theta t}}{1+\frac{1}{2}\sigma^2 t^2 - i\frac{\sigma}{\sqrt{2}}(\frac{1}{\kappa}-\kappa)t}$
$\theta = 0$ $\sigma \geq 0$ $\kappa > 0$	Asymm. Laplace	$\mathcal{AL}^*(0, \kappa, \sigma)$, $\mathcal{AL}^*(\kappa, \sigma)$	$\frac{1}{1+\frac{1}{2}\sigma^2 t^2 - i\frac{\sigma}{\sqrt{2}}(\frac{1}{\kappa}-\kappa)t}$
$\theta \in \mathbb{R}$ $\sigma \geq 0$ $\kappa = 1$	Symm. Laplace	$\mathcal{AL}^*(\theta, 1, \sigma)$, $\mathcal{L}(\theta, \sigma)$	$\frac{e^{i\theta t}}{1+\frac{1}{2}\sigma^2 t^2}$
$\theta = 0$ $\sigma = 1$ $\kappa > 0$	Standard AL	$\mathcal{AL}^*(0, \kappa, 1)$, $\mathcal{AL}^*(\kappa)$	$\frac{1}{1+\frac{1}{2}t^2 - i\frac{1}{\sqrt{2}}(\frac{1}{\kappa}-\kappa)t}$
$\theta \in \mathbb{R}$ $\sigma = 0$ $\kappa = 1$	Degenerated		$e^{i\theta t}$

Table 3.2: Special cases and notation for an asymmetric Laplace distribution in the $\mathcal{AL}^*(\theta, \kappa, \sigma)$ parametrization.

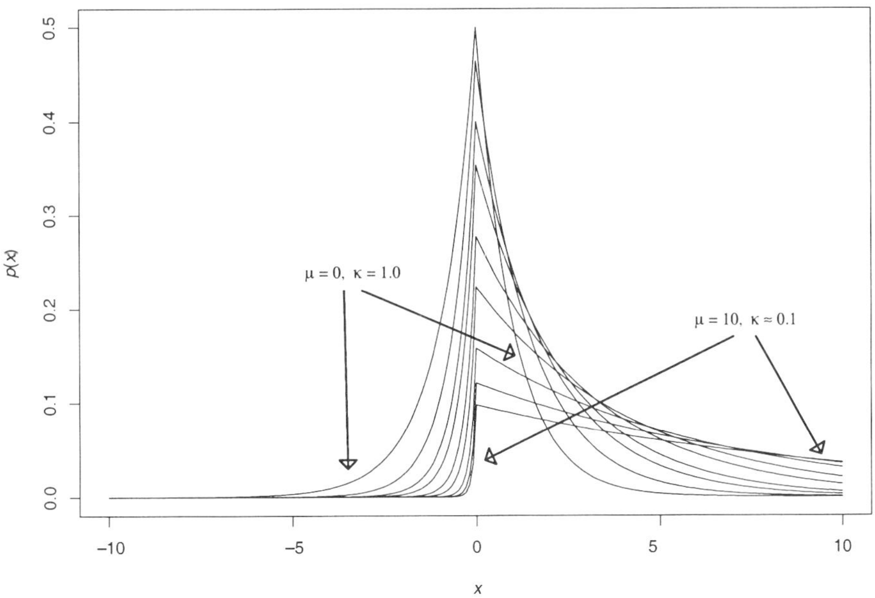

Figure 3.1: Asymmetric Laplace densities with $\sigma = \sqrt{2}$ and $\mu = 0, 0.8, 1.5, 2, 3, 4, 6, 8, 10$ that correspond to $\kappa \approx 1.0, 0.68, 0.50, 0.41, 0.30, 0.24, 0.16, 0.12, 0.1$.

Consequently, the parameter κ controls the probability assigned to each side of θ. Clearly, for $\kappa = 1$, the two probabilities are equal and the distribution is symmetric about θ.

Remark 3.1.6 Our skewed Laplace distribution with density (3.1.10), defined by its characteristic function, may be obtained formally by following a general procedure of obtaining a skewed distribution from a symmetric one, proposed by Fernández and Steel (1998). Let f be any p.d.f. which is unimodal (say about zero) and symmetric. The method of transforming the symmetric distribution given by f into a skewed one consists of introducing inverse scale factors for the positive and negative parts of the distribution, leading to density (3.0.7) discussed in the introduction. The Laplace distribution demonstrates that such distributions may appear quite naturally.

Remark 3.1.7 Every AL density can be written as a mixture of two exponential densities with means $\mu_1 = \sigma/(\kappa\sqrt{2})$ and $\mu_2 = -\sigma\kappa/\sqrt{2}$,

$$f_{\theta,\kappa,\sigma}(x) = p_\kappa \frac{1}{\mu_1} e^{-|x-\theta|/\mu_1} \mathbb{I}_{[\theta,\infty)}(x) + q_\kappa \frac{1}{|\mu_2|} e^{(x-\theta)/|\mu_2|} \mathbb{I}_{(-\infty,\theta)}(x), \tag{3.1.14}$$

with q_κ and p_κ defined by (3.1.12) and (3.1.13), respectively ($\mathbb{I}_A(x)$ is the indicator function equal to 1 if x belongs to the set A and equal to zero otherwise).

Remark 3.1.8 Since the AL density is increasing on $(-\infty, \theta)$ and decreasing on (θ, ∞), the distribution is unimodal with the mode equal to θ. The value of the density at the mode is

$$f_{\theta,\kappa,\sigma}(\theta) = \frac{\sqrt{2}}{\sigma} \frac{\kappa}{1+\kappa^2}$$

in the $\mathcal{AL}^*(\theta, \kappa, \sigma)$ parametrization and

$$f_{\theta,\mu,\sigma}(\theta) = \frac{1}{\sqrt{\mu^2 + 2\sigma^2}}$$

in the $\mathcal{AL}(\theta, \mu, \sigma)$ parametrization. This value can be located anywhere in the interval $(0, \infty)$. Further, we have

$$\lim_{\mu\to 0} f_{\theta,\mu,\sigma}(\theta) = \frac{1}{\sqrt{2}\sigma}, \quad \lim_{\sigma\to 0^+} f_{\theta,\mu,\sigma}(\theta) = \frac{1}{|\mu|}, \quad \lim_{\mu,\sigma\to 0} f_{\theta,\mu,\sigma}(\theta) = \infty. \tag{3.1.15}$$

Further properties of AL densities are discussed in the exercises.

3.1.4 Moment and cumulant generating functions. We can obtain the moment generating function of an AL distribution either by a straightforward integration utilizing the AL density (3.1.10) or from the representation (3.1.9).

Proposition 3.1.3 *If $Y \sim \mathcal{AL}^*(\theta, \kappa, \sigma)$, then the moment generating function of Y is*

$$M_{\theta,\kappa,\sigma}(t) = E[e^{tY}] = \frac{e^{\theta t}}{1 - \frac{1}{2}\sigma^2 t^2 - \frac{\sigma}{\sqrt{2}}\left(\frac{1}{\kappa} - \kappa\right)t}, \quad -\frac{\sqrt{2}}{\sigma\kappa} < t < \frac{\sqrt{2}\kappa}{\sigma}. \tag{3.1.16}$$

Proof. By the representation (3.1.9) we have

$$M_{\theta,\kappa,\sigma}(t) = E[e^{tY}] = e^{\theta t} E[e^{\frac{\sigma}{\sqrt{2}}\frac{1}{\kappa}tW_1}] E[e^{-\frac{\sigma}{\sqrt{2}}\kappa t W_2}],$$

where W_1 and W_2 are i.i.d. standard exponential variables with moment generating function

$$M_{W_i}(s) = E[e^{sW_i}] = \frac{1}{1-s}, \qquad s < 1.$$

Thus we have

$$M_{\theta,\kappa,\sigma}(t) = \frac{e^{\theta t}}{(1 - \frac{\sigma}{\sqrt{2}}\frac{1}{\kappa}t)(1 + \frac{\sigma}{\sqrt{2}}\kappa t)}, \tag{3.1.17}$$

where we must have

$$\frac{t\sigma}{\sqrt{2}\kappa} < 1 \text{ and } -\frac{t\sigma\kappa}{\sqrt{2}} < 1. \tag{3.1.18}$$

Now (3.1.17) and (3.1.18) produce (3.1.16), concluding the proof. □

Remark 3.1.9 In the $\mathcal{AL}(\theta, \mu, \sigma)$ parametrization the moment generating function is

$$M_{\theta,\mu,\sigma}(t) = \frac{e^{\theta t}}{1 - \frac{1}{2}\sigma^2 t^2 - \mu t}, \qquad -\frac{2}{\sqrt{2\sigma^2 + \mu^2} - \mu} < t < \frac{2}{\sqrt{2\sigma^2 + \mu^2} + \mu}. \tag{3.1.19}$$

In case $\mu = 0$ we obtain the moment generating function (2.1.10) of the classical Laplace distribution $\mathcal{CL}(\theta, s)$ with $s = \sigma/\sqrt{2}$ (the $\mathcal{L}(\theta, \sigma)$ distribution).

By Proposition 3.1.3 we can now write the cumulant generating function, $\log M_{\theta,\kappa,\sigma}(t)$, corresponding to the $\mathcal{AL}^*(\theta, \kappa, \sigma)$ distribution:

$$\log M_{\theta,\kappa,\sigma}(t) = \theta t - \log\left(1 - \frac{\sigma}{\sqrt{2}}\frac{1}{\kappa}t\right) - \log\left(1 + \frac{\sigma}{\sqrt{2}}\kappa t\right), \qquad -\frac{\sqrt{2}}{\sigma\kappa} < t < \frac{\sqrt{2}\kappa}{\sigma}. \tag{3.1.20}$$

Note that in the symmetric case ($\kappa = 1$) we obtain the cumulant generating function (2.1.11) of the classical Laplace distribution $\mathcal{CL}(\theta, s)$ with $s = \sigma/\sqrt{2}$.

3.1.5 Moments and related parameters.

3.1.5.1 *Cumulants.* The cumulants of a general $\mathcal{AL}^*(\theta, \kappa, \sigma)$ r.v. Y are the coefficients of $t^n/n!$ in the Taylor series (about $t = 0$) of the corresponding cumulant generating function (3.1.20). Thus the nth cumulant κ_n is equal to the nth derivative of the cumulant generating function at $t = 0$. The calculation of the derivatives is straightforward. For $n = 1$ we have

$$\frac{d}{dt}\log M_{\theta,\kappa,\sigma}(t) = \theta + \frac{\sigma}{\sqrt{2}}\left\{\frac{1/\kappa}{1 - \frac{\sigma}{\sqrt{2}}\frac{1}{\kappa}t} - \frac{\kappa}{1 + \frac{\sigma}{\sqrt{2}}\kappa t}\right\}, \tag{3.1.21}$$

while for $n > 1$ we obtain

$$\frac{d^n}{dt^n}\log M_{\theta,\kappa,\sigma}(t) = (n-1)!\left(\frac{\sigma}{\sqrt{2}}\right)^n \left\{\left(\frac{1/\kappa}{1 - \frac{\sigma}{\sqrt{2}}\frac{1}{\kappa}t}\right)^n + \left(\frac{-\kappa}{1 + \frac{\sigma}{\sqrt{2}}\kappa t}\right)^n\right\}. \tag{3.1.22}$$

Now substituting $t = 0$ into (3.1.21) and (3.1.22), we obtain the following expressions for the cumulants of an $\mathcal{AL}^*(\theta, \kappa, \sigma)$ r.v. Y:

$$\kappa_n(Y) = \begin{cases} \theta + \frac{\sigma}{\sqrt{2}}(\kappa^{-1} - \kappa) & \text{if } n = 1, \\ (n-1)!\left(\frac{\sigma}{\sqrt{2}}\right)^n(\kappa^{-n} - \kappa^n) & \text{if } n > 1 \text{ is odd}, \\ (n-1)!\left(\frac{\sigma}{\sqrt{2}}\right)^n(\kappa^{-n} + \kappa^n) & \text{if } n \text{ is even}. \end{cases} \tag{3.1.23}$$

Note that in the symmetric case ($\kappa = 1$) the cumulants of odd order greater than one vanish, and we obtain cumulants (2.1.13) of the classical Laplace distribution $\mathcal{CL}(\theta, s)$ with $s = \sigma/\sqrt{2}$. Observe also that the mean and variance of Y, which coincide with the first and second cumulants, respectively, are

$$E[Y] = \theta + \frac{\sigma}{\sqrt{2}}\left(\frac{1}{\kappa} - \kappa\right) = \theta + \mu, \qquad \mathrm{Var}[Y] = \frac{\sigma^2}{2}\left(\frac{1}{\kappa^2} + \kappa^2\right) = \mu^2 + \sigma^2. \tag{3.1.24}$$

3.1.5.2 *Moments.* Let $Y \sim \mathcal{AL}^*(\theta, \kappa, \sigma)$. For any integer $n > 0$, the nth moment of Y about θ, $E(Y - \theta)^n$, is

$$\int_{-\infty}^{\theta} (y-\theta)^n \frac{\sqrt{2}}{\sigma}\frac{\kappa}{1+\kappa^2} e^{\frac{\sqrt{2}}{\sigma\kappa}(y-\theta)} dy + \int_{\theta}^{\infty} (y-\theta)^n \frac{\sqrt{2}}{\sigma}\frac{\kappa}{1+\kappa^2} e^{\frac{\sqrt{2}\kappa}{\sigma}(\theta-y)} dy.$$

The substitution of $x = \theta - y$ in the first integral and $x = y - \theta$ in the second integral leads to

$$\frac{(-1)^n\kappa^2}{1+\kappa^2}\int_0^{\infty} x^n \frac{\sqrt{2}}{\sigma\kappa} e^{-\frac{\sqrt{2}}{\sigma\kappa}x} dx + \frac{1}{1+\kappa^2}\int_0^{\infty} x^n \frac{\sqrt{2}\kappa}{\sigma} e^{-\frac{\sqrt{2}\kappa}{\sigma}x} dx.$$

Thus

$$E(Y-\theta)^n = n!\left(\frac{\sigma}{\sqrt{2}\kappa}\right)^n \frac{1 + (-1)^n\kappa^{2(n+1)}}{1+\kappa^2}, \tag{3.1.25}$$

since for any $u > 0$ and $a > -1$ we have

$$\int_0^{\infty} x^a u e^{-ux} dx = \frac{1}{u^a}\int_0^{\infty} x^a e^{-x} dx = \frac{\Gamma(a+1)}{u^a}.$$

In the symmetric case ($\kappa = 1$) we obtain the moments (2.1.14) of the classical Laplace distribution with $s = \sigma/\sqrt{2}$.

3.1.5.3 *Absolute moments.* To obtain absolute moments of an AL distribution, we follow essentially the calculation leading to the moment formula (3.1.25), obtaining

$$E[|Y-\theta|^a] = \left(\frac{\sigma}{\sqrt{2}\kappa}\right)^a \Gamma(a+1)\frac{1+\kappa^{2(a+1)}}{1+\kappa^2}, \quad a > -1. \tag{3.1.26}$$

3.1.5.4 *Mean deviation.* Let Y have an $\mathcal{AL}^*(\theta, \kappa, \sigma)$ distribution with density $f_{\theta,\kappa,\sigma}$ given by (3.1.10). Then by (3.1.24), the mean deviation of Y is

$$E|Y - E[Y]| = \int_{-\infty}^{\infty} |y - \theta - \frac{\sigma}{\sqrt{2}}(1/\kappa - \kappa)| f_{\theta,\kappa,\sigma}(y) dy.$$

After a straightforward but tedious integration, we obtain

$$E|Y - E[Y]| = \frac{2\sigma}{\kappa(1+\kappa^2)} e^{(\kappa^2-1)}, \tag{3.1.27}$$

which equals $\sigma/\sqrt{2}$ for the symmetric case with $\mu = 0$; cf. (2.1.19). Further, since the standard deviation of Y is

$$\sqrt{\mathrm{Var}(Y)} = \sqrt{\sigma^2 + \frac{\sigma^2}{2}\left(\frac{1}{\kappa} - \kappa\right)^2} = \frac{\sigma\sqrt{1+\kappa^4}}{\sqrt{2}\kappa},$$

we have

$$\frac{\text{mean deviation}}{\text{standard deviation}} = \frac{2e^{\kappa^2-1}}{(1+\kappa^2)\sqrt{1+\kappa^4}}.$$

For the symmetric Laplace distribution ($\kappa = 1$), the above ratio is equal to $1/\sqrt{2}$, as previously derived in (2.1.20).

3.1.5.5 *Coefficient of Variation.* For an r.v. X with the mean not equal to zero, the *coefficient of variation* is defined as

$$\frac{\sqrt{\text{Var}(X)}}{|EX|}.$$

For $Y \sim \mathcal{AL}(\theta, \mu, \sigma)$ with $\theta \neq -\mu$, the mean of Y is nonzero and the coefficient of variation is equal to

$$\frac{\sqrt{\mu^2+\sigma^2}}{|\theta+\mu|}. \tag{3.1.28}$$

For $\theta = 0$ and $\mu \neq 0$, we obtain

$$\sqrt{\frac{\sigma^2}{\mu^2}+1} = \frac{\sqrt{1/\kappa^2+\kappa^2}}{1/\kappa-\kappa}. \tag{3.1.29}$$

Note that in this case the absolute value of the mean is less than or equal to the standard deviation, and thus the coefficient of variation is always greater than or equal to one.

3.1.5.6 *Coefficients of skewness and kurtosis.* The coefficient of skewness, defined in (2.1.21), is a measure of symmetry that is independent of scale. For the symmetric Laplace distribution its value is zero, as it is for any symmetric distribution with finite third moment and standard deviation greater than zero. For an $\mathcal{AL}^*(\theta, \kappa, \sigma)$ distribution, the coefficient of skewness is nonzero, unless $\kappa = 1$ ($\mu = 0$). In terms of κ, its value is

$$\gamma_1 = 2\frac{1/\kappa^3-\kappa^3}{(1/\kappa^2+\kappa^2)^{3/2}}. \tag{3.1.30}$$

It follows from (3.1.30) that the absolute value of γ_1 is bounded by two, and as κ increases within the interval $(0, \infty)$, then the corresponding value of γ_1 decreases monotonically from to 2 to -2.

Let us now study the peakedness of AL distributions. We saw in Section 2.1 that a symmetric Laplace distribution is leptokurtic, as its coefficient of kurtosis (adjusted), defined in (2.1.22), is equal to three. For an $\mathcal{AL}^*(\theta, \kappa, \sigma)$ distribution, we have

$$\gamma_2 = 6 - \frac{12}{(1/\kappa^2+\kappa^2)^2}. \tag{3.1.31}$$

Thus the distribution is leptokurtic and γ_2 varies from 3 [the least value for the symmetric Laplace distribution with $\kappa = 1$, see (2.1.23)] to 6 (the greatest value attained for the limiting exponential distribution when $\kappa \to 0$).

3.1.5.7 *Quantiles.* Since the distribution function of an asymmetric Laplace distribution is given in closed form, calculation of quantiles, including the median, is quite straightforward. Let ξ_q be the qth quantile of an AL r.v. with distribution function given by (3.1.11). Then we have

$$\xi_q = \begin{cases} \theta + \frac{\sigma\kappa}{\sqrt{2}}\log\left\{\frac{1+\kappa^2}{\kappa^2}q\right\} & \text{for } q \in (0, \frac{\kappa^2}{1+\kappa^2}], \\ \theta - \frac{\sigma}{\sqrt{2}\kappa}\log\{(1+\kappa^2)(1-q)\} & \text{for } q \in (\frac{\kappa^2}{1+\kappa^2}, 1). \end{cases} \tag{3.1.32}$$

Note that for $\kappa = 1$ we obtain the quantiles (2.1.24) of the symmetric Laplace distribution. Setting $q = 1/2$, we obtain the median m:

$$m = \xi_{1/2} = \begin{cases} \theta + \frac{\sigma}{\sqrt{2}\kappa} \log\left\{\frac{2}{1+\kappa^2}\right\} & \text{for } \kappa \le 1, \\ \theta - \frac{\sigma\kappa}{\sqrt{2}} \log\left\{\frac{2\kappa^2}{1+\kappa^2}\right\} & \text{for } \kappa > 1. \end{cases} \tag{3.1.33}$$

By setting $q = 1/4$ and $q = 3/4$ we obtain the first and third quartiles, Q_1 and Q_3, as well as the interquartile range, equal to

$$Q_3 - Q_1 = \begin{cases} \frac{\sigma \log 3}{\sqrt{2}\kappa} & \text{for } \kappa \le 1/\sqrt{3}, \\ \frac{\sigma}{\sqrt{2}\kappa} \log\left\{\frac{4}{1+\kappa^2}\right\} - \frac{\sigma\kappa}{\sqrt{2}} \log\left\{\frac{1+\kappa^2}{4\kappa^2}\right\} & \text{for } 1/\sqrt{3} < \kappa < \sqrt{3}, \\ \frac{\sigma\kappa \log 3}{\sqrt{2}} & \text{for } \kappa \ge \sqrt{3}. \end{cases} \tag{3.1.34}$$

In particular, we have

$$Q_1 = \theta \text{ and } Q_3 = \theta + \sigma\sqrt{\frac{3}{2}} \log 3 \text{ for } \kappa = \frac{1}{\sqrt{3}}$$

and

$$Q_1 = \theta - \sigma\sqrt{\frac{3}{2}} \log 3 \text{ and } Q_3 = \theta \text{ for } \kappa = \sqrt{3}.$$

Remark 3.1.10 If $\kappa = 1$ ($\mu = 0$), the relation (3.1.33) yields $m = \theta$, which is the median of a symmetric Laplace distribution. Similarly, for $\sigma = \theta = 0$, we get $m = \mu \log 2$, which is the median of an exponential distribution with mean μ (to which the asymmetric Laplace law is simplified in this case).

Remark 3.1.11 One can show that for $\kappa \ne 1$, the mode, median, and mean of an AL distribution satisfy the following inequalities:

$$\begin{array}{ll} \text{If } \kappa < 1 & \text{then } Mode < Median < Mean, \\ \text{If } \kappa > 1 & \text{then } Mode > Median > Mean. \end{array} \tag{3.1.35}$$

All three measures of location are equal to θ when $\kappa = 1$ ($\mu = 0$), in which case we obtain the symmetric Laplace distribution.

In Table 3.3 below we summarize the moments and related parameters of AL r.v.'s.

3.2 Representations

In this section we present the representations and characterizations of AL distributions that are generalizations of the corresponding properties of the symmetric Laplace distributions, as presented in Section 2.2.

3.2.1 Mixture of normal distributions. A symmetric Laplace r.v. can be regarded (informally) as a normal r.v. with mean zero and variance that is an exponentially distributed random variable (see Proposition 2.2.1). AL r.v.'s admit a similar interpretation, where the mean is a random variable as well. We state it more formally in the following result.

Parameter	Definition	Value
Absolute moment	$E\|Y\|^a, a > -1$	$\left(\frac{\sigma}{\sqrt{2}\kappa}\right)^a \Gamma(a+1)\frac{1+\kappa^{2(a+1)}}{1+\kappa^2}$
nth moment	EY^n	$n!\left(\frac{\sigma}{\sqrt{2}\kappa}\right)^n \frac{1+(-1)^n\kappa^{2(n+1)}}{1+\kappa^2}$
nth cumulant	$\frac{d^n}{dt^n}\log M_{\theta,\kappa,\sigma}(t)\Big\|_{t=0}$	$(n-1)!\left(\frac{\sigma}{\sqrt{2}\kappa}\right)^n(1+(-1)^n\kappa^{2n})$
Mean	EY	$\frac{\sigma}{\sqrt{2}}(\frac{1}{\kappa}-\kappa)=\mu$
Variance	$E(Y-EY)^2$	$\mu^2+\sigma^2$
Mean deviation	$E\|Y-EY\|$	$\frac{\sqrt{2}\sigma e^{(\kappa^2-1)}}{\kappa(1+\kappa^2)}$
Coefficient of Variation	$\frac{\sqrt{\mathrm{Var}(Y)}}{\|EY\|}$	$\sqrt{\frac{\sigma^2}{\mu^2}+1}=\frac{\sqrt{1/\kappa^2+\kappa^2}}{1/\kappa-\kappa}$
Coeffcient of Skewness	$\gamma_1=\frac{E(Y-EY)^3}{(E(Y-EY)^2)^{3/2}}$	$2\frac{1/\kappa^3-\kappa^3}{(1/\kappa^2+\kappa^2)^{3/2}}$
Kurtosis (adjusted)	$\gamma_2=\frac{E(Y-EY)^4}{(\mathrm{Var}(Y))^2}-3$	$6-\frac{12}{(1/\kappa^2+\kappa^2)^2}$
Median	$m=F^{-1}_{0,\kappa,\sigma}(1/2)$	$\begin{cases} -\frac{\sigma}{\sqrt{2}\kappa}\log\frac{1+\kappa^2}{2}, & \kappa\le 1 \\ \frac{\sigma\kappa}{\sqrt{2}}\log\frac{1+\kappa^2}{2\kappa^2}, & \kappa>1\end{cases}$

Table 3.3: Moments and related parameters of $Y \sim \mathcal{AL}^*(\theta,\kappa,\sigma)$ with $\theta=0$.

Proposition 3.2.1 *An $\mathcal{AL}(\theta,\mu,\sigma)$ random variable Y with ch.f.* (3.1.1) *admits the representation*

$$Y \stackrel{d}{=} \theta+\mu W+\sigma\sqrt{W}Z, \tag{3.2.1}$$

where Z is standard normal and W is standard exponential.

Proof. Let W have an exponential distribution with p.d.f. e^{-w}. Conditioning on W, we can express the ch.f. of the right-hand side of (3.2.1) as follows:

$$E[e^{it(\theta+\mu W+\sigma\sqrt{W}Z)}]=\int_0^\infty e^{it\theta+it\mu w}E[e^{it\sigma\sqrt{w}Z}]e^{-w}dw.$$

Note that

$$E[e^{it\sigma\sqrt{w}Z}]=\phi_Z(t\sigma\sqrt{w})=e^{-\frac{1}{2}t^2\sigma^2 w},$$

where $\phi_Z(s)=e^{-s^2/2}$ is the ch.f. of a standard normal r.v. Z. Thus

$$E[e^{it(\theta+\mu W+\sigma\sqrt{W}Z)}]=\int_0^\infty e^{it\theta}e^{-w(1+\frac{1}{2}t^2\sigma^2-i\mu t)}dw,$$

which produces the ch.f. (3.1.1) and the result follows. □

Note that in the symmetric case ($\mu = 0$) we obtain the representation of the classical Laplace r.v. discussed in Proposition 2.2.1 and the remarks following it.

3.2.2 Convolution of exponential distributions. We now formally state representation (3.1.9) in the following result.

Proposition 3.2.2 *An $\mathcal{AL}^*(\theta, \kappa, \sigma)$ random variable Y with ch.f.* (3.1.3) *admits representation* (3.1.9), *where W_1 and W_2 are i.i.d. standard exponential random variables.*

Note that for $\kappa = 1$ we obtain the representation of the classical Laplace distribution discussed in Proposition 2.2.2 and the remarks following it.

Remark 3.2.1 Denoting $H_i = 2W_i, i = 1, 2$, we have

$$Y \stackrel{d}{=} \theta + \frac{\sigma}{2\sqrt{2}}\left(\frac{1}{\kappa}H_1 - \kappa H_2\right), \tag{3.2.2}$$

where H_1 and H_2 are i.i.d. chi-square r.v.'s with two degrees of freedom.

Remark 3.2.2 Since a standard exponential r.v. W has the same distribution as $-\log(U)$, where U is standard uniform variable, we have the following representation of Y in terms of two i.i.d. standard uniform variables U_1 and U_2:

$$Y \stackrel{d}{=} \theta + \frac{\sigma}{\sqrt{2}}\log\left(\frac{U_1^{\kappa}}{U_2^{1/\kappa}}\right). \tag{3.2.3}$$

It generalizes a similar representation of the classical Laplace distribution with $\kappa = 1$.

Remark 3.2.3 Similarly, we can express an AL r.v. in terms of two i.i.d. Pareto Type I r.v.'s, P_1 and P_2, with density $f(x) = 1/x^2, x \geq 1$. Indeed, as already mentioned in Section 2.2.3, a standard exponential r.v. W has the same distribution as $\log(P_1)$, so that by (3.1.9) we have

$$Y \stackrel{d}{=} \theta + \frac{\sigma}{\sqrt{2}}\log\left(\frac{P_1^{1/\kappa}}{P_2^{\kappa}}\right). \tag{3.2.4}$$

A similar representation of the classical Laplace distribution was obtained in Proposition 2.2.4.

Remark 3.2.4 The representation of Proposition 3.2.2 may be expressed alternatively as follows:

$$Y \stackrel{d}{=} \theta + \frac{\sigma}{\sqrt{2}}IW, \tag{3.2.5}$$

where the r.v.'s I and W are independent, W is a standard exponential variable, and I takes on the values $-\kappa$ and $1/\kappa$ with probabilities $\kappa^2/(1+\kappa^2)$ and $1/(1+\kappa^2)$, respectively. In the symmetric case with $\kappa = 1$ ($\mu = 0$), the random variable I takes on the values ∓ 1 with probabilities $1/2$, and (3.2.5) reduces to the representation (2.2.10) of the symmetric Laplace r.v. with the scale parameter $s = \sigma/\sqrt{2}$.

3.2.3 Self-decomposability. We saw in Section 2.4.3 that all symmetric Laplace random variables Y are self-decomposable, that is for every $c \in (0, 1)$ we have

$$Y \stackrel{d}{=} cY + X,$$

where X and Y are independent variables. Ramachandran (1997) shows that all AL distributions are self-decomposable as well. In fact, we have the following explicit representation.

Proposition 3.2.3 *Let $Y \sim \mathcal{AL}^*(\theta, \kappa, \sigma)$. Then Y is self-decomposable and for any $c \in [0, 1]$ we have*

$$Y \stackrel{d}{=} cY + (1-c)\theta + \frac{\sigma}{\sqrt{2}}\left(\frac{1}{\kappa}\delta_1 W_1 - \kappa\delta_2 W_2\right), \tag{3.2.6}$$

where δ_1, δ_2 are dependent Bernoulli r.v.'s taking on values of either zero or one with the probabilities

$$P(\delta_1 = 0, \delta_2 = 0) = c^2, \quad P(\delta_1 = 1, \delta_2 = 1) = 0,$$
$$P(\delta_1 = 1, \delta_2 = 0) = (1-c)\left(c + \frac{1-c}{1+\kappa^2}\right),$$
$$P(\delta_1 = 0, \delta_2 = 1) = (1-c)\left(c + \frac{(1-c)\kappa^2}{1+\kappa^2}\right),$$

W_1 and W_2 are standard exponential variables, and Y, W_1, W_2, and (δ_1, δ_2) are mutually independent.

Proof. Representation (3.2.6) follows directly from the following equality for ch.f.'s:

$$\frac{(1+i\frac{\sigma}{\sqrt{2}}c\kappa t)(1-i\frac{\sigma}{\sqrt{2}}c\kappa^{-1}t)}{(1+i\frac{\sigma}{\sqrt{2}}\kappa t)(1-i\frac{\sigma}{\sqrt{2}}\kappa^{-1}t)} = c^2 + (1-c)\left(c+\frac{1-c}{1+\kappa^2}\right)\frac{1}{1-i\frac{\sigma}{\sqrt{2}}\kappa^{-1}t}$$
$$+ (1-c)\left(c+\frac{(1-c)\kappa^2}{1+\kappa^2}\right)\frac{1}{1+i\frac{\sigma}{\sqrt{2}}\kappa t}.$$

□

Remark 3.2.5 Note that in the symmetric case $\kappa = 1$, representation (3.2.6) reduces to that of a symmetric Laplace distribution with $s = \sigma/\sqrt{2}$ (see Proposition 2.4.4).

Remark 3.2.6 Note the following version of the representation

$$Y \stackrel{d}{=} cY + (1-c)\theta + \left(\frac{\delta_1}{\kappa} - \delta_2\kappa\right)\frac{\sigma}{\sqrt{2}}W,$$

where δ_i's are as before, W has the standard exponential distribution, and Y, W, (δ_1, δ_2) are independent.

By taking $c = 0$ in (3.2.6) we obtain the representation of an AL r.v. Y as a mixture of exponentially distributed random variables:

$$Y \stackrel{d}{=} \theta + \frac{\sigma}{\sqrt{2}}\left(\frac{1}{\kappa}\delta_1 W_1 - \kappa\delta_2 W_2\right), \tag{3.2.7}$$

where the zero-one variables δ_1 and δ_2, $\delta_1 + \delta_2 = 1$, assume one with probabilities $1/(1+\kappa^2)$ and $\kappa^2/(1+\kappa^2)$, respectively, and are independent of i.i.d. exponential variables W_1 and W_2. This is essentially the representation from Proposition 3.2.2.

3.2.4 Relation to 2×2 normal determinants. We have the following extension of Proposition 2.2.5 to the case of an AL random variable.

Proposition 3.2.4 *Let $Y \sim \mathcal{AL}^*(\theta, \kappa, \sigma)$ with $\theta = 0$ and $\sigma = 1$, and let (X_1, X_2) and (X_3, X_4) be i.i.d. bivariate normal r.v.'s with vector mean zero and variance-covariance matrix*

$$\mathbf{\Sigma} = \frac{1}{2\kappa} \begin{bmatrix} 1+\kappa^2, & 1-\kappa^2 \\ 1-\kappa^2, & 1+\kappa^2 \end{bmatrix}. \tag{3.2.8}$$

Then

$$Y \stackrel{d}{=} X_1 X_2 + X_3 X_4. \tag{3.2.9}$$

Note that if Y is symmetric Laplace ($\kappa = 1$), then $\mathbf{\Sigma}$ is an identity matrix, so all four variables X_1, X_2, X_3, X_4 are i.i.d. standard normal (see Proposition 2.2.5). For this case the representation (3.2.9) was derived in Mantel and Pasternack (1966) by an appropriate representation in terms of chi-square random variables [see also Farebrother (1986)], and in Mantel (1973) by calculating the appropriate characteristic functions [see also comments in Mantel (1987) and Missiakoulis and Darton (1985)]. Here we prove our generalization for asymmetric Laplace distribution using appropriate representations in terms of random variables.

Proof. Let Z_1, Z_2, Z_3, Z_4 be i.i.d. standard normal r.v.'s. Note that X_i's have the representation

$$(X_1, X_2) \stackrel{d}{=} \left(\frac{Z_1 - \kappa Z_3}{\sqrt{2\kappa}}, \frac{Z_1 + \kappa Z_3}{\sqrt{2\kappa}} \right), \tag{3.2.10}$$

$$(X_3, X_4) \stackrel{d}{=} \left(\frac{Z_2 - \kappa Z_4}{\sqrt{2\kappa}}, \frac{Z_2 + \kappa Z_4}{\sqrt{2\kappa}} \right). \tag{3.2.11}$$

Indeed, to see (3.2.10), note that the linear combinations of Z_i's are normal with

$$\mathrm{Var}\left(\frac{Z_1 - \kappa Z_3}{\sqrt{2\kappa}} \right) = \mathrm{Var}\left(\frac{Z_1 + \kappa Z_3}{\sqrt{2\kappa}} \right) = \frac{1}{2\kappa}(1+\kappa^2) \tag{3.2.12}$$

and

$$\mathrm{Cov}\left(\frac{Z_1 - \kappa Z_3}{\sqrt{2\kappa}}, \frac{Z_1 + \kappa Z_3}{\sqrt{2\kappa}} \right) = \frac{1}{2\kappa}(1-\kappa^2), \tag{3.2.13}$$

which correspond to the entries of $\mathbf{\Sigma}$ given by (3.2.8). Similar arguments apply to (3.2.11). Next, write

$$\begin{aligned} X_1 X_2 + X_3 X_4 &\stackrel{d}{=} \frac{1}{2\kappa} \{(Z_1 - \kappa Z_3)(Z_1 + \kappa Z_3) + (Z_2 - \kappa Z_4)(Z_2 + \kappa Z_4)\} \\ &= \frac{1}{2\kappa}(Z_1^2 - \kappa^2 Z_3^2 + Z_2^2 - \kappa^2 Z_4^2) = \frac{1}{2\kappa}(H_1 - \kappa^2 H_2), \end{aligned} \tag{3.2.14}$$

where

$$H_1 = Z_1^2 + Z_2^2 \text{ and } H_2 = Z_3^2 + Z_4^2 \tag{3.2.15}$$

are two i.i.d. χ^2 r.v.'s with two degrees of freedom. Finally, note that $H_i \stackrel{d}{=} 2W_i, i = 1, 2$, where W_i's are i.i.d. standard exponential variables, so (3.2.14) reduces to (3.1.9) and the result follows. □

Table 3.4 summarizes the representations studied in this section.

Representation	Variables
$\mu W + \sqrt{W} Z$	Z — standard normal r.v. W — exponentially distributed r.v.
$\frac{1}{\sqrt{2}}(\frac{1}{\kappa} W_1 - \kappa W_2)$	W_1, W_2 — standard exponential r.v.'s
$\frac{1}{2\sqrt{2}}(\frac{1}{\kappa} H_1 - \kappa H_2)$	H_1, H_2 — χ^2 r.v.'s with two degrees of freedom
$\frac{1}{\sqrt{2}} I W$	I takes on values $-\kappa$ and $\frac{1}{\kappa}$ with probabilities $\frac{\kappa^2}{1+\kappa^2}$ and $\frac{1}{1+\kappa^2}$ W — standard exponential r.v.
$\frac{1}{\sqrt{2}} \log(P_1^{1/\kappa} / P_2^{\kappa})$	P_1, P_2 — Pareto r.v.'s with p.d.f. $f(p) = 1/p^2, p > 1$
$\frac{1}{\sqrt{2}} \log(U_1^{\kappa} / U_2^{1/\kappa})$	U_1, U_2 — r.v.'s uniformly distributed on [0, 1]
$X_1 X_2 + X_3 X_4$	(X_1, X_2) and (X_3, X_4) are bivariate normal with mean 0 and covariance given by (3.2.8)
$\frac{1}{\sqrt{2}}(\frac{1}{\kappa}\delta_1 W_1 - \kappa \delta_2 W_2)$	W_1, W_2 — standard exponential r.v.'s (δ_1, δ_2) assumes values $(1, 0)$ and $(0, 1)$ with probabilities $\frac{1}{1+\kappa^2}$ and $\frac{\kappa^2}{1+\kappa^2}$

Table 3.4: Summary of the representations of the standard $\mathcal{AL}(0, \mu, 1)$ [or $\mathcal{AL}^*(0, \kappa, 1)$] random variables. All random variables (or vectors) in each representation are mutually independent.

3.3 Simulation

Random variate generation from an AL distribution is straightforward. Since the AL distribution function has closed form and so does its inverse, the inversion method can be applied [see, e.g., Devroye (1986)]. Alternatively, we can use any of the representations discussed in Section 3.2. Representation (3.2.3) in terms of two i.i.d. uniform variables seems to be most suitable for simulation, as these can be obtained directly. Here is an AL generator based on this representation.

$\mathcal{AL}^*(\theta, \kappa, \sigma)$ generator

- Generate a uniform [0, 1] random variate U_1.
- Generate a uniform [0, 1] random variate U_2, independent of U_1.
- Set $Y \leftarrow \theta + \frac{\sigma}{\sqrt{2}} \log \frac{U_1^{\kappa}}{U_2^{1/\kappa}}$.
- RETURN Y.

Remark 3.3.1 To generate an $\mathcal{AL}(\theta, \mu, \sigma)$ variate, first compute the parameter κ using relation (3.1.4), and then apply the above algorithm.

3.4 Characterizations and further properties

3.4.1 Infinite divisibility. In Section 2.4.1 of Chapter 2 we discussed a fundamental concept of infinite divisibility and showed that all symmetric Laplace laws are infinitely divisible. Similarly, all AL distributions are infinitely divisible as well, as their ch.f. ψ given by (3.1.3) can be factored as

$$\psi(t) = \left\{ e^{i\theta t/n} \left(\frac{1}{1 - i\frac{\sigma}{\sqrt{2}\kappa}t} \right)^{1/n} \left(\frac{1}{1 + i\frac{\sigma\kappa}{\sqrt{2}}t} \right)^{1/n} \right\}^n = [\psi_n(t)]^n \tag{3.4.1}$$

for each integer $n \geq 1$. The ch.f. ψ_n corresponds to the random variable

$$\frac{\theta}{n} + \frac{\sigma}{\sqrt{2}} \left(\frac{1}{\kappa} G_1 - \kappa G_2 \right), \tag{3.4.2}$$

where G_1 and G_2 are i.i.d. gamma $\Gamma(1/n, 1)$ random variables with density

$$f(x) = \frac{1}{\Gamma(1/n)} x^{1/n-1} e^{-x}, \quad x > 0. \tag{3.4.3}$$

Generalizations of Laplace distribution such as (3.4.2), whose characteristic functions are powers of the AL ch.f., are known as Bessel function distributions and are subject of Section 4.1 of Chapter 4.

The following result summarizes our discussion.

Proposition 3.4.1 *Let $Y \sim \mathcal{AL}^*(\theta, \kappa, \sigma)$. Then Y is infinitely divisible, admitting for each integer $n \geq 1$ the representation*

$$Y \stackrel{d}{=} \sum_{i=1}^{n} X_{ni}, \tag{3.4.4}$$

where the X_{ni}'s are i.i.d. variables given by (3.4.2).

The next result reveals the Lévy–Khinchine representation of an AL characteristic function, which was derived in Takano (1989, 1990).

Proposition 3.4.2 *The ch.f. ψ of $Y \sim \mathcal{AL}^*(\theta, \kappa, \sigma)$ r.v. admits the Lévy–Khinchine representation*

$$\psi(t) = \exp\left(it\theta + \int_R (e^{itu} - 1)\lambda(u)du \right), \tag{3.4.5}$$

where

$$\lambda(u) = \frac{1}{|u|} \begin{cases} e^{-\frac{\sqrt{2}\kappa}{\sigma}|u|}, & \text{for } u > 0 \\ e^{\frac{\sqrt{2}}{\kappa\sigma}u}, & \text{for } u < 0. \end{cases} \tag{3.4.6}$$

Proof. Recall that the Lévy measure of exponential distribution with parameter $\beta > 0$ has density $e^{-\beta u}/u$, $u > 0$, i.e.,

$$\frac{1}{1 - it/\beta} = \exp\left(\int_0^\infty (e^{itu} - 1) \frac{1}{u} e^{-\beta u} du \right), \quad \beta > 0, t \in \mathbb{R}.$$

Consequently,

$$\frac{1}{1 - i\frac{\sigma\kappa}{\sqrt{2}}t} = \exp\left(\int_0^\infty (e^{ity} - 1) \frac{1}{y} e^{-\frac{\sqrt{2}}{\sigma\kappa}y} dy \right), \quad t \in \mathbb{R}, \tag{3.4.7}$$

and

$$\frac{1}{1-i\frac{\sigma}{\sqrt{2}\kappa}t} = \exp\left(\int_0^\infty (e^{itu}-1)\frac{1}{u}e^{-\frac{\sqrt{2}\kappa}{\sigma}u}du\right), \quad t \in \mathbb{R}. \tag{3.4.8}$$

Replacing in (3.4.7) t with $-t$ and substituting $y=-u$ we obtain

$$\frac{1}{1+i\frac{\sigma\kappa}{\sqrt{2}}t} = \exp\left(\int_{-\infty}^0 (e^{itu}-1)\frac{1}{|u|}e^{-\frac{\sqrt{2}}{\sigma\kappa}|u|}du\right), \quad t \in \mathbb{R}. \tag{3.4.9}$$

The multiplication of the corresponding sides of (3.4.8) and (3.4.9), coupled with (3.1.3), produces (3.4.5)–(3.4.6). □

In Figure 3.2, we see graphs of the Lévy densities for various specifications of the parameter μ ($\sigma=\sqrt{2}$).

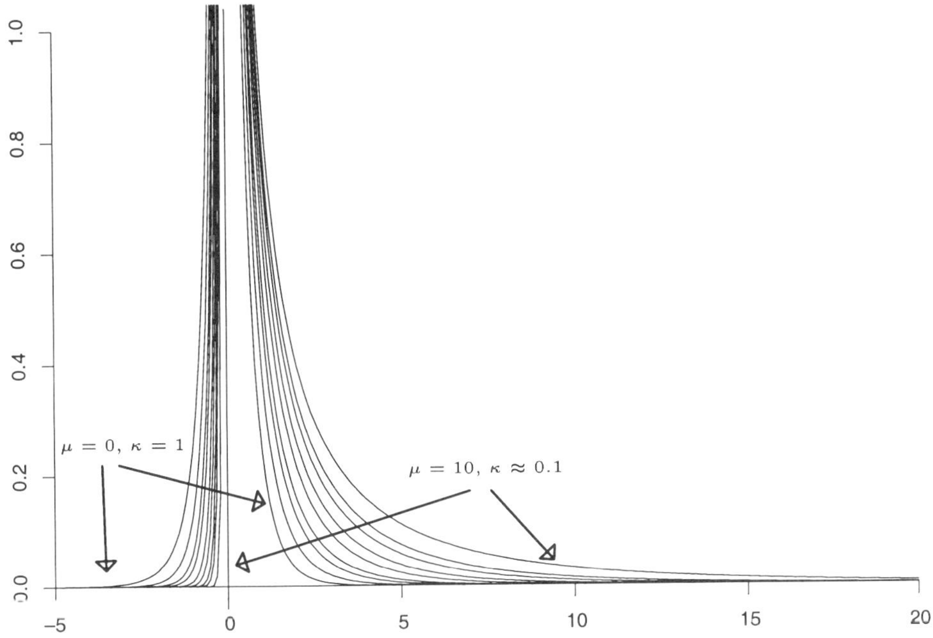

Figure 3.2: Densities of the Lévy measures for asymmetric Laplace distributions with $\sigma=\sqrt{2}$ and $\mu = 0, 0.8, 1.5, 2, 3, 4, 6, 8, 10$ that correspond to $\kappa \approx 1.0, 0.68, 0.50, 0.41, 0.30, 0.24, 0.16, 0.12, 0.1$. (The densities of these distributions are illustrated in Figure 3.1.)

3.4.2 Geometric infinite divisibility. In Section 2.4.2 of Chapter 2 we discussed the class of geometric infinitely divisible laws, and showed that all symmetric Laplace distributions with mean zero belong to this group. More generally, all AL laws with mode equal to zero are geometrically infinitely divisible as well, as shown by the following result.

Proposition 3.4.3 *If* $Y \sim \mathcal{AL}(0,\mu,\sigma)$, *then* Y *is geometrically infinitely divisible and for all* $p \in (0,1)$ *we have*

$$Y \stackrel{d}{=} \sum_{i=1}^{\nu_p} Y_p^{(i)}, \tag{3.4.10}$$

where ν_p is a geometric r.v. with mean $1/p$, the r.v.'s $Y_p^{(i)}$ are i.i.d. $\mathcal{AL}(0, p\mu, \sqrt{p}\sigma)$ for each p, and ν_p and $(Y_p^{(i)})$ are independent.

Proof. Let f_p be the ch.f. of $Y_p^{(i)}$. Conditioning on ν_p, we find the ch.f. of the right-hand side of (3.4.10) to be

$$E[e^{it\sum_{i=1}^{\nu_p} Y_p^{(i)}}] = \sum_{n=1}^{\infty} E[e^{it\sum_{i=1}^{n} Y_p^{(i)}}](1-p)^{n-1}p = \frac{pf_p(t)}{1-(1-p)f_p(t)}. \tag{3.4.11}$$

When we now substitute

$$f_p(t) = \frac{1}{1+\frac{1}{2}p\sigma^2t^2 - i\mu pt},$$

which is the ch.f. of the $\mathcal{AL}(0, p\mu, \sqrt{p}\sigma)$ distribution, into (3.4.11), we obtain the ch.f. of Y given by (3.1.1) with $\theta = 0$. □

Remark 3.4.1 If $Y \sim \mathcal{AL}^*(0, \kappa, \sigma)$, then (3.4.10) holds with $Y_p^{(i)}$ having the $\mathcal{AL}^*(0, \kappa_p, \sqrt{p}\sigma)$ distribution, where

$$\kappa_p = \frac{\sqrt{p(1/\kappa - \kappa)^2 + 4} - \sqrt{p}(1/\kappa - \kappa)}{2} \tag{3.4.12}$$

(see Exercise 3.6.17).

3.4.3 Distributional limits of geometric sums. We saw in Section 2.2.7 of Chapter 2 that the class of symmetric Laplace distributions with zero mean coincides with the class of distributional limits as $p \to 0$ of (appropriately normalized) geometric sums

$$X_1 + \cdots + X_{\nu_p},$$

where $X_1, X_2, \ldots$ are nondegenerate and symmetric i.i.d. r.v.'s with finite variance and ν_p is a geometric r.v. with mean $1/p$, independent of the X_i's. It turns out that if we omit the assumption of symmetry, then the limiting class coincides with the family of AL distributions.

Proposition 3.4.4 *The class of AL distributions with mode equal to zero coincides with the class of nondegenerate distributional limits of*

$$S_p = a_p \sum_{i=1}^{\nu_p}(X_i + b_p) \tag{3.4.13}$$

as $p \to 0$, where $X_1, X_2, \ldots$ are nondegenerate i.i.d. r.v.'s with finite variance, and ν_p is a geometric r.v. with mean $1/p$, independent of the X_i's. Moreover, if $EX_i = \mu$ and $\mathrm{Var}(X_i) = \sigma^2$, then the normalizing sequences in (3.4.13) may be taken as

$$a_p = p^{1/2}, \qquad b_p = \mu(p^{1/2} - 1), \tag{3.4.14}$$

in which case S_p converges in distribution to the $\mathcal{AL}(0, \mu, \sigma)$ random variable.

Proof. First, we shall show that if $Y \sim \mathcal{AL}(0, \mu, \sigma)$, then Y is the distributional limit of S_p, where X_i's are i.i.d. r.v.'s with $EX_i = \mu$ and $\text{Var}(X_i) = \sigma^2$, while the normalizing sequences are given by (3.4.14). Thus we need to show the convergence

$$p^{1/2} \sum_{j=1}^{\nu_p} (X_j - \mu + p^{1/2}\mu) \stackrel{d}{\to} Y, \tag{3.4.15}$$

where Y is an AL r.v. with ch.f. ψ given by (3.1.1) with $\theta = 0$. Writing (3.4.15) in terms of ch.f.'s, we obtain

$$\frac{pe^{ip\mu t}\phi(p^{1/2}t)}{1-(1-p)e^{ip\mu t}\phi(p^{1/2}t)} \to \psi(t), \tag{3.4.16}$$

where ϕ is the ch.f. of $X_j - \mu$. Taking reciprocals, we can express (3.4.16) as

$$\frac{1-(1-p)e^{ip\mu t}\phi(p^{1/2}t)}{pe^{ip\mu t}\phi(p^{1/2}t)} \to 1 + \frac{1}{2}\sigma^2 t^2 - i\mu t. \tag{3.4.17}$$

Note that the factor $\phi(p^{1/2}t)$ tends to one as p converges to zero, so we can write equivalently (splitting the numerator)

$$\frac{e^{-ip\mu t}-1}{p} + \frac{1-(1-p)\phi(p^{1/2}t)}{p} = I + II \to 1 + \frac{1}{2}\sigma^2 t^2 - i\mu t. \tag{3.4.18}$$

First, we show that $I \to -i\mu t$. Indeed, we have

$$\frac{e^{-ip\mu t}-1}{p} = -i\mu t \frac{\sin(p\mu t)}{p\mu t} + \frac{\cos(p\mu t)-1}{p\mu t} p\mu t \to -i\mu t + 0.$$

To establish the convergence

$$II = \frac{1-(1-p)\phi(p^{1/2}t)}{p} \to 1 + \frac{1}{2}\sigma^2 t^2 \tag{3.4.19}$$

use Theorem 8.44 from Breiman (1993). Since $W_j = X_j - \mu$ has finite first two moments, the ch.f. of W_j can be written as

$$\phi(u) = 1 + iuEW_j + \frac{(iu)^2}{2}(EX_j^2 + \delta(u)),$$

where δ denotes a bounded function of u such that $\lim_{u\to 0} \delta(u) = 0$. Since $EW_j = E[X_j - \mu] = 0$ and $EW_j^2 = E(X_j - \mu)^2 = \sigma^2$, we apply the above with $u = p^{1/2}t$ to the left-hand side of (3.4.19) to obtain

$$\frac{t^2}{2}(\sigma^2 + \delta(p^{1/2}t)) + 1 - \frac{pt^2}{2}(\sigma^2 + \delta(p^{1/2}t)),$$

which converges to $1 + t^2\sigma^2/2$ as $p \to 0$. Thus we have shown the first part of the proposition.

Let us now assume that the variables (3.4.13) converge in distribution to an r.v. Y with ch.f. ψ. Our goal is to show that the r.v. Y has an AL distribution. First, note that being a limit of geometric compounds (3.4.13), the r.v. Y is geometrically infinitely divisible, and thus also infinitely divisible [see, e.g., Mohan et al. (1993)]. Thus its ch.f. ψ does not vanish. Expressing the convergence in terms of ch.f.'s, we have

$$\frac{pf_p(t)}{1-(1-p)f_p(t)} \to \psi(t) \text{ for } t \in \mathbb{R}, \tag{3.4.20}$$

where

$$f_p(t) = e^{ita_p b_p}\phi(a_p t)$$

and ϕ is the ch.f. of the X_j's. Since the fraction in (3.4.20) converges to a nonzero limit while its numerator converges to zero (since f_p is bounded), we must have

$$f_p(t) \to 1 \text{ for } t \in \mathbb{R}. \tag{3.4.21}$$

We now rewrite (3.4.20) equivalently as

$$\frac{1}{1+\frac{1}{p}(\frac{1}{f_p(t)}-1)} \to \psi(t) \text{ for } t \in \mathbb{R}, \tag{3.4.22}$$

so

$$\frac{1}{p}\left(\frac{1}{f_p(t)}-1\right) \to \frac{1}{\psi(t)} - 1 \text{ for } t \in \mathbb{R}. \tag{3.4.23}$$

In view of (3.4.21), we have

$$\frac{1}{p}(f_p(t)-1) \to 1 - \frac{1}{\psi(t)} \text{ for } t \in \mathbb{R}. \tag{3.4.24}$$

We now let $p = 1/n$ and denote $a_n = a_{1/n}$, $b_n = b_{1/n}$, so (3.4.24) takes the form

$$n(f(a_n t)e^{ita_n b_n} - 1) \to 1 - \frac{1}{\psi(t)} \text{ for } t \in \mathbb{R}, \tag{3.4.25}$$

where the limit is a continuous function. Thus by Feller (1971, XVII, Theorem 1) we conclude that

$$(f(a_n t)e^{ita_n b_n})^n \to \exp\left(1 - \frac{1}{\psi(t)}\right) \text{ for } t \in \mathbb{R}. \tag{3.4.26}$$

But the left-hand side of (3.4.26) is the ch.f. of

$$a_n \sum_{i=1}^{n} (X_i + b_n), \tag{3.4.27}$$

where the X_i's are i.i.d. with finite variance, and consequently the limit in (3.4.26) must be a normal characteristic function,

$$\exp\left(1 - \frac{1}{\psi(t)}\right) = \exp\left(i\mu t - \frac{1}{2}\sigma^2 t^2\right), \tag{3.4.28}$$

where $\mu \in \mathbb{R}$ and $\sigma > 0$ are some constants. Solving (3.4.28) for $\psi(t)$, we obtain the AL ch.f. (3.1.1) with $\theta = 0$. The result has been proved. □

Remark 3.4.2 If the X_i's are $\mathcal{AL}(\mu, \sigma)$, then they have mean μ and variance $\mu^2+\sigma^2$. Consequently, we have the convergence to the $\mathcal{AL}(\mu, \sqrt{\mu^2+\sigma^2})$ law under the normalization (3.4.14).

3.4.4 Stability with respect to geometric summation. As we saw in Section 2.2.6 of Chapter 2, an AL r.v. Y with $\mu = 0$ (symmetric Laplace) is the only symmetric r.v. with a finite second moment satisfying the relation

$$Y \stackrel{d}{=} a_p \sum_{i=1}^{\nu_p} (Y_i + b_p), \tag{3.4.29}$$

where ν_p is geometrically distributed with mean $1/p$, Y_i's are i.i.d. copies of Y, and ν_p and Y_i's are independent. More generally, all AL r.v.'s satisfy a similar relation when equality in distribution is replaced by convergence in distribution. The following result, which we include here without proof, follows from a more general characterization of geometric stable distributions [see Kozubowski (1994b, Theorem 3.1)].

Proposition 3.4.5 *Let Y be a random variable with finite variance, and let $Y_1, Y_2, \ldots$ be i.i.d. copies of Y. Then the following statements are equivalent:*

(i) $Y \sim \mathcal{AL}(0, \mu, \sigma)$ *with* $\mu^2 + \sigma^2 > 0$.

(ii) *There exist* $a_p > 0$ *and* $b_p \in \mathbb{R}$ *such that*

$$a_p \sum_{i=1}^{\nu_p} (Y_i + b_p) \stackrel{d}{\to} Y, \tag{3.4.30}$$

where ν_p is a geometric r.v. with mean $1/p$, independent of the Y_i's.

Moreover, the normalizing sequences must have the form

$$a_p = Cp^{1/2}[1 + \delta(p)], \qquad b_p = [p\eta(p) + (p - a_p)\mu]/a_p, \tag{3.4.31}$$

where

$$C = \sqrt{\frac{\sigma^2}{\sigma^2 + \mu^2}} \tag{3.4.32}$$

and the sequences $\delta(p)$ and $\eta(p)$ converge to zero as $p \to 0$.

3.4.5 Maximum entropy property. In this section we characterize AL laws in terms of their entropy, defined in Section 2.4.5. Let us derive the entropy of X having an AL distribution with density (3.1.10).

Proposition 3.4.6 *Let X have an $\mathcal{AL}^*(\theta, \kappa, \sigma)$ distribution with density f given by* (3.1.10). *Then the entropy of X is given by*

$$H(X) = E[-\log f(X)] = 1 + \log\sigma + \log\left(\frac{1}{\kappa} + \kappa\right) - \frac{1}{2}\log 2. \tag{3.4.33}$$

Proof. The calculation is straightforward. Since the value of entropy is not affected by translation, we can assume that $\theta = 0$. By definition, the entropy of X is equal to

$$-\int_{-\infty}^{0} (\log C + \frac{\sqrt{2}}{\kappa\sigma}x) Ce^{\frac{\sqrt{2}}{\kappa\sigma}x} dx - \int_{0}^{\infty} (\log C - \frac{\sqrt{2}\kappa}{\sigma}x) Ce^{-\frac{\sqrt{2}\kappa}{\sigma}x} dx, \tag{3.4.34}$$

where

$$C = \frac{\sqrt{2}}{\sigma} \frac{\kappa}{1+\kappa^2}. \tag{3.4.35}$$

Recalling that for any $a > 0$ we have

$$\int_0^\infty a e^{-ax} dx = 1 \text{ and } \int_0^\infty x a e^{-ax} dx = \frac{1}{a}, \tag{3.4.36}$$

we obtain the following expression after integration in (3.4.34)

$$H(X) = -C\frac{\sigma}{\sqrt{2}}\kappa \log C + C\frac{\sigma}{\sqrt{2}}\kappa - C\frac{\sigma}{\sqrt{2}\kappa} \log C + C\frac{\sigma}{\sqrt{2}\kappa}. \tag{3.4.37}$$

The substitution of (3.4.35) into (3.4.37) produces (3.4.33). □

Remark 3.4.3 Note that for $\kappa = 1$, for which the AL distribution becomes a symmetric Laplace distribution, formula (3.4.33) simplifies to

$$H(X) = 1 + \log \sigma + \frac{1}{2} \log 2, \tag{3.4.38}$$

which was derived for symmetric Laplace distribution in Section 2.1.3 of Chapter 2.

We saw in Section 2.4.5 that the classical Laplace distribution maximizes the entropy among all distributions with a given first absolute moment and $(-\infty, \infty)$ support. It turns out that under the additional stipulation that the mean be also given, the distribution that maximizes the entropy is AL, as shown by Kotz et al. (2000a).

Proposition 3.4.7 *Consider the class $\mathcal{C}$ of all continuous random variables with nonvanishing densities on $(-\infty, \infty)$ such that*

$$EX = c_1 \in \mathbb{R} \text{ and } E|X| = c_2 > 0 \text{ for } X \in \mathcal{C}, \tag{3.4.39}$$

where

$$|c_1| < c_2. \tag{3.4.40}$$

Then the maximum entropy is attained for the AL r.v. X^* with density (3.1.10), where $\theta = 0$,

$$\kappa = \left(\frac{c_2 - c_1}{c_2 + c_1}\right)^{1/4}, \tag{3.4.41}$$

and

$$\sigma = \frac{1}{\sqrt{2}}(c_2^2 - c_1^2)^{1/4}(\sqrt{c_2 + c_1} + \sqrt{c_2 - c_1}). \tag{3.4.42}$$

Moreover, the maximum entropy is

$$\max_{X \in \mathcal{C}} H(X) = H(X^*) = 2 \log \left\{ \frac{\sqrt{c_2 + c_1} + \sqrt{c_2 - c_1}}{\sqrt{2}} \right\} + 1. \tag{3.4.43}$$

Proof. Applying Proposition 2.4.6 with $a = -\infty, b = \infty, h_1(x) = x$, and $h_2(x) = |x|$, we find that the maximum entropy is attained by density

$$p(x) = e^{a_0} e^{a_1 x + a_2 |x|} = e^{a_0} \begin{cases} e^{(a_1+a_2)x}, & \text{if } x \geq 0 \\ e^{(a_1-a_2)x}, & \text{if } x < 0 \end{cases}, \tag{3.4.44}$$

provided that the function (3.4.44) integrates to one on $(-\infty, \infty)$ and satisfies the constraints (3.4.39). Thus it is enough to find the constants a_0, a_1 and a_2 for which the constraints are satisfied. To this end, first note that the integrability of p implies the following restrictions on a_1 and a_2:

$$a_1 + a_2 < 0 \text{ and } a_1 - a_2 > 0, \tag{3.4.45}$$

implying that $a_2 < 0$. Write

$$a_1 = \frac{1}{\sqrt{2}\sigma}\left(\frac{1}{\kappa} - \kappa\right) \in \mathbb{R} \text{ and } a_2 = -\frac{1}{\sqrt{2}\sigma}\left(\frac{1}{\kappa} + \kappa\right) \in (-\infty, 0) \tag{3.4.46}$$

for some $\sigma > 0$ and $\kappa > 0$ so that the density (3.4.44) takes the form

$$p(x) = e^{a_0} \begin{cases} e^{-\frac{\sqrt{2}\kappa}{\sigma}x}, & \text{if } x \geq 0 \\ e^{\frac{1}{\sqrt{2}\kappa\sigma}x}, & \text{if } x < 0. \end{cases} \tag{3.4.47}$$

Comparing (3.4.47) with (3.1.10), we conclude that p must be an AL density, so that

$$e^{a_0} = \frac{\sqrt{2}}{\sigma}\frac{\kappa}{1+\kappa^2}. \tag{3.4.48}$$

Next, using the formulas for the mean and the first absolute moment of the AL distribution with density (3.1.10) with $\theta = 0$, we write the conditions (3.4.39) as

$$EX = \frac{\sigma}{\sqrt{2}\kappa}\frac{1-\kappa^4}{1+\kappa^2} = c_1 \tag{3.4.49}$$

and

$$E|X| = \frac{\sigma}{\sqrt{2}\kappa}\frac{1+\kappa^4}{1+\kappa^2} = c_2. \tag{3.4.50}$$

Divide the sides of (3.4.50) into the corresponding sides of (3.4.49) to obtain

$$\frac{1-\kappa^4}{1+\kappa^4} = \frac{c_1}{c_2}. \tag{3.4.51}$$

Solving the above equation for κ produces (3.4.41). Finally, the substitution of κ given by (3.4.41) into (3.4.50) and solving for σ produces (3.4.42). We thus conclude that the entropy is maximized by the AL law with $\theta = 0$ and κ and σ as specified by (3.4.41)–(3.4.42). The actual value of the maximal entropy follows from Proposition 3.4.6. □

Remark 3.4.4 Note that if the mean is zero, then $\kappa = 1$ and $\sigma = \sqrt{2}c_2$ so that the entropy is maximized by the classical Laplace r.v. with density $\frac{1}{2c_2}e^{-|x|/c_2}$. In this case the maximal entropy (3.4.43) reduces to (3.4.38).

Remark 3.4.5 If in Proposition 3.4.7 the absolute deviation *about the mean* is prescribed instead of $E|X|$, then the entropy is maximized by the symmetric Laplace distribution (Exercise 3.6.18).

3.5 Estimation

In this section we study the problem of estimating the parameters of an AL distribution. Note that our distributions are essentially convolutions of exponential random variables of different signs, and common estimation procedures for mixtures of positive exponential distributions [see, e.g., Mendenhall and Hader (1958), Rider (1961)] are not applicable in this case. We shall focus on the method of maximum likelihood, leaving the discussion of other methods of estimation (i.e., the method of moments) to exercises.

Let us start with the derivation of the Fisher information matrix, $I(\theta, \kappa, \sigma)$, corresponding to an $\mathcal{AL}^*(\theta, \kappa, \sigma)$ distribution. We have

$$I(\theta, \kappa, \sigma) = \left[\ E\left\{\frac{\partial}{\partial \gamma_i} \log f_{\theta,\kappa,\sigma}(X) \cdot \frac{\partial}{\partial \gamma_j} \log f_{\theta,\kappa,\sigma}(X)\right\}\ \right]_{i,j=1}^{3},$$

where X has an $\mathcal{AL}^*(\theta, \kappa, \sigma)$ distribution with the vector-parameter

$$\gamma = (\theta, \kappa, \sigma)$$

and density $f_{\theta,\kappa,\sigma}$. Routine calculations (Exercise 3.6.23) produce the matrix

$$I(\theta, \kappa, \sigma) = \begin{bmatrix} \frac{2}{\sigma^2} & -\frac{\sqrt{2}}{\sigma}\frac{2}{1+\kappa^2} & 0 \\ -\frac{\sqrt{2}}{\sigma}\frac{2}{1+\kappa^2} & \frac{1}{\kappa^2} + \frac{4}{(1+\kappa^2)^2} & -\frac{1}{\sigma\kappa}\frac{1-\kappa^2}{1+\kappa^2} \\ 0 & -\frac{1}{\sigma\kappa}\frac{1-\kappa^2}{1+\kappa^2} & \frac{1}{\sigma^2} \end{bmatrix}. \tag{3.5.1}$$

3.5.1 Maximum likelihood estimation. Let $X_1, \ldots, X_n$ be an i.i.d. random sample from an $\mathcal{AL}^*(\theta, \kappa, \sigma)$ distribution with the density $f_{\theta,\sigma,\kappa}$ given by (3.1.10), and let $x_1, \ldots, x_n$ be their particular realization. Then the likelihood function takes the form

$$L(\theta, \kappa, \sigma) = \frac{2^{n/2}}{\sigma^n}\frac{\kappa^n}{(1+\kappa^2)^n} \exp\left\{-\frac{\sqrt{2}\kappa}{\sigma}\sum_{j=1}^n (x_j-\theta)^+ - \frac{\sqrt{2}}{\kappa\sigma}\sum_{j=1}^n (x_j-\theta)^-\right\}, \tag{3.5.2}$$

where

$$(x_i-\theta)^+ = \begin{cases} x_i - \theta & \text{if } x_i \geq \theta \\ 0 & \text{if } x_i \leq \theta \end{cases} \tag{3.5.3}$$

and

$$(x_i-\theta)^- = \begin{cases} \theta - x_i & \text{if } x_i \leq \theta \\ 0 & \text{if } x_i \geq \theta. \end{cases} \tag{3.5.4}$$

Thus the log-likelihood function is

$$\log L(\theta, \kappa, \sigma) = \frac{n}{2}\log 2 - n\log\sigma + n\log\frac{\kappa}{1+\kappa^2} - \frac{\sqrt{2}}{\sigma}D, \tag{3.5.5}$$

where

$$D = D(\theta, \kappa) = \kappa\sum_{j=1}^n (x_j-\theta)^+ + \frac{1}{\kappa}\sum_{j=1}^n (x_j-\theta)^-. \tag{3.5.6}$$

We follow our approach to the symmetric case and consider several cases.

3.5.1.1 *Case 1: The values of κ and σ are known.* Here the likelihood function will be maximized by the value of θ that minimizes the function

$$Q(\theta) = \kappa \sum_{i=1}^{n} (x_i - \theta)^+ + \frac{1}{\kappa} \sum_{i=1}^{n} (x_i - \theta)^-. \tag{3.5.7}$$

Let $X_{1:n} \le \cdots \le X_{n:n}$ be the order statistics connected with a random sample of size n from the $\mathcal{AL}^*(\theta, \kappa, \sigma)$ distribution, and let $x_{1:n} \le \cdots \le x_{n:n}$ be their particular realization. Consider the set of $n + 1$ intervals $\{I_0, \ldots, I_n\}$, where

$$I_0 = (-\infty, x_{1:n}], \qquad I_n = [x_{n:n}, \infty), \tag{3.5.8}$$

and

$$I_j = [x_{j:n}, x_{j+1:n}], \quad j = 1, 2, \ldots n - 1. \tag{3.5.9}$$

It can be shown that the function Q is continuous on $\mathbb{R}$ and linear on each of the intervals I_j, $j = 0, 1, \ldots, n$ (Exercise 3.6.19). Further, the function Q is decreasing on I_0, increasing on I_n, while on any I_j with $1 \le j \le n - 1$ it is

$$\begin{cases} \text{decreasing} & \text{if } \frac{j}{n-j} < \kappa^2, \\ \text{constant} & \text{if } \frac{j}{n-j} = \kappa^2, \\ \text{increasing} & \text{if } \frac{j}{n-j} > \kappa^2. \end{cases} \tag{3.5.10}$$

Thus, if the parameter κ is such that

$$\kappa^2 = \frac{j}{n-j} \quad \text{for some } j = 1, 2, \ldots, n - 1, \tag{3.5.11}$$

then the function Q is minimized by any value of θ within the interval $[x_{j:n}, x_{j+1:n}]$. Consequently, any statistic of the form

$$pX_{j:n} + (1 - p)X_{j+1:n}, \quad p \in [0, 1], \tag{3.5.12}$$

may be taken as an MLE of the parameter θ in this case. If the condition (3.5.11) does not hold, the function Q attains its global minimum value at the *unique* $\hat{\theta}_n$ given by

$$\hat{\theta}_n = \begin{cases} X_{1:n} & \text{if } \kappa^2 < \frac{1}{n-1}, \\ X_{j:n} & \text{if } \frac{j-1}{n-(j-1)} < \kappa^2 < \frac{j}{n-j}, \quad j = 2, 3, \ldots, n - 1, \\ X_{n:n} & \text{if } \kappa^2 > n - 1. \end{cases} \tag{3.5.13}$$

We see that, as in the case of symmetric Laplace distribution, the problem of estimating the location parameter θ of the $\mathcal{AL}^*(\theta, \kappa, \sigma)$ distribution admits an explicit solution.

Observe that for large values of n we will have

$$\frac{j-1}{n-(j-1)} \le \kappa^2 < \frac{j}{n-j} \quad \text{for some } j = 2, 3, \ldots, n - 1, \tag{3.5.14}$$

so that consistently with the relations (3.5.12)–(3.5.13) the statistic $X_{j:n}$ may be taken as the MLE of θ. Solving the inequalities (3.5.14) for j we obtain the relation

$$n\frac{\kappa^2}{1+\kappa^2} < j \le 1 + n\frac{\kappa^2}{1+\kappa^2}, \tag{3.5.15}$$

which is satisfied uniquely by

$$j = j(n) = [[n\kappa^2/(1+\kappa^2)]] + 1 \tag{3.5.16}$$

(the bracketed $[[x]]$ denotes the integral part of x). The resulting MLE of θ, given by the order statistic

$$\hat{\theta}_n = X_{j(n):n}, \tag{3.5.17}$$

is consistent and asymptotically normal.

Proposition 3.5.1 *Let $X_1, \dots, X_n$ be i.i.d. from the $\mathcal{AL}^*(\theta, \kappa, \sigma)$ distribution with an unknown value of θ. Then the MLE of θ given by* (3.5.17) *is*

(i) *consistent;*

(ii) *asymptotically normal, i.e.,*

$$\sqrt{n}(\hat{\theta}_n - \theta) \xrightarrow{d} N(0, \sigma^2/2); \tag{3.5.18}$$

(iii) *asymptotically efficient.*

Proof. It is well known [see, e.g., David (1981)] that for a continuous distribution with density f the sample quantile

$$\hat{\xi}_{\lambda,n} = X_{[[\lambda n]]+1:n}, \quad 0 < \lambda < 1,$$

converges to the corresponding population quantile ξ_λ and the asymptotic distribution of

$$\sqrt{n}(\hat{\xi}_{\lambda,n} - \xi_\lambda)$$

is normal with mean zero and variance

$$\frac{\lambda(1-\lambda)}{(f(\xi_\lambda))^2}. \tag{3.5.19}$$

In our case the MLE is a sample quantile with $\lambda = \frac{\kappa^2}{1+\kappa^2}$, the corresponding population quantile ξ_λ is equal to θ (since the above λ coincides with the probability that the relevant asymmetric Laplace variable is less than θ), and

$$f(\xi_\lambda) = f_{\theta,\kappa,\sigma}(\theta) = \frac{\sqrt{2}}{\sigma}\frac{\kappa}{1+\kappa^2}. \tag{3.5.20}$$

Thus the consistency and asymptotic normality (3.5.18) follow. To establish asymptotic efficiency note that the asymptotic variance coincides with the inverse of the Fisher information $I(\theta) = 2/\sigma^2$ [cf. (3.5.1)]. □

The specific form of the MLE for the location parameter provides a characterization of our class of asymmetric Laplace laws. Buczolich and Székely (1989), already mentioned in a remark following Proposition 2.6.3, considered the question of when the statistic $\sum_{i=1}^n a_i X_{i:n}$, where $a_i \ge 0$ and $\sum_{i=1}^n a_i = 1$, can be the MLE of the location parameter θ for a sample $X_1, \dots, X_n$ from a distribution given by a density $f(x)$. A proof of the following result may be found in Buczolich and Székely (1989).

Theorem 3.5.1 *The weighted sum* $\sum_{i=1}^{n} a_i X_{i:n}$*, where* $n \geq 3$*,* $a_i \geq 0$*, and* $\sum_{i=1}^{n} a_i = 1$*, can be the MLE for the location parameter* θ *if and only if one of the following cases holds:*

(i) $a_i = 1/n$ *for all* $i = 1, \dots, n$.

(ii) $a_1 = p$ *and* $a_n = 1 - p$ *for some* $p \in (0, 1)$.

(iii) $a_j = p$ *and* $a_{j+1} = 1 - p$ *for some* $p \in (0, 1)$ *and some* $j = 1, \dots, n - 1$.

(iv) $a_j = 1$ *for some* $j = 1, \dots, n$.

In the first case the distribution is necessarily Gaussian centered at zero (and the estimator is a sample mean).

In the second case, the distribution is uniform on the interval $[-c(1-p), cp]$ *for some* $c > 0$ *(and the estimator is the midrange).*

In the third case, the distribution is necessarily asymmetric Laplace with the skewness parameter $\kappa^2 = j/(n-j)$.

In the fourth case, there is no parametric class to which the density f *belongs for the case when* n *is fixed. However, if the hypothesis holds for infinitely many sample sizes* $n = n_r$ *and for* $j = j_r$ *such that* j_r/n_r *converges to* α*, then the distribution is necessarily asymmetric Laplace with the skewness parameter* $\kappa^2 = \alpha/(1-\alpha)$.

3.5.1.2 *Case 2: The values of* θ *and* κ *are known.* Here the log-likelihood (3.5.5) leads to the following function of σ to be maximized:

$$Q(\sigma) = C - n \log \sigma - \frac{\sqrt{2}}{\sigma} D, \tag{3.5.21}$$

where the quantities $C = \frac{n}{2} \log 2 + n \log \frac{\kappa}{1+\kappa^2}$ and D given by (3.5.6) do not depend of σ. By differentiating, we find that Q attains its maximum value at the unique point

$$\hat{\sigma}_n = \frac{\sqrt{2}}{n} \left\{ \kappa \sum_{j=1}^{n} (x_j - \theta)^+ + \frac{1}{\kappa} \sum_{j=1}^{n} (x_j - \theta)^- \right\}, \tag{3.5.22}$$

which is the MLE of σ. Note that the distribution of $\hat{\sigma}_n$ coincides with that of the sample mean

$$\hat{\sigma}_n = \frac{1}{n} \sum_{j=1}^{n} Y_i, \tag{3.5.23}$$

where the Y_i's are i.i.d. exponential variables with mean σ and variance σ^2. This follows from the fact that if $X \sim \mathcal{AL}^*(\theta, \kappa, \sigma)$, then the variable $Y = g(X)$, where

$$g(x) = \begin{cases} \sqrt{2}\kappa(x - \theta) & \text{if } x \geq \theta \\ -\frac{\sqrt{2}}{\kappa}(x - \theta) & \text{for } x < \theta, \end{cases} \tag{3.5.24}$$

is exponentially distributed with the above mean and variance (Exercise 3.6.20).

The representation (3.5.23) immediately leads to the strong consistency and asymptotic normality of $\hat{\sigma}_n$, as the variables Y_i have finite variance. Since the asymptotic variance coincides with the reciprocal of the Fisher information $I(\sigma) = 1/\sigma^2$ [cf. (3.5.1)], the MLE is also asymptotically efficient.

Proposition 3.5.2 *Let $X_1, \ldots, X_n$ be i.i.d. r.v.'s from the $\mathcal{AL}^*(\theta, \kappa, \sigma)$ distribution, where the value of σ is unknown. Then the MLE of σ is given by* (3.5.22) *and is*

(i) *unbiased;*

(ii) *strongly consistent;*

(iii) *asymptotically normal, where*

$$\sqrt{n}(\hat{\sigma}_n - \sigma) \stackrel{d}{\rightarrow} N(0, \sigma^2); \tag{3.5.25}$$

(iv) *asymptotically efficient.*

3.5.1.3 *Case* 3: *The values of θ and σ are known.* Here, by (3.5.5), we need to maximize the function

$$g(y, \alpha, \beta) = \log y - \log(1 + y^2) - \alpha y - \frac{\beta}{y} \tag{3.5.26}$$

with respect to $y \in (0, \infty)$, where

$$\alpha = \alpha(\theta) = \frac{\sqrt{2}}{\sigma}\frac{1}{n}\sum_{j=1}^{n}(x_j - \theta)^+, \quad \beta = \beta(\theta) = \frac{\sqrt{2}}{\sigma}\frac{1}{n}\sum_{j=1}^{n}(x_j - \theta)^-. \tag{3.5.27}$$

For any fixed $\alpha, \beta > 0$, the derivative of g with respect to y is

$$h(y, \alpha, \beta) = \frac{\partial}{\partial y} g(y, \alpha, \beta) = \frac{1}{y} - \frac{2y}{1 + y^2} + \frac{\beta}{y^2} - \alpha. \tag{3.5.28}$$

To find the MLE of κ, we shall study the solutions of the equation

$$h(y, \alpha, \beta) = 0. \tag{3.5.29}$$

The relevant properties of the function h are presented in the following lemma.

Lemma 3.5.1 *For any fixed $\alpha, \beta > 0$ the function h defined in* (3.5.28) *is strictly decreasing on* $(0, \infty)$ *with*

$$\lim_{y \to 0^+} h(y, \alpha, \beta) = \infty \quad \text{and} \quad \lim_{y \to \infty} h(y, \alpha, \beta) = -\alpha < 0,$$

so that there exists a unique solution $y_0 \in (0, \infty)$ of the equation (3.5.29). *Moreover, we have*

$$\begin{array}{ll} \sqrt{\beta/\alpha} \leq y_0 \leq 1 & \text{in case } \beta \leq \alpha, \\ 1 \leq y_0 \leq \sqrt{\beta/\alpha} & \text{in case } \beta \geq \alpha \end{array} \tag{3.5.30}$$

Proof. Fix $\alpha, \beta > 0$ and write

$$h(y, \alpha, \beta) = h_1(y) + h_2(y), \tag{3.5.31}$$

where

$$h_1(y) = \frac{1}{y} - \frac{2y}{1 + y^2} \quad \text{and} \quad h_2(y) = \frac{\beta}{y^2} - \alpha. \tag{3.5.32}$$

Since

$$\frac{d}{dy}h_1(y) = -\frac{1}{y^2} - 2\frac{1-y^2}{(1+y^2)^2}, \tag{3.5.33}$$

it is easy to see that the function h_1 is decreasing on the interval $(0, y^*)$ and increasing on the interval (y^*, ∞), where

$$y^* = \sqrt{2+\sqrt{5}} > 1. \tag{3.5.34}$$

In addition,

$$\lim_{y\to 0^+} h_1(y) = \infty, \quad h_1(1) = 0, \quad h_1(y^*) < 0, \quad \lim_{y\to\infty} h_1(y) = 0. \tag{3.5.35}$$

On the other hand, the function h_2 is decreasing on $(0, \infty)$ and

$$\lim_{y\to 0^+} h_2(y) = \infty, \quad h_2(\sqrt{\beta/\alpha}) = 0, \quad h_2(1) = \beta - \alpha, \quad \lim_{y\to\infty} h_2(y) = -\alpha < 0. \tag{3.5.36}$$

Assume first that $\alpha = \beta$. Then $h(1) = 0$ and

$$h(y) = h_1(y) + h_2(y) > 0 \tag{3.5.37}$$

for $y \in (0, 1)$, while

$$h(y) = h_1(y) + h_2(y) < 0 \tag{3.5.38}$$

for $y \in (1, \infty)$. Consequently, $y_0 = 1$ is the unique solution of equation (3.5.29) satisfying (3.5.30).

Next, assume that $\beta < \alpha$. By the above properties of h_1 and h_2, we deduce that there must exist a unique

$$y_0 \in (\sqrt{\beta/\alpha}, 1) \tag{3.5.39}$$

such that relations (3.5.37)–(3.5.38) hold for $y \in (0, y_0)$ and $y \in (y_0, \infty)$, respectively. This y_0 must be a unique solution of the equation (3.5.29) satisfying (3.5.30).

Finally, if $\beta > \alpha$, then the result follows from the relation

$$h(y, \alpha, \beta) = h(1/y, \beta, \alpha) \tag{3.5.40}$$

and the application of the previous case. □

Remark 3.5.1 It is easy to see that the conclusions of Lemma 3.5.1 remain valid if either α or β is equal to zero (which occurs when all the observations are located on one side of θ). It is interesting that in this case we still get the MLE's and the corresponding two-tailed AL distribution. On the other hand, we shall see in Case 5 (when θ is known) that under this condition the maximum likelihood approach would produce an exponential distribution (a one-tailed AL law).

In view of Lemma 3.5.1, we conclude that the likelihood function (3.5.26) is maximized at a unique value of y (the MLE of κ), which can be obtained by solving equation (3.5.29). The solution does not admit a closed form and must be found numerically. The properties of the MLE are presented in the following result.

Proposition 3.5.3 *Let $X_1, \ldots, X_n$ be i.i.d. r.v.'s from an $\mathcal{AL}^*(\theta, \kappa, \sigma)$ distribution where the values of θ and σ are known. Then the MLE of κ is the unique solution $\hat{\kappa}_n$ of the equation* (3.5.29), *where the function h is defined in* (3.5.28) *and α, β are given in* (3.5.27). *The MLE $\hat{\kappa}_n$ is*

(i) *consistent;*

(ii) *asymptotically normal and efficient:*

$$\sqrt{n}(\hat{\kappa}_n - \kappa) \xrightarrow{d} N(0, \sigma_\kappa^2), \tag{3.5.41}$$

where the asymptotic variance

$$\sigma_\kappa^2 = \frac{\kappa^2(1+\kappa^2)^2}{(1+\kappa^2)^2 + 4\kappa^2} \tag{3.5.42}$$

coincides with the reciprocal of the Fisher information $I(\kappa)$.

Moreover, for any integer $n \geq 1$ we have

$$\begin{array}{ll} \sqrt{\beta/\alpha} \leq \hat{\kappa}_n \leq 1 & \text{in case } \beta \leq \alpha, \\ 1 \leq \hat{\kappa}_n \leq \sqrt{\beta/\alpha} & \text{in case } \beta \geq \alpha. \end{array} \tag{3.5.43}$$

Proof. Consider auxiliary random vectors

$$\mathbf{Z}^{(i)} = [Z_1^{(i)}, Z_2^{(i)}]', \quad i = 1, 2, \ldots, n, \tag{3.5.44}$$

where $Z_1^{(i)} = (X_i - \theta)^+$ and $Z_2^{(i)} = (X_i - \theta)^-$, so

$$X_i - \theta = [1, -1]\mathbf{Z}^{(i)}. \tag{3.5.45}$$

The above $\mathbf{Z}^{(i)}$'s admit the representation

$$\mathbf{Z}^{(i)} \stackrel{d}{=} \begin{bmatrix} \delta_{1,i} E_{1,i} \\ \delta_{2,i} E_{2,i} \end{bmatrix},$$

where the $E_{1,i}$'s are i.i.d. distributed as $\frac{\sigma}{\sqrt{2}} \frac{1}{\kappa} W$, and the $E_{2,i}$'s are i.i.d. distributed as $\frac{\sigma}{\sqrt{2}} \kappa W$, where W is a standard exponential variable and the δ_{1i}, δ_{2i} are the 0–1 random variables that appear in representation (3.2.7). The random vectors $\mathbf{Z}^{(i)}$ are i.i.d. with the mean

$$\mathbf{m}_\mathbf{Z} = \begin{bmatrix} m_{1,\mathbf{Z}} \\ m_{2,\mathbf{Z}} \end{bmatrix} = \frac{\sigma\kappa}{\sqrt{2}(1+\kappa^2)} \begin{bmatrix} 1/\kappa^2 \\ \kappa^2 \end{bmatrix} \tag{3.5.46}$$

and the covariance matrix

$$\boldsymbol{\Sigma}_\mathbf{Z} = \frac{\sigma^2\kappa^2}{2\left(1+\kappa^2\right)^2} \begin{bmatrix} (1/\kappa^2+1)^2 - 1 & -1 \\ -1 & (\kappa^2+1)^2 - 1 \end{bmatrix}. \tag{3.5.47}$$

Clearly, the sequence $\{\mathbf{Z}^{(i)}\}$ obeys the Law of Large Numbers and the Central Limit Theorem so that

$$\lim_{n\to\infty} \bar{\mathbf{Z}}^{(n)} \stackrel{\text{a.s.}}{=} \mathbf{m}_\mathbf{Z} \tag{3.5.48}$$

and

$$\lim_{n\to\infty} \sqrt{n}(\bar{\mathbf{Z}}^{(n)} - \mathbf{m_Z}) \stackrel{\mathrm{d}}{=} N(\mathbf{0}, \mathbf{\Sigma_Z}), \tag{3.5.49}$$

where

$$\bar{\mathbf{Z}}^{(n)} = \frac{1}{n}\sum_{i=1}^{n} \mathbf{Z}^{(i)} = \left[\frac{1}{n}\sum_{i=1}^{n} Z_1^{(i)}, \frac{1}{n}\sum_{i=1}^{n} Z_2^{(i)}\right]'. \tag{3.5.50}$$

Notice that the quantities α and β are related to the $\mathbf{Z}^{(i)}$'s as follows:

$$\alpha = \frac{\sqrt{2}}{\sigma}\frac{1}{n}\sum_{i=1}^{n} Z_1^{(i)} = \frac{\sqrt{2}}{\sigma}\bar{Z}_1^{(n)}, \tag{3.5.51}$$

$$\beta = \frac{\sqrt{2}}{\sigma}\frac{1}{n}\sum_{i=1}^{n} Z_2^{(i)} = \frac{\sqrt{2}}{\sigma}\bar{Z}_2^{(n)}. \tag{3.5.52}$$

Since the MLE, $\hat{\kappa}_n$, is a unique solution of equation (3.5.29), it can be written as

$$\hat{\kappa}_n = H(\alpha, \beta), \tag{3.5.53}$$

where $H(\cdot, \cdot)$ is a continuous and differentiable function satisfying the equation

$$h(H(\alpha, \beta), \alpha, \beta) = 0. \tag{3.5.54}$$

In view of (3.5.51) and (3.5.52), we have

$$\hat{\kappa}_n = H\left(\frac{\sqrt{2}}{\sigma}\bar{Z}_1^{(n)}, \frac{\sqrt{2}}{\sigma}\bar{Z}_2^{(n)}\right). \tag{3.5.55}$$

To establish the consistency of the MLE given in (3.5.55), note that by (3.5.46), (3.5.47), (3.5.48), and the continuity of H, we have

$$\hat{\kappa}_n \stackrel{d}{\to} H\left(\frac{\sqrt{2}}{\sigma}m_{1,\mathbf{Z}}, \frac{\sqrt{2}}{\sigma}m_{2,\mathbf{Z}}\right). \tag{3.5.56}$$

Substituting

$$\alpha = \frac{\sqrt{2}}{\sigma}m_{1,\mathbf{Z}} = \frac{1}{\kappa}\frac{1}{1+\kappa^2} \quad \text{and} \quad \frac{\sqrt{2}}{\sigma}m_{2,\mathbf{Z}} = \frac{\kappa^3}{1+\kappa^2} \tag{3.5.57}$$

into (3.5.29) and solving for y we obtain κ, as can be readily verified.

The asymptotic normality (3.5.41) of $\hat{\kappa}_n$ can be establish similarly. In view of (3.5.49), by the standard large sample theory results [see, e.g., Serfling (1980)] it follows that as $n \to \infty$, the variables

$$\sqrt{n}\left[H\left(\frac{\sqrt{2}}{\sigma}\bar{Z}_1^{(n)}, \frac{\sqrt{2}}{\sigma}\bar{Z}_2^{(n)}\right) - H\left(\frac{\sqrt{2}}{\sigma}m_{1,\mathbf{Z}}, \frac{\sqrt{2}}{\sigma}m_{2,\mathbf{Z}}\right)\right] \tag{3.5.58}$$

converge in distribution to a $N\left(0, 2\mathbf{D}\boldsymbol{\Sigma}_{\mathbf{Z}}\mathbf{D}'/\sigma^2\right)$ variable, where $\mathbf{D}$ is the vector of partial derivatives of H:

$$\mathbf{D} = \left[\frac{\partial H}{\partial \alpha}, \frac{\partial H}{\partial \beta}\right]_{[\alpha,\beta]=\sqrt{2}\mathbf{m}_{\mathbf{Z}}/\sigma}. \tag{3.5.59}$$

A straightforward but laborious calculation of the derivatives produces

$$\mathbf{D} = \left[-\frac{\kappa^2(1+\kappa^2)^2}{(1+\kappa^2)^2+4\kappa^2}, \frac{(1+\kappa^2)^2}{(1+\kappa^2)^2+4\kappa^2}\right], \tag{3.5.60}$$

and we obtain (3.5.41) and (3.5.42). The asymptotic efficiency is obtained by noting that σ_κ^2 given in (3.5.42) is the reciprocal of the Fisher information $I(\kappa)$ given by the middle entry in the Fisher information matrix (3.5.1). □

3.5.1.4 *Case* 4: *The value of* κ *is known.* By (3.5.5), we need to maximize the function

$$Q(\theta, \sigma) = -n \log \sigma - \frac{\sqrt{2}}{\sigma} D(\theta, \kappa), \tag{3.5.61}$$

where $D(\theta, \kappa)$ is given by (3.5.6). We have already established in Case 2 that for any fixed value of $D = D(\theta, \kappa)$ the function (3.5.61) is maximized by the following value of σ:

$$\sigma(\theta) = \frac{\sqrt{2}}{n} D(\theta, \kappa). \tag{3.5.62}$$

The corresponding maximum value of Q is

$$Q(\theta, \sigma(\theta)) = -n \log \left\{\frac{\sqrt{2}}{n} D(\theta, \kappa)\right\} - n. \tag{3.5.63}$$

Since the quantity (3.5.63) is decreasing in $D(\theta, \kappa)$, we need to find the value of θ that minimizes the latter. Such value was already obtained in Case 1. Thus the MLE of θ, denoted $\hat{\theta}_n$, is given by (3.5.12) or (3.5.13), and for large n it can be taken as the order statistic $X_{j(n):n}$ with $j(n)$ given by (3.5.16). The MLE of σ is then given by (3.5.62) with $\hat{\theta}_n$ in place on θ, that is,

$$\hat{\sigma}_n = \frac{\sqrt{2}}{n}\left\{\kappa \sum_{j=1}^{n} (x_j - \hat{\theta}_n)^+ + \frac{1}{\kappa} \sum_{j=1}^{n} (x_j - \hat{\theta}_n)^-\right\}. \tag{3.5.64}$$

We observe that both estimators are linear combinations of order statistics, as was the case with the corresponding MLE's of the parameters of a symmetric Laplace distribution. Proceeding as in the classical Laplace case, one can show that the MLE $(\hat{\theta}_n, \hat{\sigma}_n)$ is consistent, asymptotically normal, and efficient, with the asymptotic covariance matrix

$$\boldsymbol{\Sigma} = \begin{bmatrix} \sigma^2/2 & 0 \\ 0 & \sigma^2 \end{bmatrix}; \tag{3.5.65}$$

cf. (3.5.1). We omit a highly technical derivation of this result, which can be found in Kotz et al. (2000c).

3.5.1.5 *Case 5: The value of θ is known.* Here we need to maximize the function

$$Q(\kappa, \sigma) = \log \kappa - \log(1 + \kappa^2) - \log(\sigma) - [\kappa, 1/\kappa]\bar{\mathbf{Z}}^{(n)}/(\sigma/\sqrt{2}),$$

where the vector $\bar{\mathbf{Z}}^{(n)}$ was defined previously in (3.5.50). We shall proceed by considering three cases:

1. $\theta \leq x_{1:n}$,
2. $\theta \geq x_{n:n}$,
3. $x_{1:n} < \theta < x_{n:n}$.

In case 1, all sample values are greater than or equal to θ, so

$$(x_i - \theta)^+ = x_i - \theta \;\text{ and }\; (x_i - \theta)^- = 0 \;\text{ for all } i = 1, 2, \ldots, n. \tag{3.5.66}$$

Thus the two components of the vector $\bar{\mathbf{Z}}^{(n)}$ are

$$\bar{Z}_1^{(n)} = \frac{1}{n}\sum_{i=1}^{n} Z_1^{(i)} = \frac{1}{n}\sum_{i=1}^{n}(x_i - \theta)^+ = \bar{x}_n - \theta, \tag{3.5.67}$$

$$\bar{Z}_2^{(n)} = \frac{1}{n}\sum_{i=1}^{n} Z_2^{(i)} = \frac{1}{n}\sum_{i=1}^{n}(x_i - \theta)^- = 0, \tag{3.5.68}$$

so the function Q takes the form

$$Q(\kappa, \sigma) = \log \kappa - \log(1 + \kappa^2) - \log(\sigma) - \frac{\sqrt{2}}{\sigma}\kappa(\bar{x}_n - \theta). \tag{3.5.69}$$

Fix $\kappa > 0$ and differentiate (3.5.69) with respect to σ to obtain

$$\frac{\partial Q(\kappa, \sigma)}{\partial \sigma} = -\frac{1}{\sigma} + \frac{\sqrt{2}}{\sigma^2}\kappa(\bar{x}_n - \theta). \tag{3.5.70}$$

It is clear that the derivative is positive for $\sigma < \sigma(\kappa)$ and negative for $\sigma > \sigma(\kappa)$, where

$$\sigma(\kappa) = \sqrt{2}\kappa(\bar{x}_n - \theta). \tag{3.5.71}$$

Consequently, for any fixed $\kappa > 0$, the function Q in (3.5.69) is maximized by $\sigma(\kappa)$. Thus, for all $\sigma, \kappa > 0$, we have

$$Q(\kappa, \sigma) \leq Q(\kappa, \sigma(\kappa)) = -\log(1 + \kappa^2) - \log\sqrt{2} - \log(\bar{x}_n - \theta) - 1. \tag{3.5.72}$$

This function of κ is strictly decreasing on $(0, \infty)$ with the least upper bound of

$$\lim_{\kappa \to 0^+} Q(\kappa, \sigma(\kappa)) = -\log\sqrt{2} - \log(\bar{x}_n - \theta) - 1, \tag{3.5.73}$$

corresponding to the values $\kappa = 0$ and $\sigma = 0$. Since these values are not admissible, formally the MLE's of κ and σ do not exist in this case. However, as

$$\kappa \to 0^+ \;\text{ and }\; \sigma(\kappa) = \sqrt{2}\kappa(\bar{x}_n - \theta) \to 0^+, \tag{3.5.74}$$

then the $\mathcal{AL}^*(\theta, \kappa, \sigma(\kappa))$ distribution converges weakly to the exponential distribution with density

$$g(y) = \begin{cases} \frac{1}{\mu} e^{-(y-\theta)/\mu} & \text{for } y \geq \theta \\ 0 & \text{otherwise,} \end{cases} \tag{3.5.75}$$

where $\mu = \bar{x}_n - \theta$ (Exercise 3.6.24). This is actually the $\mathcal{AL}(\theta, \mu, 0)$ distribution. Intuitively, it is certainly plausible to conclude that the underlying distribution is exponential if all sample values happen to be located on one side of the location parameter θ.

Similar considerations lead to the conclusion that in the second case ($\theta \geq x_{n:n}$), where we have

$$\bar{Z}_1^{(n)} = \frac{1}{n} \sum_{i=1}^{n} Z_1^{(i)} = \frac{1}{n} \sum_{i=1}^{n} (x_i - \theta)^+ = 0 \tag{3.5.76}$$

and

$$\bar{Z}_2^{(n)} = \frac{1}{n} \sum_{i=1}^{n} Z_2^{(i)} = \frac{1}{n} \sum_{i=1}^{n} (x_i - \theta)^- = \theta - \bar{x}_n, \tag{3.5.77}$$

we can choose

$$\sigma(\kappa) = \sqrt{2}\kappa^{-1}(\theta - \bar{x}_n) \tag{3.5.78}$$

to ensure that for all $\sigma, \kappa > 0$ we have

$$Q(\kappa, \sigma) \leq Q(\kappa, \sigma(\kappa)) = \log \frac{\kappa^2}{1 + \kappa^2} - \log \sqrt{2} - \log(\theta - \bar{x}_n) - 1. \tag{3.5.79}$$

This function of κ is strictly increasing on $(0, \infty)$ with the limit at infinity

$$\lim_{\kappa \to \infty} Q(\kappa, \sigma(\kappa)) = -\log \sqrt{2} - \log(\theta - \bar{x}_n) - 1. \tag{3.5.80}$$

As in the previous case, the maximum likelihood formally does not yield a solution (since the values $\kappa = \infty$ and $\sigma = 0$ are not admissible). Not surprisingly, these limiting values of the parameters do correspond to a distribution, as in the previous case, which this time is given by density

$$g(y) = \begin{cases} 0 & \text{for } y \geq \theta \\ \frac{1}{\mu} e^{-(\theta-y)/\mu} & \text{for } y \leq \theta, \end{cases} \tag{3.5.81}$$

where $\mu = \theta - \bar{x}_n$. This is so since the $\mathcal{AL}^*(\theta, \kappa, \sigma(\kappa))$ density converges to the density (3.5.81) as $\kappa \to \infty$ (Exercise 3.6.24). Again, we see that when all sample values happen to be on the left side of the location parameter θ, then the maximum likelihood approach leads to an exponential distribution.

We now move to the third case, assuming that the value of θ is strictly between $x_{1:n}$ and $x_{n:n}$, in which case both components of the vector $\bar{\mathbf{Z}}^{(n)}$ are nonzero.

Note that the likelihood function converges to zero on the boundary of its domain, so the existence and uniqueness of the MLE's is guaranteed if the following equations for the derivatives of Q have a unique solution within the domain:

$$\begin{aligned} \frac{\partial Q(\kappa,\sigma)}{\partial \sigma} &= -\frac{1}{\sigma} + \frac{\sqrt{2}}{\sigma^2}[\kappa, 1/\kappa]\bar{\mathbf{Z}}^{(n)} = 0, \\ \frac{\partial Q(\kappa,\sigma)}{\partial \kappa} &= \frac{1}{\kappa} - \frac{2\kappa}{1+\kappa^2} - \frac{\sqrt{2}}{\sigma}[1, -1/\kappa^2]\bar{\mathbf{Z}}^{(n)} = 0. \end{aligned} \tag{3.5.82}$$

These equations are equivalent to

$$[-\kappa^2, 1/\kappa^2]\bar{\mathbf{Z}}^{(n)} = 0,$$
$$\sqrt{2}\,[\kappa, 1/\kappa]\,\bar{\mathbf{Z}}^{(n)} = \sigma,$$

and lead to the following unique and explicit solution for κ and σ:

$$\widehat{\kappa}_n = \sqrt[4]{\frac{[0,1]\bar{\mathbf{Z}}^{(n)}}{[1,0]\bar{\mathbf{Z}}^{(n)}}}, \quad \widehat{\sigma}_n = \sqrt{2}\left[\sqrt[4]{\frac{[0,1]\bar{\mathbf{Z}}^{(n)}}{[1,0]\bar{\mathbf{Z}}^{(n)}}}, \sqrt[4]{\frac{[1,0]\bar{\mathbf{Z}}^{(n)}}{[0,1]\bar{\mathbf{Z}}^{(n)}}}\right]\bar{\mathbf{Z}}^{(n)}.$$

Remark 3.5.2 The corresponding MLE of the parameter μ of the $\mathcal{AL}(\theta, \mu, \sigma)$ parametrization is the sample mean

$$\widehat{\mu}_n = [1, -1]\bar{\mathbf{Z}}^{(n)} = \frac{1}{n}\sum_{i=1}^{n} X_i.$$

The above estimators can be written more explicitly as follows:

$$\widehat{\kappa}_n = \sqrt[4]{\frac{\frac{1}{n}\sum_{i=1}^{n}(x_i-\theta)^-}{\frac{1}{n}\sum_{i=1}^{n}(x_i-\theta)^+}}, \tag{3.5.83}$$

$$\widehat{\sigma}_n = \sqrt{2}\sqrt[4]{\frac{1}{n}\sum_{i=1}^{n}(x_i-\theta)^+}\sqrt[4]{\frac{1}{n}\sum_{i=1}^{n}(x_i-\theta)^-}$$
$$\times\left(\sqrt{\frac{1}{n}\sum_{i=1}^{n}(x_i-\theta)^+} + \sqrt{\frac{1}{n}\sum_{i=1}^{n}(x_i-\theta)^-}\right). \tag{3.5.84}$$

The MLE $[\hat{\kappa}_n, \hat{\sigma}_n]'$ is consistent, asymptotically normal, and efficient for the vector-parameter $[\kappa, \sigma]'$ [see, e.g., Hartley and Revankar (1974), Kozubowski and Podgórski (2000)].

Theorem 3.5.2 *Let $X_1, \dots X_n$ be i.i.d. with the $\mathcal{AL}^*(\theta, \kappa, \sigma)$ distribution where the value of θ is known. Then the MLE of $[\kappa, \sigma]$, $[\widehat{\kappa}_n, \widehat{\sigma}_n]'$, given by* (3.5.83) *and* (3.5.84) *is*

(i) *strongly consistent;*

(ii) *asymptotically bivariate normal with the asymptotic covariance matrix*

$$\mathbf{\Sigma}_{MLE} = \frac{\sigma^2}{8}(1+\kappa^2)^2\begin{bmatrix} \frac{1}{\sigma^2} & \frac{1}{\kappa\sigma}\frac{1-\kappa^2}{1+\kappa^2} \\ \frac{1}{\kappa\sigma}\frac{1-\kappa^2}{1+\kappa^2} & \frac{1}{\kappa^2}\left(1+\frac{4\kappa^2}{(1+\kappa^2)^2}\right)\end{bmatrix}, \tag{3.5.85}$$

(iii) *asymptotically efficient, namely, this asymptotic covariance matrix coincides with the inverse of the Fisher information matrix.*

Proof. The result follows from the large sample theory [see, e.g., Serfling (1980)]. Write

$$[\hat{\kappa}_n, \hat{\sigma}_n] = G(\bar{\mathbf{Z}}^{(n)}) = [G_1(\bar{Z}_1^{(n)}, \bar{Z}_2^{(n)}), G_2(\bar{Z}_1^{(n)}, \bar{Z}_2^{(n)})], \tag{3.5.86}$$

where

$$G_1(y_1, y_2) = (y_2/y_1)^{1/4} \tag{3.5.87}$$

and

$$G_2(y_1, y_2) = \sqrt{2}(y_1 y_2)^{1/4}(\sqrt{y_1} + \sqrt{y_2}). \tag{3.5.88}$$

(i) To establish the consistency of the MLE given in (3.5.86) use the continuity of G together with (3.5.48) to conclude that

$$\lim_{n\to\infty} [\hat{\kappa}_n, \hat{\sigma}_n] = G(\lim_{n\to\infty} \bar{\mathbf{Z}}^{(n)}) = G(\mathbf{m_Z}), \tag{3.5.89}$$

and then verify by substitution that

$$G(\mathbf{m_Z}) = [\kappa, \sigma]. \tag{3.5.90}$$

(ii) Similarly, we establish the asymptotic normality of the MLE with the asymptotic variance of the form $\mathbf{D\Sigma_Z D'}$, where

$$\mathbf{D} = \left[\left. \frac{\partial G_i}{\partial y_j} \right|_{(y_1,y_2)=\mathbf{m_Z}} \right]_{i,j=1}^{2} \tag{3.5.91}$$

is the matrix of partial derivatives of the vector valued function G. We skip laborious calculations leading to the asymptotic variance (3.5.85).

(iii) To prove asymptotic efficiency we need to demonstrate that $\mathbf{\Sigma}_{MLE}$ is equal to the inverse of the Fisher information matrix $I(\kappa, \sigma)$. By (3.5.1), the Fisher information matrix is

$$I(\kappa, \sigma) = \frac{1}{\sigma^2} \begin{bmatrix} \frac{\sigma^2}{\kappa^2}\left(1 + \frac{4}{(1/\kappa+\kappa)^2}\right) & -\frac{\sigma}{\kappa}\frac{1/\kappa-\kappa}{1/\kappa+\kappa} \\ -\frac{\sigma}{\kappa}\frac{1/\kappa-\kappa}{1/\kappa+\kappa} & 1 \end{bmatrix}. \tag{3.5.92}$$

Taking the inverse of this matrix, we obtain (3.5.85). □

3.5.1.6 *Case* 6: *The value of σ is known.* If the value of σ is given, then maximizing the log-likelihood function (3.5.5) is equivalent to maximizing the function

$$Q(\theta, \kappa) = \log \kappa - \log(1 + \kappa^2) - \left\{ \kappa\alpha(\theta) + \frac{1}{\kappa}\beta(\theta) \right\}, \tag{3.5.93}$$

where $\alpha(\theta)$ and $\beta(\theta)$ are as defined previously in (3.5.27). We shall proceed by maximizing (3.5.93) with respect to (θ, κ) on the sets

$$\mathbb{R} \times J_1,\ \mathbb{R} \times J_2, \ldots, \mathbb{R} \times J_n, \tag{3.5.94}$$

where

$$J_1 = \left(0, \frac{1}{n-1}\right], \qquad J_n = [n-1, \infty), \tag{3.5.95}$$

and

$$J_i = \left[\frac{i-1}{n-(i-1)}, \frac{i}{n-i}\right], \quad j = 2, 3, \ldots, n-1. \tag{3.5.96}$$

The procedure described below will result in the set of n pairs

$$(\theta_1, \kappa_1), \ldots, (\theta_n, \kappa_n), \tag{3.5.97}$$

where the ith pair maximizes the function (3.5.93) on the set $\mathbb{R} \times J_i$, $i = 1, 2, \ldots, n$. By substituting (3.5.97) into (3.5.93) and comparing the resulting values we would obtain the required MLE's of θ and κ.

The process of obtaining each of the pairs in (3.5.97) consists of two steps. First, note that by the results on estimating θ (see Case 1), the inequality

$$Q(\theta, \kappa) \leq Q(x_{i:n}, \kappa) = \log \kappa - \log(1 + \kappa^2) - \left\{ \kappa \alpha(x_{i:n}) + \frac{1}{\kappa} \beta(x_{i:n}) \right\} \tag{3.5.98}$$

holds for all $(\theta, \kappa) \in \mathbb{R} \times J_i$. We can now maximize the right-hand side of (3.5.98) with respect to $\kappa \in J_j$ using the results obtained under Case 3 (where the only unknown parameter is κ). Namely, we conclude that the right-hand side of (3.5.98) is increasing on the interval $(0, \kappa_i^0)$ and decreasing on the interval (κ_i^0, ∞), where κ_i^0 is the unique solution of the equation (3.5.29) [with $\alpha = \alpha(x_{i:n})$ and $\beta = \beta(x_{i:n})$]. Now the value κ_i that would maximize the right-hand side of (3.5.98) would be either κ_i^0 (if $\kappa_i^0 \in J_i$) or one of the endpoints of J_i (the left endpoint if it is greater than κ_i^0, or the right endpoint if it is less than κ_i^0). The algorithm below summarizes the process of obtaining the MLE's of θ and κ for this problem.

Computation of the MLE's of θ and κ when σ is known

- For $i = 1, 2, \ldots, n$, set

$$\alpha = \frac{\sqrt{2}}{\sigma} \frac{1}{n} \sum_{j=i}^{n} (x_{j:n} - x_{i:n}), \quad \beta = \frac{\sqrt{2}}{\sigma} \frac{1}{n} \sum_{j=1}^{i} (x_{i:n} - x_{j:n}). \tag{3.5.99}$$

- For $i = 1, 2, \ldots, n$, solve the equation

$$\frac{1}{\kappa} - \frac{2\kappa}{1+\kappa^2} + \frac{\beta}{\kappa^2} - \alpha = 0, \tag{3.5.100}$$

  obtaining the unique solution κ_i^0, which lies between 1 and $\sqrt{\beta/\alpha}$.

- Set

$$\kappa_1 = \begin{cases} \kappa_1^0 & \text{if } \kappa_1^0 \geq 1/(n-1), \\ \frac{1}{n-1} & \text{otherwise.} \end{cases} \tag{3.5.101}$$

$$\text{For } i = 2, 3, \ldots, n-1, \quad \kappa_i = \begin{cases} \frac{i-1}{n-(i-1)} & \text{if } \kappa_i^0 < \frac{i-1}{n-(i-1)}, \\ \kappa_i^0 & \text{if } \frac{i-1}{n-(i-1)} \leq \kappa_i^0 < \frac{i}{n-i}, \\ \frac{i}{n-i} & \text{if } \kappa_i^0 \geq \frac{i}{n-i}, \end{cases} \tag{3.5.102}$$

$$\kappa_n = \begin{cases} \kappa_n^0 & \text{if } \kappa_n^0 \geq n-1, \\ n-1 & \text{otherwise.} \end{cases} \tag{3.5.103}$$

- For $i = 1, 2, \ldots, n$, substitute the two values $\theta_i = x_{i:n}$ and κ_i given by (3.5.101)–(3.5.103) into (3.5.93) and choose the pair that results in the maximum value.

The method for estimating θ and κ is more complex compared with other cases considered so far and may be time consuming for large problems. The consistency as well as asymptotic normality and efficiency of the estimators may be obtained similarly as in the case of estimating all three parameters.

3.5.1.7 *Case 7: The values of all three parameters are unknown.* Let us start by noting that the maximum likelihood estimators and their asymptotic distributions for this case were derived in Hartley and Revankar (1974) and Hinkley and Revankar (1977), although these authors worked in the context of the log-Laplace model and under another parametrization.

We need to maximize the log-likelihood function (3.5.5) with respect to all three parameters, which is equivalent to maximizing the function

$$Q(\theta, \kappa, \sigma) = -\log \sigma + \log \frac{\kappa}{1+\kappa^2} - \frac{\sqrt{2}}{\sigma}\left\{\kappa\alpha(\theta) + \frac{1}{\kappa}\beta(\theta)\right\}, \tag{3.5.104}$$

where this time

$$\alpha(\theta) = \frac{1}{n}\sum_{j=1}^{n}(x_j - \theta)^+ \text{ and } \beta(\theta) = \frac{1}{n}\sum_{j=1}^{n}(x_j - \theta)^-. \tag{3.5.105}$$

We proceed by first fixing the value of θ and then applying the results obtained under Case 5 (when the value of θ is known).

When $\theta \le x_{1:n}$, then by the relation (3.5.72) (see Case 5) we conclude that for any $\kappa, \sigma > 0$

$$Q(\theta, \kappa, \sigma) \le -\log(1+\kappa^2) - \log\sqrt{2} - \log(\bar{x}_n - \theta) - 1. \tag{3.5.106}$$

Similarly, when $\theta \ge x_{n:n}$, then by (3.5.79), we have

$$Q(\theta, \kappa, \sigma) \le \log \frac{\kappa^2}{1+\kappa^2} - \log\sqrt{2} - \log(\theta - \bar{x}_n) - 1. \tag{3.5.107}$$

If $x_{1:n} < \theta < x_{n:n}$, then both quantities $\alpha(\theta)$ and $\beta(\theta)$ given in (3.5.105) are positive. Thus, using the results of Case 5, we have

$$Q(\theta, \kappa, \sigma) \le Q(\theta, \hat{\kappa}, \hat{\sigma}), \tag{3.5.108}$$

where the quantities $\hat{\kappa}$ and $\hat{\sigma}$ are the MLE's of κ and σ (derived under the case when the value of θ is known) given by (3.5.83) and (3.5.84). Substituting these values into the right-hand side of (3.5.108), we obtain after some algebra

$$Q(\theta, \kappa, \sigma) \le g(\theta), \tag{3.5.109}$$

where

$$g(\theta) = -\log\sqrt{2} - 2\log(\sqrt{\alpha(\theta)} + \sqrt{\beta(\theta)}) - \sqrt{\alpha(\theta)}\sqrt{\beta(\theta)}. \tag{3.5.110}$$

Note that for $\theta \in (x_{1:n}, x_{2:n})$ we have

$$\alpha(\theta) = \frac{1}{n}\sum_{j=2}^{n}(x_{j:n} - \theta) \text{ and } \beta(\theta) = \frac{1}{n}(\theta - x_{1:n}), \tag{3.5.111}$$

so

$$\lim_{\theta \to x_{1:n}^+} \alpha(\theta) = \bar{x}_n - x_{1:n} \text{ and } \lim_{\theta \to x_{1:n}^+} \beta(\theta) = 0. \tag{3.5.112}$$

Thus

$$\lim_{\theta \to x_{1:n}^+} g(\theta) = -\log \sqrt{2} - \log(\bar{x}_n - x_{1:n}). \tag{3.5.113}$$

The limit in (3.5.113) is larger than the value $Q(\theta, \kappa, \sigma)$ at any $\theta \le x_{1:n}, 0 < \kappa, 0 < \sigma$. Indeed, in view of (3.5.106), for $\theta \le x_{1:n}$ we have

$$Q(\theta, \kappa, \sigma) \le -\log \sqrt{2} - \log(\bar{x}_n - x_{1:n}) - 1, \tag{3.5.114}$$

since here the function Q attains its least upper bound for $\kappa = \sigma = 0$ and $\theta = x_{1:n}$. In view of the above, we can restrict attention to the values $\theta > x_{1:n}$ when maximizing the function $Q(\theta, \kappa, \sigma)$ over $\theta \in \mathbb{R}, 0 < \kappa, 0 < \sigma$.

Similar arguments show that

$$\lim_{\theta \to x_{n:n}^-} g(\theta) = -\log \sqrt{2} - \log(x_{n:n} - \bar{x}_n), \tag{3.5.115}$$

which is 1 larger than the supremum of the function $Q(\theta, \kappa, \sigma)$ over the values $\theta \ge x_{n:n}, 0 < \kappa, 0 < \sigma$ (the supremum is obtained by taking $\kappa \to \infty$ and $\theta = x_{n:n}$ in the right-hand side of (3.5.107)). Consequently, we can rule out the values $\theta \ge x_{n:n}$ from further consideration.

This leaves us with the problem of maximizing the function $Q(\theta, \kappa, \sigma)$ given by (3.5.104) under the conditions

$$x_{1:n} < \theta < x_{n:n}, \quad 0 < \kappa < \infty, \quad 0 < \sigma < \infty, \tag{3.5.116}$$

or equivalently, maximizing the function $g(\theta)$ in (3.5.110) on the set

$$A = \{\theta : x_{1:n} < \theta < x_{n:n}\}. \tag{3.5.117}$$

Clearly, this is equivalent to the minimization of the function

$$h(\theta) = 2\log(\sqrt{\alpha(\theta)} + \sqrt{\beta(\theta)}) + \sqrt{\alpha(\theta)}\sqrt{\beta(\theta)} \tag{3.5.118}$$

with respect to the same values of θ. It turns out that the infimum of the function h on the set A is given by one of the values

$$h(x_{j:n}), \quad j = 1, 2, \ldots, n. \tag{3.5.119}$$

This follows from the following lemma (see Exercise 3.6.25).

Lemma 3.5.2 *The function h defined in* (3.5.118) *is continuous on the closed interval* $[x_{1:n}, x_{n:n}]$ *and concave down on each of the subintervals* $(x_{j:n}, x_{j+1:n})$, $j = 1, 2, \ldots, n-1$.

Consequently, to find the MLE's of θ, κ, and σ we should proceed as follows.

Step 1: Evaluate the n values (3.5.119) and choose a positive integer $r \le n$ such that

$$h(x_{r:n}) \le h(x_{j:n}) \quad j = 1, 2, \ldots, n. \tag{3.5.120}$$

Step 2: Set $\theta = x_{r:n}$ and find the MLE's of κ and σ (derived previously under Case 5).
There are three scenarios in Step 2:

- If $r = 1$ ($\theta = x_{1:n}$), then as in Case 5 the MLE's do not exist (as the likelihood is maximized by $\kappa = \sigma = 0$), but the likelihood approach leads to (positive) exponential distribution with density (3.5.75) with $\mu = \bar{x}_n - x_{1:n}$.

- If $r = n$ ($\theta = x_{n:n}$), then again formally the MLE's do not exist, but the likelihood approach does lead to the (negative) exponential distribution with density (3.5.81) with $\mu = x_{n:n} - \bar{x}_n$.
- If $1 < r < n$, then the MLE's are

$$\begin{aligned} \hat{\theta}_n &= X_{r:n}, \\ \hat{\kappa}_n &= \sqrt[4]{\beta(\hat{\theta}_n)}/\sqrt[4]{\alpha(\hat{\theta}_n)}, \\ \hat{\sigma}_n &= \sqrt{2}\sqrt[4]{\alpha(\hat{\theta}_n)}\sqrt[4]{\beta(\hat{\theta}_n)}\left(\sqrt{\alpha(\hat{\theta}_n)} + \sqrt{\beta(\hat{\theta}_n)}\right), \end{aligned} \tag{3.5.121}$$

where

$$\alpha(\hat{\theta}_n) = \frac{1}{n}\sum_{j=1}^{n}(x_j - \hat{\theta}_n)^+ \text{ and } \beta(\hat{\theta}_n) = \frac{1}{n}\sum_{j=1}^{n}(x_j - \hat{\theta}_n)^-. \tag{3.5.122}$$

Thus the problem of estimating all three parameters of the $\mathcal{AL}^*(\theta, \kappa, \sigma)$ distribution admits a solution that can be determined with ease. The resulting MLE's are consistent, asymptotically normal, and asymptotically efficient with the asymptotic covariance matrix equal to the inverse of the Fisher information matrix (3.5.1). We refer the reader to Hartley and Revankar (1974) and Hinkley and Revankar (1977) for technical details regarding the asymptotic results on the MLE's.

3.6 Exercises

The readers may find the 26 exercises below somewhat challenging. Again we recommend that special attention will be paid to these exercises. A number of them deal with the most recent results on asymmetric Laplace distributions.

Exercise 3.6.1 Let X have an asymmetric Laplace distribution with p.d.f. (3.0.1). Derive the mean, median, mode, and variance of X.

Exercise 3.6.2 Let X have the skewed Laplace distribution with p.d.f. (3.0.3).

(a) Find the mean and the variance of X.

(b) Show that the mode of X and the α-quantile of X are both equal to θ.

(c) Show that the characteristic function of X is

$$\varphi(t) = \alpha(1-\alpha)e^{i\theta t}\left(\frac{1}{1-\alpha+it} + \frac{1}{it-\alpha}\right).$$

What is the moment generating function of X?

Exercise 3.6.3 Consider a hyperbolic distribution with density

$$f(x) = \frac{\sqrt{\alpha^2 - \beta^2}}{2\alpha\delta K_1(\delta\sqrt{\alpha^2 - \beta^2})} e^{-\alpha\sqrt{\delta^2 + (x-\theta)^2} + \beta(x-\theta)}, \quad -\infty < x < \infty, \tag{3.6.1}$$

where

$$\alpha > 0, \quad 0 \le |\beta| < \alpha, \quad -\infty < \theta < \infty, \quad \delta > 0$$

and $K_1(\cdot)$ is the modified Bessel function of the third kind with index 1 (see the appendix).

(a) Show that as

$$\delta \to \infty, \quad \frac{\delta}{\sqrt{\alpha^2 - \beta^2}} \to \sigma^2 > 0, \quad \beta \to 0,$$

density (3.6.1) converges (pointwise) to the density of the normal distribution with mean θ and variance σ^2.

(b) Show that as $\delta \to 0$, density (3.6.1) converges (pointwise) to an asymmetric Laplace density

$$g(x) = C \begin{cases} e^{-(\alpha-\beta)|x-\theta|} & \text{for } x \geq \theta, \\ e^{-(\alpha+\beta)|x-\theta|} & \text{for } x < \theta. \end{cases} \tag{3.6.2}$$

What is the normalizing constant C in (3.6.2)?

(c) Show that the density (3.6.2) corresponds to the $\mathcal{AL}^*(\theta, \kappa, \sigma)$ distribution, where

$$\sigma = \sqrt{\frac{2}{\alpha^2 - \beta^2}}, \quad \text{and } \kappa = \sqrt{\frac{\alpha - \beta}{\alpha + \beta}}.$$

Thus the latter distribution arises as a limit of hyperbolic distributions [Barndorff-Nielsen (1977)].

Exercise 3.6.4 For $a < 0 < b$ and $n \in \mathbb{N}$ consider a r.v. X_n with p.d.f.

$$f_n(x) = \frac{n+1}{b-a} \begin{cases} \left(\frac{x-a}{-a}\right)^n & \text{for } a \leq x \leq 0 \\ \left(\frac{b-x}{b}\right)^n & \text{for } 0 \leq x \leq b. \end{cases} \tag{3.6.3}$$

(a) Show that the function f_n is a genuine probability density function.

(b) Let $a = -nA$ and $b = nB$, where $A, B > 0$. Show that as $n \to \infty$, then for every $x \in \mathbb{R}$, the density $f_n(x)$ converges to

$$f(x) = \frac{1}{A+B} \begin{cases} e^{-|x|/A} & \text{for } x \leq 0 \\ e^{-|x|/B} & \text{for } x \geq 0. \end{cases} \tag{3.6.4}$$

(c) Show that the function (3.6.4) is the p.d.f. of the $\mathcal{AL}^*(\sigma, \kappa)$ distribution with

$$\sigma = \sqrt{AB} \text{ and } \kappa = \sqrt{A/B}$$

(cf. Exercise 2.7.56).

Exercise 3.6.5 Establish the relations (3.1.4) and (3.1.5). Further, show that for every $\sigma > 0$ the functions of μ and κ, given by (3.1.4) and (3.1.5), respectively, are strictly decreasing on their domains, and prove the relations given in (3.1.6).

Exercise 3.6.6 Let $f_{\theta,\kappa,\sigma}(x)$ be the density (3.1.10) of an AL distribution.

(a) Show that for any $x \in \mathbb{R}$ we have

$$f_{\theta,\kappa,\sigma}(-x) = f_{-\theta,1/\kappa,\sigma}(x). \tag{3.6.5}$$

What is the interpretation of (3.6.5) in terms of random variables?

(b) Show that for $0 < \kappa < 1$ and $x > 0$ we have

$$f_{\theta,\kappa,\sigma}(\theta + x) > f_{\theta,\kappa,\sigma}(\theta - x). \tag{3.6.6}$$

What happens for $\kappa > 1$? For $\kappa = 1$?

(c) Clearly, when $x \to \infty$, then densities on both sides of (3.6.6) converge to zero. Investigate whether they converge with the same rate, or one of them converges to zero faster than the other one.

(d) Repeat parts (a)–(c) using the $\mathcal{AL}(\theta, \mu, \sigma)$ parametrization.

Exercise 3.6.7 In this problem we investigate the derivatives of an AL density.

(a) Show that the AL densities (3.1.10) have derivatives of any order (except at $x = \theta$), which are expressed by the following formulas:

$$f^{(n)}_{\theta,\kappa,\sigma}(x) = \begin{cases} (-1)^n \left(\frac{\sqrt{2}\kappa}{\sigma}\right)^{n+1} \frac{1}{1+\kappa^2} e^{-\sqrt{2}\kappa|x-\theta|/\sigma}, & \text{if } x > \theta, \\ \left(\frac{\sqrt{2}}{\sigma\kappa}\right)^{n+1} \frac{\kappa^2}{1+\kappa^2} e^{-\sqrt{2}|x-\theta|/(\kappa\sigma)}, & \text{if } x < \theta. \end{cases} \tag{3.6.7}$$

(b) Find the limits

$$\lim_{x\to\theta+} (-1)^n f^{(n)}_{\theta,\kappa,\sigma}(x) \text{ and } \lim_{x\to\theta-} f^{(n)}_{\theta,\kappa,\sigma}(x), \tag{3.6.8}$$

check for what values of n or the parameters, if any, the two limits in (3.6.8) are equal, and give an interpretation of the equality.

(c) Show that if $0 < \kappa \le 1$ and $x \ge \sigma n/\sqrt{2}$, where n is a positive integer, then

$$(-1)^n f^{(n)}_{\theta,\kappa,\sigma}(\theta + x) \ge f^{(n)}_{\theta,\kappa,\sigma}(\theta - x). \tag{3.6.9}$$

What happens if $\kappa > 1$? If $x \le \sigma n/\sqrt{2}$?

Exercise 3.6.8 Show that the AL density f given by (3.1.10) is completely monotone on (θ, ∞) and absolutely monotone on $(-\infty, \theta)$ (that is, for any $k = 0, 1, 2, \ldots$, we have $(-1)^k f^{(k)}(x) \ge 0$ for $x > \theta$ and $f^{(k)}(x) \ge 0$ for $x < \theta$).

Exercise 3.6.9 Establish formulas (3.1.16) and (3.1.19) for the m.g.f. of an AL distribution.

Exercise 3.6.10 Let $Y \sim \mathcal{AL}^*(\theta, \kappa, \sigma)$.

(a) Show that the ath absolute moment of $Y - \theta$ is finite for any $a > -1$, and is given by (3.1.26).

(b) Show that the mean absolute deviation of Y is given by (3.1.27).

Exercise 3.6.11 Calculate the nth moment about zero of the $\mathcal{AL}(\theta, \kappa, \sigma)$ distribution.

Exercise 3.6.12 Let $Y \sim \mathcal{AL}^*(\theta, \kappa, \sigma)$.

(a) Show that the coefficients of skewness and kurtosis of Y, defined by (2.1.21) and (2.1.22), are given by (3.1.30) and (3.1.31), respectively.

(b) Show that the coefficient of skewness is bounded by 2 in absolute value and decreases monotonically from 2 to -2 as κ increases from zero to infinity.

(c) Show that the coefficient of kurtosis varies from three to six.

Exercise 3.6.13 The κ-criterion is a preliminary selection test useful in reducing the number of plausible models for a given set of data [see, e.g., Elderton (1938), Hirschberg et al. (1989)]. The κ-criterion is defined as

$$\kappa = \frac{\beta_1(\beta_2 + 3)^2}{4(4\beta_2 - 3\beta_1)(2\beta_2 - 3\beta_1 - 6)}, \tag{3.6.10}$$

where β_1 is the square of the coefficient of skewness γ_1 and β_2 is the (un-adjusted) kurtosis $\gamma_2 + 3$ [cf. (2.1.21)–(2.1.22), Chapter 2] of the underlying probability distribution. It is clear that the κ-criterion is zero for the symmetric Laplace distribution (as it is for any symmetric distribution with a finite fourth moment since in this case $\beta_1 = 0$). Derive the κ-criterion for the $\mathcal{AL}(\mu, \sigma)$ distribution (not to be confused with the *parameter* κ of the distribution). What is the range of the κ-criterion in this case?

Exercise 3.6.14 Let $Y \sim \mathcal{AL}^*(\theta, \kappa, \sigma)$. Establish the mode-median-mean inequalities (3.1.35).

Exercise 3.6.15 A common measure of skewness of a probability distribution with distribution function F is given by the limit

$$\lim_{x\to\infty} \frac{1 - F(x) - F(-x)}{1 - F(x) + F(-x)}.$$

This limit is equal to zero if the distribution is symmetric about zero. Show that for an AL distribution with distribution function (3.1.11), the limit is equal to 1 if $\kappa < 1$ $(\mu > 0)$, is equal to -1 if $\kappa > 1$ $(\mu < 0)$ and is equal to

$$\frac{e^{\frac{\sqrt{2}\theta}{\sigma}} - e^{-\frac{\sqrt{2}\theta}{\sigma}}}{e^{\frac{\sqrt{2}\theta}{\sigma}} + e^{-\frac{\sqrt{2}\theta}{\sigma}}}$$

for $\kappa = 1$. Note that for an $\mathcal{AL}(0, \mu, \sigma)$ distribution, which is a special case of a geometric stable distribution $GS_\alpha(\sigma/\sqrt{2}, \beta, \mu)$ with $\alpha = 2$ [see (4.4.7)], the above limit is equal to sign(μ). Since for geometric stable distributions this limit is equal to β [see, e.g., Kozubowski (1994a)], for consistency, we set $\beta = \text{sign}(\mu)$ for a GS law with $\alpha = 2$.

Exercise 3.6.16 Show that an $\mathcal{AL}^*(\theta, \kappa, \sigma)$ r.v. Y admits the representation (3.2.5).

Exercise 3.6.17 Let $Y \sim \mathcal{AL}^*(0, \kappa, \sigma)$, let $Y_p^{(i)}, i \geq 1$, be i.i.d. variables having the $\mathcal{AL}^*(0, \kappa_p, \sqrt{p}\sigma)$ distribution, where k_p is given by (3.4.12), and let ν_p be a geometric random variable with $P(\nu_p = n) = (1-p)^{n-1}p$, $n \geq 1$, which is independent of the $Y_p^{(i)}$'s. Show that for each $p \in (0, 1)$, representation (3.4.10) is valid.

Exercise 3.6.18 Show that if in Proposition 3.4.7 the mean and mean deviation *about the mean* are prescribed, that is, if the condition (3.4.39) is replaced by

$$EX = c_1 \in \mathbb{R} \text{ and } E|X - c_1| = c_2 > 0 \text{ for } X \in \mathcal{C},$$

then the maximum entropy is attained by the classical symmetric Laplace distribution with density $f(x) = \frac{1}{2c_2}e^{-|x-c_1|/c_2}$ [Kapur (1993)].

Exercise 3.6.19 Let $X_{1:n} \leq \cdots \leq X_{n:n}$ be the order statistics connected to a random sample of size n from the Laplace $\mathcal{AL}^*(\theta, \kappa, \sigma)$ distribution where κ and σ are known, while θ is to be estimated by the method of maximum likelihood.

(a) Show that the likelihood function is maximized by any θ that minimizes the function Q given by (3.5.7).

(b) Show that the function Q is continuous on $\mathbb{R}$ and linear on the intervals $I_0, I_1, \ldots I_n$ given by (3.5.8) and (3.5.9).

(c) Show that the function Q is decreasing on I_0, increasing on I_n, and on I_j $(1 \leq j \leq n-1)$ the behavior of Q is given by (3.5.10).

(d) Conclude that if condition (3.5.11) holds, then any statistic of the form (3.5.12) is an MLE of θ, and if it does not, then the MLE of θ is given by (3.5.13).

(e) Derive the mean and variance of the MLE. Check whether the estimator is efficient (i.e., its variance attains the Cramér–Rao lower bound).

Exercise 3.6.20 Show that if $X \sim \mathcal{AL}^*(\theta, \kappa, \sigma)$, then $Y = g(X)$, where the function g is given by (3.5.24), has an exponential distribution with mean σ and variance σ^2. This is a generalization of the fact that the r.v. $|X|$ is exponential whenever X is a symmetric Laplace variable (with mean 0).

Exercise 3.6.21 Let $X_1, \dots, X_n$ be a random sample from the $\mathcal{AL}(\theta, \mu, \sigma)$ distribution. Derive the method of moments estimators of each of the parameters assuming that the values of the other two are known. Investigate consistency and asymptotic normality of the estimators. Compare with the corresponding results for the MLE's.

Exercise 3.6.22 Let $X_1, \dots, X_n$ be a random sample from the $\mathcal{AL}(\theta, \mu, \sigma)$ distribution.

(a) Assuming that the value of θ is known (and for convenience set to zero), show that the method of moments estimators (MME's) of μ and σ are given by

$$\tilde{\mu}_n = \bar{X}_n = \frac{1}{n}\sum_{i=1}^{n} X_i, \qquad \tilde{\sigma}_n = \sqrt{\frac{1}{n}\sum_{i=1}^{n} X_i^2 - 2\bar{X}_n^2}. \tag{3.6.11}$$

Further, show that the estimator $(\tilde{\mu}_n, \tilde{\sigma}_n)'$ is strongly consistent and its asymptotic distribution is normal with (vector) mean zero and the covariance matrix

$$\mathbf{\Sigma}_{MME} = \frac{1}{4\sigma^2}\begin{bmatrix} 4\sigma^2 + 4\mu^2\sigma^2 & 2\mu\sigma^3 \\ 2\mu\sigma^3 & 4\mu^4 + 8\mu^2\sigma^2 + 5\sigma^5 \end{bmatrix}. \tag{3.6.12}$$

[Kozubowski and Podgórski (2000)].

Hint: Consider an auxiliary sequence of bivariate i.i.d. random vectors $\mathbf{V}_i = (X_i, X_i^2)'$. Show that the vector mean and covariance matrix of $\mathbf{V}_i$ are

$$\mathbf{m}_{\mathbf{V}} = \begin{bmatrix} \mu \\ 2\mu^2 + \sigma^2 \end{bmatrix}, \quad \mathbf{\Sigma}_{\mathbf{V}} = \begin{bmatrix} \sigma^2 + \mu^2 & 5\mu\sigma^2 + 4\mu^3 \\ 5\mu\sigma^2 + 4\mu^3 & 20\mu^4 + 32\mu^2\sigma^2 + 5\sigma^4 \end{bmatrix}.$$

Then use the fact that the Law of Large Numbers and the Central Limit Theorem are valid for the sequence $\{\mathbf{V}_i\}$.

(b) Derive the MME's for the remaining pairs of the parameters (assuming that the value of the remaining parameter is known) and study their consistency and asymptotic normality.

(c) Investigate the method of moments estimation of all three parameters.

Exercise 3.6.23 Show that the Fisher information matrix corresponding to the $\mathcal{AL}^*(\theta, \kappa, \sigma)$ distribution is given by (3.5.1).

Exercise 3.6.24 Let X have an $\mathcal{AL}^*(\theta, \kappa, \sigma)$ distribution.

(a) Suppose that $\sigma = \sigma(\kappa) = \sqrt{2}\kappa\mu$ for some $\mu > 0$. Show that when $\kappa \to 0$, then the corresponding AL density (3.1.10) converges to the exponential density (3.5.75).

(b) Suppose that $\sigma = \sigma(\kappa) = \sqrt{2}\kappa^{-1}\mu$ for some $\mu > 0$. Show that when $\kappa \to 0$, then the corresponding AL density (3.1.10) converges to the exponential density (3.5.81).

Exercise 3.6.25 Prove Lemma 3.5.2.

Hint: To establish the concavity show that $h''(\theta) < 0$ for all $\theta \in (x_{1:n}, x_{n:n})$, $j = 1, 2, \dots n$.

Exercise 3.6.26 Let $x_{1:3} < x_{2:3} < x_{3:3}$ be particular realizations of the order statistics corresponding to a random sample of size $n = 3$ from the $\mathcal{AL}^*(\theta, \kappa, \sigma)$ distribution. Derive the MLE's of all three parameters. Under what conditions on $x_{j:3}$'s the MLE of θ is $\hat{\theta}_3 = x_{2:3}$? When does the maximum likelihood approach lead to an exponential distribution?

4

Related Distributions

Symmetric Laplace distributions can be extended in various ways. As we discussed in Chapter 3, skewness may be introduced, leading to asymmetric Laplace laws. Next, one can consider a more general class of distributions whose ch.f.'s are positive powers of Laplace ch.f.'s. These are marginal distributions of the Lévy process $\{Y(t), t \geq 0\}$ with independent increments, for which $Y(1)$ has symmetric or asymmetric Laplace distribution. We term such a process the Laplace motion. Finally, one obtains a wider class of limiting distributions consisting of geometric stable laws, by allowing for infinite variance of the components in the geometric compounds (2.2.1). More generally, if the random number of components in the summation (2.2.1) is distributed according to a discrete law ν on positive integers, a wider class of *ν-stable* laws is obtained as the limiting distributions. This chapter is devoted to a discussion of all such related distributions and random variables.

Barndorff-Nielsen (1977) introduced a general class of hyperbolic distributions [see also Eberlein and Keller (1995) for applications in finance]. The Bessel function distributions discussed in this chapter could be studied through the theory of this class. However, hyperbolic distributions do not constitute a direct generalization of Laplace laws. Thus we decided not to present the following material through this alternative approach as it would take us "too far" from the classical Laplace distribution.

4.1 Bessel function distribution

If $X_1, \ldots, X_n$ are i.i.d. Laplace r.v.'s with mean zero and variance σ^2, then their sum S_n has the ch.f.

$$\psi_{S_n}(t) = \prod_{i=1}^{n} \psi_{X_i}(t) = \left(\frac{1}{1 + \frac{1}{2}\sigma^2 t^2} \right)^n, \quad -\infty < t < \infty. \tag{4.1.1}$$

By infinite divisibility of the Laplace distribution, the function (4.1.1) is a legitimate ch.f. even when n is not an integer (but is still positive). More generally, taking in (4.1.1) the ch.f. of an asymmetric Laplace distribution with the mode at zero (which is still infinitely divisible) we conclude that the function

$$\psi(t) = \left(\frac{1}{1 + \frac{1}{2}\sigma^2 t^2 - i\mu t} \right)^{\tau}, \quad -\infty < t < \infty, \tag{4.1.2}$$

is a characteristic function for any $\mu \in \mathbb{R}$ and $\sigma, \tau \geq 0$. The function (4.1.2) yields an AL ch.f. for $\tau = 1$ and symmetric Laplace ch.f. for $\tau = 1$ and $\mu = 0$ (and gamma ch.f. for $\sigma = 0$). Not surprisingly, it is known in the literature as a *generalized (asymmetric) Laplace distribution* [see, e.g., Mathai (1993), Kozubowski and Podgórski (1999c)]. Since the corresponding density function can be written in terms of the Bessel function of the third kind (defined in the appendix), *Bessel function distribution* is another name for this class [see, e.g., McKay (1932)]. The formula for the density appeared in Pearson et al. (1929) in connection with the distribution of sample covariance for a random sample drawn from a bivariate normal population [see also Pearson et al. (1932) and Bhattacharyya (1942)]. This distribution arises as a mixture of normal distributions with stochastic variance having gamma distribution, so it is also called *variance gamma* model [see, e.g., Madan and Seneta (1990)]. Such mixtures (with mean zero) were introduced in Teichroew (1957), who commented that in some practical problems the variable of interest may be normal with variance varying with time. Rowland and Sichel (1960) applied the generalized Laplace model to logarithms of the ratios of duplicate check-sampling values (of gold ore) in South African gold mines, reporting an excellent fit. Sichel (1973) applied this distribution for modeling the size of diamonds mined in southwest Africa. More recently, the variance gamma model became popular among some financial modelers, due to its simplicity, flexibility, and an excellent fit to empirical data [see, e.g., Madan and Seneta (1990), Madan et al. (1998), Levin and Tchernitser (1999), Kozubowski and Podgórski (1999ac)].

4.1.1 Definition and parametrizations. We start with a definition, terminology, and some notation. We define a general four-parameter family of distributions, although in what follows we often consider a three-parameter model with the location parameter fixed at zero.

Definition 4.1.1 *A random variable Y is said to have a generalized asymmetric Laplace (GAL) distribution if its ch.f. is given by*

$$\psi(t) = \frac{e^{i\theta t}}{(1 + \frac{1}{2}\sigma^2 t^2 - i\mu t)^{\tau}}, \quad -\infty < t < \infty, \tag{4.1.3}$$

where $\theta, \mu \in \mathbb{R}$ and $\sigma, \tau \geq 0$. We denote such distribution by $\mathcal{GAL}(\theta, \mu, \sigma, \tau)$ and write $Y \sim \mathcal{GAL}(\theta, \mu, \sigma, \tau)$.

Remark 4.1.1 The terminology for this family of distributions is not well-established and various names can be equally justified. First, in McKay (1932) and Johnson et al. (1994) we have two types of Bessel function distributions: Bessel I-function distribution (not considered here) and Bessel K-function distribution (which is an alternative name for generalized Laplace distributions). The name *Bessel K-function distribution* is then historically well justified. On the other hand, in various contexts a more compact name is handier: we prefer Laplace motion instead of Bessel K-function motion. In this book we use the terms *Bessel function distribution* and *variance-gamma distribution* interchangeably with the name *generalized Laplace distribution* used in Definition 4.1.1.

While the distribution is well-defined for every $\theta, \mu \in \mathbb{R}$ and $\sigma, \tau \geq 0$, we have the following special cases. If $\theta = \mu = \sigma = 0$, then $\psi(t) = 1$ for every $t \in \mathbb{R}$, and the distribution is degenerate at 0. For $\theta = \sigma = 0$ and $\mu > 0$, we have a gamma r.v. with the scale parameter μ and the shape parameter τ (which reduces to an exponential variable for $\tau = 1$). For $\tau = 1$, we obtain an AL distribution, which for $\mu = 0$ and $\sigma > 0$ yields a symmetric Laplace distribution with mean θ and variance σ^2.

The GAL ch.f. (4.1.3) with $\sigma > 0$ can be factored similarly as an AL ch.f.,

$$\psi(t) = e^{i\theta t}\left(\frac{1}{1+i\frac{\sqrt{2}}{2}\sigma\kappa t}\right)^{\tau}\left(\frac{1}{1-i\frac{\sqrt{2}}{2\kappa}\sigma t}\right)^{\tau}, \tag{4.1.4}$$

where the additional parameter $\kappa > 0$ is related to μ and σ as before,

$$\mu = \frac{\sigma}{\sqrt{2}}\left(\frac{1}{\kappa} - \kappa\right) \quad \text{and} \quad \kappa = \frac{\sqrt{2}\sigma}{\mu + \sqrt{2\sigma^2 + \mu^2}} = \frac{\sqrt{2\sigma^2+\mu^2} - \mu}{\sqrt{2}\sigma}. \tag{4.1.5}$$

It will be convenient to express certain properties of the GAL distributions in the $(\theta, \kappa, \sigma, \tau)$-parametrization, using the notation $\mathcal{GAL}^*(\theta, \kappa, \sigma, \tau)$ for the distribution given by (4.1.4). Analogous to the AL case, the parameter κ is scale invariant, while σ is a genuine scale parameter [in the $(\theta, \kappa, \sigma, \tau)$-parametrization].

The following result extends an analogous property of AL laws (Proposition 3.1.1).

Proposition 4.1.1 *Let $X \sim \mathcal{GAL}^*(\theta, \kappa, \sigma, \tau)$ and let c be a nonzero real constant. Then*

(i) $c + X \sim \mathcal{GAL}^*(c+\theta, \kappa, \sigma, \tau)$;

(ii) $cX \sim \mathcal{GAL}^*(c\theta, \kappa_c, |c|\sigma, \tau)$, *where* $\kappa_c = \kappa^{sign(c)}$.

Remark 4.1.2 Note that in particular, if $X \sim \mathcal{GAL}^*(\theta, \kappa, \sigma, \tau)$, then $-X \sim \mathcal{GAL}^*(-\theta, 1/\kappa, \sigma, \tau)$.

Since θ is a location parameter, we shall often assume that $\theta = 0$ and denote the corresponding distribution as either $\mathcal{GAL}(\mu, \sigma, \tau)$ or $\mathcal{GAL}^*(\kappa, \sigma, \tau)$, depending on the parametrization. For $\theta = 0$ and $\sigma = 1$ we shall refer to the GAL distribution as *standard* and write $\mathcal{GAL}(\mu, \tau)$ and $\mathcal{GAL}^*(\kappa, \tau)$, respectively, for the distributions $\mathcal{GAL}(0, \mu, 1, \tau)$ and $\mathcal{GAL}^*(0, \kappa, 1, \tau)$. Table 4.1 below contains a summary of the notation and special cases.

4.1.2 Representations. A Bessel function random variable admits certain representations analogous to those corresponding to AL random variables. First, we shall consider a mixture representation in terms of normal distribution with a stochastic mean and variance. Then we shall discuss a representation as a convolution of two gamma distributions, analogous to the previously considered representations of (symmetric and asymmetric) Laplace r.v.'s in terms of exponential r.v.'s. Finally, we shall discuss the relation between the Bessel function distribution and a sample covariance for bivariate normal random samples.

4.1.2.1 *Mixture of normal distributions.* Let Z be a standard normal random variable. Then for any $\mu \in \mathbb{R}$ and $\sigma > 0$, the r.v.

$$\mu + \sigma Z \tag{4.1.6}$$

has normal distribution with mean μ and variance σ^2. The ch.f. of the latter r.v. is

$$\phi_{\mu,\sigma}(t) = e^{i\mu t - \frac{1}{2}\sigma^2 t^2}, \quad t \in \mathbb{R}. \tag{4.1.7}$$

Now suppose that the mean and variance of this normal r.v. are multiplied by an independent and positive random variable W and let us write the resulting new random variable Y as the following function of Z and W:

$$Y = \mu W + \sigma\sqrt{W}Z. \tag{4.1.8}$$

Case	Distribution	Notation	Density
$\theta=0$ $\tau=1$ $\sigma=0$ $\mu>0$	Exponential (with mean μ)	$\mathcal{GAL}(0,\mu,0,1)$ $\mathcal{AL}(\mu,0)$ $\mathcal{GAL}(\mu,0,1)$ $G(1,\mu), \mathcal{E}(\mu)$	$\frac{1}{\mu}e^{-x/\mu}\quad (x>0)$
$\theta=0$ $\sigma=0$ $\mu>0$	Gamma with parameters $\alpha=\tau, \beta=\mu$	$\mathcal{GAL}(0,\mu,0,\tau)$ $\mathcal{GAL}(\mu,0,\tau)$ $G(\tau,\mu)$	$\frac{x^{\tau-1}e^{-x/\mu}}{\mu^\tau\Gamma(\tau)}\quad (x>0)$
$\tau=1$ $\sigma>0$ $\mu=0$	Symmetric Laplace	$\mathcal{L}(\theta,\sigma)$, $\mathcal{AL}(\theta,0,\sigma)$ $\mathcal{GAL}(\theta,0,\sigma,1)$	$\frac{1}{\sqrt{2}\sigma}e^{-\sqrt{2}\|x-\theta\|/\sigma}\quad (x\in\mathbb{R})$
$\tau=1$ $\sigma>0$ $\mu\neq0$	Asymmetric Laplace	$\mathcal{AL}(\theta,\mu,\sigma)$, $\mathcal{GAL}(\theta,\mu,\sigma,1)$	$\frac{\sqrt{2}\kappa}{\sigma(1+\kappa^2)}\begin{cases} e^{\frac{\sqrt{2}\kappa}{\sigma}(\theta-x)}, & x\geq\theta \\ e^{\frac{\sqrt{2}}{\sigma\kappa}(x-\theta)}, & x<\theta, \end{cases}$
$\theta=0$ $\sigma=0$ $\mu=0$ $\tau=0$	Degenerate at 0		

Table 4.1: Special cases and notation for the Bessel function distribution in the $\mathcal{GAL}(\theta,\mu,\sigma,\tau)$ parametrization.

Thus conditionally on $W=w$, the random variable Y has a normal distribution with mean μw and variance $w\sigma^2$. To find the marginal distribution of Y, we may find its density by integrating the product of the conditional density of $Y|W=w$ and the marginal density $f(w)$ of W. Alternatively, we may find the ch.f. of Y by conditioning on W. This is exactly how we have found mixture representations of this type for the classical as well as asymmetric Laplace distributions. We shall follow this approach to show that Y given by (4.1.8) has the Bessel function distribution when W is gamma distributed. Indeed, let W have a gamma distribution $G(\alpha=\tau,\beta=1)$ with density

$$g(x)=\frac{1}{\Gamma(\tau)}x^{\tau-1}e^{-x},\quad x>0,\ \ \tau>0. \tag{4.1.9}$$

Conditioning on W, we obtain

$$\psi_Y(t)=Ee^{itY}=E[E(e^{itY}|W)]=\int_0^\infty Ee^{it(\mu w+\sigma\sqrt{w}Z)}g(w)dw.$$

When we put the gamma density (4.1.9) and the normal ch.f. (4.1.7) into this relation, we get

$$\psi_Y(t)=\int_0^\infty \phi_{\mu w,\sigma\sqrt{w}}(t)g(w)dw=\frac{1}{\Gamma(\tau)}\int_0^\infty w^{\tau-1}e^{-w(1+\frac{1}{2}\sigma^2t^2-i\mu t)}dw.$$

The latter integral can be related to the standard gamma function to produce

$$\psi_Y(t) = \frac{1}{\Gamma(\tau)}\Gamma(\tau)\left(\frac{1}{1+\frac{1}{2}\sigma^2t^2 - i\mu t}\right)^{\tau} = \left(\frac{1}{1+\frac{1}{2}\sigma^2t^2 - i\mu t}\right)^{\tau},$$

which we recognize as the $\mathcal{GAL}(\mu, \sigma, \tau)$ characteristic function. We summarize our findings in the following result, where we consider a more general four-parameter model.

Proposition 4.1.2 *A $\mathcal{GAL}(\theta, \mu, \sigma, \tau)$ random variable Y with ch.f.* (4.1.3) *admits the representation*

$$Y \stackrel{d}{=} \theta + \mu W + \sigma\sqrt{W}Z, \tag{4.1.10}$$

where Z is standard normal and W is gamma with density (4.1.9).

Remark 4.1.3 Note that in the case $\tau = 1$, where W has the standard exponential distribution, for $\mu = 0$ (and $\sigma = \sqrt{2}, \theta = 0$) we obtain the representation (2.2.3) of the standard classical Laplace distribution, while for $\mu \neq 0$, we get the representation (3.2.1) obtained previously for asymmetric Laplace laws.

This representation produces the following result, showing that as the parameter τ converges to infinity, the corresponding Bessel random variable converges in distribution to a normal variable.

Theorem 4.1.1 *Let $Y_\tau \sim \mathcal{GAL}(\mu_\tau, \sigma_\tau, \tau)$, where*

$$\lim_{\tau\to\infty} \mu_\tau \tau = \mu_0 \quad \textit{and} \quad \lim_{\tau\to\infty} \sigma_\tau^2 \tau = \sigma_0^2.$$

Then Y_τ converges in distribution to the Gaussian r.v. with mean μ_0 and variance σ_0^2.

Proof. Let W_τ be a gamma $G(\alpha = \tau, \beta = 1)$ random variable. It follows from the form of the relevant characteristic functions that the random variables $\mu_\tau W_\tau$ and $\sigma_\tau^2 W_\tau$ converge in probability to μ_0 and σ_0^2, respectively. Thus the result follows from the representation given in Proposition 4.1.2 by invoking the independence of W and Z. □

4.1.2.2 *Relation to gamma distribution.* We now study the relation between the Bessel function and gamma distributions. Let Y have the Bessel function distribution with the ch.f. ψ given by (4.1.3). Note that in the factorization of ψ given by (4.1.4), the third factor corresponds to the r.v. $\frac{\sigma}{\sqrt{2}}\frac{1}{\kappa}G_1$, while the second factor corresponds to the r.v. $-\frac{\sigma}{\sqrt{2}}\kappa G_2$, where G_1, G_2 are i.i.d. $G(\alpha = \tau, \beta = 1)$ random variables. Thus we obtain the following result derived by Press (1967).

Proposition 4.1.3 *A $\mathcal{GAL}^*(\theta, \kappa, \sigma, \tau)$ random variable Y with ch.f.* (4.1.4) *admits the representation*

$$Y \stackrel{d}{=} \theta + \frac{\sigma}{\sqrt{2}}\left(\frac{1}{\kappa}G_1 - \kappa G_2\right), \tag{4.1.11}$$

where G_1 and G_2 are i.i.d. gamma random variables with density (4.1.9).

As before, for the special case $\tau = 1$, the representation (4.1.11) reduces to that of an AL r.v., in which case G_1 and G_2 are standard exponential variables (see Proposition 3.2.2).

Remark 4.1.4 Writing $G_i = -\log(U_i)$, where U_i's have *log-gamma distribution* on $(0, 1)$ with p.d.f.

$$f(u) = \frac{1}{\Gamma(\tau)}(-\log u)^{\tau-1}, \quad u \in (0, 1),$$

[see, e.g., Johnson et al. (1994)], we obtain the representation

$$Y \stackrel{d}{=} \theta + \frac{\sigma}{\sqrt{2}} \log\left(\frac{U_1^{\kappa}}{U_2^{1/\kappa}}\right). \tag{4.1.12}$$

For $\kappa = 1$ the U_i's are standard uniform and we obtain the representation (3.2.3) of AL random variables.

Remark 4.1.5 Similarly, writing $G_i = \log P_i$, we obtain the representation

$$Y \stackrel{d}{=} \theta + \frac{\sigma}{\sqrt{2}} \log\left(\frac{P_1^{1/\kappa}}{P_2^{\kappa}}\right). \tag{4.1.13}$$

Here the i.i.d. variables P_i have density

$$f(u) = \frac{1}{\Gamma(\tau)}\frac{1}{u^2}(\log u)^{\tau-1}, \quad u \in (1, \infty).$$

For $\kappa = 1$ the P_i's have Pareto Type I distribution and (4.1.13) reduces to the representation (3.2.4) of AL r.v.'s.

Remark 4.1.6 Recall that if G has a gamma distribution with density (4.1.9), then the r.v. $H = 2G$ has a chi-square distribution with $\nu = 2\tau$ degrees of freedom, denoted by χ^2_ν. Consequently, $Y \sim \mathcal{GAL}^*(\theta, \kappa, \sigma, \tau)$ has the following representation in terms of two i.i.d. $\chi^2_{2\tau}$-distributed r.v.'s H_1 and H_2:

$$Y \stackrel{d}{=} \theta + \frac{\sqrt{2}\sigma}{4}\left(\frac{1}{\kappa}H_1 - \kappa H_2\right). \tag{4.1.14}$$

4.1.3 Self-decomposability. As shown in Proposition 3.2.3, every $\mathcal{AL}^*(\theta, \kappa, \sigma)$ r.v. Y is self-decomposable, that is, for every $c \in (0, 1)$ it admits the representation

$$Y \stackrel{d}{=} cY + (1 - c)\theta + V, \tag{4.1.15}$$

where the r.v. V can be expressed as

$$V \stackrel{d}{=} \frac{\sigma}{\sqrt{2}}\left(\frac{1}{\kappa}\delta_1 W_1 - \kappa\delta_2 W_2\right). \tag{4.1.16}$$

Here δ_1, δ_2 are r.v.'s taking values of either zero or one with probabilities

$$P(\delta_1 = 0, \delta_2 = 0) = c^2, \qquad P(\delta_1 = 1, \delta_2 = 1) = 0,$$

$$P(\delta_1 = 1, \delta_2 = 0) = (1 - c)\left(c + \frac{1 - c}{1 + \kappa^2}\right),$$

$$P(\delta_1 = 0, \delta_2 = 1) = (1 - c)\left(c + \frac{(1 - c)\kappa^2}{1 + \kappa^2}\right),$$

W_1 and W_2 are standard exponential variables, and Y, W_1, W_2, and (δ_1, δ_2) are mutually independent. Now consider a $\mathcal{GAL}^*(\theta, \kappa, \sigma, \tau)$ r.v. X, where $\tau = n$ is a positive integer. Then

$$X \stackrel{d}{=} \theta + \sum_{i=1}^{n} Y_i, \tag{4.1.17}$$

where the Y_i's are i.i.d. $\mathcal{AL}^*(0, \kappa, \sigma)$ random variables. Consequently, since each Y_i admits the representation (4.1.15) with $\theta = 0$,

$$Y_i \stackrel{d}{=} cY_i + V_i, \tag{4.1.18}$$

where V_i's are i.i.d. copies of V given by (4.1.16), we obtain

$$X \stackrel{d}{=} \theta + \sum_{i=1}^{n} Y_i \stackrel{d}{=} \theta + c\sum_{i=1}^{n} Y_i + \sum_{i=1}^{n} V_i = c(\theta + \sum_{i=1}^{n} Y_i) + (1-c)\theta + \sum_{i=1}^{n} V_i. \tag{4.1.19}$$

Thus we conclude that X is self-decomposable as well. The following result summarizes our findings.

Proposition 4.1.4 *Let $X \sim \mathcal{GAL}^*(\theta, \kappa, \sigma, n)$, where $n \geq 1$ is a positive integer. Then X is self-decomposable and for any $c \in [0, 1]$ we have*

$$X \stackrel{d}{=} cX + (1-c)\theta + \sum_{i=1}^{n} V_i, \tag{4.1.20}$$

where the V_i's are i.i.d. variables with the representation (4.1.16).

Remark 4.1.7 The fact that a GAL r.v. with the parameters $\theta = 0, \kappa = 1, \sigma > 0$ and $\tau = n \in \mathbb{N}$ has the same distribution as the sum of n i.i.d. symmetric Laplace variables shows that this distribution is stable with respect to a random summation where the number of terms $\nu_{p,n}$ has the *Pascal distribution*

$$P(\nu_{p,n} = k) = \binom{k-1}{n-1} p^n (1-p)^{k-n}, \quad k = n, n+1, \dots, \quad 0 < p < 1. \tag{4.1.21}$$

More precisely, if X_i's are i.i.d. with the $\mathcal{GAL}^*(0, 1, \sigma, n)$ distribution and $\nu_{p,n}$ is an independent of the X_i's Pascal r.v., then the relation

$$p^{1/2} \sum_{i=1}^{\nu_{p,n}} X_i \stackrel{d}{=} X_1, \tag{4.1.22}$$

holds for all $p \in (0, 1)$. Moreover, under the symmetry and finite variance of the X_i's, the stability property (4.1.22) characterizes this class of distributions (recall that with the geometric number of terms, which corresponds to $n = 1$, we obtain the characterization of symmetric Laplace laws). In addition, the class of $\mathcal{GAL}^*(0, 1, \sigma, n)$ distributions consists of all distributional limits as $p \to 0$ of Pascal compounds

$$a_p \sum_{i=1}^{\nu_{p,n}} (Y_i - b_p) \tag{4.1.23}$$

with $b_p = 0$, where the Y_i's are symmetric and i.i.d. variables with finite variance, independent of the Pascal number of terms $\nu_{p,n}$. If the restrictions on symmetry or finite variance are relaxed, we obtain a larger class of *Pascal-stable distributions*, introduced in Janković (1993b) as the class of distributional limits of (4.1.23).

4.1.3.1 *Relation to sample covariance.* Pearson et al. (1929) showed analytically, that if (X_i, Y_i), $i = 1, \dots n$, are i.i.d. from a bivariate normal distribution with means μ_X and μ_Y, variances σ_X^2 and σ_Y^2, and correlation coefficient ρ, then the product-moment coefficient

$$p_{11} = \frac{1}{n}\sum_{i=1}^{n}(X_i - \overline{X})(Y_i - \overline{Y}) \tag{4.1.24}$$

has the Bessel function distribution. We provide an alternative derivation, utilizing appropriate representations of random variables along with convolution representation (4.1.11) of the Bessel function distribution. Without loss of generality we assume that a random sample comes from the standard normal distribution with mean zero, variances equal to one, and correlation (covariance) ρ. The following result shows that the statistic

$$T_n = n\sum_{i=1}^{n}(X_i - \overline{X})(Y_i - \overline{Y}) = n\sum_{i=1}^{n} X_i Y_i - \left(\sum_{i=1}^{n} X_i\right)\left(\sum_{i=1}^{n} X_i\right) \tag{4.1.25}$$

has a Bessel function distribution with appropriate parameters (and consequently so does the statistic p_{11} defined above).

Proposition 4.1.5 *Let X_i and Y_i, $i = 1, \dots n$, be i.i.d. bivariate normal with zero mean, unit variances, and covariance ρ. Then for any $n > 1$, the statistic T_n given by* (4.1.25) *has the Bessel function distribution $\mathcal{GAL}^*(\kappa, \sigma, \tau)$ with*

$$\sigma = \sqrt{2}n\sqrt{1-\rho^2}, \quad \kappa = \sqrt{\frac{1-\rho}{1+\rho}}, \quad \tau = \frac{n-1}{2}. \tag{4.1.26}$$

Before proving Proposition 4.1.5 we establish the following lemma.

Lemma 4.1.1 *Let $x_1, \dots, x_n$ and $y_1, \dots, y_n$ be two sets of real numbers, and let $\overline{x}$ and $\overline{y}$ be their arithmetic means. Then for any integer $n \geq 1$, we have*

$$n\sum_{i=1}^{n}(x_i - \overline{x})^2 = \sum_{1\le i<j\le n}(x_i - x_j)^2 \tag{4.1.27}$$

$$n\sum_{i=1}^{n}(x_i - \overline{x})(y_i - \overline{y}) = \sum_{1\le i<j\le n}(x_i - x_j)(y_i - y_j). \tag{4.1.28}$$

Proof. Since (4.1.27) follows from (4.1.28), we only prove the latter relation. We have the following chain of equalities:

$$\begin{aligned} n\sum_{i=1}^{n}(x_i - \overline{x})(y_i - \overline{y}) &= n\sum_{i=1}^{n} x_i y_i - \sum_{i=1}^{n} x_i \sum_{j=1}^{n} y_j \\ &= n\sum_{i=1}^{n} x_i y_i - \sum_{i=1}^{n} x_i y_i - 2\sum_{1\le i<j\le n} x_i y_j \\ &= (n-1)\sum_{i=1}^{n} x_i y_i - \sum_{1\le i<j\le n} 2x_i y_j \\ &= \sum_{1\le i<j\le n}(x_i - x_j)(y_i - y_j). \end{aligned}$$

□

We now turn to the proof of Proposition 4.1.5.

Proof. In view of the representation (4.1.11), our goal is to show that

$$T_n \stackrel{d}{=} n\sqrt{1-\rho^2}\left[\frac{1}{\kappa}G_1 - \kappa G_2\right]. \tag{4.1.29}$$

By Lemma 4.1.1 we have

$$T_n = \sum_{1\le i<j\le n} a_{i,j},$$

where $a_{i,j} = (X_i - X_j)(Y_i - Y_j)$. Write

$$a_{i,j} = \frac{1}{4}\{[b_{i,j}^+]^2 - [b_{i,j}^-]^2\},$$

where

$$b_{i,j}^\pm = (Y_i - Y_j) \pm (X_i - X_j),$$

so that

$$T_n = \frac{1}{4}\left\{\sum_{1\le i<j\le n}[b_{i,j}^+]^2 - \sum_{1\le i<j\le n}[b_{i,j}^-]^2\right\}.$$

Next, note that for all $1 \le i < j \le n$ and $1 \le k < l \le n$, the variables $b_{i,j}^+$ and $b_{k,l}^-$ are independent. Indeed, they are normally distributed and their covariance is equal to zero:

$$\begin{aligned}
\mathrm{Cov}(b_{i,j}^+, b_{k,l}^-) &= \mathrm{Cov}\{(Y_i - Y_j) + (X_i - X_j), (Y_k - Y_l) - (X_k - X_l)\}\\
&= \mathrm{Cov}(Y_i, Y_k) - \mathrm{Cov}(Y_i, Y_l) - \mathrm{Cov}(Y_i, X_k) + \mathrm{Cov}(Y_i, X_l)\\
&\quad - \mathrm{Cov}(Y_j, Y_k) + \mathrm{Cov}(Y_j, Y_l) + \mathrm{Cov}(Y_j, X_k) - \mathrm{Cov}(Y_j, X_l)\\
&\quad + \mathrm{Cov}(X_i, Y_k) - \mathrm{Cov}(X_i, Y_l) - \mathrm{Cov}(X_i, X_k) + \mathrm{Cov}(X_i, X_l)\\
&\quad - \mathrm{Cov}(X_j, Y_k) + \mathrm{Cov}(X_j, Y_l) + \mathrm{Cov}(X_j, X_k) - \mathrm{Cov}(X_j, X_l)\\
&= \delta_{ik} - \delta_{il} - \rho\delta_{ik} + \rho\delta_{il} - \delta_{jk} + \delta_{jl} + \rho\delta_{jk} - \rho\delta_{jl}\\
&\quad + \rho\delta_{ik} - \rho\delta_{il} - \delta_{ik} + \delta_{il} - \rho\delta_{jk} + \rho\delta_{jl} + \delta_{jk} - \delta_{jl} = 0,
\end{aligned}$$

since δ_{ij} is equal to one if $i = j$ and zero otherwise. Next, write

$$T_n = \frac{1}{4}(W^+ - W^-),$$

where

$$W^+ = \sum_{1\le i<j\le n}[b_{i,j}^+]^2 \quad\text{and}\quad W^- = \sum_{1\le i<j\le n}[b_{i,j}^-]^2$$

are independent random variables. Further, we have

$$b_{i,j}^\pm = (Y_i \pm X_i) - (Y_j \pm X_j) = Z_i^\pm - Z_j^\pm,$$

where

$$Z_i^\pm = (Y_i \pm X_i), \quad i = 1, \ldots n.$$

Note that the Z_i^+'s are i.i.d. normal with mean zero and variance $2(1+\rho)$, since

$$\mathrm{Var}(Y_i + X_i) = \mathrm{Var}(Y_i) + \mathrm{Var}(X_i) + 2\mathrm{Cov}(Y_i, X_i) = 2(1+\rho).$$

Similarly, the Z_i^-'s are i.i.d. normal with mean zero and variance $2(1-\rho)$. We now express T_n in terms of the $Z_i^{\pm}$'s as

$$T_n = \frac{1}{4}\left\{\sum_{1\le i<j\le n}[Z_i^+ - Z_j^+]^2 - \sum_{1\le i<j\le n}[Z_i^- - Z_j^-]^2\right\},$$

and apply Lemma 4.1.1 to conclude that

$$W^+ = n\sum_{i=1}^n [Z_i^+ - \overline{Z}^+]^2 \text{ and } W^- = n\sum_{i=1}^n [Z_i^- - \overline{Z}^-]^2,$$

where $\overline{Z}^+$ and $\overline{Z}^-$ denote the arithmetic means of Z_i^+'s and Z_i^-'s, respectively. Since the Z_i^+'s are i.i.d. normal with mean zero and variance $\sigma_+^2 = 2(1+\rho)$, we conclude that the statistic

$$H_1 = \frac{1}{n}\frac{W^+}{\sigma_+^2} = \frac{\sum_{i=1}^n [Z_i^+ - \overline{Z}^+]^2}{2(1+\rho)}$$

has a chi-square distribution with $n-1$ degrees of freedom. Similarly, the same distribution has the statistic

$$H_2 = \frac{1}{n}\frac{W^-}{\sigma_-^2} = \frac{\sum_{i=1}^n [Z_i^- - \overline{Z}^-]^2}{2(1-\rho)},$$

which is independent of W_1. Finally, we can write

$$T_n = \frac{1}{4}\{2n(1+\rho)H_1 - 2n(1-\rho)H_2\} = \frac{n}{2}\{(1+\rho)H_1 - (1-\rho)H_2\},$$

which is equivalent to (4.1.29) by the relation between chi-square and gamma distributions. The result has been proved. □

Remark 4.1.8 For the special case $n = 3$ we obtain $\tau = 1$ so that the statistic T_3 has an asymmetric Laplace distribution $\mathcal{AL}^*(\kappa, \sigma)$ with parameters as in (4.1.26). Equivalently, an $\mathcal{AL}^*(\kappa, 1)$ r.v. Y admits a representation

$$Y \stackrel{d}{=} \frac{(X_1 - \overline{X})(Y_1 - \overline{Y}) + (X_2 - \overline{X})(Y_2 - \overline{Y}) + (X_3 - \overline{X})(Y_3 - \overline{Y})}{\sqrt{2}\sqrt{1-\rho^2}},$$

where ρ and κ are related as in (4.1.26) and (X_i, Y_i), $i = 1, 2, 3$, are i.i.d. bivariate normal variables with vector mean zero, unit variances, and correlation ρ.

4.1.4 Densities. To derive the p.d.f. of a GAL random variable we can either apply the inversion formula to the GAL ch.f. (4.1.2) or exploit the representations (4.1.10) and (4.1.11). Actually, we have already done the latter (for the case $\sigma = 1$) in Lemma 2.3.1, where we were dealing with functions of Laplace random variables. Thus the density of a $\mathcal{GAL}^*(\theta, \kappa, \sigma, \tau)$ r.v. has the following form for $x \neq \theta$:

$$h(x) = \frac{\sqrt{2}e^{\frac{\sqrt{2}}{2\sigma}(1/\kappa-\kappa)(x-\theta)}}{\sqrt{\pi}\sigma^{\tau+1/2}\Gamma(\tau)}\left(\frac{\sqrt{2}|x-\theta|}{\kappa+1/\kappa}\right)^{\tau-\frac{1}{2}} K_{\tau-1/2}\left(\frac{\sqrt{2}}{2\sigma}\left(\frac{1}{\kappa}+\kappa\right)|x-\theta|\right), \tag{4.1.30}$$

where K_λ is the modified Bessel function of the third kind with the index λ, given in the appendix. A standard GAL density is obtained for $\theta = 0$ and $\sigma = 1$. This density, derived by a variety of methods and under various parametrizations, has appeared in several papers, including Pearson et al. (1929), McKay (1932), Madan et al. (1998), Levin and Tchernitser (1999), Kozubowski and Podgórski (1999a). In Figure 4.1, we present a variety of standard GAL densities. Note the behavior of the densities at zero, which is the subject of Theorem 4.1.2.

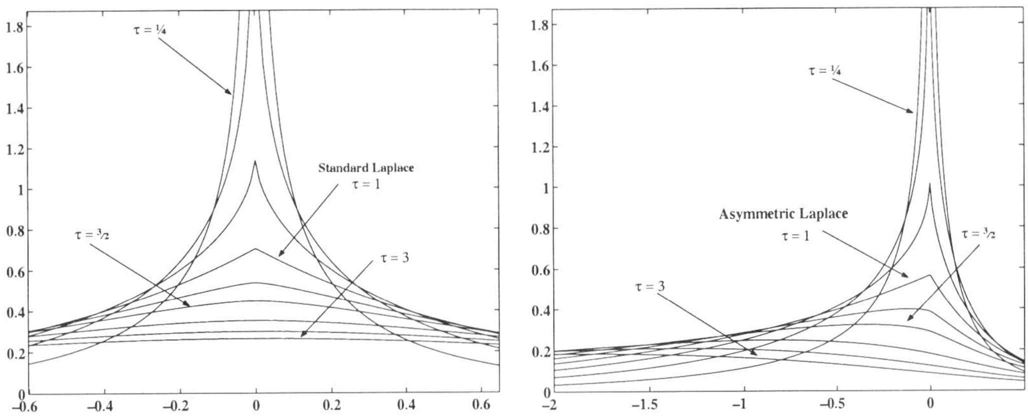

Figure 4.1: Densities of the standard generalized Laplace distributions with $\tau = 1/4, 1/2, 3/4, 1, 5/4, 3/2, 2, 5/2$, and 3. *Left*: $\kappa = 1$—the symmetric case. *Right*: $\kappa = 2$—an asymmetric case.

Let us note several special cases.

4.1.4.1 *Asymmetric Laplace laws.* Consider a standard density GAL density with $\tau = 1$. Here the Bessel function has index 1/2, so it admits a closed form given by (A.0.11) in the appendix. Thus the density (4.1.30) takes the form

$$\begin{aligned} h(x) &= \frac{\sqrt{2}}{\Gamma(1)\sqrt{\pi}}\left(\frac{\sqrt{2}|x|}{\kappa+1/\kappa}\right)^{1/2} e^{\frac{\sqrt{2}}{2}(1/\kappa-\kappa)x} K_{1/2}\left(\frac{\sqrt{2}}{2}\left(\frac{1}{\kappa}+\kappa\right)|x|\right) \\ &= \frac{\sqrt{2}}{\sqrt{\pi}}\frac{|x|^{1/2}}{(\kappa+1/\kappa)^{1/2}}e^{\frac{\sqrt{2}}{2}(1/\kappa-\kappa)x}\frac{\sqrt{\pi}}{(\kappa+1/\kappa)^{1/2}|x|^{1/2}}e^{-\frac{\sqrt{2}}{2}(1/\kappa+\kappa)|x|} \\ &= \frac{\sqrt{2}}{\kappa+1/\kappa}e^{\frac{\sqrt{2}}{2}(1/\kappa-\kappa)x-\frac{\sqrt{2}}{2}(1/\kappa+\kappa)|x|}, \end{aligned} \tag{4.1.31}$$

which we recognize as the density of the standard $\mathcal{AL}^*(0, \kappa, 1)$ distribution. Further, in the symmetric case $\kappa = 1$, (4.1.31) reduces to the density of the standard Laplace distribution.

4.1.4.2 *Symmetric case.* When $\kappa = 1$ and $\theta = 0$, the distribution is symmetric (about zero) since the corresponding characteristic function is real. In this case, the density is given by the following even function of x:

$$h(x) = \frac{\sqrt{2}}{\sigma^{\tau+1/2}\Gamma(\tau)\sqrt{\pi}} \left(\frac{|x|}{\sqrt{2}}\right)^{\tau-1/2} K_{\tau-1/2}(\sqrt{2}|x|/\sigma), \quad x \neq 0. \tag{4.1.32}$$

This particular distribution arises as a mixture of normal distributions with mean zero and (stochastic) variance $\sigma^2 W$, where W has the gamma distribution with density (4.1.9) [see e.g., Teichroev (1957), Madan and Seneta (1990), McLeish (1982)].

In our next result we summarize some properties of the densities of the symmetric generalized Laplace distributions. In particular, we show that they are all unimodal for $\tau \geq 1$, and study their behavior at the mode.

Theorem 4.1.2 *Let $h(x; \tau)$ be the density of a symmetric generalized Laplace distribution $\mathcal{GAL}^*(0, 1, 1, \tau)$. Then $h(x; \tau)$ has the following asymptotic behavior as $x \to 0^+$:*

$$h(x;\tau) = \begin{cases} \frac{1}{2^\tau\sqrt{\pi}}\frac{\Gamma(1/2-\tau)}{\Gamma(\tau)}x^{2\tau-1} + o(x^{2\tau-1}) & \text{for} \quad \tau \in (0, 1/2), \\ -\frac{\sqrt{2}}{\pi}\log x + o(\log x) & \text{for} \quad \tau = 1/2, \\ \frac{1}{2\pi}\frac{\Gamma(\tau-1/2)}{\Gamma(\tau)} + o(1) & \text{for} \quad \tau > 1/2. \end{cases}$$

Moreover, for $x > 0$, we have

$$\frac{\partial}{\partial x}h(x;\tau) = -\frac{1}{\tau-1}\frac{\sqrt{2}}{2}h(x;\tau-1), \quad \tau > 1,$$

and

$$\frac{\partial}{\partial x}h(x;\tau) = \begin{cases} \frac{2\tau-1}{2^\tau\sqrt{\pi}}\frac{\Gamma(1/2-\tau)}{\Gamma(\tau)}x^{2\tau-2} + o(x^{2\tau-2}) & \text{for} \quad \tau \in (0, 1/2), \\ -\frac{\sqrt{2}}{\pi}x^{-1} + o(x^{-1}) & \text{for} \quad \tau = 1/2, \\ -\frac{1}{\sin(\pi(\tau-1/2))\Gamma(2\tau-1)}x^{2\tau-2} + o(x^{2\tau-2}) & \text{for} \quad \tau \in (1/2, 1), \\ -1 + o(1) & \text{for} \quad \tau = 1, \\ -\frac{t-1/2}{\sin(\pi(\tau-1/2))\Gamma(2\tau)}x^{2\tau-2} + o(x^{2\tau-2}) & \text{for} \quad \tau \in (1, 3/2), \\ \frac{\sqrt{2}}{2\pi}x\log(\sqrt{2}x) + o(x\log(\sqrt{2}x)) & \text{for} \quad \tau = 3/2, \\ -\sqrt{\frac{2}{\pi}}\frac{\Gamma(\tau-1/2)}{(\tau-3/2)\Gamma(\tau)}x + o(x) & \text{for} \quad \tau \in (n-1/2, n+1/2), \\ -\frac{\sqrt{2}(n-2)(2n)!}{\pi n!}x + o(x) & \text{for} \quad \tau = n+1/2, \end{cases}$$

where in the last two relations $n \in \mathbb{N} + 1$.

Proof. Let $H(x;\tau) = x^{\tau-1/2}K_{\tau-1/2}(x)$. We have the following relation, which follows from the form of the density (4.1.32):

$$h(x;\tau) = \frac{2^{1-\tau}}{\sqrt{\pi}\Gamma(\tau)}H(\sqrt{2}x;\tau), \quad x > 0.$$

The result follows from Properties 6, 9, and 10 of the functions $H(x;\tau)$ and K_λ given in the appendix. The behavior of the density at zero follows from Property 6 (and also Property 10 for $\tau < 1/2$). The recurrent relation follows from Property 9. The behavior of the derivative of $h(x;\tau)$ follows from all three properties. □

A direct consequence of Theorem 4.1.2 is the following.

Corollary 4.1.1 *The density of a symmetric $\mathcal{GAL}^*(0, 1, \sigma, \tau)$ distribution with $\tau > 1$ is unimodal with the mode at zero.*

Proof. The recurrent relation of Theorem 4.1.2 implies that the derivative of the density is negative for positive arguments. Thus the density is a decreasing function that does not have any maximum, except possibly at zero. □

The graphs of the densities in the symmetric case illustrating their behavior at zero, which is studied in this theorem, are presented in Figure 4.1 (the left-hand side picture). The influence of the parameters on the shape of the densities is perhaps better illustrated by Figure 4.2.

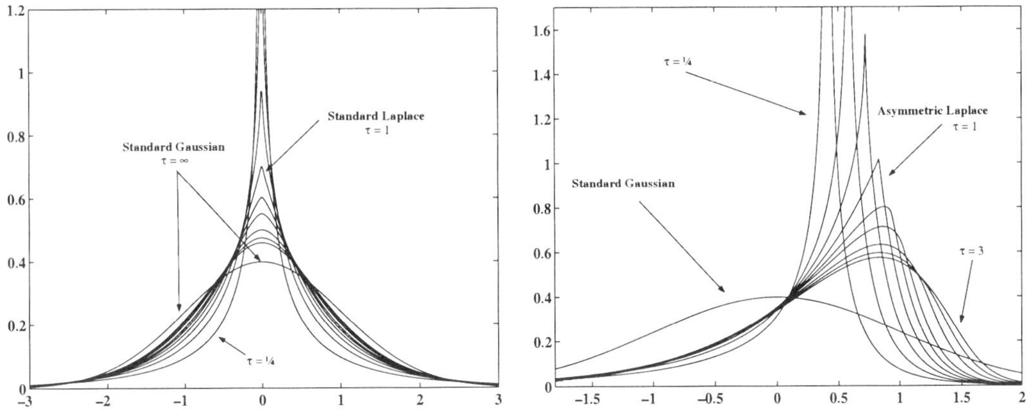

Figure 4.2: Comparison of the standardized generalized Laplace densities and normal standard density. Both pictures contain densities with $\tau = 1/4, 1/2, 3/4, 1, 5/4, 3/2, 2, 5/2$, and 3. *Left*: The symmetric case, $\kappa = 1$. *Right*: An asymmetric case, $\kappa = 2$. All densities have the mean equal to zero and variance equal to one.

4.1.4.3 *An integer value of* τ. We already know that when $\tau = n$ is a nonnegative integer, then the corresponding GAL r.v. is a sum of n i.i.d. AL random variables with the same parameters σ and μ (or κ). In this case the Bessel function $K_{n-1/2}$ admits a closed form (see (A.0.10) in the appendix), and so does the corresponding standard GAL density with the parameter $\tau = n \geq 1$:

$$h(x) = \frac{1}{(n-1)!} \sum_{j=0}^{n-1} \frac{(n-1+j)!}{(n-1-j)!j!} \frac{2^{(n-j)/2}|x|^{n-1-j}}{(\kappa + 1/\kappa)^{n+j}} \begin{cases} e^{-\sqrt{2}\kappa|x|}, & \text{for } x \geq 0, \\ e^{-\sqrt{2}\frac{1}{\kappa}|x|}, & \text{for } x < 0 \end{cases} \tag{4.1.33}$$

[see, e.g., Press (1967), Levin and Tchernitser (1999), Kozubowski and Podgórski (1999a)]. Note that in the symmetric case ($\kappa = 1$), this density simplifies to (2.3.25) considered previously in connection with the distribution of the sum of n i.i.d. Laplace r.v.'s [see also Teichroev (1957), McLeish (1982)]. Also observe that (4.1.33) coincides with (4.1.31) if $\tau = 1$, which is the AL case. Further, here the density (4.1.33) is a mixture of n densities on $(-\infty, \infty)$. For $j = 0, \ldots, n-1$, the jth density has the form

$$f_{n,j}(x) = p_{n,j} g_{n-j,1/\kappa}(x) \mathbb{I}_{[0,\infty)}(x) + q_{n,j} g_{n-j,\kappa}(-x) \mathbb{I}_{(0,\infty)}(-x), \tag{4.1.34}$$

where $g_{\alpha,\beta}$ stands for the gamma $G(\alpha, \beta)$ density, and

$$p_{n,j} = \frac{p^n q^j}{p^n q^j + p^j q^n}, \qquad q_{n,j} = 1 - p_{n,j} = \frac{p^j q^n}{p^n q^j + p^j n q^n}, \tag{4.1.35}$$

with $p = 1/(1+\kappa^2)$ and $q = \kappa^2/(1+\kappa^2)$. Under this notation, the $\mathcal{GAL}^*(0, \kappa, 1, n)$ density is

$$h(x) = \sum_{j=0}^{n-1} \frac{(n+j-1)!}{j!(n-1)!} 2^{(n-j)/2} (p^n q^j + p^j q^n) f_{n,j}(x). \tag{4.1.36}$$

This result, taken from Kozubowski and Podgórski (1999a), is a generalization of the exponential mixture representation discussed previously for the AL random variables.

4.1.5 Moments. Exploiting representations of the K-Bessel function random variables it is easy to find their moments. This is done in the following result.

Proposition 4.1.6 *The moments of a $\mathcal{GAL}(\mu, \sigma, \tau)$ random variable Y are given by the relations*

$$E(Y^n) = \frac{1}{\sqrt{\pi}\Gamma(\tau)} \sum_{k=0}^{[[n/2]]} \binom{n}{2k} \sigma^{2k} \mu^{n-2k} 2^k \Gamma(1/2+k)\Gamma(\tau+n-k).$$

In particular, if $\mu = 0$ (symmetric case), then

$$E(Y^{2m}) = \sigma^{2m} \prod_{i=0}^{m-1} [(\tau+i)(2i+1)].$$

Proof. We exploit representation (4.1.8) and the following formulas for the moments of a gamma variable W with parameter $\alpha = \tau$ and a standard normal random variable Z:

$$E(W^s) = \frac{\Gamma(\tau+s)}{\Gamma(\tau)}, \qquad E(Z^{2k}) = 2^k \frac{\Gamma(1/2+k)}{\Gamma(1/2)} = \prod_{i=0}^{k-1}(2i+1).$$

Since odd moments of the standard normal random variable vanish, we obtain

$$\begin{aligned} E(Y^n) &= \sum_{l=0}^{n} \binom{n}{l} \sigma^l \mu^{n-l} E(Z^l) E(W^{n-l/2}) \\ &= \sum_{k=0}^{[[n/2]]} \binom{n}{2k} \sigma^{2k} \mu^{n-2k} E(Z^{2k}) E(W^{n-k}), \end{aligned}$$

and the formula follows from a direct application of the expressions for the moments of W and Z.

In the symmetric case all terms except the last one in this sum vanish and the conclusion follows from the identity

$$\Gamma(\tau+k) = \Gamma(\tau) \prod_{i=0}^{k-1} (\tau+i), \quad k \in \mathbb{N}. \qquad \square$$

Corollary 4.1.2 *The mean of a $\mathcal{GAL}(\mu, \sigma, \tau)$ random variable Y is equal to*

$$E(Y) = \tau\mu,$$

and the variance is

$$\mathrm{Var}(Y) = \tau(\mu^2 + \sigma^2).$$

4.2 Laplace motion

In this section we study Laplace motion — a stochastic process that plays the same role in the Laplacian domain as the Brownian motion does among Gaussian processes. The Laplace motions have several interesting properties that distinguish them from their famous Gaussian counterpart. We study here only the most fundamental ones, leaving more extensive investigation for future work on processes generated by the Laplace distribution.

Laplace motions are special cases of Lévy processes. The latter are defined through the class of infinitely divisible distributions to which Laplace distributions belong. Although Laplace motions share some common properties with Brownian motions, including the finite second (or any order) moments, independence and stationarity of increments, their observed features are essentially different. First, their trajectories (paths) are discontinuous at any point and, in fact, they are purely jump functions. In general, they can be asymmetric, including properties of their paths. The space scale is not exchangeable with the time scale which, even in the symmetric case, requires two different parameters for these scales.

Laplace motions have several representations that relate them to other processes. First, they can be written as Brownian motion evaluated at random time, the latter being the gamma process. In other words they are Brownian motions subordinated to the gamma process. Alternatively, the Laplace motion can be obtained as a difference of two independent gamma processes. Finally, using a general representation of Lévy processes, we can write them as compound Poisson processes with independent and random jumps having a special form of the distribution (given by so-called Lévy density). The last characterization gives an insight into the structure of the trajectories and sizes of jumps, the latter completely characterizing trajectories of pure jumps processes.

The finiteness of their moments and their convenient characterizations make Laplace motion an interesting object for future investigation and for developing the theory of Laplacian processes more or less in the same spirit as the theory of Gaussian processes is developed based on Brownian motion.

4.2.1 Symmetric Laplace motion. As we already know, Laplace distributions are infinitely divisible (see, for example, Subsection 2.4.1 in Chapter 2). Thus it is a direct consequence of the general theory of infinitely divisible distributions and processes that we can define the following subclass of Lévy processes [cf. Ferguson and Klass (1972)].

Definition 4.2.1 *A stochastic process $L(t)$ is called a symmetric Laplace motion with the space scale parameter σ and the time scale parameter ν (in short, $\mathcal{LM}(\sigma, \nu)$ process) if*

1. *it starts at the origin, i.e.,*
$$L(0) = 0;$$
2. *it has independent and stationary (homogeneous) increments;*
3. *the increments by the time scale unit have a symmetric Laplace distribution with the parameter σ, i.e.,*
$$L(t+\nu) - L(t) \stackrel{d}{=} \mathcal{L}(\sigma).$$

The symmetric Laplace motion $\mathcal{LM}(1, 1)$ is called the standard Laplace motion *or simply the Laplace motion.*

A symmetric Laplace motion $Y(t)$ with drift *m is a $\mathcal{LM}(\sigma, \nu)$ process $L(t)$ shifted by a linear function mt, i.e.,*
$$Y(t) = mt + L(t).$$

Remark 4.2.1 This definition, along with the properties of infinitely divisible distributions, imply the following characteristic function for the increment $L(s+t) - L(s)$ of $\mathcal{LM}(\sigma, \nu)$:

$$\phi_t(u) = \frac{1}{(1 + \sigma^2 u^2/2)^{t/\nu}},$$

i.e., the increment has the generalized symmetric Laplace distribution (the symmetric K-Bessel function distribution) with the parameters σ and $\tau = t/\nu$, which is denoted by $\mathcal{GAL}(0, 0, \sigma, \tau)$.

Remark 4.2.2 Recall that standard Brownian motion $\{B(t), t > 0\}$ is self-similar with index $H = 1/2$, that is,

$$\{B(at), t > 0\} \stackrel{d}{=} \{a^H B(t), t > 0\} \quad \text{for all} \quad a > 0. \tag{4.2.1}$$

In contrast with Brownian motion, for Laplace motion, the time scale and the space scale are no longer exchangeable, and the process is not self-similar. Indeed, for any $a > 0$ and $H > 0$, we have

$$a^H L(t) \stackrel{d}{=} \mathcal{GAL}(0, 0, a^H \sigma, t/\nu)$$

and

$$L(at) \stackrel{d}{=} \mathcal{GAL}(0, 0, \sigma, at/\nu),$$

so the self-similarity property (4.2.1) cannot hold for the Laplace motion $L(t)$.

Remark 4.2.3 As expected, a general Laplace motion with a drift can be defined through the standard Laplace motion L by the expression

$$mt + \sigma L(t/\nu).$$

Let us start a more detailed discussion of the properties of Laplace motion with the derivation of their moments.

Proposition 4.2.1 *Let $L(t)$ be a $\mathcal{LM}(\sigma, \nu)$ Laplace motion with drift m. Then*

$$E[L(t)] = mt, \qquad \text{Var}[L(t)] = t\sigma^2/\nu.$$

Proof. The result follows from Remark 4.2.1 and Corollary 4.1.2. □

It follows immediately that fixing the variance and the mean is not enough to define a Laplace motion completely. Therefore, there are infinitely many Laplace motions $\mathcal{LM}(\sigma, \nu)$ each with $\sigma^2/\nu = 1$, having the same covariance structure as standard Brownian motion characterized by unit variance at the time equal to one. In Figure 4.3, we present trajectories of the processes with the same covariance structure. We see that sample properties differ significantly for these processes.

4.2.2 Representations. There are several important representations of Laplace motion. Most of the results presented here were discussed and partially proved in Madan and Seneta (1990).

The first representation relates Laplace motion to Brownian motion evaluated at an independent random time distributed according to a gamma process. Recall that a stochastic process Γ_t is called a gamma process if it starts at zero, has independent and homogeneous increments, and the distribution of the increment $\Gamma_{t+s} - \Gamma_t$ is given by the standard gamma distribution with the shape parameter s/ν. If $\nu = 1$, we refer to such a process as the standard gamma process.

Theorem 4.2.1 *Let $B(t)$ be a Brownian motion with the scale parameter σ and let Γ_t be a gamma process with parameter ν independent of B_t. Then the process*

$$L(t) = B(\Gamma_t), t > 0,$$

is $\mathcal{LM}(\sigma, \nu)$.

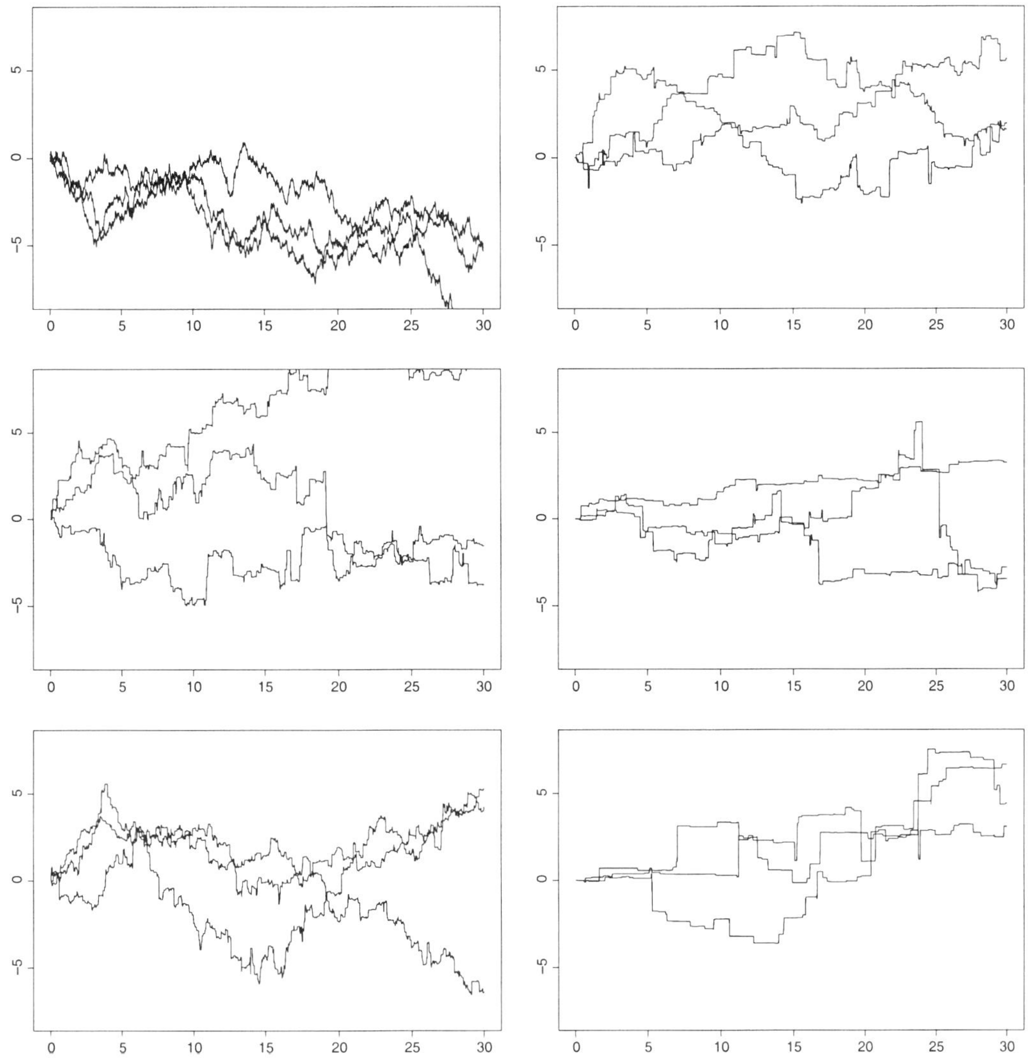

Figure 4.3: Trajectories of Laplace motions and Brownian motion (three paths for each process). All processes have the same covariance structure characterized by unit variance at time $t = 1$. This requirement for the Laplace motion $\mathcal{LM}(\sigma, \nu)$ is satisfied by setting $\sigma = \sqrt{\nu}$. *Top*: Standard Brownian motion vs. standard Laplace motion ($\nu = 1$). *Middle*: $\mathcal{LM}(\sigma = \sqrt{2}, \nu = 2)$ and $\mathcal{LM}(\sigma = \sqrt{2}/2, \nu = 1/2)$. *Bottom*: $\mathcal{LM}(\sigma = \sqrt{5}, \nu = 5)$ and $\mathcal{LM}(\sigma = 1/2, \nu = 1/4)$.

Proof. That the process $L(t)$ starts at the origin is obvious. The distribution of $L(t)$ can be obtained from the characteristic function

$$\begin{aligned}\phi_{L(t)}(\xi) &= Ee^{iL(t)\xi} = E(E(e^{iB(\Gamma_t)\xi}|\Gamma_t)) \\ &= Ee^{-\Gamma_t\sigma^2\xi^2/2} = \frac{1}{(1+\sigma^2\xi^2/2)^{t/\nu}},\end{aligned}$$

which corresponds to the $\mathcal{GAL}(0, 0, \sigma^2, t/\nu)$ distribution. The proof then follows from the general property stating that the composition of two independent processes with independent and homogenous increments (in this case Brownian motion and gamma process) is again a process with independent and homogenous increments [see Bertoin (1996)]. □

Another simple representation of Laplace motion is given in the following theorem.

Theorem 4.2.2 *Let Γ_t and $\tilde{\Gamma}_t$ be two independent gamma processes with the same parameter ν. Then the process defined by*

$$L(t) = \frac{\sqrt{2}}{2}\sigma(\Gamma_t - \tilde{\Gamma}_t), t > 0,$$

is $\mathcal{LM}(\sigma, \nu)$.

Proof. The process obviously starts at zero and has independent and homogeneous increments since Γ_t and $\tilde{\Gamma}_t$ are such processes. Thus the thesis follows from Proposition 4.1.3 applied to $G_1 = \Gamma_t$ and $G_2 = \tilde{\Gamma}_t$ for $\kappa = 1$. □

The last representation we want to discuss here follows from an application of the general result of Ferguson and Klass (1972). It is sometimes described as a Poisson approximation of independent increments processes.

Recall first the Lévy–Khinchine representation of a symmetric process $X(t)$ with independent and homogeneous increments with no Gaussian component (Laplace motions are examples of such processes):

$$\phi_{X(t)}(u) = \exp\left[\int_{-\infty}^{\infty}[\cos(uz) - 1]d\Lambda_t(z)\right],$$

where $\Lambda_t = t\Lambda$ and Λ is the Lévy measure of $X(1)$.

Consider the standard Laplace motion $L(t)$, i.e., with $\nu = 1$ and $\sigma = 1$. By the Lévy–Khinchine representation derived in Proposition 2.4.2, the above representation holds with Λ defined through

$$\Lambda([-u, u]^c) = 2E_1(\sqrt{2}u) = 2\int_{\sqrt{2}u}^{\infty}\frac{1}{x}e^{-x}dx.$$

Here E_1 stands for the exponential integral function [see, e.g., Abramowitz and Stegun (1965)]. In the following series representation we restrict ourselves to a standard Laplace motion and to time interval [0, 1].

Theorem 4.2.3 *Let $L(t)$ be a standard Laplace motion. Assume that (δ_i) is a Rademacher sequence (i.i.d. symmetric signs), (U_i) is an i.i.d. sequence of random variables distributed uniformly on* [0, 1], *(Γ_i) are arrival times in a standard Poisson process. We assume that all three sequences, (δ_i), (U_i), and (Γ_i), are independent. Then the following representation holds for $L(t)$:*

$$L(t) \stackrel{d}{=} \sum_{i=1}^{\infty}\delta_i J_i \mathbb{I}_{[0,t)}(U_i),$$

where the series is absolutely convergent with probability one, $J_i = \frac{1}{\sqrt{2}}E_1^{-1}(\Gamma_i)$, and $\mathbb{I}_{[0,t)}(U_i)$ is the indicator function of the interval $[0, t)$ evaluated at U_i.

Proof. The proof is a direct consequence of the theorem of Ferguson and Klass (1972, p. 1640). The absolute convergence follows from the fact that $\int_0^1 z d\Lambda(z)$ is finite. Consequently, no centering of the terms of the series is needed. By adding random signs to the representation, we obtain symmetry of the process. □

Remark 4.2.4 From this representation, one can derive properties of trajectories of Laplace motions. First of all, sample paths are pure jump functions (a function is a jump function if its value is equal the sum of the jumps, or in other words, if it is increasing only at the jumps). The absolute values of the jumps are given by J_i's, and are ordered. The largest jump is represented by $J_1 = \frac{1}{\sqrt{2}} E_1^{-1}(\Gamma_1)$, and its distribution is given by

$$P(J_1 \le x) = e^{-E_1(\sqrt{2}x)}, x > 0.$$

Since $E_1(x)$ converges to infinity when x approaches zero, the distribution of the first jump is continuous on $[0, \infty)$ and has density

$$f_{J_1}(x) = e^{-E_1(\sqrt{2}x)} e^{-\sqrt{2}x}/x.$$

Using the probability structure of the arrivals of a Poisson process one can easily derive the conditional distribution of the next jump given the previous ones. Namely, the distribution of J_n given that $J_1 = x_1, \dots, J_{n-1} = x_{n-1}$ has the c.d.f.

$$F(x|x_1, \dots, x_{n-1}) = e^{-E_1(\sqrt{2}x) + E_1(\sqrt{2}x_{n-1})}, \quad x > x_{n-1} > \cdots > x_1.$$

4.2.3 Asymmetric Laplace motion. The definition and properties of Laplace motion extend naturally to the asymmetric case. The fact that the asymmetric Laplace distribution $\mathcal{AL}(\mu, \sigma)$ is infinitely divisible justifies the following definition.

Definition 4.2.2 *A stochastic process $L(t)$ is called an asymmetric Laplace motion with the space scale parameter σ, the time scale parameter ν, and centered at μ (and denoted by $\mathcal{ALM}(\mu, \sigma, \nu)$) if*

1. *it starts at the origin, i.e.,*
$$L(0) = 0;$$
2. *it has independent and stationary (homogeneous) increments;*
3. *the increments by the time scale unit have an asymmetric Laplace distribution with the parameters μ and σ, i.e.,*
$$L(t + \nu) - L(t) \stackrel{d}{=} \mathcal{AL}(\mu, \sigma).$$

An asymmetric Laplace motion with drift *m is an $\mathcal{ALM}(\mu, \sigma, \nu)$ process $L(t)$ shifted by a linear function mt, i.e.,*

$$Y(t) = mt + L(t).$$

Remark 4.2.5 This definition and the properties of infinitely divisible distributions lead to the following characteristic function of the increment $L(s + t) - L(s)$ of the $\mathcal{ALM}(\mu, \sigma, \nu)$ process:

$$\phi_{L(t)}(u) = \frac{1}{(1 - i\mu u + \sigma^2 u^2/2)^{t/\nu}},$$

i.e., the increment has the generalized asymmetric Laplace distribution (the asymmetric Bessel function distribution) with the parameters μ, σ, and $\tau = t/\nu$, denoted $\mathcal{GAL}(\mu, \sigma, \tau)$.

Proposition 4.2.2 *Let $L(t)$ be an $\mathcal{ALM}(\mu, \sigma, \nu)$ Laplace motion with a drift m. Then*

$$E[L(t)] = mt + \mu t/\nu, \qquad \mathrm{Var}[L(t)] = t(\mu^2 + \sigma^2)/\nu.$$

Proof. The result follows from Remark 4.2.5 and Corollary 4.1.2. □

In Figure 4.4, we present trajectories of asymmetric Laplace motions with the same covariance structure as the symmetric processes of Figure 4.3.

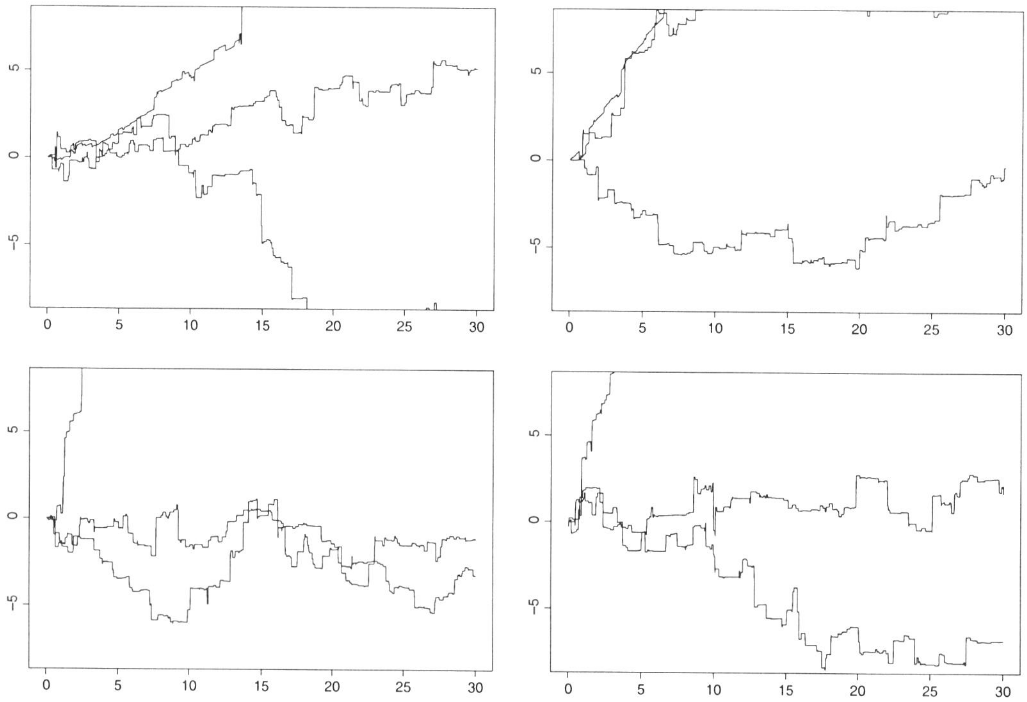

Figure 4.4: Trajectories of asymmetric Laplace motions with centering drifts (three paths for each process). All processes are asymmetric, but have the same covariance structure characterized by unit variance at time $t = 1$ and the mean zero, i.e., the same as for the symmetric processes of Figure 4.3. This requirement for asymmetric Laplace motion $\mathcal{ALM}(\mu, \sigma, \nu)$ with a drift m is satisfied by setting $m = -\mu/\nu$ and $\sigma = \sqrt{\nu - \mu^2}$, where $\mu^2 < \nu$. *Top*: Laplace motions with $\nu = 1$ and $\mu = 0.4$ (*left*) $\mu = 0.8$ (*right*). *Bottom*: Laplace motions with $\nu = 4$ and $\mu = 1$ (*left*) $\mu = 1.5$ (*right*).

Below we list representations of the $\mathcal{ALM}(\mu, \sigma, \nu)$ process, which are direct extensions of the ones obtained for the symmetric Laplace motions.

4.2.3.1 *Subordinated Brownian motion.* Assume that $B(t)$ is a Brownian motion with scale σ and with drift μ, and that Γ_t is a gamma process with the parameter ν independent of $B(t)$. Then the following representation for the $\mathcal{ALM}(\mu, \sigma, \nu)$ process $L(t)$ holds:

$$L(t) \stackrel{d}{=} B(\Gamma_t), \quad t > 0. \tag{4.2.2}$$

4.2.3.2 *Difference of gamma processes.* Let Γ_t and $\tilde{\Gamma}_t$ be two independent gamma processes with parameter ν. Let $\kappa = \sqrt{2}\sigma/(\mu + \sqrt{2\sigma^2 + \mu^2})$. Then we have the following representation of the $\mathcal{ALM}(\mu, \sigma, \nu)$ process $L(t)$:

$$L(t) \stackrel{d}{=} \frac{\sqrt{2}}{2}\sigma\left(\frac{1}{\kappa}\Gamma_t - \kappa\tilde{\Gamma}_t\right), \quad t > 0. \tag{4.2.3}$$

4.2.3.3 *Compound Poisson approximation.* The series representation of an $\mathcal{ALM}(\mu, \sigma, \nu)$ process is a direct generalization of the symmetric case, and involves a series that is absolutely convergent almost surely. Let us recall that the Lévy measure Λ of the asymmetric Laplace distribution $\mathcal{AL}(\mu, \sigma)$ is given by

$$\Lambda(u, \infty) = E_1(\sqrt{2}\kappa u/\sigma), \qquad \Lambda(-\infty, -u) = E_1(\sqrt{2}u/(\sigma\kappa)), \quad u > 0.$$

Let us now define $\Lambda_-(x) = E_1(\sqrt{2}x/(\sigma\kappa))$ and $\Lambda_+(x) = E_1(\sqrt{2}x\kappa/\sigma)$, $x > 0$.

Let $L(t)$ be an asymmetric Laplace motion $\mathcal{ALM}(\mu, \sigma, 1)$. Assume that (δ_i) is a Rademacher sequence of i.i.d. symmetric signs, (U_i) is an i.i.d. sequence of random variables distributed uniformly on $[0, 1]$, and (Γ_i) is a sequence of the arrival times in a standard Poisson process. We assume that all three sequences, (δ_i), (U_i), and (Γ_i), are independent. Then the following representation in distribution holds for $L(t)$:

$$L_t \stackrel{d}{=} \sum_{i=1}^{\infty} \delta_i J_i \mathbb{I}_{[0,t)}(U_i), \tag{4.2.4}$$

where the series is absolutely convergent with probability one and $J_i = \Lambda_{\delta_i}^{-1}(\Gamma_i)$.

4.3 Linnik distribution

The univariate symmetric Linnik distribution with index $\alpha \in (0, 2]$ and scale parameter $\sigma > 0$ is given by the characteristic function

$$\psi_{\alpha,\sigma}(t) = \frac{1}{1 + \sigma^\alpha |t|^\alpha}, \quad t \in \mathbb{R}, \tag{4.3.1}$$

and is named after Ju.V. Linnik, who showed that the function (4.3.1) is a bona fide ch.f. for any $\alpha \in (0, 2]$ [see Linnik (1953)]. Since for $\alpha = 2$ we obtain symmetric Laplace distribution, the distribution is also known as α-Laplace [see, e.g., Pillai (1985)]. We write $L_{\alpha,\sigma}$ to denote the distribution given by (4.3.1).

Linnik laws are special cases of strictly geometric stable distributions, introduced in Klebanov et al. (1984). A random variable Y (and its probability distribution) is called strictly geometric stable, if for any $p \in (0, 1)$ there is an $a_p > 0$ such that

$$a_p \sum_{i=1}^{\nu_p} Y_i \stackrel{d}{=} Y_1, \tag{4.3.2}$$

where ν_p is a geometric r.v. with mean $1/p$, while the Y_i's are i.i.d. copies of Y, independent of ν_p. Strictly geometric stable laws are a special case of geometric stable laws discussed in Subsection 4.4.4; they have ch.f. (4.4.7) with either $\mu = 0$ and $\alpha \neq 1$ or $\beta = 0$ and $\alpha = 1$. Thus strictly geometric stable laws form a three-parameter family, and their ch.f. can be written as

$$\psi_{\alpha,\sigma,\tau}(t) = \frac{1}{1 + \sigma^\alpha |t|^\alpha \exp(-i\pi\alpha\tau \operatorname{sign}(t)/2)}, \quad t \in \mathbb{R}, \tag{4.3.3}$$

where α and σ are as before, and τ is such that $|\tau| \leq \min(1, 2/\alpha - 1)$. Since for $\tau = 0$ we obtain the symmetric Linnik distribution (4.3.1), some authors refer to (4.3.3) as a *nonsymmetric Linnik distribution* [see, e.g., Erdogan and Ostrovskii (1997)]. As we shall see in this section, Linnik distributions share some, but not all the properties of the symmetric Laplace distribution. Like symmetric Laplace distributions, Linnik laws are stable with respect to geometric summation and appear as limit laws of geometric compounds when the summands are symmetric and have an infinite variance. We shall discuss their various characterizations in Section 4.3.1. In Section 4.3.2, while discussing representations of Linnik laws, we shall show that they are mixtures of stable laws as well as exponential mixtures and scale mixtures of normal distributions. These representations lead directly to integral representations of the Linnik densities, which are discussed in Section 4.3.3, devoted to Linnik densities and distribution functions. Although a closed-form expression for the Linnik density seems to be unavailable, as it is the case for stable laws, asymptotic results have been investigated by Kotz et al. (1995). In Section 4.3.4, we shall study moments and the tail behavior of the Linnik laws. We shall show that their tail probabilities are no longer exponential, and the moments are governed by the parameter α. Unlike Laplace laws, although analogous to stable distributions, the Linnik laws have an infinite variance, while the mean is finite only for $1 < \alpha < 2$. In Section 4.3.5, we shall list properties of the Linnik laws, which include unimodality, geometric and classical infinite divisibility, and self-decomposability. Sections 4.3.6 and 4.3.7 are devoted to the problems of simulation and estimation, respectively. For the Linnik laws, the standard methods (which are based on explicit forms of the relevant distribution functions and densities) are not practical. We shall show that the problem of simulation is easily handled by the mixture representations of Linnik laws and discuss some recent advances in the estimation problem. Section 4.3.8 is devoted to the extension of the Linnik distribution.

4.3.1 Characterizations. In this section we present characterizations of Linnik laws related mostly to geometric summation. Many results are consequences of the fact that Linnik laws are special cases of strictly geometric stable distributions.

4.3.1.1 *Stability with respect to geometric summation.* We saw in Section 2.2.6 that within the class of symmetric r.v.'s with a finite variance, the classical Laplace r.v. is characterized by the stability property (4.3.2). Anderson (1992) observed that the Linnik distribution is closed under geometric compounding as well, so that (4.3.2) holds with $L_{\alpha,\sigma}$ distributed Y_i's and $a_p = p^{1/\alpha}$. In the case $\alpha = 1$ this result is due to Arnold (1973) and it serves as a foundation for the development of Anderson's (1992) multivariate Linnik distribution. In the subsequent result we show that stability property (4.3.2) actually characterizes symmetric Linnik distributions within the class of symmetric r.v.'s (not necessarily with finite variance).

Proposition 4.3.1 *Let $Y, Y_1, Y_2, \ldots$ be symmetric, i.i.d. random variables and let ν_p be a geometric random variable with mean $1/p$, independent of the Y_i's. Then the following statements are equivalent:*

(i) *Y is stable with respect to geometric summation,*

$$a_p \sum_{i=1}^{\nu_p} (Y_i + b_p) \stackrel{d}{=} Y \quad \text{for all} \quad p \in (0, 1), \tag{4.3.4}$$

where $a = a_p > 0$ and $b = b_p \in \mathbb{R}$.

(ii) *Y has a symmetric Linnik distribution.*

Moreover, the constants a_p and b_p are necessarily of the form: $a_p = p^{1/\alpha}$, $b_p = 0$.

Proof. First, we show that the Linnik r.v. with ch.f. (4.3.1) satisfies the relation (4.3.4) with a_p and b_p given above. Using the typical conditional argument we write the ch.f. of the variable on the left-hand side of (4.3.4) in the form

$$\frac{p}{1 + p\sigma^\alpha |t|^\alpha} \frac{1}{1 - (1-p)(1 + p\sigma^\alpha |t|^\alpha)^{-1}}, \tag{4.3.5}$$

and note that it simplifies to (4.3.1), which is the ch.f. of the right-hand side of (4.3.4). To prove the converse, use the corresponding characterization of strictly geometric stable laws [see, e.g., Kozubowski (1994b), Theorem 3.2] and conclude that if an r.v. Y_1 satisfies (4.3.4), it then must be a strictly geometric stable r.v. with ch.f. (4.3.3) and the normalizing constants must be as specified in the statement of the proposition. Since Y_1 is assumed to be symmetric, its ch.f. must be real, implying that the parameter τ in (4.3.3) equals zero, leading to the Linnik ch.f. (4.3.1). This concludes the proof. □

What happens if relation (4.3.4) holds only for one particular value of p? Then the solution of (4.3.4) consists of a larger class than the class of strictly geometric stable laws; see Lin (1994) for details. However, under certain additional tail conditions, relation (4.3.4) with one particular value of p characterizes symmetric Linnik distributions as well. Specifically, assuming that ψ satisfies the condition

$$\lim_{t\to 0}(1-\psi(t))/|t|^{\alpha} = \gamma \quad \text{for some } \gamma > 0 \text{ and } 0 < \alpha \le 2, \tag{4.3.6}$$

we have the following result.

Proposition 4.3.2 *Let $Y, Y_1, Y_2, \ldots$ be i.i.d. r.v.'s whose ch.f. ψ satisfies condition* (4.3.6). *Let $p \in (0,1)$ and let ν_p be a geometric r.v. with mean $1/p$, independent of the sequence (Y_i). Then*

$$a_p \sum_{i=1}^{\nu_p} Y_i \stackrel{d}{=} Y \tag{4.3.7}$$

for some $a_p > 0$ if and only if $a_p = p^{1/\alpha}$ and Y has a symmetric Linnik distribution.

See Lin (1994) for a proof and also for a similar characterization of Mittag–Leffler distributions. The result also appeared in Kakosyan et al. (1984) under the additional assumptions that $a_p = p^{1/\alpha}$ and the distribution of Y is nondegenerate and symmetric.

The following characterization of the Linnik distribution is also proved in Lin (1994) [as well as in Kakosyan et al. (1984) under the additional assumptions that $a_p = (p/q)^{1/\alpha}$ and the distribution of Y is nondegenerate and symmetric].

Proposition 4.3.3 *Let $Y_1, Y_2, \ldots$ be i.i.d. r.v.'s whose ch.f. ψ satisfies condition* (4.3.6). *Let $p, q \in (0,1)$, where $p \neq q$, and let ν_p and ν_q be geometric r.v.'s with means $1/p$ and $1/q$, respectively, independent of (Y_i). Then*

$$a_p \sum_{i=1}^{\nu_p} Y_i \stackrel{d}{=} \sum_{i=1}^{\nu_q} Y_i \tag{4.3.8}$$

with some $a_p \neq 0$ if and only if $|a_p|^{\alpha} = p/q$ and Y has a symmetric Linnik distribution.

We conclude this section by noting that relation (4.3.7) remains valid under the *randomization* of parameter p. More precisely, let $Y, Y_1, Y_2, \ldots$ be i.i.d. symmetric and nondegenerate r.v.'s whose ch.f. ψ satisfies condition (4.3.6). Let ν_p be a geometric r.v. with mean $1/p$, independent of the sequence (Y_i), where $p \in (0,1)$. Further, assume that the parameter p is itself an r.v. with a probability distribution on (0, 1). Then relation (4.3.7) holds with $a_p = p^{1/\alpha}$ if and only if Y has symmetric Linnik distribution. In addition, if (4.3.7) holds with nonnegative r.v.'s and $a_p = p$, then Y must have an exponential distribution; see Kakosyan et al. (1984) for proofs and further details.

4.3.1.2 *Distributional limits of geometric sums.* We saw in Section 2.2.7 that the classical Laplace distribution arises as the only possible limit of a geometric sum with symmetric i.i.d. components with finite variance. If the condition of finite variance is omitted, we then obtain a characterization of symmetric Linnik distributions.

Proposition 4.3.4 *The class of symmetric Linnik distributions coincides with the class of distributional limits of*

$$S_p = c_p \sum_{i=1}^{\nu_p} X_i \tag{4.3.9}$$

as $p \to 0$, where $c_p > 0$, the X_i's are symmetric i.i.d. random variables, and ν_p is a geometric random variable with mean $1/p$, independent of the X_i's.

Proof. First, note that by Proposition 4.3.1, a symmetric Linnik r.v. X is *equal* in distribution to the r.v. S_p given by (4.3.9), where $a_p = p^{1/\alpha}$ and X_i's are i.i.d. copies of X. So it is a distributional limit of S_p as well. Thus it remains to show that if geometric compounds (4.3.9) with i.i.d. and symmetric X_i's converge in distribution to an r.v. Y, then the latter must have a symmetric Linnik distribution. Our proof consists of showing that the r.v. Y is symmetric and stable with respect to geometric summation (i.e., (4.3.2) holds), and thus it must have a symmetric Linnik distribution by Proposition 4.3.1. First, note that as the r.v.'s X_i are symmetric, their ch.f. is real, so the ch.f. of S_p must be real, implying that the ch.f. of the limiting r.v. Y is real as well. Consequently, Y has a symmetric distribution. If Y is degenerate at zero, it is (a degenerate) Linnik (with $\sigma = 0$) and the result is valid. Assume now that the distribution of Y is not concentrated at zero. It then follows that Y cannot have a degenerate distribution (concentrated at some constant not equal to zero), since its ch.f. would not be real.

Next, fix an arbitrary $p' \in (0, 1)$ and for any $p \in (0, p')$ define $p'' = p/p'$. Then the geometric r.v. ν_p admits the representation

$$\nu_p = \sum_{i=1}^{\nu_{p'}} \nu_{p''}^{(i)}, \tag{4.3.10}$$

where $\nu_{p''}^{(i)}$'s are i.i.d. geometric r.v.'s with mean $1/p''$ while $\nu_{p'}$ is a geometric r.v. with mean $1/p'$, independent of the $\nu_{p''}^{(i)}$'s (Exercise 4.5.15). This allows us to express S_p in the following manner:

$$S_p = c_p \sum_{i=1}^{\nu_p} X_i \stackrel{d}{=} \frac{c_p}{c_{p''}} \sum_{i=1}^{\nu_{p'}} W_{p''}^{(i)}, \tag{4.3.11}$$

where the $W_{p''}^{(i)}$'s are i.i.d. r.v.'s equal in distribution to $S_{p''} = c_{p''} \sum_{i=1}^{\nu_{p''}} X_i$. Now as $p \to 0$, we note that $p'' = p/p'$ also converges to zero (p' being fixed!), so by assumption we have

$$W_{p''}^{(i)} = S_{p''} = c_{p''} \sum_{i=1}^{\nu_{p''}} X_i \stackrel{d}{\to} Y_i, \quad i = 1, 2, \ldots, \tag{4.3.12}$$

where the Y_i's are independent copies of Y. Thus we have the convergence

$$\sum_{i=1}^{\nu_{p'}} W_{p''}^{(i)} \stackrel{d}{\to} \sum_{i=1}^{\nu_{p'}} Y_i \tag{4.3.13}$$

(see Exercise 4.5.16). Since by the assumption $S_p \xrightarrow{d} Y$, where Y is nondegenerate, in view of (4.3.11) and (4.3.13) we conclude that the sequence $c_p/c_{p''}$ must converge to a limit (which may depend on p') denoted by $a_{p'}$, and we must have

$$a_{p'} \sum_{i=1}^{\nu_{p'}} Y_i \stackrel{d}{=} Y. \tag{4.3.14}$$

Consequently, by Proposition 4.3.1, Y must have symmetric Linnik distribution, as p' is an arbitrary real number in (0, 1). The result has been proved. □

4.3.1.3 *Stability with respect to deterministic summation.* We saw in Section 2.2.8 that within the class of symmetric distributions with finite variance, the classical Laplace distribution can be characterized by means of the stability property under deterministic summation and random normalization. Omitting the condition of finite variance leads to a characterization of symmetric Linnik laws.

Proposition 4.3.5 *Let the variables B_n, where $n > 0$, have a* Beta$(1, n)$ *distribution given by* (2.2.45). *Let $0 < \alpha \le 2$, and let $\{Y_i\}$ be a sequence of symmetric i.i.d. random variables. Then the following statements are equivalent:*

(i) *For all $n \ge 2$, $Y_1 \stackrel{d}{=} B_{n-1}^{1/\alpha}(Y_1 + \cdots + Y_n)$.*

(ii) *Y_1 has a symmetric Linnik distribution.*

Proof. The proof is very similar to that of Proposition 2.2.11 for the symmetric Laplace case. Write the right-hand side of the representation in (i) in the form $U_n V_n$, where

$$U_n = (nB_{n-1})^{1/\alpha} \text{ and } V_n = \frac{\sum_{i=1}^n Y_i}{n^{1/\alpha}}, \tag{4.3.15}$$

and let $n \to \infty$. Then U_n converges in distribution to a random variable $W^{1/\alpha}$, where the variable W has a standard exponential distribution. Further, since the product $U_n V_n$ as well as the sequence U_n are convergent, while V_n has a symmetric distribution, we conclude that the sequence V_n must be convergent as well. Moreover, if X is the limit of V_n, then it must have a symmetric stable distribution with ch.f. (4.3.20). Since by the assumption U_n is independent of V_n, the limit of the product $U_n V_n$ is the product of the limits, so

$$U_n V_n \xrightarrow{d} W^{1/\alpha} X. \tag{4.3.16}$$

But this is representation (4.3.19) of Linnik random variables discussed in the next section. The implication (i) ⇒ (ii) follows, since Y_1 must have the same distribution as the limit in (4.3.16).

We now turn to the proof of the implication (ii) ⇒ (i). Multiply both sides of (4.3.21) from Proposition 4.3.8 by $B_{n-1}^{1/\alpha}$ (which is independent of all other r.v.'s) to obtain

$$B_{n-1}^{1/\alpha}(Y_1 + \cdots + Y_n) \stackrel{d}{=} (G_n B_{n-1})^{1/\alpha} X \tag{4.3.17}$$

(with X as above). By Lemma 2.2.2, the product $G_n B_{n-1}$ has the same distribution as a standard exponential r.v. W, so the right-hand side of (4.3.17) has a Linnik distribution by the representation (4.3.19). The proof is thus complete. □

We conclude our discussion on stability with another characterization of symmetric Linnik laws, derived in Pillai (1985) for a larger class of semi-α-Laplace distributions, a class that includes all strictly geometric stable laws.

Proposition 4.3.6 *Let $Y, Y_1, Y_2,$ and Y_3 be i.i.d. symmetric Linnik variables $L_{\alpha,\sigma}$. Let $p \in (0, 1)$, and let I be an indicator random variable, independent of Y, Y_1, Y_2, Y_3, with $P(I = 1) = p$ and $P(I = 0) = 1 - p$. Then the following equality in distribution is valid for any $p \in (0, 1)$:*

$$Y \stackrel{d}{=} p^{1/\alpha} I Y_1 + (1 - I)(Y_2 + p^{1/\alpha} Y_3). \tag{4.3.18}$$

Proof. The result follows by writing the ch.f. of the right-hand side in (4.3.18) conditioning on the distribution of the r.v. I. □

4.3.2 Representations. Representations of Linnik random variables were studied by Devroye (1990), Anderson (1992), Anderson and Arnold (1993), Kotz and Ostrovskii (1996), and Kozubowski (1998). Devroye (1990) derived the following fundamental representation of a Linnik r.v. in terms of independent exponential and symmetric stable random variables, which is analogous to the representation (2.2.3) of the Laplace distribution.

Proposition 4.3.7 *A Linnik r.v. Y with the ch.f.* (4.3.1) *admits the representation*

$$Y \stackrel{d}{=} W^{1/\alpha} X, \tag{4.3.19}$$

where X is symmetric stable variable with ch.f.

$$\phi(t) = \exp(-\sigma^\alpha |t|^\alpha) \tag{4.3.20}$$

and W is a standard exponential r.v., independent of X.

This representation is a special case with $n = 1$ of the next result, which describes the distribution of the sum of n i.i.d. Linnik random variables. It generalizes similar representation for the case of symmetric Laplace random variables; see Proposition 2.2.10.

Proposition 4.3.8 *Let $Y_1, Y_2, \ldots$ be i.i.d. Linnik r.v.'s with ch.f.* (4.3.1). *Then*

$$Y_1 + \cdots + Y_n \stackrel{d}{=} G_n^{1/\alpha} X, \tag{4.3.21}$$

where X is symmetric stable with ch.f. (4.3.20) *and G_n has gamma $G(n, 1)$ distribution.*

Proof. The result follows by computing the ch.f.'s on both sides of (4.3.21). By conditioning on G_n, we calculate the ch.f. of $G_n^{1/\alpha} X$ as follows:

$$Ee^{itG_n^{1/\alpha} X} = \int_0^\infty Ee^{itz^{1/\alpha} X} \frac{z^{n-1}}{\Gamma(n)} e^{-z} dz = \int_0^\infty \phi(tz^{1/\alpha}) \frac{z^{n-1}}{\Gamma(n)} e^{-z} dz,$$

where ϕ is the symmetric stable ch.f. (4.3.20). Since

$$\phi(tz^{1/\alpha}) e^{-z} = e^{-z(\sigma^\alpha |t|^\alpha + 1)},$$

a straightforward integration results in the Linnik ch.f. (4.3.1). □

Representation (4.3.19) allows for obtaining properties of Linnik distributions from those of stable laws. However, its value for certain applications may be limited. For instance, this representation is not very convenient for simulating Linnik random variates, since stable distributions do not admit densities or distribution functions in closed form and require mixture representations themselves for simulation. Kotz and Ostrovskii (1996) and Kozubowski (1998) have studied alternative mixture representations of the Linnik distribution which allow efficient generation of the corresponding random variates. Kotz and Ostrovskii (1996) observe that for any $0 < \alpha < \alpha' \le 2$, the ch.f.'s of the Linnik distributions $L_{\alpha,1}$ and $L_{\alpha',1}$ satisfy the equation

$$\psi_{\alpha,1}(t) = \int_0^\infty \psi_{\alpha',1}(t/s)g(s;\alpha,\alpha')ds, \tag{4.3.22}$$

where

$$g(s;\alpha,\alpha') = \frac{\alpha'}{\pi} \sin\frac{\pi\alpha}{\alpha'} \frac{s^{\alpha-1}}{1+s^{2\alpha}+2s^\alpha\cos\frac{\pi\alpha}{\alpha'}} \tag{4.3.23}$$

is the density of a nonnegative r.v. $U_{\alpha,\alpha'}$. Kozubowski (1998) notes the representation

$$\psi_{\alpha,1}(t) = \int_0^\infty \psi_{\alpha',1}(ts)g(s;\alpha,\alpha')ds, \tag{4.3.24}$$

using this notation. Representations (4.3.22)–(4.3.24) lead to the conclusion that the corresponding Linnik r.v.'s $Y_{\alpha,1}$ and $Y_{\alpha',1}$ obey the representations

$$Y_{\alpha,1} \stackrel{d}{=} Y_{\alpha',1} \cdot U_{\alpha,\alpha'} \stackrel{d}{=} Y_{\alpha'}/U_{\alpha,\alpha'}. \tag{4.3.25}$$

Kozubowski (1998) modifies representations (4.3.25) by introducing an r.v. $W_\rho = U^\alpha_{\alpha,\alpha'}$, where $\rho = \alpha/\alpha' < 1$, with a folded Cauchy density g_ρ on $(0,\infty)$ given by

$$g_\rho(x) = \frac{\sin(\pi\rho)}{\pi\rho[x^2+2x\cos(\pi\rho)+1]}. \tag{4.3.26}$$

Note that the definition of W_ρ can be extended to the cases $\rho = 0$ and $\rho = 1$ as well by taking weak limits as $\rho \to 0^+$ and $\rho \to 1^-$, thus arriving at the density $g_0(x) = (1+x)^{-2}$ for W_0 and $W_1 = 1$ (see Exercise 4.5.19). The following result is a restatement of (4.3.25) in terms of the r.v. W_ρ [see Kozubowski (1998)].

Proposition 4.3.9 *Let $0 < \alpha < \alpha' \le 2$ and $\rho = \alpha/\alpha' < 1$. Let W_ρ be a nonnegative r.v. with density* (4.3.26), *and let $Y_{\alpha',\sigma}$ be a Linnik $L_{\alpha',\sigma}$ r.v., independent of W_ρ. Then an r.v. $Y_{\alpha,\sigma}$ with the Linnik $L_{\alpha,\sigma}$ distribution admits the representations*

$$Y_{\alpha,\sigma} \stackrel{d}{=} Y_{\alpha',\sigma} \cdot W_\rho^{1/\alpha} \stackrel{d}{=} Y_{\alpha',\sigma}/W_\rho^{1/\alpha}. \tag{4.3.27}$$

The fact that the representations involve both division and multiplication follows from the reciprocal property of the r.v. W_ρ (see Exercises 4.5.20 and 4.5.21).

Taking $\alpha' = 2$, we arrive at the classical Laplace r.v. and the representation provides a direct method of simulating Linnik random variates discussed in section 4.3.6. Thus a Linnik $L_{\alpha,\sigma}$ r.v. can be thought of as a Laplace variable with a stochastic variance and also as a normal variable with a stochastic variance (since a Laplace distribution is a scale mixture of normal distributions). In addition, the Laplace r.v. corresponding to $\alpha' = 2$ has the representation σIW in accordance with Proposition 2.2.3. Consequently, we obtain the following *exponential mixture* representation of the Linnik r.v. $L_{\alpha,\sigma}$.

Proposition 4.3.10 *Let $Y_{\alpha,\sigma}$ be a Linnik $L_{\alpha,\sigma}$ r.v. with any $0 < \alpha \le 2$, and let W_ρ be a nonnegative r.v. with density* (4.3.26) *for $\rho = \alpha/2 \le 1$. Then*

$$Y_{\alpha,\sigma} \stackrel{d}{=} \sigma \cdot I \cdot W \cdot W_\rho^{1/\alpha} \stackrel{d}{=} \sigma \cdot I \cdot W/W_\rho^{1/\alpha}, \tag{4.3.28}$$

where I is an indicator r.v. taking values ± 1 with probabilities $1/2$, W is standard exponential variable, and all the variables are independent.

Taking $\alpha = 2$, this representation reduces to representation (2.2.10) of the Laplace distribution, as $W_1 = 1$.

Remark 4.3.1 Choosing $\alpha = 1$ and $\alpha' = 2$ and noting that in this case the r.v. $W_{1/2}$ has a folded standard Cauchy distribution, we arrive at the representation

$$Y_{1,1} \stackrel{d}{=} \text{exponential} \cdot \text{Cauchy} \stackrel{d}{=} |\text{Cauchy}| \cdot \text{Laplace}, \qquad (4.3.29)$$

which is essentially a restatement of the well-known result that the density of Cauchy variable is of the same form as the characteristic function of the Laplace while the characteristic function of the Cauchy variable is of the same functional form as the density of the Laplace.

Remark 4.3.2 Nonsymmetric Linnik distributions with ch.f. (4.3.3) and more general geometric stable r.v.'s admit similar mixture representations [see Erdogan and Ostrovskii (1998a), Kozubowski (2000a), and Belinskiy and Kozubowski (2000) for further details].

4.3.3 Densities and distribution functions. Here we study Linnik distribution functions and densities. There are no closed-form expressions for Linnik distribution functions and densities, except for $\alpha = 2$, which corresponds to the Laplace distribution. However, the mixture representations of Section 4.3.2 lead to integral as well as asymptotic and convergent series representations of Linnik densities and distribution functions, which we present here.

4.3.3.1 *Integral representations.* Representation (4.3.19) leads to the representations of Linnik densities and distribution functions through their stable counterparts. Let $p_{\alpha,\sigma}$ and $F_{\alpha,\sigma}$ denote the density and distribution function of the Linnik $L_{\alpha,\sigma}$ distribution given by ch.f. (4.3.1). Similarly, let $g_{\alpha,\sigma}$ and $G_{\alpha,\sigma}$ denote the density and distribution function of the corresponding stable law specified in Proposition 4.3.7.

Proposition 4.3.11 *Every Linnik distribution with $0 < \alpha \le 2$ is absolutely continuous and*

$$F_{\alpha,\sigma}(x) = \int_0^\infty G_{\alpha,\sigma}\left(\frac{x}{z^{1/\alpha}}\right) e^{-z} dz, \qquad (4.3.30)$$

$$p_{\alpha,\sigma}(x) = \int_0^\infty z^{-1/\alpha} g_{\alpha,\sigma}\left(\frac{x}{z^{1/\alpha}}\right) e^{-z} dz. \qquad (4.3.31)$$

These representations, which are dealt with in Exercise 4.5.22, appeared in Kozubowski (1994a) and Lin (1994). Note that in case $\alpha = 2$, equations (4.3.30) and (4.3.31) produce the distribution function and density of a symmetric Laplace distribution.

Now we express the exponential mixture representation (4.3.28) in terms of the corresponding densities and distribution functions (see Exercise 4.5.23).

Proposition 4.3.12 *The distribution function and density of the Linnik $L_{\alpha,1}$ distribution with $0 < \alpha < 2$ admit the following representations for $x > 0$:*

$$F_{\alpha,1}(x) = 1 - \frac{\sin\frac{\pi\alpha}{2}}{\pi} \int_0^\infty \frac{v^{\alpha-1}\exp(-vx)dv}{1 + v^{2\alpha} + 2v^\alpha \cos\frac{\pi\alpha}{2}} \qquad (4.3.32)$$

and

$$p_{\alpha,1}(x) = \frac{\sin\frac{\pi\alpha}{2}}{\pi} \int_0^\infty \frac{v^\alpha \exp(-v|x|)dv}{1 + v^{2\alpha} + 2v^\alpha \cos\frac{\pi\alpha}{2}}. \qquad (4.3.33)$$

For $x < 0$, use $F_{\alpha,1}(x) = 1 - F_{\alpha,1}(-x)$ and $p_{\alpha,1}(x) = p_{\alpha,1}(-x)$.

This representation appears in Erdogan (1995) and, for the case $1 < \alpha < 2$, in Klebanov et al. (1996). Note that the density (4.3.33) can be written equivalently in the form

$$p_{\alpha,1}(x) = \frac{\sin\frac{\pi\alpha}{2}}{\pi}\int_0^\infty \frac{v^\alpha \exp(-v|x|)dv}{|1+v^\alpha \exp(i\pi\alpha/2)|^2}, \qquad x \neq 0, \tag{4.3.34}$$

in which it was originally first derived (by the inversion formula and the Cauchy theorem for complex variables) in Linnik (1953). Indeed, since for real x we have $\exp(ix) = \cos x + i\sin x$, the denominator under the integral in (4.3.34) is equal to

$$\left|1 + v^\alpha \cos\frac{\pi\alpha}{2} + iv^\alpha \sin\frac{\pi\alpha}{2}\right|^2 = \left(1 + v^\alpha\cos\frac{\pi\alpha}{2}\right)^2 + \left(v^\alpha \sin\frac{\pi\alpha}{2}\right)^2$$

and coincides with that in (4.3.33).

Remark 4.3.3 Hayfavi (1998) derived another representation of the Linnik density $p_{\alpha,1}$ by a contour integral: for any $\delta \in (0,1)$ and $\alpha \in [\delta, 2-\delta]$, we have

$$p_{\alpha,1}(x) = \frac{1}{x}\frac{i}{4\alpha}\int_{L(\delta)} \frac{e^{z\log x}dz}{\Gamma(z)\sin\frac{\pi z}{\alpha}\cos\frac{\pi z}{2}}, \qquad x > 0,$$

where $L(\delta)$ is the boundary of the region

$$\{z : |z| > \delta/2, |\arg z| < \pi/4\}.$$

Note that

$$\lim_{x\to 0^+} p_{\alpha,1}(x) = \frac{\sin\frac{\pi\alpha}{2}}{\pi}\int_0^\infty \frac{v^\alpha dv}{|1+v^\alpha\exp(i\pi\alpha/2)|^2}. \tag{4.3.35}$$

The integral is divergent for $0 < \alpha \leq 1$ and it is convergent for $1 < \alpha < 2$. In the latter case

$$p_{\alpha,1}(0) = \lim_{x\to 0^+} p_{\alpha,1}(x) = (\alpha\sin(\pi\alpha))^{-1}. \tag{4.3.36}$$

Thus the limit of $p_{\alpha,1}(x)$ as $x \to 0^+$ is finite for $1 < \alpha < 2$ and infinite for $0 < \alpha \leq 1$, in which case the densities have an infinite peak at $x = 0$. On the interval $(0, \infty)$, the function $p_{\alpha,1}(x)$ is decreasing and its kth derivative satisfies the relations

$$\lim_{x\to 0^+} (-1)^k p_{\alpha,1}^{(k)}(x) = \infty, \quad k = 1, 2, \ldots$$

and

$$(-1)^k p_{\alpha,1}^{(k)}(x) \geq 0, \quad k = 1, 2, \ldots.$$

The latter property implies *complete monotonicity* of the Linnik density on $(0, \infty)$ [see, e.g., Kotz et al. (1995)]. Since the characteristic function is real for all $t \in \mathbb{R}$, the density $p_{\alpha,1}(x)$ is an even function of x. Finally, since the integral on the right-hand side of (4.3.34) is a continuous function of α for any fixed x, the density $p_{\alpha,1}(x)$ is a continuous function of $\alpha \in (0, 2)$. Figure 4.5 presents graphs of several selected Linnik densities.

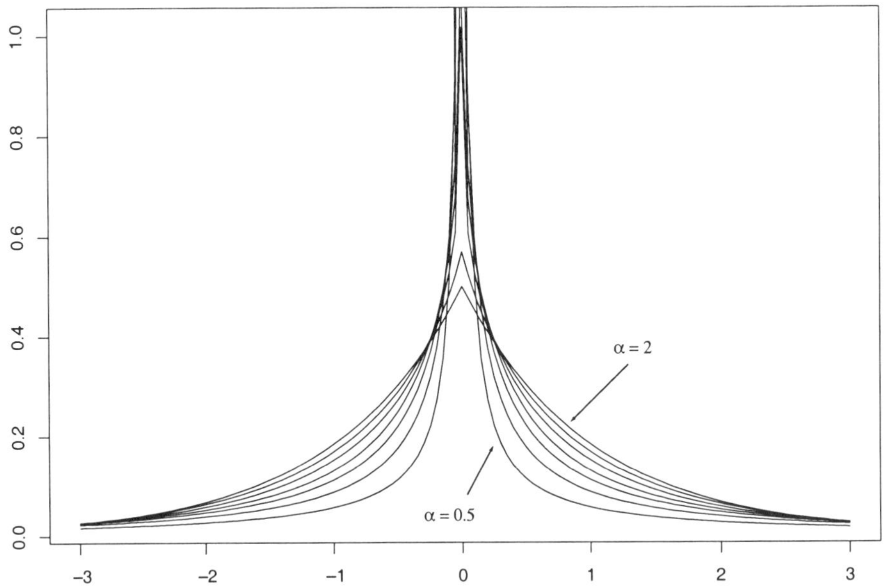

Figure 4.5: Densities of Linnik distributions with $\sigma = 1$ and α's equal to 0.5, 0.75, 1.00, 1.25, 1.50, 1.75, 2.00.

4.3.3.2 *Series expansions.* We briefly discuss asymptotic and convergent series representations of Linnik distribution functions and densities. We start with the asymptotic expansions at infinity, due to Kozubowski (1994a), Erdogan (1995), and Kotz et al. (1995). Let $p_\alpha = p_{\alpha,1}$ be the density and let $F_\alpha = F_{\alpha,1}$ be the distribution function corresponding to the Linnik characteristic function (4.3.1) with $\sigma = 1$. Consider the densities first. The following asymptotic relation is valid as $x \to \infty$:

$$p_\alpha(\pm x) \sim \frac{1}{\pi} \sum_{k=1}^{\infty} (-1)^{k+1} \Gamma(k\alpha + 1) \sin(k\pi\alpha/2) x^{-1-k\alpha}. \tag{4.3.37}$$

This asymptotic relation can be written alternatively as follows.

Proposition 4.3.13 *The density p_α of a Linnik $L_{\alpha,1}$ distribution has the following representation for $x > 0$:*

$$\forall n > 0 \quad p_\alpha(\pm x) = \frac{1}{\pi} \sum_{k=1}^{n} c_k x^{-k\alpha-1} + R_n(x), \tag{4.3.38}$$

where

$$c_k = (-1)^{k+1} \Gamma(k\alpha + 1) \sin(k\pi\alpha/2),$$

$$|R_n(x)| \le \frac{\alpha \Gamma(\alpha(n+1)+1)}{\pi |\sin(\pi\alpha/2)|} x^{-\alpha(n+1)-1}.$$

See Kozubowski (1994a) for the proof of Proposition 4.3.13 and Belinskiy and Kozubowski (2000) for its extension to geometric stable laws.

The approximation of $p_\alpha(x)$ with the finite sum (4.3.38) should be used for large values of x, since for fixed n the remainder $|R_n(x)|$ converges to zero as $x \to \infty$ (with the rate of $O(\frac{1}{x^{(n+1)\alpha+1}})$).

In particular, for $n = 1$, we have the following asymptotic expansion:

$$p_\alpha(\pm x) \sim \frac{1}{\pi}\Gamma(1+\alpha)\sin(\pi\alpha/2)x^{-1-\alpha}, \quad x \to \infty, \tag{4.3.39}$$

with the absolute value of the remainder $R_1(x)$ bounded by

$$b_1(x,\alpha) = \frac{\alpha\Gamma(2\alpha+1)}{\pi \sin\frac{\pi\alpha}{2}} x^{-2\alpha-1}. \tag{4.3.40}$$

As an illustration of asymptotic expansion (4.3.39), in Table 4.2 we present the values of the approximation, along with the corresponding values of the bound (4.3.40) and the percent error (equal to the ratio of bound (4.3.40) to approximate value (4.3.39) multiplied by 100%).

x	α	appr. of $p_\alpha(x)$	$b_1(x,\alpha)$	percent error
10	1/2	6.307831E-3	2.250791E-3	36%
10	3/2	9.461747E-4	4.051423E-4	42%
20	1/2	2.230155E-3	5.626977E-4	25%
20	3/2	1.672616E-4	2.532140E-5	15%
50	1/2	5.641896E-4	9.003163E-5	16%
50	3/2	1.692569E-5	6.482277E-7	3.83%
100	1/2	1.994711E-4	2.250791E-5	11%
100	3/2	2.992067E-6	4.051423E-8	1.35%
1000	1/2	6.307831E-6	2.250791E-7	3.57%
1000	3/2	9.461747E-9	4.051423E-12	0.04%

Table 4.2: The values of the one-term asymptotic expansion of $p_\alpha(x)$, along with the values of the error bound $b_1(x,\alpha)$ and the corresponding maximal percent error, for selected values of α and x.

Next, we turn to distribution functions. Their asymptotic expansions are obtained by integration of the corresponding series for the densities. We have the following asymptotic relation as $x \to \infty$:

$$1 - F_\alpha(x) \sim \frac{1}{\pi}\sum_{k=1}^{\infty}(-1)^{k+1}\Gamma(k\alpha)\sin(k\pi\alpha/2)x^{-k\alpha}. \tag{4.3.41}$$

Similarly, we get the behavior of the Linnik c.d.f. at $-\infty$:

$$F_\alpha(-x) \sim \frac{1}{\pi}\sum_{k=1}^{\infty}(-1)^{k+1}\Gamma(k\alpha)\sin(k\pi\alpha/2)x^{-k\alpha}, \quad x \to \infty. \tag{4.3.42}$$

More precisely, we have the following result:

Proposition 4.3.14 *The distribution function F_α of a Linnik distribution $L_{\alpha,1}$ admits the following representation for $x > 0$:*

$$\forall n > 0 \quad 1 - F_\alpha(x) = \frac{1}{\pi}\sum_{k=1}^{n} b_k x^{-k\alpha} + R_n^*(x), \tag{4.3.43}$$

where

$$b_k = (-1)^{k+1}\Gamma(k\alpha)\sin(k\pi\alpha/2),$$

$$|R_n^*(x)| \le \frac{\alpha\Gamma(\alpha(n+1))}{\pi|\sin(\pi\alpha/2)|}x^{-\alpha(n+1)}.$$

See Kozubowski (1994a) for the proof of Proposition 4.3.14.

We now turn our attention to series expansions and asymptotics at zero for Linnik densities, which were theoretically thoroughly studied by Kotz et al. (1995). We add here some numerical results. The structure of such series representations depends on the arithmetic nature of the parameter α. Three cases ought to be investigated:

(i) $1/\alpha$ is an integer.

(ii) $1/\alpha$ is a noninteger rational number.

(iii) α is an irrational number.

In case (i) we have the following representation.

Proposition 4.3.15 *Let p_α be the density of a Linnik distribution $L_{\alpha,1}$, where $0 < \alpha = \frac{1}{n} < 2$ and n is a positive integer. Then*

$$p_\alpha(\pm x) = \frac{1}{2}\sum_{k=1, k/n\in\mathbb{Q}\setminus\mathbb{N}}^{\infty} \frac{(-1)^{k+1}x^{k/n-1}}{\Gamma(k/n)\cos\frac{k\pi}{2n}} + \frac{(-1)^{n+1}}{\pi}\cos x\cdot\log\frac{1}{x} + \frac{1}{2}\sin x + \frac{(-1)^{n+1}}{\pi}\sum_{k=0}^{\infty}(-1)^k\frac{\Gamma'(2k+1)}{\Gamma(2k+1)}x^{2k}, \quad x>0. \tag{4.3.44}$$

See Erdogan (1995) and Kotz et al. (1995) for the proofs. The series representation leads to the asymptotic formula for each $n \geq 2$

$$p_\alpha(\pm x) = \frac{1}{2}\sum_{k=1}^{n-1}\frac{(-1)^{k+1}x^{k/n-1}}{\Gamma(k/n)\cos\frac{k\pi}{2n}} + \frac{(-1)^{n+1}}{\pi}\log\frac{1}{x} + (-1)^n\frac{\gamma}{\pi} + \frac{(-1)^{n+1}nx^{1/n}}{2\Gamma(1/n)\sin\frac{\pi}{2n}} + O(|x|^{2/n}), \quad x\to 0, \tag{4.3.45}$$

where γ is the Euler constant. Let us note the following two special cases. For $\alpha = 1$, which corresponds to ch.f. $\psi_{1,1}(t) = [1+|t|]^{-1}$, we obtain the representation

$$p_1(\pm x) = \frac{1}{\pi}\cos x\cdot\log\frac{1}{x} + \frac{1}{2}\sin x + \frac{1}{\pi}\sum_{k=0}^{\infty}(-1)^k\frac{\Gamma'(2k+1)}{\Gamma(2k+1)}x^{2k}, \quad x>0, \tag{4.3.46}$$

and the corresponding asymptotic formula

$$p_1(\pm x) = \frac{1}{\pi}\log\frac{1}{x} - \frac{\gamma}{\pi} + \frac{1}{2}x - \frac{1}{2\pi}x^2\log\frac{1}{x} + O(x^2), \quad x\to 0. \tag{4.3.47}$$

For $\alpha = 1/2$, we obtain

$$p_{1/2}(x) = \frac{1}{\sqrt{2x}}\sum_{k=0}^{\infty}\frac{(-1)^{[[\frac{k+1}{2}]]}|x|^k}{\Gamma(k+\frac{1}{2})} - \frac{\cos x}{\pi}\cdot\log\frac{1}{|x|} + \frac{\sin|x|}{2} - \frac{1}{\pi}\sum_{k=0}^{\infty}(-1)^k\frac{\Gamma'(2k+1)}{\Gamma(2k+1)}|x|^{2k}, \tag{4.3.48}$$

corresponding to $\phi_{1/2,1}(t) = [1+|t|^{1/2}]^{-1}$.

In case (ii), things are getting a little more complicated, as the expansion includes several series.

Proposition 4.3.16 *Let p_α be the density of a Linnik distribution $L_{\alpha,1}$, where $0 < \alpha = \frac{m}{n} < 2$ and m and n are relatively prime integers greater than one. Then*

$$\begin{aligned} p_\alpha(\pm x) = & \sum_{k=1, k/n \in \mathbb{Q}\setminus\mathbb{N}}^{\infty} \frac{(-1)^{k+1}}{\Gamma(k\alpha)} \frac{\sin(k\pi\alpha/2)}{\sin(k\pi\alpha)} x^{k\alpha-1} \\ & + \frac{1}{\pi} \log \frac{1}{x} \sum_{t=1}^{\infty} \frac{(-1)^{(m+n)t}}{\Gamma(mt)} \sin(t\pi n\alpha/2) x^{mt-1} \\ & + \frac{1}{2} \sum_{t=1}^{\infty} \frac{(-1)^{(m+n)t-1}}{\Gamma(mt)} \cos(t\pi n\alpha/2) x^{mt-1} \\ & + \frac{1}{\alpha} \sum_{j=1, j/m \in \mathbb{Q}\setminus\mathbb{N}}^{\infty} \frac{(-1)^{j-1}}{\Gamma(j)} \frac{\sin(j\pi/2)}{\sin(j\pi/\alpha)} x^{j-1} \\ & + \frac{1}{\pi} \sum_{t=1}^{\infty} (-1)^{(m+n)t} \frac{\Gamma'(mt)}{\Gamma^2(mt)} \sin(t\pi n\alpha/2) x^{mt-1}, \quad x > 0. \end{aligned} \tag{4.3.49}$$

See Erdogan (1995) and Kotz et al. (1995) for proofs. Rather remarkably, under the additional assumption that the number m is even, the series expansion for $p_{m/n}$ simplifies to

$$p_\alpha(\pm x) = \frac{1}{2} \sum_{k=1}^{\infty} \frac{(-1)^{k+1} x^{k\alpha-1}}{\Gamma(k\alpha)\cos(k\pi\alpha/2)} + \frac{1}{\alpha} \sum_{k=0}^{\infty} \frac{(-1)^k x^{2k}}{\Gamma(2k+1)\sin(\pi(2k+1)/\alpha)}, \quad x > 0, \tag{4.3.50}$$

where the series on the right-hand side is absolutely convergent. We note here that the expansion for $\alpha = 1/n$ given in Proposition 4.3.15 follows from the one with $\alpha = m/n$ by setting $m = 1$ in (4.3.49) (see Exercise 4.5.24).

To obtain asymptotic formulas for $x \to 0$ describing the behavior up to $O(|x|^N)$, it is necessary to select from the right-hand side of (4.3.49) the terms involving powers of $|x|$ that are less than N and to add the term containing $\log(1/|x|)$, if available. For example, for $\alpha = 3/2$, we have $m = 3$, $n = 2$, and

$$p_{3/2}(\pm x) = \frac{4}{3\sqrt{3}} - \sqrt{\frac{2}{\pi}} \cdot |x|^{1/2} + \frac{1}{2\pi} x^2 \log \frac{1}{x} + \frac{\Gamma'(3)}{4\pi} x^2 + O(|x|^{7/2}) \tag{4.3.51}$$

as $x \to 0$. Another remarkable result is that under case (iii), where α is irrational, the representation of p_α is similar to (4.3.50) rather than to (4.3.49)! Indeed, if $\alpha \in (0, 2)$ is not rational of the form $\alpha = m/n$ with an odd m, we have the representation

$$p_\alpha(x) = \frac{1}{x} \lim_{s\to\infty} \left\{ \frac{1}{2} \sum_{k=1}^{s} \frac{(-1)^{k+1} |x|^{k\alpha}}{\Gamma(k\alpha)\cos\frac{k\pi\alpha}{2}} + \frac{1}{\alpha} \sum_{k\in A_s} \frac{(-1)^k |x|^{2k+1}}{\Gamma(2k+1)\sin\frac{\pi(2k+1)}{\alpha}} \right\}, \tag{4.3.52}$$

where A_s denotes the set of positive integers k satisfying the relation $1 \le 2k + 1 < \alpha(s + 1/2)$. In addition, the limit on the right-hand side is uniform with respect to x on any compact subset of $\mathbb{R}$. Moreover, for almost all (but not all) irrational values of α, representation (4.3.50) remains valid and the series converges absolutely and uniformly on any compact set. More precisely, the "lucky" set of irrational α's is the set $(0, 2) \setminus L$, where L is the set of the so-called Liouville numbers — namely, numbers β such that for any $r = 2, 3, 4, \ldots$ there exists a pair of integers $p, q \ge 2$ such that

$$0 < |\beta - p/q| < q^{-r}.$$

It is well known that these numbers are transcendental and the set of all Liouville numbers has Lebesgue measure zero. We thus have the following proposition [see Kotz et al. (1995)].

Proposition 4.3.17 *The density p_α of a Linnik distribution $L_{\alpha,1}$, where $0 < \alpha < 2$ is irrational and not Liouville, admits representation* (4.3.50). *Moreover, both series converge absolutely and uniformly on any compact set.*

To construct an α for which both series in (4.3.50) are divergent, we have to construct a sequence of very rapidly growing integers by the recurrence relation

$$q_{s+1} = (q_s!)^2 q_s, \quad s = 1, 2, \ldots,$$

and set

$$\alpha = \sum_{k=1}^{\infty} \frac{1}{q_k}.$$

Evidently, since $q_s > 2^s$ for $s \geq 2$ and $\alpha \in (1/2, 1)$, it is not difficult to show that these α's are Liouville numbers and the terms of the form

$$\frac{(-1)^{k+1} x^{k\alpha-1}}{\Gamma(\alpha k)\cos(\pi\alpha k/2)}$$

with index $k = q_s$ diverge to ∞ as $s \to \infty$.

4.3.4 Moments and tail behavior. The asymptotic representation (4.3.43) shows that Linnik distributions have regularly varying tails with index α. More precisely, if the r.v. Y_α have the Linnik distribution $L_{\alpha,1}$, then we have

$$\lim_{x\to\infty} x^\alpha P(Y_\alpha > x) = \frac{\Gamma(\alpha)\sin\frac{\pi\alpha}{2}}{\pi}. \tag{4.3.53}$$

Consequently, as noticed by Lin (1994), the absolute moments of positive order p, $e(p) = E|Y_\alpha|^p$, are finite for $p < \alpha$ and infinite for $p \geq \alpha$. The following computational formula for $e(p)$ is useful for estimating the parameters of Linnik distribution [see Kozubowski and Panorska (1996), Proposition 5.3].

Proposition 4.3.18 *Let $Y \sim L_{\alpha,\sigma}$ with $0 < \alpha \leq 2$. Then for every $0 < p < \alpha$, we have*

$$e(p) = E|Y|^p = \frac{p(1-p)\sigma^p\pi}{\alpha\Gamma(2-p)\sin\frac{\pi p}{\alpha}\cos\frac{\pi p}{2}}. \tag{4.3.54}$$

In case $p = 1$, we need to set $(1-p)/\cos\frac{\pi p}{2}$ to its limiting value when $p \to 1$, which is equal to $2/\pi$. Note that for $\alpha = 2$, we obtain a familiar expression for the moments of the symmetric Laplace distribution, $E|Y|^p = \sigma^p\Gamma(p+1)$ (see Exercise 4.5.25). In particular, the first absolute moment ($p = 1$) is equal to σ for the Laplace distribution, and

$$\frac{2\sigma}{\alpha\sin\frac{\pi}{\alpha}} \tag{4.3.55}$$

for the Linnik $L_{\alpha,\sigma}$ distribution. We list a few selected values of $E|Y|$ for the latter distribution with $\sigma = 1$ in Table 4.3 (the corresponding value of the standard classical Laplace distribution is equal to 1). We can clearly see the increase in $E|Y|$ as the parameter α approaches 1. In fact, for each given

α	1.01	1.025	1.05	1.10	1.25	1.50	1.75	2
$E\|Y\|$	63.67	25.49	12.78	6.45	2.72	1.54	1.17	1

Table 4.3: Selected values of $E|Y|$, where Y has the Linnik distribution with $\sigma = 1$ and various α's.

$\sigma > 0$, the function of α given by (4.3.55) is strictly decreasing on $(1, 2]$ and converges to infinity as $\alpha \to 1^+$. For $\alpha = 1$, the first absolute moment of Linnik distribution is infinite, while for $\alpha = 2$ it coincides with its counterpart of the standard classical Laplace distribution.

Since Linnik distribution $L_{\alpha,\sigma}$ has the tails $P(Y_{\alpha,\sigma} > x)$ asymptotically equivalent to the power function $x^{-\alpha}$, it is in the domain of attraction of stable distribution with index α. Indeed, for a given sequence $X_1, X_2, \ldots$ of i.i.d. Linnik $L_{\alpha,1}$ random variables, as $n \to \infty$, the sum

$$S_n = n^{-1/\alpha} \sum_{i=1}^{n} X_i$$

converges in distribution to the stable law with characteristic function $\phi(t) = \exp(-|t|^\alpha)$:

$$\lim_{n\to\infty} E[e^{itS_n}] = \lim_{n\to\infty} (1 + |t|^\alpha/n)^{-n} = \exp(-|t|^\alpha).$$

We conclude this section with the result on the asymptotic behavior of absolute fractional moments of Linnik distribution, which follows from the tail behavior of geometric stable distributions [see Kozubowski and Panorska (1996)].

Proposition 4.3.19 *Let $Y \sim L_{\alpha,\sigma}$ with $0 < \alpha \leq 2$. Then*

$$\lim_{r\to\alpha^-} (\alpha - r)E|Y|^r = \frac{2\alpha\Gamma(\alpha)\sigma^\alpha \sin\frac{\pi\alpha}{2}}{\pi}. \tag{4.3.56}$$

4.3.5 Properties. In this section, we collect (somewhat fragmented) further results on symmetric Linnik distributions.

4.3.5.1 *Self-decomposability.* In Section 2.4.3 we discussed the class L of self-decomposable distributions and showed that symmetric Laplace distributions belong to this class. It was shown in Lin (1994) that this property is shared by Linnik distributions as well.

Proposition 4.3.20 *All symmetric Linnik distributions are in class L, that is, for all $c \in (0, 1)$ the Linnik characteristic function $\psi_{\alpha,\sigma}$ given by (4.3.1) can be written as*

$$\psi_{\alpha,\sigma}(t) = \psi_{\alpha,\sigma}(ct)\phi_c(t), \tag{4.3.57}$$

where ϕ_c is a characteristic function.

Proof. Lukacs (1970) has shown that if $p > 1$ and g is a ch.f., then the function $(p-1)/(p-g(t))$ is also a characteristic function. Since

$$\frac{\psi_{\alpha,\sigma}(t)}{\psi_{\alpha,\sigma}(ct)} = \frac{p-1}{p-\psi_{\alpha,\sigma}(ct)} = \phi_c(t),$$

where $p = (1-c^\alpha)^{-1} > 1$, we conclude that ϕ_c is a characteristic function. □

Remark 4.3.4 We note also that strictly geometric stable laws are self-decomposable as well [see, e.g., Kozubowski (1994a)], while geometric stable r.v.'s with $0 < \alpha < 2$ and $\mu \neq 0$, in general, are not [see Ramachandran (1997)].

As shown by Yamazato (1978), self-decomposability implies unimodality, so Linnik distributions are unimodal (with the mode at zero). The unimodality of Linnik distributions was also proved in Laha (1961). In conclusion, we note that although general geometric stable laws may not belong to class L, they are all unimodal (with the mode at zero), as recently shown by Belinskiy and Kozubowski (2000).

4.3.5.2 *Infinite divisibility.* We saw in Section 2.4.1 that symmetric Laplace distribution is infinitely divisible, and its characteristic function admits a Lévy–Khinchine representation with an explicit expression for the Lévy measure. Linnik distributions are infinitely divisible as well, although the Lévy measure can be no longer written explicitly. Their Lévy–Khinchine representation follows from Lemma 7, VI.2 of Bertoin (1996) and the fact that a Linnik random variable $Y_{\alpha,1}$ can be written as $Y = S(W)$, where W is standard exponential variable and $S(t)$ is a stable process with independent increments, independent of W, and $S(1)$ has the stable law with the characteristic function

$$\phi(t) = e^{-|t|^\alpha}. \tag{4.3.58}$$

Proposition 4.3.21 *The ch.f.* (4.3.1) *of the Linnik distribution* $L_{\alpha,\sigma}$ *admits the representation*

$$\psi(t) = \exp\left(\int_{\mathbb{R}} (e^{itu} - 1)d\Lambda(u)\right), \tag{4.3.59}$$

where

$$\frac{d\Lambda}{du}(u) = \frac{\alpha}{2|u|} E \exp\left(-\left|\frac{u}{\sigma X}\right|^\alpha\right) = \frac{1}{\sigma}\int_0^\infty g_\alpha\left(\frac{u}{\sigma w^{1/\alpha}}\right)\frac{e^{-w}}{w^{1+1/\alpha}}dw,$$

where X has the stable distribution (4.3.58) *and g_α is the density of X.*

Remark 4.3.5 See Kozubowski et al. (1998) for a more detailed discussion on the Linnik and the more general geometric stable Lévy measure and their asymptotics at zero.

Remark 4.3.6 If $\alpha = 2$, the Linnik distribution $L_{\alpha,\sigma}$ reduces to the classical Laplace distribution $\mathcal{CL}(0,\sigma)$ with mean zero and variance $2\sigma^2$. In this case the stable random variable X has ch.f. e^{-t^2}, which corresponds to the normal distribution with mean zero and variance equal to two. Consequently, the density of the Lévy measure is

$$\frac{d\Lambda}{du}(u) = \frac{2}{2|u|} E \exp\left(-\left|\frac{u}{\sigma X}\right|^2\right) = \frac{1}{|u|}\int_{-\infty}^\infty e^{-\frac{u^2}{\sigma^2 x^2}}\frac{1}{2\sqrt{\pi}}e^{-\frac{1}{4}x^2}dx. \tag{4.3.60}$$

Noting that the integral in (4.3.60) is an even function of x, we obtain after some algebra

$$\frac{d\Lambda}{du}(u) = \frac{1}{\sqrt{\pi}}\frac{1}{|u|}\int_0^\infty t^{1/2-1}e^{-(t+\frac{u^2}{\sigma^2}\frac{1}{4t})}dt. \tag{4.3.61}$$

Relating the integral in (4.3.61) to the modified Bessel function $K_{-1/2}$, defined in (A.0.4) (see the appendix), we obtain

$$\frac{d\Lambda}{du}(u) = \frac{1}{\sqrt{\pi}}\frac{1}{|u|}K_{-1/2}\left(\frac{|u|}{\sigma}\right)\cdot 2\cdot\frac{|u|^{1/2}}{\sqrt{2}}\sigma^{-1/2}. \tag{4.3.62}$$

Finally, the application of Properties 5 and 10 of the function K_λ results in

$$\frac{d\Lambda}{du}(u) = \frac{1}{|u|}e^{-|u|/\sigma}, \tag{4.3.63}$$

which is the density obtained previously for the classical Laplace distribution (see Proposition 2.4.2).

4.3.6 Simulation. Devroye's representation (4.3.19) allows us to generate Linnik distributions from independent stable and exponential variates. However, the generation of stable distributions requires nonstandard methods, as their distribution functions are not given explicitly [see, e.g., Weron (1996)]. An alternative computer simulation of Linnik random variables is obtained through representation (4.3.27) with $\alpha' = 2$. Here the r.v.'s that appear in the representation have explicit distribution functions and thus can conveniently be generated by the inversion method. Indeed, the Laplace distribution function is given in Section 2.1.1, while the distribution function of the r.v. W_ρ has the form

$$F_\rho(x) = \frac{1}{\pi\rho}\left[\arctan\left(\frac{x}{\sin\pi\rho} + \cot\pi\rho\right) - \frac{\pi}{2}\right] + 1. \tag{4.3.64}$$

Since the inverse function of F_ρ has an explicit form,

$$F_\rho^{-1}(x) = \sin(\pi\rho)\cot(\pi\rho(1-x)) - \cos(\pi\rho), \tag{4.3.65}$$

the r.v. W_ρ can be generated by the inversion method. Here is a generator of a symmetric Linnik $L_{\alpha,\sigma}$ distribution given by the ch.f. (4.3.1).

Linnik $L_{\alpha,\sigma}$ generator

- Generate random variate Z from the $L_{2,1}$ distribution (standard Laplace with location 0 and scale 1).
- Generate uniform [0,1] variate U, independent of Z.
- Set $\rho \leftarrow \alpha/2$.
- Set $W \leftarrow \sin(\pi\rho)\cot(\pi\rho U) - \cos(\pi\rho)$.
- Set $Y \leftarrow \sigma Z W^{1/\alpha}$.
- RETURN Y.

More details on generation variates from the Linnik laws and the more general geometric stable laws can be found in Kozubowski (2000b).

4.3.7 Estimation. This section is devoted to the problem of estimating the parameters α and σ of the Linnik distribution $L_{\alpha,\sigma}$. Since densities and distribution functions of Linnik laws cannot in general be written in closed form, most estimation methods for Linnik laws suggested in the literature are based on the characteristic function and its empirical counterpart. Recall that if $X_1, X_2, \ldots, X_n$ are i.i.d. random variables with characteristic function ψ, then the *empirical characteristic function* (sample ch.f.) is defined as

$$\widehat{\psi}_n(t) = \frac{1}{n}\sum_{j=1}^{n} e^{itX_j}. \tag{4.3.66}$$

This function is the characteristic function of the *empirical distribution* of the data, which assigns probability $1/n$ to each observation. By definition and the strong LLN, it follows that

$$E[\widehat{\psi}_n(t)] = \psi(t) \text{ and } \widehat{\psi}_n(t) \stackrel{a.s.}{\to} \psi(t) \text{ as } n \to \infty. \tag{4.3.67}$$

Consequently, estimators based on the sample characteristic function are usually strongly consistent.

Below we present several estimation procedures for Linnik parameters, based on the random sample $X_1, X_2, \ldots, X_n$ from the Linnik $L_{\alpha,\sigma}$ distribution given by the ch.f. $\psi = \psi_{\alpha,\sigma}$ as specified by (4.3.1). Here the characteristic function is real and the distribution is symmetric about zero. Thus the real part of the empirical characteristic function,

$$\widehat{\eta}_n(t) = \frac{1}{n}\sum_{j=1}^{n} \cos(tX_j), \tag{4.3.68}$$

can be used in estimation.

4.3.7.1 *Method of moments type estimators.* The first method is a special case of the estimation procedure for geometric stable parameters suggested by Anderson (1992) and Kozubowski (1993). The method is based on the sample characteristic function (4.3.68) for the symmetric case and produces computationally simple, consistent, and asymptotically normal estimators. For convenience, we set $\lambda = \sigma^\alpha$ to be consistent with the notation used in Kozubowski (1993). Since

$$1/\psi(t) = 1 + \lambda \mid t \mid^\alpha,$$

we have

$$\upsilon(t_i) = \lambda |t_i|^\alpha, \quad i = 1, 2, \tag{4.3.69}$$

where $\upsilon(t) = |1/\psi(t) - 1]$ and $t_1 \neq t_2$, are both greater than 0. Solving equations (4.3.69) for α and λ we obtain

$$\alpha = \frac{\log[\upsilon(t_1)/\upsilon(t_2)]}{\log[t_1/t_2]}, \quad \lambda = \exp\left\{\frac{\log \mid t_1 \mid \log[\upsilon(t_2)] - \log \mid t_2 \mid \log[\upsilon(t_1)]}{\log[t_1/t_2]}\right\}. \tag{4.3.70}$$

Substituting the sample ch.f. $\widehat{\eta}_n(t)$ for $\psi(t)$ into (4.3.70) we get estimators of α and λ:

$$\widehat{\alpha} = \frac{\log[\widehat{\upsilon}_n(t_1)/\widehat{\upsilon}_n(t_2)]}{\log[t_1/t_2]},$$
$$\widehat{\lambda} = \exp\left\{\frac{\log \mid t_1 \mid \log[\widehat{\upsilon}_n(t_2)] - \log \mid t_2 \mid \log[\widehat{\upsilon}_n(t_1)]}{\log[t_1/t_2]}\right\},$$

where $\widehat{\upsilon}_n(t) = |1/\widehat{\eta}_n(t) - 1|$ is the sample counterpart of $\upsilon(t)$. Since $\widehat{\eta}_n(t) \stackrel{a.s.}{\to} \psi(t)$, also $\widehat{\upsilon}_n(t) \stackrel{a.s.}{\to} \upsilon(t)$, and the estimators are consistent.

Remark 4.3.7 See Jacques et al. (1999) for an extension of the method to the case of *generalized Linnik laws* given by the ch.f.

$$\psi_{\alpha,\sigma,\beta}(t) = \left(\frac{1}{1 + \sigma^\alpha |t|^\alpha}\right)^\beta, \quad t \in \mathbb{R}.$$

4.3.7.2 *Least-squares estimators.* Another estimation procedure based on the sample ch.f. is the regression-type estimation of Koutrouvelis (1980) adapted to the Linnik case, which was discussed in Kozubowski (1993) in the more general setting of geometric stable laws. Again, set $\lambda = \sigma^\alpha$. Taking the logarithms of both sides in the relation

$$\mid 1/\psi(t) - 1 \mid = \lambda \mid t \mid^\alpha \tag{4.3.71}$$

results in

$$\log \mid 1/\psi(t) - 1 \mid = \log \lambda + \alpha \log \mid t \mid . \tag{4.3.72}$$

We can now estimate λ and α using the regression of $y = \log \mid 1/\widehat{\eta}_n(t) - 1 \mid$ on $x = \log \mid t \mid$ via the model

$$y_i = \delta + \alpha x_i + \epsilon_i, \quad i = 1, \dots, K, \tag{4.3.73}$$

where $\{t_i\}$, $i = 1, \dots, K$, is a suitable sequence of real numbers, $\delta = \log \lambda$, and ϵ_i is an error term. Denote these estimators by $\widetilde{\alpha}$ and $\widetilde{\lambda}$.

Like the method of moments procedure, the regression-type estimation presented here produces consistent estimators and is computationally straightforward. However, optimality properties for estimators are lacking, and the methods may not be robust with respect to the choice of the required constants.

4.3.7.3 *Minimal distance method.* Anderson and Arnold (1993) discuss another estimation method for Linnik parameters, based on empirical characteristic function (4.3.66). They consider estimation of the parameter α of the Linnik distribution with $\sigma = 1$, although the procedure can be generalized to include the scale parameter as well. The method is based on minimization of the objective function

$$I_L(\alpha) = \int_{-\infty}^{\infty} |\widehat{\psi}(t) - (1 + |t|^{\alpha})^{-1}|^2 e^{-t^2} dt, \tag{4.3.74}$$

where $\widehat{\psi}$ is the empirical characteristic function (4.3.66) based on the random sample $X_1, X_2, \dots, X_n$ from the Linnik $L_{\alpha,1}$ distribution. Again, since the distribution is symmetric, the real part of $\widehat{\psi}$ given by (4.3.68) can be used, in which case the objective function becomes

$$I_L(\alpha) = \int_{-\infty}^{\infty} |\widehat{\eta}(t) - (1 + |t|^{\alpha})^{-1}|^2 e^{-t^2} dt. \tag{4.3.75}$$

The weights e^{-t^2} are incorporated mainly for mathematical convenience, as integrals of the form

$$\int_{-\infty}^{\infty} f(t) e^{-t^2} dt \tag{4.3.76}$$

can be well approximated by the sum

$$\sum_{i=1}^{m} \omega_i f(z_i) + R_m \tag{4.3.77}$$

(via so-called Hermite integration). Here the weights are

$$\omega_i = \frac{2^{m-1} m! \sqrt{m}}{(m H_{m-1}(z_i))^2}, \tag{4.3.78}$$

and z_i is the ith zero of the mth degree Hermite polynomial $H_m(z)$. The values of z_i, ω_i, and $\omega_i e^{z_i^2}$ are presented in Abramowitz and Stegun (1965, p. 924). They reproduce tables of zeroes and weight factors of the first 20 Hermite polynomials from Salzer et al. (1952).

The objective function in the symmetric case can be well approximated by

$$\widehat{I_L}(\alpha) = \sum_{i=1}^{m} \omega_i (\widehat{\eta}(z_i) - (1 + |z_i|^{\alpha})^{-1})^2. \tag{4.3.79}$$

The values of $\widehat{\alpha}_L$ that minimize $\widehat{I_L}(\alpha)$ are strongly consistent estimators of α. Anderson and Arnold (1993) carried out extensive simulations that indicate that this approach provides reasonable estimators.

4.3.7.4 *Fractional moment estimation.* Here we present the approach to estimation based on fractional moments of Section 4.3.4, which was considered in Kozubowski (1999). The basis for the method is formula (4.3.54), which expresses the fractional moment $E|Y|^p$ in terms of the parameters α and σ. We can substitute sample fractional moments and solve the resulting equations for the parameters. As noted in Kozubowski (1999), the method is computationally simple, requires minimal implementation effort, and provides accurate estimates even for small sample sizes.

Consider $0 < p < \alpha \le 2$, and let $e(p) = E|Y_1|^p$ denote the pth absolute moment of $L_{\alpha,\sigma}$. Next, choose two values of p, say p_1 and p_2, replace $e(p_k)$ in the fractional moment formula (4.3.54) with its sample counterpart $\hat{e}(p_k) = \frac{1}{n}\sum |Y_i|^{p_k}$, $k = 1, 2$, and solve the resulting equations for α and σ.

As an illustration, assume $1 < \alpha \le 2$ and take $p_1 = 1/2$ and $p_2 = 1$ so that by (4.3.54), we have

$$\hat{e}(1/2) = \frac{1}{n}\sum |Y_i|^{1/2} = \sqrt{\frac{\pi\sigma}{2}\frac{1}{\alpha \sin\frac{\pi}{2\alpha}}} \tag{4.3.80}$$

and

$$\hat{e}(1) = \frac{1}{n}\sum |Y_i| = \frac{2\sigma}{\alpha \sin\frac{\pi}{\alpha}}. \tag{4.3.81}$$

Next, eliminate σ from (4.3.80) and (4.3.81) by squaring both sides of (4.3.80) and dividing the two sides of the resulting equation into the corresponding sides of equation (4.3.81). This results in the equation for α,

$$\frac{\hat{e}(1)}{(\hat{e}(1/2))^2} = \frac{4\alpha \sin^2\frac{\pi}{2\alpha}}{\pi \sin\frac{\pi}{\alpha}}. \tag{4.3.82}$$

As remarked by Kozubowski (1999), finding a numerical solution of (4.3.82) is straightforward, since the right-hand side of (4.3.82) is strictly decreasing in α. Now we can substitute $\hat{\alpha}$ into either (4.3.80) or (4.3.81) and solve the resulting equations for $\hat{\sigma}_1$ and $\hat{\sigma}_2$, obtaining

$$\hat{\sigma}_1 = \frac{2}{\pi}\hat{\alpha}^2 \sin^2\frac{\pi}{2\hat{\alpha}}[\hat{e}(1/2)]^2, \tag{4.3.83}$$

$$\hat{\sigma}_2 = \frac{1}{2}\hat{\alpha} \sin\frac{\pi}{\hat{\alpha}}\hat{e}(1). \tag{4.3.84}$$

One can compute the average $\hat{\sigma} = (\hat{\sigma}_1 + \hat{\sigma}_2)/2$ to estimate σ. As reported in Kozubowski (1999), these estimators perform well on simulated data. The results are most accurate when α is close to 2, and generally improve as n increases. The procedure provides quite satisfactory results even for sample sizes as small as 100, and can easily be adapted to the general strictly geometric stable case as well.

4.3.8 Extensions. We have already seen that symmetric Linnik distributions form a subclass of strictly geometric stable laws given by ch.f. (4.3.3). Distributions from this three-parameter family share many properties of the Linnik laws [see, for example, Kozubowski (1994ab), Erdogan (1995)]. In turn, strictly geometric stable laws form a subclass of geometric stable laws, defined in Section 4.4.4. The latter is a four-parameter family of distributions that are limiting laws for (normalized) geometric sums with i.i.d. components. More information on geometric stable laws can be found in Kozubowski and Rachev (1999ab).

Since the Linnik distribution is infinitely divisible, any positive power of the Linnik ch.f. (4.3.1) is a well-defined ch.f. corresponding to real-valued (and symmetric) random variable. The resulting distributions are called *generalized Linnik laws* [see, e.g., Devroye (1993), Pakes (1998), Erdogan and Ostrovskii (1998b), and Jacques et al. (1999) for more details].

Nonnegative r.v.'s with Laplace–Stieltjes transform

$$f_{\alpha,c}(s) = \frac{1}{1+cs^{\alpha}}, \quad s \ge 0, \alpha \in (0,1], c > 0, \tag{4.3.85}$$

are the Mittag–Leffler distributions, introduced by Pillai (1990). Pakes (1995) considered a more general class of distributions with Laplace–Stieltjes transform

$$f_{\alpha,c,\beta}(s) = \left(\frac{1}{1+cs^{\alpha}}\right)^{\beta}, \quad s \ge 0, \quad \alpha \in (0,1], \quad c > 0, \quad \beta > 0, \tag{4.3.86}$$

and referred to them as the *positive Linnik* laws. Note that the functions (4.3.85) and (4.3.86) ought to be restricted to the case $\alpha \in (0, 1]$, since otherwise they are not completely monotone, and hence cannot serve as Laplace–Stieltjes transforms.

Replacing s in (4.3.86) by $1 - z$, we obtain the function

$$g_{\alpha,c,\beta}(z) = \left(\frac{1}{1+c(1-z)^{\alpha}}\right)^{\beta}, \quad |z| \ge 1, \quad \alpha \in (0,1], \quad c > 0, \quad \beta > 0, \tag{4.3.87}$$

which is a probability-generating function of a nonnegative integer-valued r.v. with the *discrete Linnik distribution*, studied by Devroye (1990) for $c = 1$ and Pakes (1995) for $c > 0$. For $\beta = 1$, we obtain here the *discrete Mittag–Leffler distribution* [see Pillai (1990) and Jayakumar and Pillai (1995)]. Letting $\beta \to \infty$, we arrive in the limit at the probability-generating function

$$h_{\alpha,c}(z) = e^{-c(1-z)^{\alpha}}, \quad |z| \le 1, \quad \alpha \in (0,1], \quad c > 0, \tag{4.3.88}$$

which represents a *discrete stable* distributed r.v. [see Steutel and van Horn (1979) and Christoph and Schreiber (1998a)]. We refer the interested reader to Christoph and Schreiber (1998abc) for more information on and further references for these discrete distributions.

4.4 Other cases

4.4.1 Log-Laplace distribution. By analogy with the lognormal, S_U, and S_B systems of distributions [see, e.g., Johnson et al. (1994), Chapters 12 and 14], Johnson (1954) considered the system

$$X = \begin{cases} \theta + s \log Y & (S'_L \text{ system}), \\ \theta + s \sinh^{-1} Y & (S'_U \text{ system}), \\ \theta + s \log\left(\frac{Y}{1-Y}\right) & (S'_B \text{ system}), \end{cases} \tag{4.4.1}$$

where Y has the standard classical Laplace distribution. The S'_L system of distributions is known as the *log-Laplace distributions* (in analogy with the log-normal distributions) [see Uppuluri (1981), Chipman (1985), Kotz et al. (1985), and Johnson et al. (1994) for further discussion on log-Laplace distributions].

4.4.2 Generalized Laplace distribution. The following generalization of the Laplace distribution was proposed by Subbotin (1923):

$$f_p(x) = [2p^{1/p}\sigma_p\Gamma(1+1/p)]^{-1}\exp(-(p\sigma_p^p)^{-1}|x-\mu|^p), \tag{4.4.2}$$

where $\mu = E(X)$ is the location parameter, $\sigma_p = [E(|X-\mu|^p)]^{1/p}$ is the scale parameter, and $p > 0$ is the shape parameter. The distributions with these densities form a family called *exponential power function* distributions, and they are also called *generalized Laplace distributions*, as for $p = 1$ they reduce to the standard Laplace laws. The estimation of the parameters was treated in a number of papers, for example the MLE's and their properties were derived in Agró (1995) [see also Zeckhauser and Thompson (1970)]. The distribution is widely used in Bayesian inference [see, e.g., Box and Tiao (1962), Tiao and Lund (1970)]. Other related papers include Jakuszenkow (1979), Sharma (1984), and Taylor (1992).

4.4.3 Sargan distribution. Consider a symmetric Bessel function distribution $\mathcal{GAL}(0, \sigma, \tau)$, where $\tau = n + 1$ is an integer. Here the Bessel function $K_{\tau-1/2} = K_{n+1/2}$ admits a closed form (A.0.10) given in the appendix, and density (4.1.32) becomes

$$f(x) = \frac{1}{2} e^{-|x|} \sum_{j=0}^{n} \gamma_j |x|^j, \tag{4.4.3}$$

where

$$\gamma_j = \frac{(2n-j)!2^{j-2n}}{n!j!(n-j)!} \tag{4.4.4}$$

[cf. equation (4.1.33)]. This distribution corresponds to the sum of $n+1$ i.i.d. standard Laplace r.v.'s (for $n = 0$ we obtain the standard Laplace density (2.1.2)).

More generally, if $Y_1, \ldots Y_{n+1}$ are i.i.d. with general Laplace distribution (2.1.1), then the sample mean, $\overline{Y}$, has density

$$f(x) = \frac{K\alpha}{2} e^{-\alpha|x-\theta|} \sum_{j=0}^{n} \gamma_j \alpha^j |x-\theta|^j, \tag{4.4.5}$$

where $K = 1$, γ_j are as above, and $\alpha = (n+1)/\sigma$ [see, e.g., Weida (1935)].

The function (4.4.5) is a special case of *Sargan densities* of order n, which for $\theta = 0$ are given by (4.4.5) with

$$\gamma_j \geq 0, \quad \gamma_0 = 1, \quad \alpha > 0, \quad K = \left(\sum_{j=0}^{n} \gamma_j j!\right)^{-1}. \tag{4.4.6}$$

Sargan densities have been suggested as an alternative to normal distributions in some econometric models, where it is desirable that the relevant *distribution* function be similar to normal but computable in closed form [see, e.g., Goldfeld and Quandt (1981), Missiakoulis (1983) (who observes that the density of the arithmetic mean of $n+1$ independent Laplace variables is an nth order Sargan density), Kafaei and Schmidt (1985), and Tse (1987)].

4.4.4 Geometric stable laws. If the random variables in (2.2.1) have infinite variance, then the geometric compounds no longer converge to an AL law given by (3.1.10) with $\theta = 0$. Instead, the limiting distributions form a broader class of *geometric stable* (GS) laws. It is a four-parameter family denoted by $GS_\alpha(\sigma, \beta, \mu)$ and conveniently described in terms of the characteristic function

$$\psi(t) = [1 + \sigma^\alpha |t|^\alpha \omega_{\alpha,\beta}(t) - i\mu t]^{-1}, \tag{4.4.7}$$

where

$$\omega_{\alpha,\beta}(x) = \begin{cases} 1 - i\beta \operatorname{sign}(x) \tan(\pi\alpha/2), & \text{if } \alpha \neq 1, \\ 1 + i\beta \frac{2}{\pi} \operatorname{sign}(x) \log|x|, & \text{if } \alpha = 1. \end{cases} \tag{4.4.8}$$

The parameter $\alpha \in (0, 2]$ is the *index* that determines the tail of the distribution $P(Y > y) \sim Cy^{-\alpha}$ (as $y \to \infty$) for $0 < \alpha < 2$. For $\alpha = 2$ the tail is exponential and the distribution reduces to an AL law, since $\omega_{2,\beta} \equiv 1$. The parameter $\beta \in [-1, 1]$ is the skewness parameter, while $\mu \in \mathbb{R}$ and $\sigma \geq 0$ control, as usual, the location and scale, respectively. We provide a few comments on basic features of GS laws, referring an interested reader to Kozubowski and Rachev (1999ab) for up-to-date information and numerous references on GS laws and their particular cases.

Remark 4.4.1 Special cases of GS laws include Linnik distribution [discussed in Chapter 4, Section 4.3, where $\beta = 0$ and $\mu = 0$; see Linnik (1953)], and Mittag–Leffler distributions, which are GS with $\beta = 1$ and either $\alpha = 1$ and $\sigma = 0$ (exponential distribution) or $0 < \alpha < 1$ and $\mu = 0$. The latter are the only nonnegative GS r.v.'s [see, e.g., Pillai (1990), Fuita (1993), Jayakumar and Pillai (1993)]. For applications of Mittag–Leffler laws, see, e.g., Weron and Kotulski (1996).

Remark 4.4.2 GS laws share many, but not all, properties of the so-called *Paretian stable* distributions. In fact, Paretian stable and GS laws are related via their characteristic functions, φ and ψ, as shown in Mittnik and Rachev (1991):

$$\psi(t) = \gamma(-\log \varphi(t)), \tag{4.4.9}$$

where $\gamma(x) = 1/(1 + x)$ is the Laplace transform of the standard exponential distribution. Relation (4.4.9) produces representation (4.4.7), as well as the mixture representation of a GS random variable Y in terms of independent standardized Pareto stable and exponential r.v.'s X and W:

$$Y \stackrel{d}{=} \begin{cases} \mu W + W^{1/\alpha} \sigma X, & \alpha \neq 1, \\ \mu W + W\sigma X + \sigma W \beta (2/\pi) \log(W\sigma), & \alpha = 1. \end{cases} \tag{4.4.10}$$

Note that this representation reduces to (2.2.3) in the case $\alpha = 2$ and $\mu = 0$, as then X has the normal distribution with mean zero and variance 2.

Remark 4.4.3 The asymmetric Laplace distribution, which is GS with $\alpha = 2$, plays, among GS laws, a role analogous to that of the normal distribution among Paretian stable laws. Namely, AL are the only laws in this class with a finite variance. Also they are limits in the random summation scheme with a geometrically distributed number of terms as the normal laws are limits in the ordinary summation scheme. In contrast to normal distribution, c.d.f.'s of AL laws have explicit expressions, which makes them far easier to handle in applications.

Remark 4.4.4 Similar to Paretian stable laws, the GS laws lack explicit expressions for densities and distribution functions, which handicap their practical implementation. Moreover, they are "fat-tailed," have stability properties (with respect to random summation), and generalize the central limit theorem (being the only limiting laws for geometric compounds). *However, they are different from the stable (and normal) laws in that their densities are more "peaked"; consequently, they are similar to the Laplace type distributions, being heavy-tailed.* Unlike Paretian stable densities, GS densities "blow-up" at zero if $\alpha < 1$. Since many financial data are "peaked" and "fat-tailed," they are often consistent with a GS model [see, e.g., Kozubowski and Rachev (1994)].

4.4.5 ν-stable laws. Suppose that the random number of terms in the summation (2.2.1) is any integer-valued random variable, and, as p converges to zero, ν_p approaches infinity (in probability) while $p\nu_p$ converges in distribution to a r.v. ν with Laplace transform γ. Then the normalized compounds (2.2.1) converge in distribution to a ν-*stable* r.v., whose characteristic function is (4.4.9) [see, e.g., Gnedenko and Korolev (1996), Klebanov and Rachev (1996), Kozubowski and Panorska (1996)]. The class of ν-stable laws contains GS and generalized AL laws as special cases: if ν_p is geometric with mean $1/p$, then $p\nu_p$ converges to the standard exponential and (4.4.9) leads to (4.4.7). The tail behavior of ν-stable laws is essentially the same as that of stable and GS laws.

4.5 Exercises

Exercise 4.5.1 For any given $\sigma^2 > 0$, let the r.v. X be log-normal with the p.d.f.

$$f(x|\sigma^2) = \begin{cases} \frac{1}{\sqrt{2\pi}\sigma x} e^{-(\log x)^2/(2\sigma^2)} & \text{for } x > 0, \\ 0 & \text{otherwise,} \end{cases}$$

so that given σ^2, the r.v. $\log X$ is $N(0, \sigma^2)$. Show that if the quantity σ^2 is a random variable with the standard exponential distribution, then X has the log-Laplace distribution with the p.d.f.

$$g(x) = \frac{1}{\sqrt{2}} \begin{cases} x^{\sqrt{2}-1} & \text{for } 0 < x < 1, \\ x^{-1-\sqrt{2}} & \text{for } x \geq 1, \end{cases}$$

so that the r.v. $\log X$ is standard Laplace $\mathcal{L}(0, 1)$.

Exercise 4.5.2 Using the results on symmetric generalized Laplace densities, demonstrate that asymmetric generalized Laplace densities are unimodal. Is the mode for those distributions always at zero?

Exercise 4.5.3 Recall that if X has the standard symmetric Bessel function distribution $\mathcal{GAL}^*(0, 1, \sqrt{2}, n)$ with ch.f. is $(1 + t^2)^{-n}$, then X has the same distribution as the sum of n i.i.d. standard classical Laplace random variables. Thus the variable X admits the random sum representation discussed in Proposition 2.3.2. Investigate whether a skewed Bessel r.v. $\mathcal{GAL}^*(0, \kappa, \sigma, n)$ admits a similar representation.

Exercise 4.5.4 Using Theorem 4.1.1, show that under the conditions of this theorem, the corresponding generalized Laplace densities converge to a normal density.

Exercise 4.5.5 Derive the coefficient of skewness and kurtosis for the K-Bessel function distribution, and compare them with the corresponding values for the Laplace and AL laws.

Exercise 4.5.6 Derive estimators of the K-Bessel function distribution parameters by the method of moments, and study their asymptotic properties. You may want to consider several cases as to which of the four parameters are unknown.

Exercise 4.5.7 Consider a sequence of stochastic processes $\{L_n(t)\}$ and a process $B(t)$. We say that $\{L_n(t)\}$ has finite dimensional distributions convergent to the finite dimensional distributions of $B(t)$, if for each $N \in \mathbb{N}$ and $t_1, \ldots, t_N$, the sequence of the random vectors $(L_n(t_1), \ldots, L_n(t_N))$ converges in distribution to $(B(t_1), \ldots, B(t_N))$. Let $L_n(t)$ be $\mathcal{LM}(1/\sqrt{\tau_n}, 1/\tau_n)$, where τ_n converges to infinity, and let $B(t)$ be a standard Brownian motion. Show that the convergence of finite dimensional distributions holds in this case.

Exercise 4.5.8 Let $X_1, \dots, X_n$ be i.i.d. with the exponential power function density

$$g(x) = \frac{k}{2s\Gamma(1/k)} e^{-(|x|/s)^k}, \quad -\infty < x < \infty, \quad s, k > 0, \tag{4.5.1}$$

where k is assumed to be known (for $k = 1$ we obtain the Laplace distribution).

(a) Show that the method of moments estimator of the parameter s^2 is

$$\delta_1 = \frac{\Gamma(1/k)}{\Gamma(3/k)} \sum_{i=1}^{n} X_i^2 \tag{4.5.2}$$

[Jakuszenkow (1979)]. Derive the mean and the variance of δ_1. Show that δ_1 is unbiased and consistent for s^2. Is δ_1 an efficient estimator for s^2, i.e., does the variance of δ_1 coincide with the Cramér–Rao lower bound?

(b) Show that the MLE of the parameter s^k is

$$\delta_2 = \frac{k}{n} \sum_{i=1}^{n} |X_i|^k \tag{4.5.3}$$

[Jakuszenkow (1979)]. Show that δ_2 is unbiased and consistent for s^k. Is δ_2 an efficient estimator for s^k?

(c) Show that among all estimators of the form

$$\delta = \alpha \sum_{i=1}^{n} X_i^2, \quad \alpha > 0, \tag{4.5.4}$$

the one that minimizes the expected value of the loss function

$$L(\delta, s^2) = f(s^2)(\delta - s^2)^2, \tag{4.5.5}$$

where f is an arbitrary positive function, corresponds to

$$\alpha^* = \frac{\Gamma(3/k)\Gamma(1/k)}{\Gamma(5/k)\Gamma(1/k) + (n-1)\Gamma^2(3/k)} \tag{4.5.6}$$

[Jakuszenkow (1979)]. Is the resulting estimator unbiased for s^2?

(d) Note that the estimator considered in part (c) is not a function of the complete and sufficient statistic $T = \sum_{i=1}^{n} |X_i|^k$. To improve the estimator, consider the class of estimators of the form $\alpha T^{2/k}, \alpha > 0$, and show that the best estimator (with respect to the loss function (4.5.5)) is obtained for

$$\alpha^* = \frac{\Gamma((n+2)/k)}{\Gamma((n+4)/k)} \tag{4.5.7}$$

[Sharma (1984)].

Exercise 4.5.9 Extend Theorem 4.2.3 to an arbitrary symmetric Laplace motion $\mathcal{LM}(\sigma, \nu)$ defined over the interval $[0, T]$.

Exercise 4.5.10 It is well-known that there exist essentially different stochastic processes having the same distribution at any fixed time point. Consider the following two processes:

$$\tilde{L}_t = \sqrt{\Gamma_t}\sigma B_t + \Gamma_t \mu + mt$$

and

$$\tilde{\tilde{L}}_t = \sqrt{\tilde{\Gamma}_t}\sigma B_t + \tilde{\Gamma}_t \mu + mt,$$

where Γ_t is a gamma process independent of a Brownian motion B_t, while $\tilde{\Gamma}_t$ is a gamma white noise, i.e., for each $n \in \mathbb{N}$ and $t_1, \dots, t_n \in \mathbb{R}$ the variables $\tilde{\Gamma}_{t_1}, \dots, \tilde{\Gamma}_{t_n}$ are independent gamma distributed with the shape parameters $t_1/\nu, \dots, t_n/\nu$, respectively.

Let L_t be $\mathcal{ALM}(\mu, \sigma, \nu)$ with a drift m. Show that for each fixed t

$$L_t \stackrel{d}{=} \tilde{L}_t \stackrel{d}{=} \tilde{\tilde{L}}_t.$$

Are $\tilde{L}_t$ and $\tilde{\tilde{L}}_t$ Laplace motions? Why?

Hint: Use the representation given in Proposition 4.1.2 to show the first part.

Exercise 4.5.11 Prove the representation 4.2.2 of $\mathcal{ALM}(\mu, \sigma, \nu)$.

Exercise 4.5.12 Prove the representation 4.2.3 of $\mathcal{ALM}(\mu, \sigma, \nu)$.

Exercise 4.5.13 Prove the representation 4.2.4 of $\mathcal{ALM}(\mu, \sigma, \nu)$.

Exercise 4.5.14 Show that the function (4.3.1) is a genuine characteristic function for any $0 < \alpha < 1$.

Hint: Proceed by showing the following:

(i) $\psi_{\alpha,\sigma}(t) = \psi_{\alpha,\sigma}(-t), t > 0$.

(ii) $\psi_{\alpha,\sigma}(0) = 1$.

(iii) $\lim_{t\to\infty} \psi_{\alpha,\sigma}(t) = 0$.

(iv) $\psi''_{\alpha,\sigma}(t) > 0$ for $t > 0$ so that $\psi_{\alpha,\sigma}$ is convex on $(0, \infty)$.

Thus $\psi_{\alpha,\sigma}$ is a Polya-type ch.f. [see, e.g., Lukacs (1970)].

Exercise 4.5.15 For any $p \in (0, 1)$, let ν_p denote a geometric r.v. with mean $1/p$ and probability function

$$P(\nu_p = k) = p(1-p)^{k-1}, \quad k = 1, 2, \dots.$$

Let $p, q \in (0, 1)$, and consider a sequence $(\nu_p^{(i)})$ of i.i.d. geometric random variables with mean $1/p$ and another geometric r.v. ν_q independent of the sequence. Show that the geometric sum $\sum_{i=1}^{\nu_q} \nu_p^{(i)}$ has the same probability distribution as ν_{pq} (a geometric r.v. with mean $1/(pq)$).

Hint: Write the ch.f. of the geometric sum conditioning on the ν_q.

Exercise 4.5.16 For each $n \geq 1$, let $Z_n^{(1)}, Z_n^{(2)}, \dots$ be a sequence of i.i.d. r.v.'s. Assume that for each i we have the convergence $Z_n^{(i)} \stackrel{d}{\to} Z^{(i)}$ as $n \to \infty$, where the $Z^{(i)}$'s are independent and identically distributed variables. Let ν be any integer-valued r.v. independent of all the other r.v.'s involved. Show that, as $n \to \infty$, the random sum $\sum_{i=1}^{\nu} Z_n^{(i)}$ converges in distribution to the random sum $\sum_{i=1}^{\nu} Z^{(i)}$.

Exercise 4.5.17 Prove Proposition 4.3.6.

Exercise 4.5.18 For any $0 < \rho < 1$, let f_ρ be the Cauchy density on $(-\infty, \infty)$, defined as

$$f_\rho(x) = \frac{\sin(\pi\rho)}{\pi[(x+\cos(\pi\rho))^2 + \sin^2(\pi\rho)]}, \quad x \in \mathbb{R}. \tag{4.5.8}$$

Show that $\int_0^\infty f_\rho(x)dx = \rho$, so that $g_\rho(x) = \frac{1}{\rho} f_\rho(x)$ is a density on $(0, \infty)$.

Exercise 4.5.19 For any $0 < \rho < 1$, let W_ρ be a positive r.v. with the density g_ρ defined in Exercise 4.5.18. Show that as $\rho \to 0^+$, the distribution of W_ρ converges weakly to the distribution given by the density $g_0(x) = (1+x)^{-2}$, while as $\rho \to 1^+$, the distribution of W_ρ converges weakly to a distribution of a unit mass at 1, namely, $W_1 \equiv 1$.

Exercise 4.5.20 For any $0 < \rho < 1$, let W_ρ be a positive r.v. with the density g_ρ defined in Exercise 4.5.18. Show that W_ρ has the reciprocal property $W_\rho \stackrel{d}{=} 1/W_\rho$.

Exercise 4.5.21 Show that if X is the Pareto Type I random variable with the p.d.f.

$$f(x) = \frac{1}{x}\frac{1}{\log b - \log a}, \quad 0 < a < x < b,$$

then $Y = 1/X$ has a distribution of the same type.

Exercise 4.5.22 Prove Proposition 4.3.11.

Exercise 4.5.23 Prove Proposition 4.3.12

Exercise 4.5.24 Show that setting $m = 1$ in (4.3.49) produces (4.3.44).

Exercise 4.5.25 Using the well-known identity for noninteger values of z,

$$\gamma(z)\Gamma(1-z) = \frac{\pi}{\sin \pi z},$$

show that for $\alpha = 2$, the fractional absolute moments of the Linnik distribution given by (4.3.54) coincide with $\sigma^\alpha\Gamma(p+1)$, which are the moments of the symmetric Laplace distribution.

Exercise 4.5.26 Show that the Sargan density (4.4.5) with restrictions (4.4.6) is a bona fide probability density function on $(-\infty, \infty)$.

Part II

Multivariate Distributions

Introduction

In this part we discuss current results on *multivariate Laplace distributions* and their generalizations. The field is relatively unexplored, and the subject matter is quite fresh and somewhat fragmented; thus our account is intentionally concise. In our opinion, some period of digestion is required to put these results in a proper perspective. Hopefully, a separate monograph will be available on this burgeoning area of statistical distributions in the not-too-distant future.

Multivariate generalizations of the Laplace laws have been considered on various occasions by various authors. The term *multivariate Laplace law* is still somewhat ambiguous, but at present it applies most often to the class of symmetric, elliptically contoured distributions for which the characteristic function is of the form

$$\Phi(\mathbf{t}) = \frac{1}{1 + \frac{1}{2}\mathbf{t}'\mathbf{\Sigma}\mathbf{t}}. \tag{II.1}$$

Recall that an r.v. in $\mathbb{R}^d$ has an elliptically contoured distribution if its ch.f. has the form

$$\Phi(\mathbf{t}) = e^{i\mathbf{t}'\mathbf{m}}\phi(\mathbf{t}'\mathbf{\Sigma}\mathbf{t}) \tag{II.2}$$

for some function ϕ, where $\mathbf{m}$ is a $d \times 1$ vector in $\mathbb{R}^d$ and $\mathbf{\Sigma}$ is a $d \times d$ nonnegative definite matrix [see, e.g., Fang et al. (1990)].

Probably the simplest multivariate generalization of Laplace distribution is the distribution of a vector of independent Laplace random variables [see, e.g., Osiewalski and Steel (1993), Marshall and Olkin (1993)]. However not many properties of univariate laws can be extended to this class of distributions. Moreover, it is not invariant on rotations (see, for example, the graph of bivariate density in Figure 8.7).

Transforming a bivariate normal distribution, Ulrich and Chen (1987) obtained another bivariate distribution with Laplace marginals, noting that there were no "naturally occurring" bivariate Laplace distributions. Much earlier, McGraw and Wagner (1968) in their seminal paper provided a number of examples of bivariate elliptically contoured distributions, including the multivariate Laplace distribution (II.1) and their generalizations [see also Johnson and Kotz (1972), Table 3, p. 297 and equation (69), p. 301, and Johnson (1987)]. This multivariate Laplace law also appears in Anderson (1992) as a special case of the multivariate Linnik distribution [also known as the semi-α-Laplace distribution; see Pillai (1985)].

Ernst (1998) introduced yet another multivariate extension of symmetric Laplace distributions again via an elliptic contouring. In the one-dimensional case his class reduces to the univariate symmetric Laplace laws.

Barndorff-Nielsen (1977) introduced the class of so-called hyperbolic distributions, which was later extended to the multivariate case in Blaesid (1981). With an appropriate passage to the limit of their parameters, one can obtain a multivariate and asymmetric extension of the Laplace laws. This class is studied here on its own, independently of the theory of hyperbolic and inverse Gaussian distributions.

This part of the monograph is organized so that special cases — bivariate and symmetric distributions — are discussed (albeit rather briefly) prior to the more general cases of multivariate and asymmetric distributions. We believe that this exposition, despite the fact that formally most of the properties follow from the results derived for the general case, allows for a faster reference to the special important cases without the need to absorb the more cumbersome notation and description of the general multivariate asymmetric Laplace distributions. Thus the symmetric (elliptically contoured) multivariate distributions are discussed before the general asymmetric ones and the bivariate cases precede the general multivariate ones. On the other hand, we present proofs for the general setting, omitting explicit proofs in particular cases unless they provide a better insight.

While discussing the multivariate Laplace distributions we always consider them to be centered at zero. One can add the location parameter in a natural manner and thus consider, as we did in the previous chapters, a more general class of asymmetric Laplace distributions. However, this complicates the already cumbersome notation in the multivariate case without adding substantially to deeper understanding.

5

Symmetric Multivariate Laplace Distribution

In this chapter we discuss a natural extension of the univariate symmetric Laplace distribution to the multivariate setting. The material discussed here has not — to the best of our knowledge — appeared before in book literature. A comparison with the commonly used multivariate normal distribution would be most instructive.

5.1 Bivariate case

5.1.1 Definition. As in the univariate case, the most direct and simple way to introduce the bivariate symmetric Laplace distributions is through their characteristic functions. Thus the *bivariate symmetric Laplace* distributions constitute a three-parameter family of two-dimensional distributions with the characteristic functions given by

$$\psi(t_1, t_2) = \left(1 + \frac{\sigma_1^2 t_1^2}{2} + \rho\sigma_1\sigma_2 t_1 t_2 + \frac{\sigma_2^2 t_2^2}{2}\right)^{-1},$$

where the three parameters σ_1, σ_2, and ρ satisfy

$$\sigma_1 \geq 0, \sigma_2 \geq 0, \rho \in [0, 1].$$

We shall use $\mathcal{BSL}(\sigma_1, \sigma_2, \rho)$ instead of the lengthy expression to describe membership in this family.

Note that in this definition, as well as in all others in this part of the book, we do not take into account the location of the distribution, always centering it at zero. The word "symmetric" in our terminology represents the fact that our distribution is actually obtained from a one-dimensional distribution spread uniformly along an ellipsoid in the two dimensions. Formally, this means that the characteristic function depends on its argument $\mathbf{t} = (t_1, t_2)'$ through $\mathbf{t}'\mathbf{\Sigma}\mathbf{t}$, where $\mathbf{\Sigma}$ is a certain nonnegative definite matrix, in this case

$$\mathbf{\Sigma} = \begin{bmatrix} \sigma_1^2 & \sigma_1\sigma_2\rho \\ \sigma_1\sigma_2\rho & \sigma_2^2 \end{bmatrix}. \tag{5.1.1}$$

In general, for this type of distribution the name *elliptically contoured* is used, and more appropriately the distribution under consideration should be called the elliptically contoured Laplace distribution.

The following property follows immediately from the definition.

Proposition 5.1.1 *A linear combination $a_1 Y_1 + a_2 Y_2$ of the coordinates of a $\mathcal{BSL}(\sigma_1, \sigma_2, \rho)$ random vector $\mathbf{Y} = (Y_1, Y_2)'$ has a one-dimensional symmetric Laplace distribution $\mathcal{L}(0, \sigma)$, where*

$$\sigma = \sqrt{\sigma_1^2 a_1^2 + 2\rho\sigma_1\sigma_2 a_1 a_2 + \sigma_2^2 a_2^2)}.$$

In particular, the marginal distributions of a $\mathcal{BSL}$ distribution are symmetric Laplace distributions.

The case when $\sigma_1 = \sigma_2 = 1$ and $\rho = 0$ will be distinguished, and the corresponding distribution will be referred to as the *standard bivariate Laplace distribution*.

5.1.2 Moments. The moments of the Laplace distribution are easily obtained by differentiating its characteristic function. In particular, we have the following formulas for the mean vector and variance-covariance matrix of a $\mathcal{BSL}(\sigma_1, \sigma_2, \rho)$ random vector $\mathbf{Y}$:

$$E\mathbf{Y} = \mathbf{0}; \qquad \mathbf{Cov}(\mathbf{Y}) = E(\mathbf{Y}\mathbf{Y}') = \begin{bmatrix} \sigma_1^2 & \sigma_1\sigma_2\rho \\ \sigma_1\sigma_2\rho & \sigma_2^2 \end{bmatrix}.$$

Note that if $\mathbf{Y}$ is uncorrelated ($\rho = 0$), Y_1 and Y_2 are not independent (unlike the situation in the case of bivariate normal distribution).

Remark 5.1.1 One can consider a vector of two independent Laplace random variables and its distribution. By the above property, such a vector does not belong to the multivariate Laplace family. An example of the density for such a random vector can be seen in Figure 8.7.

5.1.3 Densities. The formula for densities is taken from the general case, considered in Section 6.5 of Chapter 6, equation (6.5.3). Namely, assuming that the distribution is nonsingular, we have

$$g(x, y) = \frac{1}{\pi\sigma_1\sigma_2\sqrt{1-\rho^2}} \cdot K_0\left(\sqrt{\frac{2(x^2/\sigma_1^2 - 2\rho xy/(\sigma_1\sigma_2) + y^2/\sigma_2^2)}{1-\rho^2}}\right),$$

where K_0 is the Bessel function of the third kind given by (A.0.4) or (A.0.5) in the appendix.

In particular, the standard bivariate Laplace distribution is given by

$$\frac{1}{\pi} K_0\left(\sqrt{2(x^2 + y^2)}\right). \tag{5.1.2}$$

To compare the Gaussian and Laplace distributions we present in Figures 5.1 and 5.2 bivariate AL and Gaussian densities. Figure 5.1 deals with uncorrelated distributions with the two covariance matrices $\mathbf{\Sigma}$ given by

$$\begin{bmatrix} 1 & 0 \\ 0 & 1 \end{bmatrix} \text{ and } \begin{bmatrix} 1 & 0 \\ 0 & 0.5 \end{bmatrix}. \tag{5.1.3}$$

The graphs present contour lines at the levels in the interval $(0, 0.5)$. The densities were cut off above the level of 0.5 (the Laplace densities are unbounded around zero). To illustrate the tails and behavior around zero, the contour levels were chosen differently in two different subintervals. From the subinterval $(0, 0.005)$ we chose 10 equally spaced levels to show contours representing tails of a distribution and from the subinterval $(0.005, 0.5)$ we selected 50 equally spaced levels to present contours of a distribution at its center.

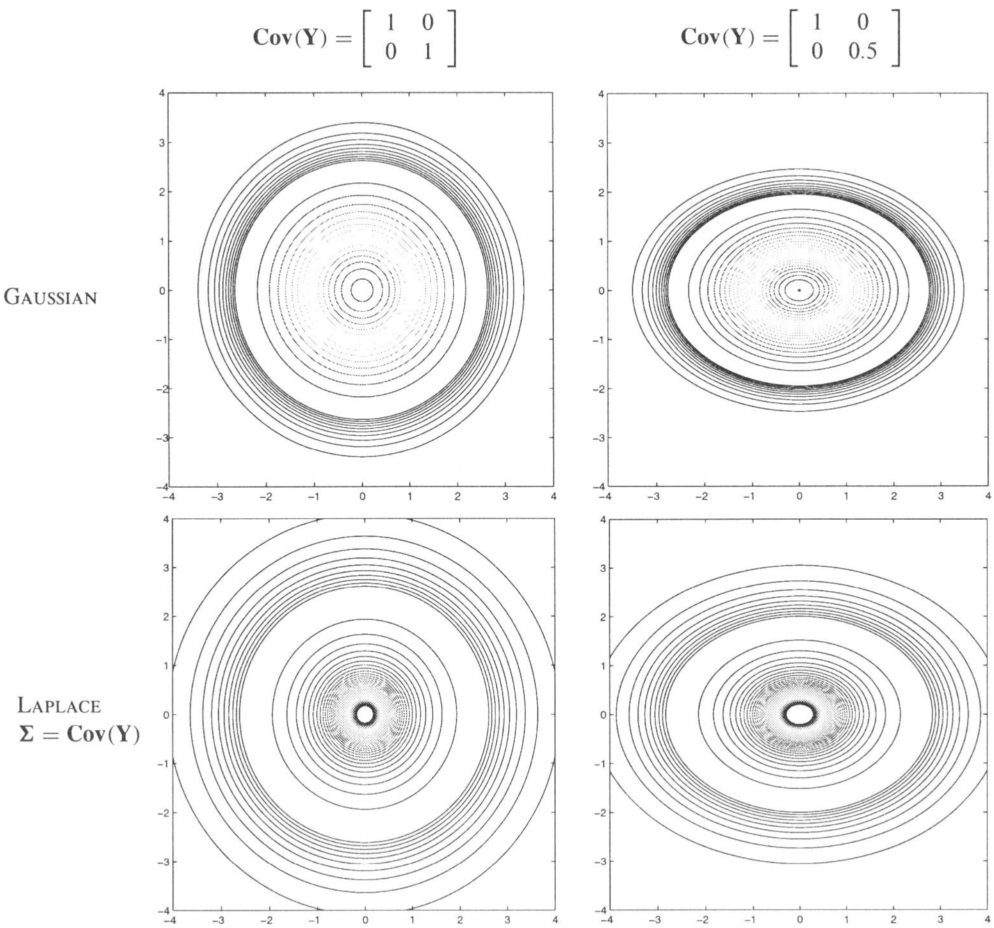

Figure 5.1: Laplace and Gaussian bivariate densities corresponding to the uncorrelated distributions.

The first two drawings represent Gaussian densities of the distributions with the covariance matrices specified by the values of $\mathbf{\Sigma}$. The third and fourth drawings represent densities of the Laplace random variables — symmetric and having the same covariance matrices. The bivariate parameters of these two distributions are given by $\mathbf{\Sigma} = \mathbf{Cov}(\mathbf{Y})$. The one on the left-hand side corresponds to the *bivariate standard Laplace* random variable with the density (5.1.2) for which $\mathbf{\Sigma}$ is the identity matrix.

In Figure 5.2, we present the correlated version of the graphs in Figure 5.1. Namely, we consider the covariance matrices $\mathbf{\Sigma}$ given by

$$\begin{bmatrix} 1 & 0.5 \\ 0.5 & 1 \end{bmatrix} \quad \text{and} \quad \begin{bmatrix} 1 & 0.5 \\ 0.5 & 0.5 \end{bmatrix}. \tag{5.1.4}$$

In the the top drawings, the Gaussian distributions are presented with covariance matrices given by (5.1.4). In the bottom drawings, the corresponding Laplace densities are provided.

5.1.4 Simulation of bivariate Laplace variates. The general algorithm for simulation of asymmetric multivariate Laplace variables is derived in Section 6.4 of the next chapter. We present here its version for the bivariate symmetric case.

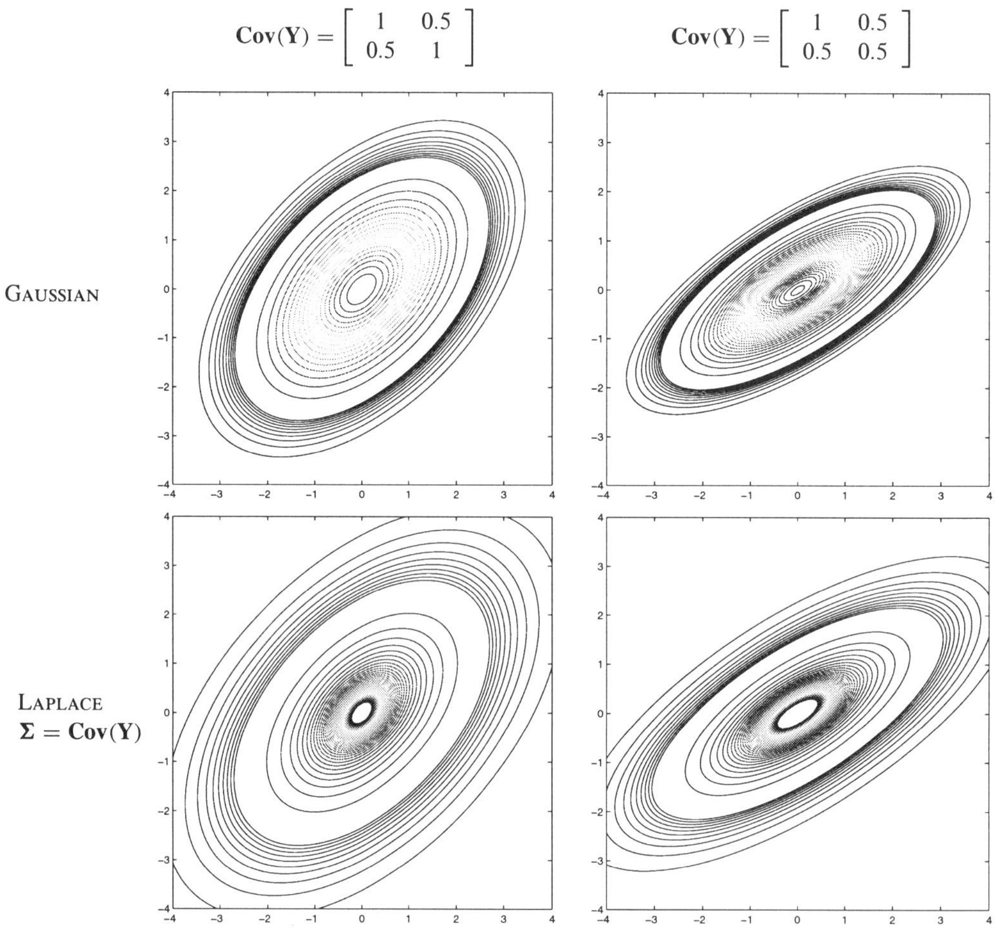

Figure 5.2: Laplace and Gaussian bivariate densities corresponding to the correlated distributions.

$\mathcal{BSL}(\sigma_1, \sigma_2, \rho)$ *generator*

- Generate a bivariate normal variable $\mathbf{X}$ with mean zero and covariance matrix $\mathbf{\Sigma}$ given by (5.1.1).
- Generate a standard exponential variable W.
- Set $\mathbf{Y} \leftarrow \sqrt{W} \cdot \mathbf{X}$.
- RETURN $\mathbf{Y}$.

In Figures 5.3 and 5.4 below we have used this method implemented in the S-Plus package to simulate samples from the distributions that are given by the densities presented on Figures 5.1 and 5.2.

5.2 General symmetric multivariate case

5.2.1 Definition. A multivariate symmetric Laplace distribution is a direct generalization of the bivariate case. As before, the word "symmetric" refers to elliptically contoured or elliptically sym-

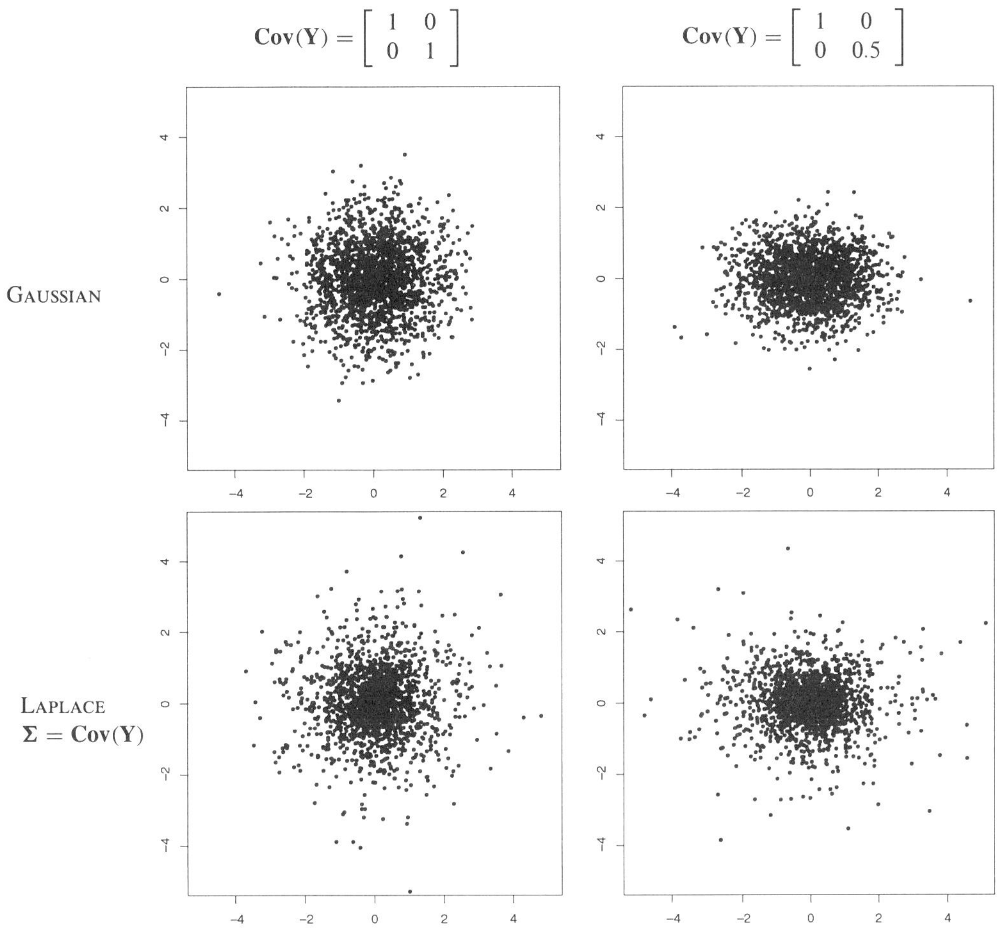

Figure 5.3: Uncorrelated Laplace and Gaussian random samples. Monte-Carlo simulation is based on the described algorithm. (The sample size equals 2000.)

metric distributions and means that the distributions possess the characteristic function that depends on its variables only through a quadratic form.

Let $\boldsymbol{\Sigma}$ be an $d \times d$ nonnegative definite symmetric matrix. We shall say that a d-dimensional distribution is multivariate symmetric Laplace with the parameter $\boldsymbol{\Sigma}$, denoted $\mathcal{SL}_d(\boldsymbol{\Sigma})$ if its characteristic function is of the form

$$\Psi(\mathbf{t}) = \frac{1}{1 + \frac{1}{2}\mathbf{t}'\boldsymbol{\Sigma}\mathbf{t}}. \tag{5.2.1}$$

5.2.2 Moments and densities. It follows directly from the definition that the $\mathcal{SL}_d(\boldsymbol{\Sigma})$ distribution is centered at zero (the mean is zero) and its covariance matrix is given by $\boldsymbol{\Sigma}$.

From the representation of the density for the general multivariate asymmetric case, we have that the $\mathcal{SL}_d(\boldsymbol{\Sigma})$ density function (for a nonsingular distribution) is of the form

$$g(\mathbf{y}) = \frac{2}{(2\pi)^{d/2}|\boldsymbol{\Sigma}|^{1/2}} \left(\frac{\mathbf{y}'\boldsymbol{\Sigma}^{-1}\mathbf{y}}{2} \right)^{v/2} K_v \left(\sqrt{2\mathbf{y}'\boldsymbol{\Sigma}^{-1}\mathbf{y}} \right), \tag{5.2.2}$$

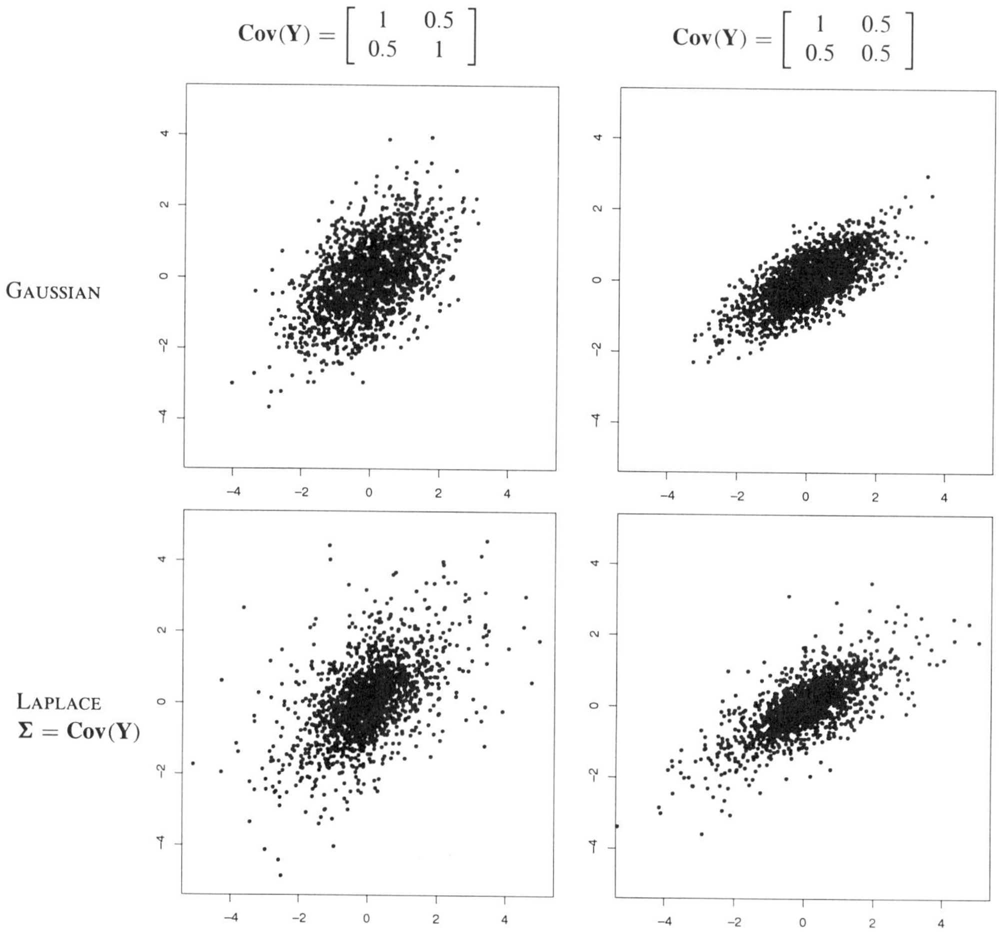

Figure 5.4: Correlated Laplace and Gaussian random samples. Monte-Carlo simulation is based on the described algorithm. (The sample size equals 2000.)

where $v = (2 - d)/2$ and $K_v(\cdot)$ is the modified Bessel function of the third kind given by (A.0.4) or (A.0.5) in the appendix. This density was derived in George and Pillai (1988) for the case $\mathbf{\Sigma} = 2\mathbf{I}_d$ and in Anderson (1992) as a special case of multivariate Linnik density (note that density (8) of Anderson (1992) contains an extra factor of $\sqrt{2Q}$). Additional properties of $\mathcal{SL}_d(\mathbf{\Sigma})$ are provided in the exercises below. They should be viewed as an integral part of this chapter.

5.3 Exercises

Exercise 5.3.1 Let $\mathbf{X} = (X_1, X_2)'$ have a standard bivariate Laplace distribution $\mathcal{BSL}(1, 1, 0)$. Show that the two random variables X_1 and X_2 are uncorrelated but not independent.

Exercise 5.3.2 Let $\mathbf{X} = (X_1, X_2)'$ have a standard bivariate Laplace distribution $\mathcal{BSL}(1, 1, 0)$. Convert to polar coordinates by setting $X_1 = R\cos\theta$, $X_2 = R\sin\theta$ $(R > 0, 0 < \theta < 2\pi)$.

(a) Derive the marginal density function of R.

(b) Derive the marginal density function of θ.

(c) Are R and θ independent?

(d) Repeat parts (a)–(c) under the assumption that X_1 and X_2 are i.i.d. with the standard Laplace $\mathcal{L}(0, 1)$ distribution.

(e) Repeat parts (a)–(c) under the assumption that X_1 and X_2 are i.i.d. with the standard normal distribution.

Exercise 5.3.3 Let $\mathbf{X} = (X_1, X_2)' \sim \mathcal{BSL}(\sigma_1, \sigma_2, \rho)$.

(a) Derive the marginal p.d.f.'s of X_1 and X_2.

(b) Derive the conditional p.d.f. of X_2 given $X_1 = x_1$.

Exercise 5.3.4* Let $\mathbf{X} = (X_1, \ldots, X_d)'$ have a symmetric multivariate Laplace distribution $\mathcal{SL}_d(\boldsymbol{\Sigma})$, and let Ψ be the ch.f. of $\mathbf{X}$.

(a) Verify that the mean vector of $\mathbf{X}$ is $\mathbf{0}$ and the covariance matrix of $\mathbf{X}$ is $\boldsymbol{\Sigma}$.

(b) Using the following expression for the kth moment of $\mathbf{X}$,

$$m_k(\mathbf{X}) = \frac{1}{i^k} \frac{\partial^k \Psi(\mathbf{t})}{\partial \mathbf{t} \partial \mathbf{t}' \ldots}\bigg|_{\mathbf{t}=\mathbf{0}}, \tag{5.3.1}$$

show that every moment of $\mathbf{X}$ of odd order vanishes.

(c)* Using the following expression for the kth cumulant of $\mathbf{X}$,

$$c_k(\mathbf{X}) = \frac{1}{i^k} \frac{\partial^k \log \Psi(\mathbf{t})}{\partial \mathbf{t} \partial \mathbf{t}' \ldots}\bigg|_{\mathbf{t}=\mathbf{0}}, \tag{5.3.2}$$

show that $c_1(\mathbf{X}) = \mathbf{0}$, $c_2(\mathbf{X}) = \boldsymbol{\Sigma}$, $c_3(\mathbf{X}) = \mathbf{0}$ and

$$c_4(\mathbf{X}) = \operatorname{vec} \boldsymbol{\Sigma} \otimes \operatorname{vec}' \boldsymbol{\Sigma} + (\mathbf{I}_{d^2} + \mathbf{K}_{dd})(\boldsymbol{\Sigma} \otimes \boldsymbol{\Sigma}) \tag{5.3.3}$$

[Kollo (2000)], where $\operatorname{vec} \mathbf{A}$ is the vec operator of matrix $\mathbf{A}$, $\mathbf{A} \otimes \mathbf{B}$ is the Kronecker product of matrices $\mathbf{A}$ and $\mathbf{B}$, and $\mathbf{K}_{dd}$ is the vec-permutation matrix [see, e.g., Harville (1997) or Magnus and Neudecker (1999) for the matrix notation]. What are the corresponding results for the multivariate normal vector $\mathbf{X}$ with vector mean zero and covariance matrix $\boldsymbol{\Sigma}$? [You may wish to consult Kotz et al. (2000).]

Exercise 5.3.5* Recall that if X is a univariate standard classical Laplace variable with density $p(x) = \frac{1}{2}e^{-|x|}$ $(-\infty < x < \infty)$, then the ordinate $p(X)$ has uniform distribution on $(0, 1/2)$, while the ordinate $p(Z)$ fails to be uniform for standard normal variable Z with density $p(x) = \frac{1}{\sqrt{2\pi}}e^{-x^2/2}$ $(-\infty < x < \infty)$ (see Exercise 2.7.9). However, show that if the variables X_1 and X_2 have bivariate normal distribution with p.d.f.

$$p(x_1, x_2) = \frac{1}{2\pi} e^{-\frac{1}{2}(x_1^2 + x_2^2)}, \quad -\infty < x_1, x_2 < \infty,$$

then the ordinate $p(X_1, X_2)$ is uniform on $(0, 2\pi)$ [Troutt (1991)]. Investigate the corresponding case of standard bivariate Laplace distribution with p.d.f.

$$p(x_1, x_2) = \frac{1}{\pi} K_0\left(\sqrt{2(x_1^2 + x_2^2)}\right), \quad (x_1, x_2) \neq (0, 0).$$

We suggest that you consult Troutt (1991), Kotz and Troutt (1996), or Kotz et al. (1997).

Exercise 5.3.6 Generalize the results of Exercises 2.7.9 and 5.3.5 by showing that if $\mathbf{X}$ is a random vector in $\mathbb{R}^d$, $d \geq 1$, with probability density function

$$f(\mathbf{x}) = c_d e^{-(\mathbf{x}'\mathbf{x})^d/2},$$

then the random variable $U = f(\mathbf{X})$ has uniform distribution on $(0, c_d)$. What is the value of c_d?

Exercise 5.3.7 Let $\mathbf{Y} = (Y_1, \ldots, Y_d)'$ have a multivariate $\mathcal{AL}_d(\mathbf{0}, \mathbf{I}_d)$ distribution in $\mathbb{R}^d$. Show that the random vector

$$\left(\frac{Y_1}{Y_d}, \ldots, \frac{Y_{d-1}}{Y_d}\right)$$

has a multivariate Cauchy distribution with density

$$\Gamma(d/2)\pi^{-d/2}\left(1 + \sum_{i=1}^{d-1}\right)^{-d/2}$$

and is independent of $||\mathbf{Y}|| = (\sum_{i=1}^{d} Y_i^2)^{1/2}$. This result is actually a characterization of spherically symmetric distributions [see George and Pillai (1988)].

6
Asymmetric Multivariate Laplace Distribution

In this chapter we present the theory of a class of multivariate laws that we term *asymmetric Laplace* (AL) distributions [see Kozubowski and Podgórski (1999bc), Kotz et al. (2000b)]. The class is an extension of both the symmetric multivariate Laplace distributions and the univariate AL distributions that were discussed in previous chapters. This extension retains the natural, asymmetric, and multivariate features of the properties characterizing these two important subclasses. In particular, the AL distributions arise as the limiting laws in a random summation scheme with i.i.d. terms having a finite second moment, where the number of terms in the summation is geometrically distributed independently of the terms themselves. This class can be viewed as a subclass of hyperbolic distributions and some of its properties are inherited from them. However, to demonstrate an elegant theoretical structure of the multivariate AL laws and also for the sake of simplicity we prefer direct derivations of the results. Thus we provide explicit formulas for the probability density and the density of the Lévy measure. The results presented also include characterizations, mixture representations, formulas for moments, a simulation algorithm, and a brief discussion of linear regression models with AL errors.

The multivariate laws discussed, unlike the laws of Ernst (1998) already mentioned, have multivariate (and univariate) Laplace marginal distributions, allow for asymmetry, and in general are not elliptically contoured. Asymmetric Laplace laws can be defined in various equivalent ways, which we express in the form of their characterizations and representations. Their significance comes from the fact that they are the only distributional limits for (appropriately normalized) random sums of i.i.d. random vectors (r.v.'s) with finite second moments

$$\mathbf{X}^{(1)} + \cdots + \mathbf{X}^{(\nu_p)}, \tag{6.0.1}$$

where ν_p has a geometric distribution with the mean $1/p$ (independent of $\mathbf{X}^{(i)}$'s):

$$P(\nu_p = k) = p(1-p)^{k-1}, \quad k = 1, 2, \ldots, \tag{6.0.2}$$

and p converges to zero [see, e.g., Mittnik and Rachev (1991)]. Thus these multivariate laws arise rather naturally. Since the sums such as (6.0.1) frequently appear in many applied problems in biology, economics, insurance mathematics, reliability, and other fields [see examples in Kalashnikov (1997) and references therein], AL distributions should have a wide variety of applications. In particular, this class seems to be suitable for modeling heavy-tailed asymmetric multivariate data for which one is reluctant to sacrifice the property of finiteness of moments. (Multivariate stable distributions are an alternative where this concession has to be made.)

From the standpoint of classical distribution theory, the AL laws form a subclass of the geometric stable distributions [see, e.g., Rachev and SenGupta (1992)]. The geometric stable laws approximate geometric compounds (6.0.1) with arbitrary components, including those with infinite means [see Kozubowski and Rachev (1999b) for references on multivariate geometric stable laws]. The geometric stable distributions, similar to stable laws, have the tail behavior governed by the index of stability $\alpha \in (0, 2]$. The AL distributions correspond to the geometric stable subclass with $\alpha = 2$. Thus they play an analogous role among geometric stable laws as Gaussian distributions do among stable laws. Like Gaussian distributions, they have finite moments of all orders, and their theory is equally elegant and straightforward. However, in spite of finiteness of moments, their tails are substantially longer than those for Gaussian laws; this coupled with the fact that they allow for asymmetry makes them more flexible and attractive for modeling heavy-tailed asymmetric data.

Incidentally, the multivariate AL laws can be obtained as a limiting case of the generalized hyperbolic distributions, introduced by Barndorff-Nielsen (1977). Consequently, certain properties of AL laws can be deduced from the corresponding properties of the generalized hyperbolic distributions and passing to the limit. However, direct proofs for AL laws are often simpler than their "hyperbolic" counterparts and in addition provide better insight into this class, and we have included them in our work. Moreover, many properties are quite specific to AL laws, such as their convolution properties in relation to the random summation model. From the latter point of view, which coincides with our main interest and motivation, the relation to the generalized hyperbolic laws, although an important one, is not crucial.

6.1 Bivariate case: Definition and basic properties

6.1.1 Definition. The *bivariate asymmetric Laplace distributions* constitute a five parameter family of two-dimensional distributions given by the characteristic function

$$\psi(t_1, t_2) = \frac{1}{1 + \frac{\sigma_1^2 t_1^2}{2} + \rho\sigma_1\sigma_2 t_1 t_2 + \frac{\sigma_2^2 t_2^2}{2} - im_1 t_1 - im_2 t_2},$$

where the five parameters $m_1, m_2, \sigma_1, \sigma_2$, and ρ satisfy

$$m_1 \in \mathbb{R}, \quad m_2 \in \mathbb{R}, \quad \sigma_1 \geq 0, \quad \sigma_2 \geq 0, \quad \rho \in [0, 1].$$

In what follows, the notation $\mathcal{BAL}(m_1, m_2, \sigma_1, \sigma_2, \rho)$ stands for the asymmetric bivariate Laplace distribution with the given parameters.

The distribution is no longer elliptically contoured (unless $m_1 = m_2 = 0$), which justifies using the term "asymmetric distributions." The following property follows immediately from the definition.

Proposition 6.1.1 *A linear combination $a_1Y_1 + a_2Y_2$ of the coordinates of a $\mathcal{BAL}(m_1, m_2, \sigma_1, \sigma_2, \rho)$ random vector $\mathbf{Y} = (Y_1, Y_2)'$ has a one-dimensional AL distribution $\mathcal{AL}(\mu, \sigma)$, where*

$$\mu = m_1 a_1 + m_2 a_2 \text{ and } \sigma = \sqrt{\sigma_1^2 a_1^2 + 2\rho\sigma_1\sigma_2 a_1 a_2 + \sigma_2^2 a_2^2)}.$$

As in the symmetric case, the marginal distributions of a $\mathcal{BAL}$ distribution are univariate asymmetric Laplace distributions.

6.1.2 Moments. The moments of the $\mathcal{BAL}$ distribution are easily obtained by differentiating their characteristic function. In particular, we have the following formulas for the means and the elements of the variance-covariance matrix of a $\mathcal{BAL}(m_1, m_2, \sigma_1, \sigma_2, \rho)$ random vector $\mathbf{Y} = (Y_1, Y_2)'$:

$$EY_1 = m_1, \quad EY_2 = m_2; \quad \text{Var}\, Y_1 = \sigma_1^2 + m_1^2, \quad \text{Var}\, Y_2 = \sigma_2^2 + m_2^2,$$
$$\text{Cov}(Y_1, Y_2) = \sigma_1\sigma_2\rho + m_1 m_2.$$

Note that as in the symmetric case, even if the components of $\mathbf{Y}$ are uncorrelated (i.e., $\sigma_1\sigma_2\rho + m_1m_2 = 0$), they are not independent. Moreover, the matrix $\mathbf{\Sigma}$ given by (5.1.1) is no longer the variance-covariance matrix of $\mathbf{Y}$ (unless $\mathbf{m} = (m_1, m_2)' = \mathbf{0}$).

6.1.3 Densities. The expression for densities is obtained from the general case considered in the next section [equation (6.5.3)]. For a nonsingular distribution, we have

$$g(x, y) = \frac{\exp\left[((m_1\sigma_2/\sigma_1 - m_2\rho)\, x + (m_2\sigma_1/\sigma_2 - m_1\rho)\, y) / (\sigma_1\sigma_2(1-\rho^2))\right]}{\pi\sigma_1\sigma_2\sqrt{1-\rho^2}}$$
$$\cdot K_0\left(C(m_1, m_2, \sigma_1, \sigma_2, \rho)\sqrt{x^2\sigma_2/\sigma_1 - 2\rho xy + y^2\sigma_1/\sigma_2}\right),$$

where

$$C(m_1, m_2, \sigma_1, \sigma_2, \rho) = \frac{\sqrt{2\sigma_1\sigma_2(1-\rho^2) + m_1^2\sigma_2/\sigma_1 - 2m_1m_2\rho + m_2^2\sigma_1/\sigma_2}}{\sigma_1\sigma_2(1-\rho^2)}.$$

In Figure 6.1, we present four different asymmetric bivariate Laplace densities for which the covariance matrix is exactly the same as for the symmetric cases of Gaussian and Laplace distributions presented in Figure 5.1. These densities are still uncorrelated but the matrix $\mathbf{\Sigma}$ is no longer diagonal.

The four graphs deal with various cases when $m_1 \neq 0$ and $m_2 \neq 0$, and thus the distributions are no longer elliptically contoured (symmetric). The values of the five parameters are as follows. The two cases in the top row of Figure 6.1 correspond to $m_1 = m_2 = 1/2$ and $\sigma_1 = \sigma_2 = \sqrt{3}/2$, $\rho = -1/3$ (left graph) and $\sigma_1 = \sqrt{3}/2$, $\sigma_2 = 1/2$, $\rho = -\sqrt{3}/3$ (right graph). The two cases in the bottom row of Figure 6.1 correspond to $m_1 = 1/2, m_2 = 1/4$ and $\sigma_1 = \sqrt{3}/2, \sigma_2 = \sqrt{15}/4$, $\rho = -\sqrt{5}/15$ (left graph) and $\sigma_1 = \sqrt{3}/2$, $\sigma_2 = \sqrt{7}/4$, $\rho = -\sqrt{21}/21$ (right graph). For the meaning of the presented contour lines, see Section 5.1.

The graphs indicate that even in the uncorrelated case, the Laplace distributions exhibit a large variety of asymmetric features, this property not shared by Gaussian distributions (compare with Figure 5.1).

Similar graphs are obtained for the correlated densities corresponding to the covariance matrices given in Section 5.1. Figure 6.2 should be compared with the symmetric case provided in Figure 5.2. In both cases, we have the same correlation structure. These graphs present densities of four asymmetric Laplace distributions with parameters specified as follows:

$$\mathbf{\Sigma} = \begin{bmatrix} 0.75 & 0.25 \\ 0.25 & 0.75 \end{bmatrix}, \qquad \mathbf{m} = (0.5, 0.5)';$$
$$\mathbf{\Sigma} = \begin{bmatrix} 0.75 & 0.25 \\ 0.25 & 0.25 \end{bmatrix}, \qquad \mathbf{m} = (0.5, 0.5)';$$
$$\mathbf{\Sigma} = \begin{bmatrix} 0.75 & 0.375 \\ 0.375 & 0.9375 \end{bmatrix}, \qquad \mathbf{m} = (0.5, 0.25)';$$
$$\mathbf{\Sigma} = \begin{bmatrix} 0.75 & 0.375 \\ 0.375 & 0.4375 \end{bmatrix}, \qquad \mathbf{m} = (0.5, 0.25)'.$$

Asymmetry of the distributions is clearly noticeable.

$\mathbf{Cov}(\mathbf{Y}) = \begin{bmatrix} 1 & 0 \\ 0 & 1 \end{bmatrix}$ $\mathbf{Cov}(\mathbf{Y}) = \begin{bmatrix} 1 & 0 \\ 0 & 0.5 \end{bmatrix}$

LAPLACE
$\mathbf{m} = (0.5, 0.5)'$
$\boldsymbol{\Sigma} \neq \mathbf{Cov}(\mathbf{Y})$

LAPLACE
$\mathbf{m} = (0.5, 0.25)'$
$\boldsymbol{\Sigma} \neq \mathbf{Cov}(\mathbf{Y})$

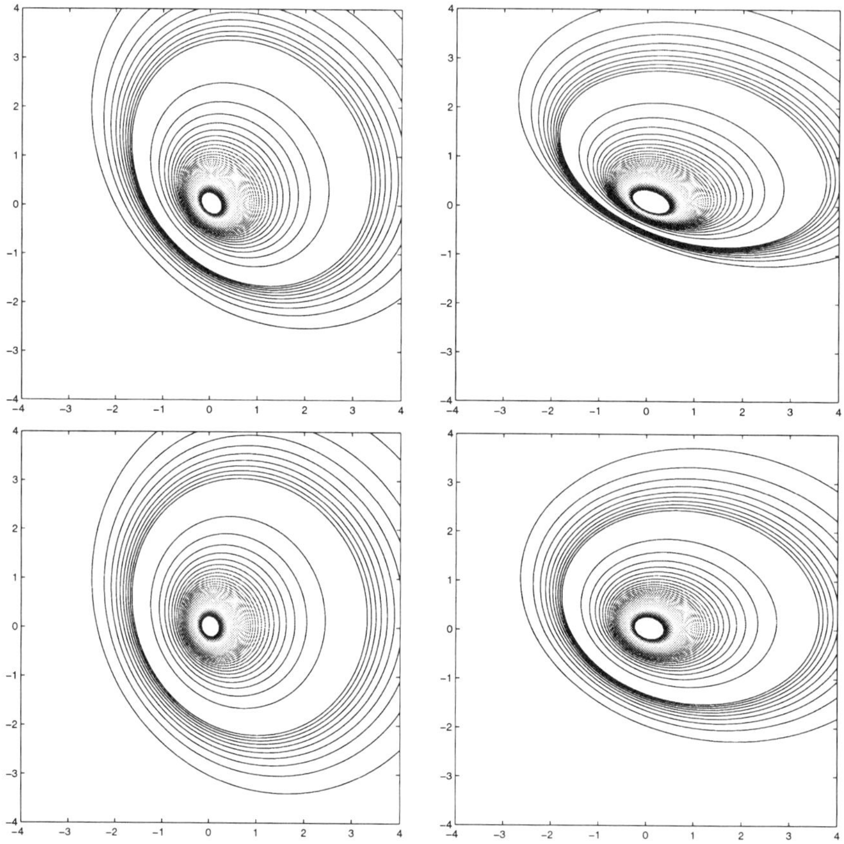

Figure 6.1: Asymmetric bivariate Laplace densities corresponding to the uncorrelated distributions. The covariances are the same as in the symmetric case in Figure 5.1.

6.1.4 Simulation of bivariate asymmetric Laplace variates. The general algorithm for simulating asymmetric multivariate Laplace variables is derived in Section 6.4 of the next chapter. In the bivariate case it takes the following form:

$\mathcal{BAL}(m_1, m_2, \sigma_1, \sigma_2, \rho)$ generator

- Generate a bivariate normal variable $\mathbf{X}$ with mean zero and covariance matrix $\boldsymbol{\Sigma}$ given by (5.1.1).
- Generate a standard exponential variable W.
- Set $\mathbf{Y} \leftarrow \sqrt{W} \cdot \mathbf{X} + \mathbf{m}W$, where $\mathbf{m} = (m_1, m_2)'$.
- RETURN $\mathbf{Y}$.

Note that compared with the corresponding algorithm for the symmetric case (see Section 5.1), here we have an extra variable $\mathbf{m}W$, which combined with $\sqrt{W}\mathbf{X}$ leads to an AL variable.

$\mathbf{Cov}(\mathbf{Y}) = \begin{bmatrix} 1 & 0.5 \\ 0.5 & 1 \end{bmatrix}$ $\mathbf{Cov}(\mathbf{Y}) = \begin{bmatrix} 1 & 0.5 \\ 0.5 & 0.5 \end{bmatrix}$

LAPLACE
$\mathbf{m} = (0.5, 0.5)'$
$\boldsymbol{\Sigma} \neq \mathbf{Cov}(\mathbf{Y})$

LAPLACE
$\mathbf{m} = (0.5, 0.25)'$
$\boldsymbol{\Sigma} \neq \mathbf{Cov}(\mathbf{Y})$

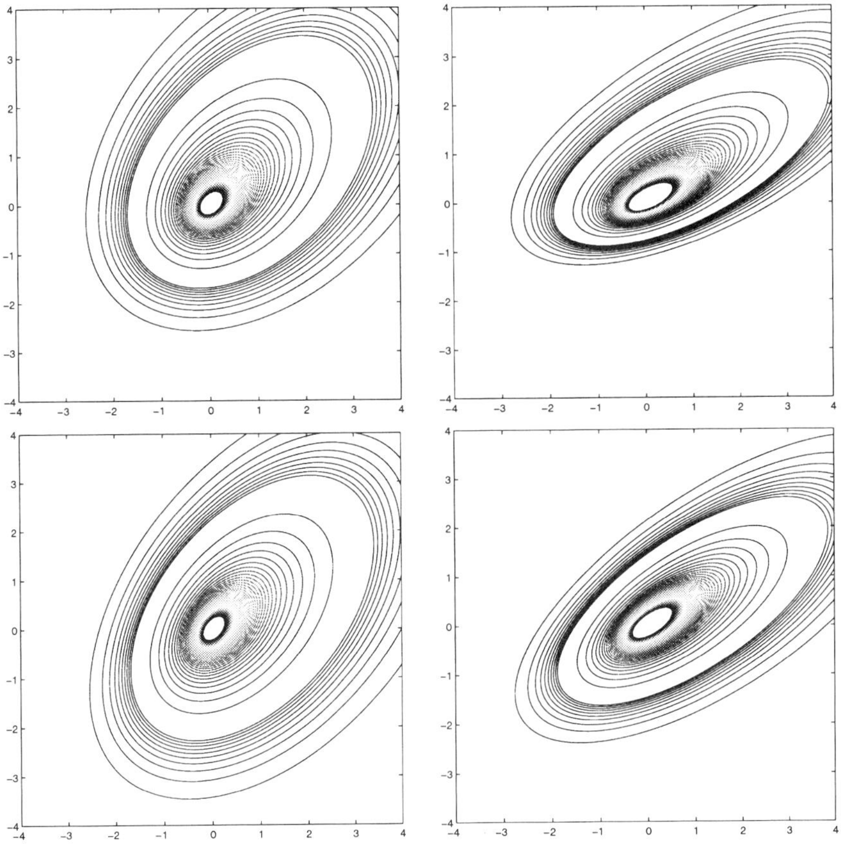

Figure 6.2: Laplace asymmetric bivariate densities corresponding to correlated distributions. The same covariances as in the symmetric case in Figure 5.2 are used.

In Figures 6.3 and 6.4, we present graphs of the same densities (based on Monte-Carlo simulation) as those presented in the graphs of the densities in Figures 6.1 and 6.2.

6.2 General multivariate asymmetric case

6.2.1 Definition. First, we provide a definition of multivariate AL laws.

Definition 6.2.1 *A random vector* $\mathbf{Y}$ *in* $\mathbb{R}^d$ *is said to have a multivariate asymmetric Laplace distribution (AL) if its characteristic function is given by*

$$\Psi(\mathbf{t}) = \frac{1}{1 + \frac{1}{2}\mathbf{t}'\boldsymbol{\Sigma}\mathbf{t} - i\mathbf{m}'\mathbf{t}}, \tag{6.2.1}$$

where $\mathbf{m} \in \mathbb{R}^d$ *and* $\boldsymbol{\Sigma}$ *is a* $d \times d$ *nonnegative definite symmetric matrix.*

We use the notation $\mathcal{AL}_d(\mathbf{m}, \boldsymbol{\Sigma})$ to denote the distribution of $\mathbf{Y}$, and write $\mathbf{Y} \sim \mathcal{AL}_d(\mathbf{m}, \boldsymbol{\Sigma})$. If the matrix $\boldsymbol{\Sigma}$ is positive-definite, the distribution is truly d-dimensional and has a probability density

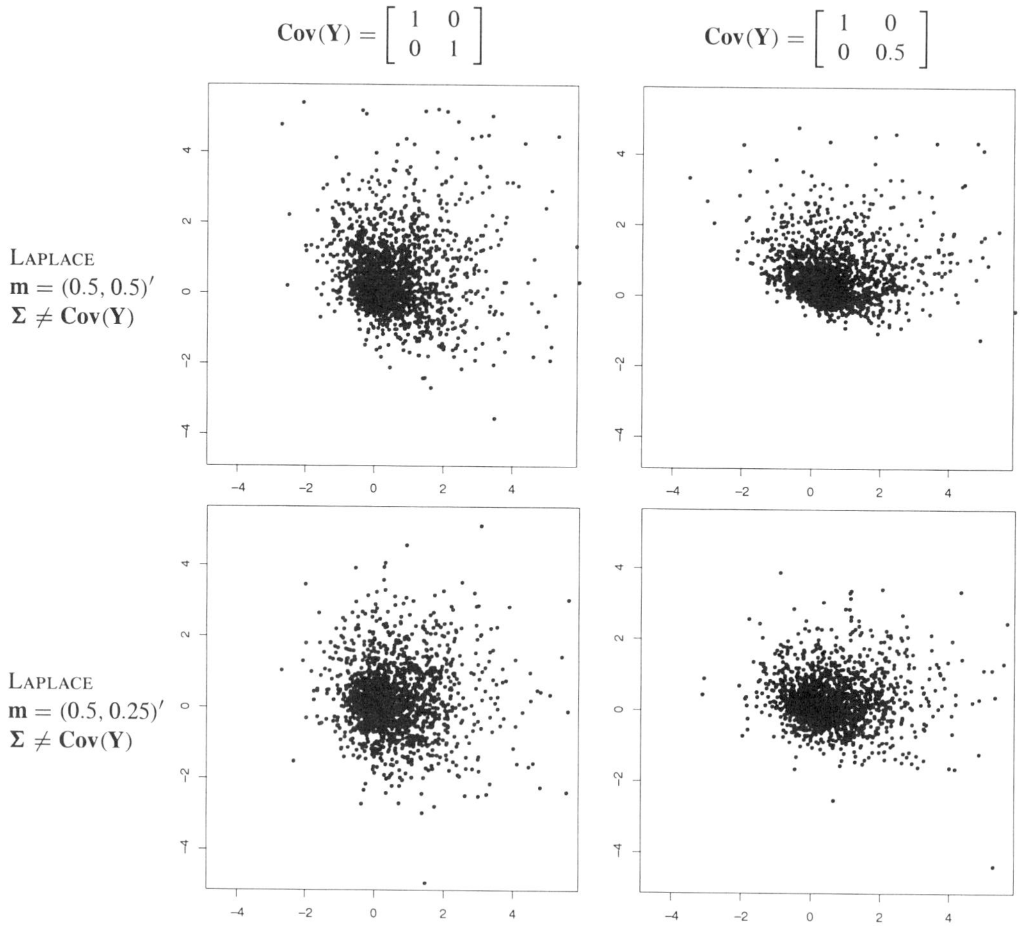

Figure 6.3: Uncorrelated asymmetric Laplace random samples. Monte-Carlo simulation based on the algorithm described in the text. (The sample size equals 2000.)

function. Otherwise, it is degenerate and the probability mass of the distribution is concentrated in a linear proper subspace of the d-dimensional space.

For $\mathbf{m} = \mathbf{0}$ the distribution $\mathcal{AL}_d(\mathbf{0}, \mathbf{\Sigma})$ reduces to the *symmetric multivariate Laplace law* $\mathcal{L}_d(\mathbf{\Sigma})$ discussed in Section 5.2 of Chapter 5 (although more appropriately it should perhaps be called an elliptically contoured Laplace law).

Remark 6.2.1 The parameter $\mathbf{m} = (m_1, \ldots, m_d)'$ appearing in (6.2.1) is not a shift parameter: if $\mathbf{Y} \sim \mathcal{AL}_d(\mathbf{m}, \mathbf{\Sigma})$ it does not follow that $\mathbf{Y} + \mathbf{n} \sim \mathcal{AL}_d(\mathbf{m} + \mathbf{n}, \mathbf{\Sigma})$. In fact, the distribution of $\mathbf{Y} + \mathbf{n}$ is not even AL (unless $\mathbf{n} = \mathbf{0}$). However, the mean of $\mathbf{Y}$ exists and equals $\mathbf{m}$.

Remark 6.2.2 The class of AL laws is not closed under summation of independent r.v.'s: if $\mathbf{X}$ and $\mathbf{Y}$ are independent AL r.v.'s, then in general $\mathbf{X} + \mathbf{Y}$ does not possess an AL law.

6.2.2 Special cases. In the following remarks we discuss some special cases of AL laws.

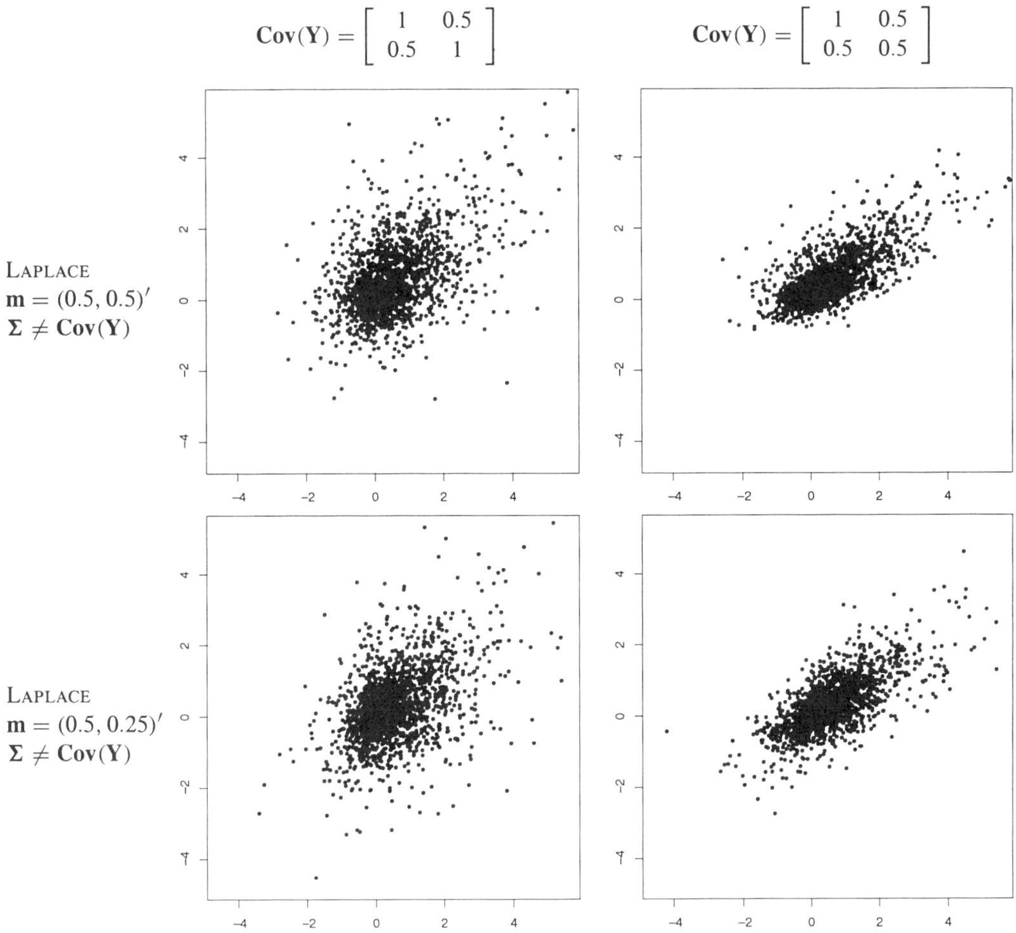

Figure 6.4: Correlated asymmetric Laplace random samples. Monte-Carlo simulation based on the algorithm described in the text. (The sample size equals 2000.)

Remark 6.2.3 For $d = 1$ we obtain a univariate $\mathcal{AL}(\mu, \sigma)$ distribution with mean μ and variance $\sigma^2 + \mu^2$.

Remark 6.2.4 For $d = 2$ the distribution $\mathcal{AL}_2(\mathbf{m}, \boldsymbol{\Sigma})$ with $\mathbf{m} = (m_1, m_2)'$ and $\boldsymbol{\Sigma}$ given by (5.1.1) reduces to $\mathcal{BAL}(m_1, m_2, \sigma_1, \sigma_2, \rho)$ distribution (and to the $\mathcal{BSL}(\sigma_1, \sigma_2, \rho)$ distribution for $\mathbf{m} = \mathbf{0}$).

Remark 6.2.5 Here is an example of a degenerate AL law in $\mathbb{R}^d$. If Y has a univariate $\mathcal{AL}(1, 1)$ law and $\mathbf{m} \in \mathbb{R}^d$, then the r.v. $\mathbf{Y} = \mathbf{m}Y$ has the ch.f.

$$\Psi_{\mathbf{Y}}(\mathbf{t}) = Ee^{i\mathbf{t}'\mathbf{Y}} = \psi_Y(\mathbf{t}'\mathbf{m}) = \frac{1}{1 + \frac{1}{2}\mathbf{t}'(\mathbf{m}\mathbf{m}')\mathbf{t} - i\mathbf{m}'\mathbf{t}}.$$

Thus $\mathbf{Y} \sim \mathcal{AL}_d(\mathbf{m}, \boldsymbol{\Sigma})$ with $\boldsymbol{\Sigma} = \mathbf{m}\mathbf{m}'$.

Remark 6.2.6 Consider an r.v. $\mathbf{Y} \sim \mathcal{AL}_d(\mathbf{m}, \mathbf{0})$, with the ch.f.

$$\Psi_{\mathbf{Y}}(\mathbf{t}) = \frac{1}{1 - i\mathbf{m}'\mathbf{t}}. \tag{6.2.2}$$

Then $\mathbf{Y}$ admits the representation $\mathbf{Y} \stackrel{d}{=} \mathbf{m}Z$, where Z is the standard exponential variable. Indeed, we have

$$\Psi_{\mathbf{Y}}(\mathbf{t}) = Ee^{i\mathbf{t}'\mathbf{Y}} = \psi_Z(\mathbf{t}'\mathbf{m}) = \frac{1}{1 - i\mathbf{m}'\mathbf{t}}.$$

This distribution is related to the Marshall–Olkin exponential distribution of the r.v.

$$\mathbf{W} = (W_1, \ldots, W_d)',$$

given by its survival function

$$P(W_1 > x_1, \ldots, W_d > x_d) = e^{-\max(x_1, \ldots, x_d)}, \quad x_i \geq 0, \quad i = 1, 2, \ldots, d.$$

Since the ch.f. of $\mathbf{W}$ is

$$\Psi_{\mathbf{W}}(\mathbf{t}) = (1 - i(t_1 + \cdots + t_d))^{-1},$$

we have $\mathbf{Y} \stackrel{d}{=} D(\mathbf{m}) \cdot \mathbf{W}$, where $D(\mathbf{m})$ is a diagonal matrix with the elements of the vector $\mathbf{m}$ on its main diagonal.

6.3 Representations

6.3.1 Basic representation. The following result follows directly from the representation of geometric stable laws discussed in Kozubowski and Panorska (1999).

Theorem 6.3.1 *Let* $\mathbf{Y} \sim \mathcal{AL}_d(\mathbf{m}, \mathbf{\Sigma})$ *and let* $\mathbf{X} \sim N_d(\mathbf{0}, \mathbf{\Sigma})$. *Let* W *be an exponentially distributed r.v. with mean* 1, *independent of* $\mathbf{X}$. *Then*

$$\mathbf{Y} \stackrel{d}{=} \mathbf{m}W + W^{1/2}\mathbf{X}. \tag{6.3.1}$$

Remark 6.3.1 More general mixtures of normal distributions, where W has a generalized inverse Gaussian distribution, were considered by Barndorff-Nielsen (1977). A generalized inverse Gaussian distribution with parameters (λ, χ, ψ), denoted $GIG(\lambda, \chi, \psi)$, has the p.d.f.

$$p(x) = \frac{(\psi/\chi)^{\lambda/2}}{2K_\lambda(\sqrt{\chi\psi})} x^{\lambda-1} e^{-\frac{1}{2}(\chi x^{-1} + \psi x)}, x > 0, \tag{6.3.2}$$

where K_λ is the modified Bessel function of the third kind (see the appendix). The range of parameters is

$$\chi \geq 0, \psi > 0, \lambda > 0; \quad \chi > 0, \psi > 0, \lambda = 0; \quad \chi > 0, \psi \geq 0, \lambda < 0.$$

Barndorff-Nielsen (1977) considered mixtures of the form

$$\mathbf{Y} \stackrel{d}{=} \boldsymbol{\mu} + \mathbf{m}W + W^{1/2}\mathbf{X}, \tag{6.3.3}$$

where $\mathbf{X}$ is as before, $\mathbf{m} = \mathbf{\Sigma}\boldsymbol{\beta}$ with some d-dimensional vector $\boldsymbol{\beta}$, and $W \sim GIG(\lambda, \chi, \psi)$. With the notation $\chi = \delta^2$, $\psi = \xi^2$, and $\alpha^2 = \xi^2 + \boldsymbol{\beta}'\mathbf{\Sigma}\boldsymbol{\beta}$, $\mathbf{Y}$ has a d-dimensional generalized hyperbolic distribution with index λ, denoted by $H_d(\lambda, \alpha, \boldsymbol{\beta}, \delta, \boldsymbol{\mu}, \boldsymbol{\Sigma})$ [a hyperbolic distribution is obtained for $\lambda = 1$; see, e.g., Blaesild (1981)]. Taking the *limiting case* $GIG(1, 0, 2)$ as a mixing distribution (which is a standard exponential) and setting $\boldsymbol{\Sigma\beta} = \mathbf{m}$ and $\boldsymbol{\mu} = \mathbf{0}$ so that $\delta^2 = 0$, $\xi^2 = 2$, and $\alpha = \sqrt{2 + \mathbf{m}'\boldsymbol{\Sigma}^{-1}\mathbf{m}}$, we obtain the mixture $W\mathbf{m} + W^{1/2}\mathbf{X}$, where $\mathbf{X}$ is $N_d(\mathbf{0}, \boldsymbol{\Sigma})$, independent of W, which has a multivariate AL distribution.

Remark 6.3.2 By Theorem 6.3.1, each component Y_i of an AL r.v. $\mathbf{Y}$ admits the representation

$$Y_i \stackrel{d}{=} m_i W + W^{1/2}\sigma_{ii} X_i, \tag{6.3.4}$$

where X_i is standard normal variable. This is the representation 3.2.1 obtained previously for univariate AL laws.

6.3.2 Polar representation. Note that AL laws with $\mathbf{m} = \mathbf{0}$ are elliptically contoured (EC), as their ch.f. depends on $\mathbf{t}$ only through the quadratic form $\mathbf{t}'\boldsymbol{\Sigma}\mathbf{t}$. The class of *elliptically symmetric* distributions consists of EC laws with nonsingular $\boldsymbol{\Sigma}$ and density

$$f(\mathbf{x}) = k_d|\boldsymbol{\Sigma}|^{-1/2} g[(\mathbf{x} - \mathbf{m})'\boldsymbol{\Sigma}^{-1}(\mathbf{x} - \mathbf{m})], \tag{6.3.5}$$

where g is a one-dimensional real-valued function (independent of d) and k_d is a proportionality constant [see, e.g., Fang et al. (1990)]. We shall denote the laws with the density (6.3.5) by $EC_d(\mathbf{m}, \boldsymbol{\Sigma}, g)$. It is well known that every r.v. $\mathbf{Y} \sim EC_d(\mathbf{0}, \boldsymbol{\Sigma}, g)$ admits the polar representation

$$\mathbf{Y} \stackrel{d}{=} R\mathbf{H}\mathbf{U}^{(d)}, \tag{6.3.6}$$

where $\mathbf{H}$ is a $d \times d$ matrix such that $\mathbf{HH}' = \boldsymbol{\Sigma}$, R is a positive r.v. independent of $\mathbf{U}^{(d)}$ (having the distribution of $\sqrt{\mathbf{Y}'\boldsymbol{\Sigma}^{-1}\mathbf{Y}}$), and $\mathbf{U}^{(d)}$ is a r.v. uniformly distributed on the sphere $\mathbb{S}_d$. Thus $\mathbf{HU}^{(d)}$ is uniformly distributed on the surface of the hyperellipsoid

$$\{\mathbf{y} \in \mathbb{R}^d : \mathbf{y}'\boldsymbol{\Sigma}^{-1}\mathbf{y} = 1\}.$$

Our next basic result identifies the distribution of R in the class of AL distributed variables $\mathbf{Y}$ [see Kotz et al. (2000b)].

Proposition 6.3.1 *Let* $\mathbf{Y} \sim \mathcal{AL}_d(\mathbf{0}, \boldsymbol{\Sigma})$, *where* $|\boldsymbol{\Sigma}| > 0$. *Then* $\mathbf{Y}$ *admits the polar representation* (6.3.6), *where* $\mathbf{H}$ *is a* $d \times d$ *matrix such that* $\mathbf{HH}' = \boldsymbol{\Sigma}$, $\mathbf{U}^{(d)}$ *is an r.v. uniformly distributed on the sphere* $\mathbb{S}_d$, *and* R *is a positive r.v. independent of* $\mathbf{U}^{(d)}$ *with density*

$$f_R(x) = \frac{2x^{d/2}K_{d/2-1}(\sqrt{2}x)}{(\sqrt{2})^{d/2-1}\Gamma(d/2)}, \quad x > 0, \tag{6.3.7}$$

where K_v *is the modified Bessel function of the third kind defined by* (A.0.4) *in the appendix.*

Proof. By Theorem 6.3.1, $\mathbf{Y}$ has the representation (6.3.1) with $\mathbf{m} = \mathbf{0}$. Write $\boldsymbol{\Sigma} = \mathbf{HH}'$, where $\mathbf{H}$ is a $d \times d$ nonsingular lower triangular matrix [see, e.g., Devroye (1986), p. 566, for a recipe for such a matrix for a given nonsingular $\boldsymbol{\Sigma}$]. Then the r.v. $\mathbf{X} \sim N_d(\mathbf{0}, \boldsymbol{\Sigma})$ in (6.3.1) has the representation $\mathbf{X} = \mathbf{HN}$, where $\mathbf{N} \sim N_d(\mathbf{0}, \mathbf{I}_d)$. Further, the r.v. $\mathbf{N}$, which is EC, has the well-known representation $\mathbf{N} \stackrel{d}{=} R_{\mathbf{N}}\mathbf{U}^{(d)}$, where $R_{\mathbf{N}}$ and $\mathbf{U}^{(d)}$ are independent, $\mathbf{U}^{(d)}$ is uniformly distributed on $\mathbb{S}_d$, while $R_{\mathbf{N}}$ is positive with density

$$f_{R_{\mathbf{N}}}(x) = \frac{d \cdot x^{d-1}\exp(-x^2/2)}{2^{d/2}\Gamma(d/2+1)}, \quad x > 0. \tag{6.3.8}$$

(It is distributed as the square root of a chi-square r.v. with d degrees of freedom.) Therefore, it is sufficient to show that $W^{1/2}R_{\mathbf{N}}$ has density (6.3.7). To this end, apply standard transformation theorem to write the density of $W^{1/2}R_{\mathbf{N}}$ as

$$f_{W^{1/2}R_{\mathbf{N}}}(y) = d \cdot y \int_0^\infty \frac{x^{d/2-2}\exp(-\frac{1}{2}(x^2+2y^2/x))}{2^{d/2}\Gamma(d/2+1)}dx. \tag{6.3.9}$$

Let $f_{\lambda,\chi,\psi}$ be the GIG density (6.3.2) with $\psi = 1$, $\chi = 2y^2$, and $\lambda = d/2 - 1$. Then relation (6.3.9) becomes

$$f_{W^{1/2}R_{\mathbf{N}}}(y) = \frac{d \cdot yK_\lambda(\sqrt{2}y)}{2^{d/2}\Gamma(d/2+1)(\chi)^{-\lambda/2}} \int_0^\infty f_{\lambda,\chi,\psi}(x)dx, \tag{6.3.10}$$

which yields (6.3.7) since the function $f_{\lambda,\chi,\psi}$ integrates to one. □

Remark 6.3.3 In case $d = 1$, where the AL law has ch.f. $\psi(t) = (1+\sigma_{11}t^2/2)^{-1}$, the r.v. $U^{(1)}$ takes on values ± 1 with probabilities $1/2$, while the Bessel function simplifies to

$$K_{1/2}(\sqrt{2}y) = \sqrt{\pi/2}\exp(-\sqrt{2}y)/(\sqrt{2}y)^{1/2}$$

(see formula (A.0.11) in the appendix). Thus $R \stackrel{d}{=} (1/\sqrt{2})W$, where W is a standard exponential variable. Consequently, the right-hand side of (6.3.6) becomes $\sqrt{\sigma_{11}/2} \cdot WU^{(1)}$, and we obtain the representation of symmetric Laplace r.v.'s already discussed in Section 2.2 of Chapter 2.

6.3.3 Subordinated Brownian motion. All AL r.v.'s can be interpreted as values of a subordinated Gaussian process. More precisely, if $\mathbf{Y} \sim \mathcal{AL}_d(\mathbf{m}, \mathbf{\Sigma})$, then

$$\mathbf{Y} \stackrel{d}{=} \mathbf{X}(W),$$

where $\mathbf{X}$ is a d-dimensional Gaussian process with independent increments, $\mathbf{X}(0) = \mathbf{0}$, and $\mathbf{X}(1) \sim N_d(\mathbf{m}, \mathbf{\Sigma})$. This follows immediately from evaluating the characteristic function on the right-hand side through conditioning on the exponential random variable W. Consequently, AL distributions may be studied via the theory of (stopped) Lévy processes [see Bertoin (1996)].

6.4 Simulation algorithm

The problem of random number generation for symmetric Laplace laws was posed in Devroye (1986) and reiterated in Johnson (1987): "Variate generation has not been explicitly worked out for (the bivariate Laplace and generalized Laplace distributions) in the literature." However, simulation of generalized hyperbolic random variables was studied earlier by Atkinson (1982). The algorithms were based on the normal mixture representations of the distributions under consideration. In this sense, in principle the problem of simulations for multivariate AL distributions was resolved. However, the solution cannot be considered to be an explicit one, since the fact that AL distributions can be obtained as the limiting case of hyperbolic distributions is not commonly known.

To state simulation algorithm for the general multivariate AL distributions, we use representation (6.3.1). The approach is quite straightforward [see Kozubowski and Podgórski (1999b)], as both exponential and multivariate normal variates are relatively easy to generate and appropriate procedures are by now implemented in all standard statistical packages.

$\mathcal{AL}_d(\mathbf{m}, \mathbf{\Sigma})$ *generator*

- Generate a standard exponential variate W.
- Independently of W, generate multivariate normal $N_d(\mathbf{0}, \mathbf{\Sigma})$ variate $\mathbf{N}$.
- Set $\mathbf{Y} \leftarrow \mathbf{m} \cdot W + \sqrt{W} \cdot \mathbf{N}$.
- RETURN $\mathbf{Y}$.

This algorithm and pseudorandom samples of normal and exponential random variables obtained from the S-Plus package were used to produce the graphs of bivariate Laplace distributions in Figures 5.3, 5.4, 6.3, and 6.4.

6.5 Moments and densities

6.5.1 Mean vector and covariance matrix. The relation between the mean vector $E\mathbf{Y}$, the covariance matrix $\mathbf{Cov}(\mathbf{Y})$ and the parameters $\mathbf{m}$ and $\mathbf{\Sigma}$ can easily be obtained from the representation (6.3.4). We have $EY_i = m_i$, so

$$E(\mathbf{Y}) = \mathbf{m}.$$

Furthermore, the variance-covariance matrix of $\mathbf{Y}$ is

$$\mathbf{Cov}(\mathbf{Y}) = \mathbf{\Sigma} + \mathbf{m}\mathbf{m}'.$$

Indeed, since $E(X_i X_j) = \sigma_{ij}$ and $EW^2 = 2$, we have

$$\begin{aligned} E(Y_i Y_j) &= E[(m_i W + W^{1/2} X_i)(m_j + W^{1/2} X_j)] = m_i m_j E W^2 + E(W) E(X_i X_j) \\ &= 2 m_i m_j + \sigma_{ij}. \end{aligned}$$

Thus

$$\mathrm{Cov}(Y_i, Y_j) = E(Y_i Y_j) - E(Y_i)E(Y_j) = 2m_i m_j + \sigma_{ij} - m_i m_j = m_i m_j + \sigma_{ij}.$$

6.5.2 Densities in the general case. In this section we study AL densities (assuming that the distribution is nonsingular). The representation given in Theorem 6.3.1, coupled with conditioning on the exponential variable W, produces a relation between the distribution functions and the densities of AL and multivariate normal random vectors. Let $G(\cdot)$ and $F(\cdot)$ be the c.d.f.'s of $\mathcal{AL}_d(\mathbf{m}, \mathbf{\Sigma})$ and $N_d(\mathbf{0}, \mathbf{\Sigma})$ r.v.'s, respectively, and let $g(\cdot)$ and $f(\cdot)$ be the corresponding densities.

Corollary 6.5.1 *Let* $\mathbf{Y} \sim \mathcal{AL}_d(\mathbf{m}, \mathbf{\Sigma})$. *The distribution function and the density (if it exists) of* $\mathbf{Y}$ *can be expressed as follows:*

$$\begin{aligned} G(\mathbf{y}) &= \int_0^\infty F(z^{-1/2}\mathbf{y} - z^{1/2}\mathbf{m}) e^{-z} dz \\ g(\mathbf{y}) &= \int_0^\infty f(z^{-1/2}\mathbf{y} - z^{1/2}\mathbf{m}) z^{-d/2} e^{-z} dz. \end{aligned} \tag{6.5.1}$$

We can express an AL density in terms of the modified Bessel function of the third kind (see the definition in the appendix). By (6.5.1), the density of $\mathbf{Y} \sim \mathcal{AL}_d(\mathbf{m}, \mathbf{\Sigma})$ becomes

$$g(\mathbf{y}) = (2\pi)^{-d/2} |\mathbf{\Sigma}|^{-1/2} \int_0^\infty \exp\left(-\frac{(\mathbf{y} - z\mathbf{m})'\mathbf{\Sigma}^{-1}(\mathbf{y} - z\mathbf{m})}{2z} - z \right) z^{-d/2} dz. \tag{6.5.2}$$

For $\mathbf{y} = \mathbf{0}$, we arrive at

$$g(\mathbf{0}) = (2\pi)^{-d/2}|\boldsymbol{\Sigma}|^{-1/2} \int_0^\infty \exp\left(-z(\frac{1}{2}\mathbf{m}'\boldsymbol{\Sigma}^{-1}\mathbf{m} + 1)\right) z^{-d/2} dz,$$

so the density blows up at zero unless $d = 1$. For $\mathbf{y} \neq \mathbf{0}$, we can simplify the exponential part of the integrand and substitute $w = z(1 + \mathbf{m}'\boldsymbol{\Sigma}^{-1}\mathbf{m}/2)$ in (6.5.2) to obtain

$$g(\mathbf{y}) = \frac{e^{\mathbf{y}'\boldsymbol{\Sigma}^{-1}\mathbf{m}}(1 + \frac{1}{2}\mathbf{m}'\boldsymbol{\Sigma}^{-1}\mathbf{m})^{d/2-1}}{(2\pi)^{d/2}|\boldsymbol{\Sigma}|^{1/2}} \int_0^\infty \exp\left(-\frac{a^2}{4z} - z\right) z^{-(d-2)/2-1} dz,$$

where $a = \sqrt{(2 + \mathbf{m}'\boldsymbol{\Sigma}^{-1}\mathbf{m})(\mathbf{y}'\boldsymbol{\Sigma}^{-1}\mathbf{y})}$. Taking into account the integral representation (A.0.4) of the corresponding Bessel functions (see the appendix), we finally obtain the following basic result.

Theorem 6.5.1 *The density of* $\mathbf{Y} \sim \mathcal{AL}_d(\mathbf{m}, \boldsymbol{\Sigma})$ *can be expressed as*

$$g(\mathbf{y}) = \frac{2e^{\mathbf{y}'\boldsymbol{\Sigma}^{-1}\mathbf{m}}}{(2\pi)^{d/2}|\boldsymbol{\Sigma}|^{1/2}} \left(\frac{\mathbf{y}'\boldsymbol{\Sigma}^{-1}\mathbf{y}}{2 + \mathbf{m}'\boldsymbol{\Sigma}^{-1}\mathbf{m}}\right)^{v/2} K_v\left(\sqrt{(2 + \mathbf{m}'\boldsymbol{\Sigma}^{-1}\mathbf{m})(\mathbf{y}'\boldsymbol{\Sigma}^{-1}\mathbf{y})}\right), \quad (6.5.3)$$

where $v = (2-d)/2$ *and* $K_v(u)$ *is the modified Bessel function of the third kind given by* (A.0.4) *or* (A.0.5) *in the appendix.*

Remark 6.5.1 This density is a limiting case of a generalized hyperbolic density

$$\frac{\xi^\lambda \exp(\boldsymbol{\beta}'(\mathbf{x}-\boldsymbol{\mu}))K_{d/2-\lambda}(\alpha\sqrt{\delta^2 + (\mathbf{x}-\boldsymbol{\mu})'\boldsymbol{\Sigma}^{-1}(\mathbf{x}-\boldsymbol{\mu})})}{(2\pi)^{d/2}|\boldsymbol{\Sigma}|^{1/2}\delta^\lambda K_\lambda(\delta\xi)[\sqrt{\delta^2 + (\mathbf{x}-\boldsymbol{\mu})'\boldsymbol{\Sigma}^{-1}(\mathbf{x}-\boldsymbol{\mu})}/\alpha]^{d/2-\lambda}} \quad (6.5.4)$$

with $\lambda = 1$, $\xi^2 = 2$, $\delta^2 = 0$, $\boldsymbol{\mu} = \mathbf{0}$, $\boldsymbol{\beta} = \boldsymbol{\Sigma}^{-1}\mathbf{m}$, and $\alpha = \sqrt{2 + \mathbf{m}'\boldsymbol{\Sigma}^{-1}\mathbf{m}}$ (see the remarks following Theorem 6.3.1). Note that in case $\delta = 0$ we use the asymptotic relation (A.0.12) given in the appendix.

6.5.3 Densities in the symmetric case. In the symmetric case ($\mathbf{m} = \mathbf{0}$), we obtain the density (5.2.2) of the $\mathcal{SL}_d(\boldsymbol{\Sigma})$ distribution

$$g(\mathbf{y}) = 2(2\pi)^{-d/2}|\boldsymbol{\Sigma}|^{-1/2}\left(\mathbf{y}'\boldsymbol{\Sigma}^{-1}\mathbf{y}/2\right)^{v/2} K_v\left(\sqrt{2\mathbf{y}'\boldsymbol{\Sigma}^{-1}\mathbf{y}}\right).$$

6.5.4 Densities in the one-dimensional case. If $d = 1$, we have $\boldsymbol{\Sigma} = \sigma_{11} = \sigma$ and the ch.f. corresponds to a univariate $\mathcal{AL}(\mu, \sigma)$ distribution with $\sigma^2 = \boldsymbol{\Sigma}$ and $\mu = \mathbf{m}$. In this case we have $v = 1/2$, and the Bessel function is simplified as in (A.0.11). Consequently, the density becomes

$$g(y) = \frac{1}{\gamma} e^{-\frac{|y|}{\sigma^2}(\gamma - \mu\cdot\mathrm{sign}(y))},$$

where $\gamma = \sqrt{\mu^2 + 2\sigma^2}$, and coincides with the density of a univariate AL distribution given by (3.1.10) with $\theta = 0$. In the symmetric case ($\mu = 0$), we obtain the density of a univariate Laplace distribution with mean zero and variance σ^2.

6.5.5 Densities in the case of odd dimension. If d is odd, the density can be written in a closed form. Indeed, suppose $d = 2r+3$, where $r = 0, 1, 2, \ldots$, so $v = (2-d)/2 = -r-1/2$. Since $K_v(u) = K_{-v}(u)$ and the Bessel function K_v with $v = r+1/2$ has an explicit form (A.0.10) given in the appendix, the AL density (6.5.3) becomes

$$g(\mathbf{y}) = \frac{C^r e^{\mathbf{y}'\boldsymbol{\Sigma}^{-1}\mathbf{m} - C\sqrt{\mathbf{y}'\boldsymbol{\Sigma}^{-1}\mathbf{y}}}}{(2\pi\sqrt{\mathbf{y}'\boldsymbol{\Sigma}^{-1}\mathbf{y}})^{r+1}|\boldsymbol{\Sigma}|^{1/2}} \sum_{k=0}^{r} \frac{(r+k)!}{(r-k)!k!}(2C\sqrt{\mathbf{y}'\boldsymbol{\Sigma}^{-1}\mathbf{y}})^{-k}, \quad \mathbf{y} \neq \mathbf{0},$$

where $v = (2-d)/2$ and $C = \sqrt{2 + \mathbf{m}'\boldsymbol{\Sigma}^{-1}\mathbf{m}}$.

The density has a particularly simple form in three-dimensional space ($d = 3$), where we have $r = 0$ and

$$g(\mathbf{y}) = \frac{e^{\mathbf{y}'\boldsymbol{\Sigma}^{-1}\mathbf{m} - C\sqrt{\mathbf{y}'\boldsymbol{\Sigma}^{-1}\mathbf{y}}}}{2\pi\sqrt{\mathbf{y}'\boldsymbol{\Sigma}^{-1}\mathbf{y}}|\boldsymbol{\Sigma}|^{1/2}}, \quad \mathbf{y} \neq \mathbf{0}.$$

6.6 Unimodality

6.6.1 Unimodality. We already know that all univariate AL distributions are unimodal with the mode at zero. There are many nonequivalent notions of unimodality for probability distributions in $\mathbb{R}^d$ [see, e.g., Dharmadhikari and Joag-Dev (1988)]. A natural extension of univariate unimodality is *star unimodality* in $\mathbb{R}^d$, which for a distribution with continuous density f requires that f be nonincreasing along rays emanating from zero. Here is an exact criterion for star unimodality due to Dharmadhikari and Joag-Dev (1988).

Criterion 1 *A distribution P with continuous density f on $\mathbb{R}^d$ is star unimodal about zero if and only if whenever*

$$0 < t < u < \infty \quad \textit{and} \quad \mathbf{x} \neq \mathbf{0},$$

then

$$f(u\mathbf{x}) \leq f(t\mathbf{x}).$$

It is clear from its statement that the criterion remains valid for densities discontinuous at zero as well. We show that all truly d-dimensional AL laws are star unimodal about zero.

Proposition 6.6.1 *Let $\mathbf{Y} \sim \mathcal{AL}_d(\mathbf{m}, \boldsymbol{\Sigma})$ with $|\boldsymbol{\Sigma}| > 0$. Then the distribution of $\mathbf{Y}$ is star unimodal about $\mathbf{0}$.*

Proof. Assume that $d > 1$ and let $\mathbf{x} \neq \mathbf{0}$. For $t > 0$ define $h(t) = \log g(t\mathbf{x})$, where g is the density of $\mathbf{Y}$ given by (6.5.3). Write

$$h(t) = \log C_1 + C_2 t + v \log t + \log K_v(C_3 t),$$

where $v = 1 - d/2$ and the constants C_1, C_2, and C_3 are given by

$$C_1 = \frac{2(\mathbf{x}'\boldsymbol{\Sigma}^{-1}\mathbf{x})^{v/2}}{(2\pi)^{d/2}|\boldsymbol{\Sigma}|^{1/2}(2 + \mathbf{m}'\boldsymbol{\Sigma}^{-1}\mathbf{m})^{v/2}} > 0,$$

$$C_2 = \mathbf{m}'\boldsymbol{\Sigma}^{-1}\mathbf{x} \in \mathbb{R},$$

$$C_3 = \sqrt{2 + \mathbf{m}'\boldsymbol{\Sigma}^{-1}\mathbf{m}}\sqrt{\mathbf{x}'\boldsymbol{\Sigma}^{-1}\mathbf{x}} > 0.$$

Showing that h is an nonincreasing function of t is required. The derivative of h with respect to t is

$$\frac{d}{dt}h(t) = C_2 + \frac{v}{t} + \frac{K_v'(C_3t)}{K_v(C_3t)}C_3. \tag{6.6.1}$$

Use properties (A.0.8)–(A.0.9) of Bessel function K_v (listed in the appendix) to write (6.6.1) as

$$\frac{d}{dt}h(t) = C_2 - \frac{K_{v-1}(C_3t)}{K_v(C_3t)}C_3. \tag{6.6.2}$$

If $C_2 < 0$, then (6.6.2) implies that $h'(t) \le 0$ since the Bessel function K_v is always positive and $C_3 > 0$. Otherwise, write $\mathbf{\Sigma}^{-1} = \mathbf{Q}'\mathbf{Q}$ and use the Cauchy–Schwarz inequality to conclude that

$$|C_2| = |(\mathbf{Qm})'(\mathbf{Qx})| \le ||\mathbf{Qm}|| \cdot ||\mathbf{Qx}|| = \sqrt{\mathbf{m}'\mathbf{\Sigma}^{-1}\mathbf{m}}\sqrt{\mathbf{x}'\mathbf{\Sigma}^{-1}\mathbf{x}} < C_3.$$

Thus the conclusion $h'(t) \le 0$ follows if we show that the ratio $\frac{K_{v-1}(C_3t)}{K_v(C_3t)}$ is greater or equal to one. Since, for any v, $K_v(x) = K_{-v}(x)$, this is equivalent to showing that

$$K_{-v}(C_3t) \le K_{-v+1}(C_3t).$$

This is indeed true since $-v \ge 0$ (as $d > 1$) and by using Property 3 of Bessel functions listed in the appendix, we obtain the desired inequality. □

Remark 6.6.1 Any AL r.v. $\mathbf{Y}$ is *linear unimodal* about $\mathbf{0}$ in the sense that every linear combination $\mathbf{c}'\mathbf{Y}$ is univariate unimodal about zero [see Definition 2.3 of Dharmadhikari and Joag-Dev (1988)]. This follows from part (iii) of Corollary 6.8.1 since all univariate AL laws are unimodal about zero.

6.6.2 A related representation. A univariate r.v. Y is unimodal about zero if and only if it has the representation $Y \stackrel{d}{=} UX$, where U and X are independent and U is uniformly distributed on (0, 1) [see, e.g., Shepp (1962)]. Similarly, every star unimodal (about $\mathbf{0}$) r.v. in $\mathbb{R}^d$ has the representation $\mathbf{Y} \stackrel{d}{=} U^{1/d}\mathbf{X}$, where U is as before and is independent from $\mathbf{X}$ [see Dharmadhikari and Joag-Dev (1988), Theorem 2.1]. Below we identify the distribution of $\mathbf{X}$ in case of a symmetric AL r.v. $\mathbf{Y}$. Let $\mathbf{Y} \sim \mathcal{AL}_d(\mathbf{0}, \mathbf{\Sigma})$ with $|\mathbf{\Sigma}| > 0$. From the proof of Proposition 6.3.1 we have the representation $\mathbf{Y} \stackrel{d}{=} W^{1/2}R_{\mathbf{N}}\mathbf{H}\mathbf{U}^{(d)}$, where $\mathbf{H}$ is a matrix satisfying $\mathbf{\Sigma} = \mathbf{HH}'$, $\mathbf{U}^{(d)}$ is uniform on the unit sphere $\mathbb{S}_d$, W is standard exponential, $R_{\mathbf{N}}$ has the density (6.3.8), and all variables are independent. Note that $R_{\mathbf{N}} \stackrel{d}{=} V^{1/d}$, where V has density

$$f_V(x) = \frac{\exp(-x^{2/d}/2)}{2^{d/2}\Gamma(d/2+1)}, \quad x > 0.$$

The density of V is unimodal; hence by Shepp (1962) it has the representation $V \stackrel{d}{=} US$ for some S (where U is standard uniform and independent of S). It can be shown by routine calculations that the density of S is

$$f_S(x) = \frac{x^{2/d}\exp(-x^{2/d}/2)}{d2^{d/2}\Gamma(d/2+1)}, \quad x > 0.$$

Thus we have $\mathbf{Y} \stackrel{d}{=} U^{1/d}(W^{1/2}S^{1/d}\mathbf{H}\mathbf{U}^{(d)})$. The density of $W^{1/2}S^{1/d}$ is readily obtained also as

$$f_{W^{1/2}S^{1/d}}(x) = \frac{2x^{d/2+1}K_{d/2}(\sqrt{2}x)}{\sqrt{2}^{d/2}\Gamma(d/2+1)}, \quad x > 0. \tag{6.6.3}$$

The following statement summarizes this discussion.

Theorem 6.6.1 *Let* $\mathbf{Y} \sim \mathcal{AL}_d(\mathbf{0}, \boldsymbol{\Sigma})$, *where* $|\boldsymbol{\Sigma}| > 0$ *and* $\boldsymbol{\Sigma} = \mathbf{HH}'$. *Then* $\mathbf{Y}$ *admits the representation*

$$\mathbf{Y} \stackrel{d}{=} U^{1/d}\mathbf{X},$$

where U *and* $\mathbf{X}$ *are independent,* U *is uniform on* $(0, 1)$ *while* $\mathbf{X}$ *is elliptically symmetric with the representation* $\mathbf{X} \stackrel{d}{=} R_{\mathbf{X}}\mathbf{HU}^{(d)}$, *where* $\mathbf{U}^{(d)}$ *is uniform on* $\mathbb{S}_d$ *while* $R_{\mathbf{X}}$ *has density* (6.6.3).

6.7 Conditional distributions

6.7.1 Conditional distributions. We obtain conditional distributions of $\mathbf{Y} \sim \mathcal{AL}_d(\mathbf{m}, \boldsymbol{\Sigma})$ with a nonsingular $\boldsymbol{\Sigma}$. The derivation is similar to that for the case of multivariate generalized hyperbolic distribution [see Blaesild (1981)]. It turns out that the conditional laws are not AL, but generalized hyperbolic ones. However, the conditional distributions can be AL if $\mathbf{Y}$ has multivariate K-Bessel function distribution (6.9.1), discussed in Section 6.9. The conditional distributions of a multivariate AL laws are given in the following result [Kotz et al. (2000b)].

Theorem 6.7.1 *Let* $\mathbf{Y} \sim \mathcal{GAL}_d(\mathbf{m}, \boldsymbol{\Sigma}, s)$ *have ch.f.* (6.9.1) (*see Section* 6.9) *with nonsingular* $\boldsymbol{\Sigma}$. *Let* $\mathbf{Y}' = (\mathbf{Y}_1', \mathbf{Y}_2')$ *be a partition of* $\mathbf{Y}$ *into* $r \times 1$ *and* $k \times 1$*-dimensional subvectors, respectively. Let* $(\mathbf{m}_1', \mathbf{m}_2')$ *and*

$$\boldsymbol{\Sigma} = \begin{pmatrix} \boldsymbol{\Sigma}_{11} & \boldsymbol{\Sigma}_{12} \\ \boldsymbol{\Sigma}_{21} & \boldsymbol{\Sigma}_{22} \end{pmatrix}$$

be the corresponding partitions of $\mathbf{m}$ *and* $\boldsymbol{\Sigma}$, *where* $\boldsymbol{\Sigma}_{11}$ *is an* $r \times r$ *matrix.*

(i) *If* $s = 1$ (*so that* $\mathbf{Y}$ *is AL*), *then the conditional distribution of* $\mathbf{Y}_2$ *given* $\mathbf{Y}_1 = \mathbf{y}_1$ *is the generalized* k*-dimensional hyperbolic distribution* $H_k(\lambda, \alpha, \boldsymbol{\beta}, \delta, \boldsymbol{\mu}, \boldsymbol{\Delta})$ *having density*

$$p(\mathbf{y}_2|\mathbf{y}_1) = \frac{\xi^\lambda \exp(\boldsymbol{\beta}'(\mathbf{y}_2 - \boldsymbol{\mu}))K_{k/2-\lambda}\left(\alpha\sqrt{\delta^2 + (\mathbf{y}_2 - \boldsymbol{\mu})'\boldsymbol{\Delta}^{-1}(\mathbf{y}_2 - \boldsymbol{\mu})}\right)}{(2\pi)^{k/2}|\boldsymbol{\Delta}|^{1/2}\delta^\lambda K_\lambda(\delta\xi)\left[\sqrt{\delta^2 + (\mathbf{y}_2 - \boldsymbol{\mu})'\boldsymbol{\Delta}^{-1}(\mathbf{y}_2 - \boldsymbol{\mu})}/\alpha\right]^{k/2-\lambda}}, \tag{6.7.1}$$

where $\lambda = 1 - r/2$, $\alpha = \sqrt{\xi^2 + \boldsymbol{\beta}'\boldsymbol{\Delta}\boldsymbol{\beta}}$, $\boldsymbol{\beta} = \boldsymbol{\Delta}^{-1}(\mathbf{m}_2 - \boldsymbol{\Sigma}_{21}\boldsymbol{\Sigma}_{11}^{-1}\mathbf{m}_1)$, $\delta = \sqrt{\mathbf{y}_1'\boldsymbol{\Sigma}_{11}^{-1}\mathbf{y}_1}$, $\boldsymbol{\mu} - \boldsymbol{\Sigma}_{21}\boldsymbol{\Sigma}_{11}^{-1}\mathbf{y}_1$, $\boldsymbol{\Delta} = \boldsymbol{\Sigma}_{22} - \boldsymbol{\Sigma}_{21}\boldsymbol{\Sigma}_{11}^{-1}\boldsymbol{\Sigma}_{12}$, *and* $\xi = \sqrt{2 + \mathbf{m}_1'\boldsymbol{\Sigma}_{11}^{-1}\mathbf{m}_1}$.

(ii) *If* $\mathbf{m}_1 = \mathbf{0}$, *then the conditional distribution of* $\mathbf{Y}_2$ *given* $\mathbf{Y}_1 = \mathbf{0}$ *is* $\mathcal{GAL}_k(\mathbf{m}_{2\cdot1}, \boldsymbol{\Sigma}_{2\cdot1}, s_{2\cdot1})$, *where*

$$s_{2\cdot1} = s - r/2, \quad \boldsymbol{\Sigma}_{2\cdot1} = \boldsymbol{\Sigma}_{22} - \boldsymbol{\Sigma}_{21}\boldsymbol{\Sigma}_{11}^{-1}\boldsymbol{\Sigma}_{12}, \quad \mathbf{m}_{2\cdot1} = \mathbf{m}_2.$$

Proof. We shall sketch the proof of part (i); the proof for part (ii) is similar. By part (i) of Corollary 6.8.1 with $n = r$, the r.v. $\mathbf{Y}_1$ is $\mathcal{AL}_r(\mathbf{m}_1, \boldsymbol{\Sigma}_{11})$. Write the densities of $\mathbf{Y}$ and $\mathbf{Y}_1$ according to (6.5.3) and simplify the ratio of the densities utilizing the familiar relations from the classical multivariate analysis:

$$\begin{aligned}
\mathbf{Y}'\boldsymbol{\Sigma}^{-1}\mathbf{m} &= \mathbf{Y}_1'\boldsymbol{\Sigma}_{11}^{-1}\mathbf{m}_1 + (\mathbf{m}_2 - \boldsymbol{\Sigma}_{21}\boldsymbol{\Sigma}_{11}^{-1}\mathbf{m}_1)'\boldsymbol{\Delta}^{-1}(\mathbf{y}_2 - \boldsymbol{\Sigma}_{21}\boldsymbol{\Sigma}_{11}^{-1}\mathbf{y}_1),\\
\mathbf{Y}'\boldsymbol{\Sigma}^{-1}\mathbf{Y} &= \mathbf{Y}_1'\boldsymbol{\Sigma}_{11}^{-1}\mathbf{Y}_1 + (\mathbf{Y}_2 - \boldsymbol{\Sigma}_{21}\boldsymbol{\Sigma}_{11}^{-1}\mathbf{Y}_1)'\boldsymbol{\Delta}^{-1}(\mathbf{y}_2 - \boldsymbol{\Sigma}_{21}\boldsymbol{\Sigma}_{11}^{-1}\mathbf{y}_1),\\
\mathbf{m}'\boldsymbol{\Sigma}^{-1}\mathbf{m} &= \mathbf{m}_1'\boldsymbol{\Sigma}_{11}^{-1}\mathbf{m}_1 + (\mathbf{m}_2 - \boldsymbol{\Sigma}_{21}\boldsymbol{\Sigma}_{11}^{-1}\mathbf{m}_1)'\boldsymbol{\Delta}^{-1}(\mathbf{m}_2 - \boldsymbol{\Sigma}_{21}\boldsymbol{\Sigma}_{11}^{-1}\mathbf{m}_1),\\
|\boldsymbol{\Sigma}| &= |\boldsymbol{\Sigma}_{22} - \boldsymbol{\Sigma}_{21}\boldsymbol{\Sigma}_{11}^{-1}\boldsymbol{\Sigma}_{12}| \cdot |\boldsymbol{\Sigma}_{11}|.
\end{aligned}$$

Finally, verify that $\alpha^2 = \boldsymbol{\beta}'\boldsymbol{\Delta}\boldsymbol{\beta} + \xi^2$. □

Remark 6.7.1 Note that in view of part (i) of the theorem, the parameter λ cannot equal one. Hence, in case of a multivariate AL distribution no conditional law can be AL. However, in part (ii) we might have $s - r/2 = 1$, in which case we do obtain a conditional AL law for a multivariate *generalized* AL distribution.

6.7.2 Conditional mean and covariance matrix. Since the conditional distributions of an AL r.v. are generalized hyperbolic distributions, we can derive expressions for the conditional mean vector and the covariance matrix via the theory of hyperbolic distributions.

Proposition 6.7.1 *Let* $\mathbf{Y}$ *have a GAL law* (6.9.1) *with a nonsingular* $\mathbf{\Sigma}$. *Let* $\mathbf{Y}$, $\mathbf{m}$, *and* $\mathbf{\Sigma}$ *be partitioned as in Theorem* 6.7.1. *Then*

$$E(\mathbf{Y}_2|\mathbf{Y}_1 = \mathbf{y}_1) = \mathbf{\Sigma}_{21}\mathbf{\Sigma}_{11}^{-1}\mathbf{y}_1 + (\mathbf{m}_2 - \mathbf{\Sigma}_{21}\mathbf{\Sigma}_{11}^{-1}\mathbf{m}_1)\frac{Q(\mathbf{y}_1)}{C}R_{1-r/2}(CQ(\mathbf{y}_1))$$

and

$$\begin{aligned}\mathrm{Cov}(\mathbf{Y}_2|\mathbf{Y}_1 = \mathbf{y}_1) &= \frac{Q(\mathbf{y}_1)}{C}(\mathbf{\Sigma}_{22} - \mathbf{\Sigma}_{21}\mathbf{\Sigma}_{11}^{-1}\mathbf{\Sigma}_{12})R_{1-r/2}(CQ(\mathbf{y}_1)) \\ &\quad + (\mathbf{m}_2 - \mathbf{\Sigma}_{21}\mathbf{\Sigma}_{11}^{-1}\mathbf{m}_1)(\mathbf{m}_2 - \mathbf{\Sigma}_{21}\mathbf{\Sigma}_{11}^{-1}\mathbf{m}_1)'\left(\frac{Q(\mathbf{y}_1)}{C}\right)^2 G(\mathbf{y}_1),\end{aligned}$$

where $C = \sqrt{2 + \mathbf{m}_1'\mathbf{\Sigma}_{11}^{-1}\mathbf{m}_1}$, $Q(\mathbf{y}_1) = \sqrt{\mathbf{y}_1'\mathbf{\Sigma}_{11}^{-1}\mathbf{y}_1}$, $R_s(x) = K_{s+1}(x)/K_s(x)$, *and*

$$G(\mathbf{y}_1) = (R_{1-r/2}(CQ(\mathbf{y}_1))R_{2-r/2}(CQ(\mathbf{y}_1)) - R_{1-r/2}^2(CQ(\mathbf{y}_1))).$$

Proof. Our outline of the proof follows Kotz et al. (2000b). Apply Theorem 6.7.1 and utilize the representation (6.3.3) of the generalized hyperbolic distribution to conclude that $E(\mathbf{Y}_2|\mathbf{Y}_1 = \mathbf{y}_1) = \boldsymbol{\mu} + \mathbf{\Delta}\boldsymbol{\beta}E(W)$ and $\mathrm{Cov}(\mathbf{Y}_2|\mathbf{Y}_1 = \mathbf{y}_1) = \mathbf{\Delta}\boldsymbol{\beta}(\mathbf{\Delta}\boldsymbol{\beta})'\mathrm{Var}(W) + \mathbf{\Delta}E(W)$, where W has the $GIG(s, \delta^2, \xi^2)$ distribution (6.3.2) and $\boldsymbol{\mu}$, $\boldsymbol{\beta}$, $\mathbf{\Delta}$, δ, and ξ are as given in Theorem 6.7.1. Then apply the well-known formulas for the moments of W, $E(W^r) = (\delta/\xi)^r K_{s+r}(\delta\xi)/K_s(\delta\xi)$ [see, e.g., Barndorff-Nielsen and Blaesild (1981)]. □

Remark 6.7.2 If $\mathbf{m}_1'\mathbf{\Sigma}_{11}^{-1}\mathbf{\Sigma}_{12} = m_d$, then by Theorem 6.7.1, the conditional distribution of Y_d given $(Y_1, \ldots, Y_{d-1})$ is generalized hyperbolic and symmetric about $\boldsymbol{\mu} = \mathbf{\Sigma}_{21}\mathbf{\Sigma}_{11}^{-1}\mathbf{y}_1$ (since $\beta = 0$ in this case), which must be the mean of the conditional distribution. This provides an alternative way for proving the result on linear regression to be discussed in the next section.

6.8 Linear transformations

6.8.1 Linear combinations. In this section we discuss the distribution of AL vectors under the linear transformations. The next proposition shows that if $\mathbf{Y} \sim \mathcal{AL}_d(\mathbf{m}, \mathbf{\Sigma})$, then all linear combinations of components of $\mathbf{Y}$ are jointly AL.

Proposition 6.8.1 *Let* $\mathbf{Y} = (Y_1, \ldots, Y_d)' \sim \mathcal{AL}_d(\mathbf{m}, \mathbf{\Sigma})$. *Let* $\mathbf{A}$ *be an* $l \times d$ *real matrix. Then the random vector* $\mathbf{AY}$ *is* $\mathcal{AL}_l(\mathbf{m}_\mathbf{A}, \mathbf{\Sigma}_\mathbf{A})$, *where* $\mathbf{m}_\mathbf{A} = \mathbf{Am}$ *and* $\mathbf{\Sigma}_\mathbf{A} = \mathbf{A\Sigma A}'$.

Proof. The assertion follows from the general relation

$$\Psi_{\mathbf{AY}}(\mathbf{t}) = Ee^{i(\mathbf{AY})'\mathbf{t}} = Ee^{i\mathbf{Y}'\mathbf{A}'\mathbf{t}} = \Psi_\mathbf{Y}(\mathbf{A}'\mathbf{t})$$

and the fact that the matrix $\mathbf{A\Sigma A}'$ is nonnegative definite whenever $\mathbf{\Sigma}$ is. □

Remark 6.8.1 Note that the proof is quite general, and applies to any multivariate distribution whose ch.f. depends on $\mathbf{t}$ only through the quadratic form $\mathbf{t}\boldsymbol{\Sigma}\mathbf{t}'$ and linear function $\mathbf{m}'\mathbf{t}$. Thus it applies to all elliptically contoured distributions with ch.f. (II.2) as well as to the ν-stable laws with ch.f.'s of the form $g(1+\mathbf{t}\boldsymbol{\Sigma}\mathbf{t}' - i\mathbf{m}'\mathbf{t})$, where g is a Laplace transform of a positive random variable [see, e.g., Kozubowski and Panorska (1998)].

It follows that all univariate and multivariate marginals, as well as linear combinations of the components of a multivariate AL vector, are AL.

Corollary 6.8.1 *Let* $\mathbf{Y} = (Y_1, \ldots, Y_d)' \sim \mathcal{AL}_d(\mathbf{m}, \boldsymbol{\Sigma})$, *where* $\boldsymbol{\Sigma} = (\sigma_{ij})_{i,j=1}^d$.

(i) *For all* $n \le d$, $(Y_1, \ldots, Y_n) \sim \mathcal{AL}_n(\tilde{\mathbf{m}}, \tilde{\boldsymbol{\Sigma}})$, *where* $\tilde{\mathbf{m}} = (m_1, \ldots, m_n)'$ *and* $\tilde{\boldsymbol{\Sigma}}$ *is a* $n \times n$ *matrix with* $\tilde{\sigma}_{ij} = \sigma_{ij}$ *for* $i, j = 1, \ldots, n$.

(ii) *For any* $\mathbf{b} = (b_1, \ldots b_d)' \in \mathbb{R}^d$, *the r.v.* $Y_\mathbf{b} = \sum_{k=1}^d b_k Y_k$ *is univariate* $\mathcal{AL}(\mu, \sigma)$ *with* $\sigma = \sqrt{\mathbf{b}'\boldsymbol{\Sigma}\mathbf{b}}$ *and* $\mu = \mathbf{m}'\mathbf{b}$. *Further, if* $\mathbf{Y}$ *is symmetric AL, then so is* $Y_\mathbf{b}$.

(iii) *For all* $k \le d$, $Y_k \sim \mathcal{AL}(\mu, \sigma)$ *with* $\sigma = \sqrt{\sigma_{kk}}$ *and* $\mu = m_k$.

Proof. Here is an outline of the proof. For part (i), apply Proposition 6.8.1 with $n \times d$ matrix $\mathbf{A} = (a_{ij})$ such that $a_{ii} = 1$ and $a_{ij} = 0$ for $i \ne j$. For part (ii), apply Proposition 6.8.1 with $l = 1$ and compare the resulting ch.f. with the characteristic function of the univariate asymmetric Laplace distribution. For part (iii), apply part (ii) to the standard base vectors in $\mathbb{R}^d$. □

Remark 6.8.2 Corollary 6.8.1 part (ii) implies that the sum $\sum_{k=1}^d Y_k$ has an AL distribution if all Y_k's are components of a multivariate AL r.v. (and thus all Y_k's are univariate AL r.v.'s). This is in contrast with a sum of i.i.d. AL r.v.'s, which generally does not have an AL distribution.

Remark 6.8.3 Note that if $\mathbf{Y}$ has a nonsingular AL law (that is $\boldsymbol{\Sigma}$ is positive definite) and the matrix $\mathbf{A}$ is such that $\mathbf{AA}'$ is positive-definite, then the vector $\mathbf{AY}$ has a nonsingular AL law as well. In particular, this holds if $\mathbf{A}$ is a nonsingular square matrix.

We have shown in Corollary 6.8.1, part (ii), that if $\mathbf{Y}$ is an AL r.v. in $\mathbb{R}^d$, then all linear combinations of its components are univariate AL r.v.'s. A natural question is whether the converse is true. As of now, we do not have a complete answer to this question. The following result provides a partial answer for the case where all linear combinations are univariate $\mathcal{AL}(\mu, \sigma)$ with either $\mu = 0$ (symmetric Laplace distribution) or $\sigma = 0$ (exponential distribution).

Theorem 6.8.1 *Let* $\mathbf{Y} = (Y_1, \ldots, Y_d)'$ *be an r.v. in* $\mathbb{R}^d$. *If all linear combinations* $\sum_{k=1}^d c_k Y_k$ *have either symmetric Laplace or exponential distribution, then* $\mathbf{Y}$ *has an* $\mathcal{AL}_d(\mathbf{m}, \boldsymbol{\Sigma})$ *distribution with either* $\boldsymbol{\Sigma} = \mathbf{0}$ *or* $\mathbf{m} = \mathbf{0}$.

Proof. The proof follows from the corresponding result for GS laws [see Kozubowski (1997), Theorem 3.3] and the fact that $\mathcal{AL}_d(\mathbf{m}, \boldsymbol{\Sigma})$ distributions with either $\boldsymbol{\Sigma} = \mathbf{0}$ or $\mathbf{m} = \mathbf{0}$ are strictly geometric stable. □

6.8.2 Linear regression. Interestingly enough the conditions for linearity of the regression of Y_d on $Y_1, \ldots, Y_{d-1}$, where $\mathbf{Y} = (Y_1, \ldots, Y_d)'$ is AL, coincide with those for multivariate normal laws.

Proposition 6.8.2 *Let* $\mathbf{Y} = (Y_1, \ldots, Y_d)' \sim \mathcal{AL}_d(\mathbf{m}, \boldsymbol{\Sigma})$. *Let*

$$\mathbf{m}_1 = (m_1, \ldots, m_{d-1})'$$

and let

$$\mathbf{\Sigma} = \begin{pmatrix} \mathbf{\Sigma}_{11} & \mathbf{\Sigma}_{12} \\ \mathbf{\Sigma}_{21} & \mathbf{\Sigma}_{22} \end{pmatrix}$$

be a partition of $\mathbf{\Sigma}$ *such that* $\mathbf{\Sigma}_{11}$ *is a* $d-1 \times d-1$ *matrix. Then*

$$E(Y_d|Y_1,\ldots,Y_{d-1}) = a_1Y_1 + \cdots + a_{d-1}Y_{d-1} \quad (a.s.) \tag{6.8.1}$$

if and only if

$$\mathbf{\Sigma}_{11}\mathbf{a} = \mathbf{\Sigma}_{12} \ \ and \ \ \mathbf{m}_1'\mathbf{a} = m_d. \tag{6.8.2}$$

Moreover, in case $|\mathbf{\Sigma}| > 0$, *condition* (6.8.2) *is equivalent to*

$$\mathbf{m}_1'\mathbf{\Sigma}_{11}^{-1}\mathbf{\Sigma}_{12} = m_d \ \ and \ \ \mathbf{a} = (a_1,\ldots,a_{d-1})' = \mathbf{\Sigma}_{11}^{-1}\mathbf{\Sigma}_{12}.$$

Proof. It is well known that for an r.v. $\mathbf{Y}$ with a finite mean, the condition (6.8.1) holds if and only if

$$\left.\frac{\partial\Psi(\mathbf{t})}{\partial t_d}\right|_{t_d=0} = a_1\left.\frac{\partial\Psi(\mathbf{t})}{\partial t_1}\right|_{t_d=0} + \cdots + a_{d-1}\left.\frac{\partial\Psi(\mathbf{t})}{\partial t_{d-1}}\right|_{t_d=0},$$

where Ψ is the ch.f. of $\mathbf{Y}$ [see, e.g., Miller (1978)]. Substitution of the AL ch.f. (6.2.1) into this equation followed by differentiation results in (6.8.2). In case $|\mathbf{\Sigma}| > 0$, the solution of the first equation in (6.8.2) is $\mathbf{a} = \mathbf{\Sigma}_{11}^{-1}\mathbf{\Sigma}_{12}$, which solves the second equation in (6.8.2) if and only if $\mathbf{m}_1'\mathbf{\Sigma}_{11}^{-1}\mathbf{\Sigma}_{12} = m_d$. □

Remark 6.8.4 The regression is always linear for $\mathbf{m} = \mathbf{0}$.

6.9 Infinite divisibility properties

6.9.1 Infinite divisibility. The following result establishes infinite divisibility of multivariate AL laws and identifies their Lévy measure.

Theorem 6.9.1 *Let* $\mathbf{Y}$ *have a nondegenerate d-dimensional* $\mathcal{AL}_d(\mathbf{m}, \mathbf{\Sigma})$ *law. Then the ch.f. of* $\mathbf{Y}$ *is of the form*

$$\Psi(\mathbf{t}) = \exp\left(\int_{R^n}\left(e^{i\mathbf{t}\cdot\mathbf{x}} - 1\right)\Lambda(d\mathbf{x})\right)$$

with

$$\frac{d\Lambda}{d\mathbf{x}}(\mathbf{x}) = \frac{2\exp(\mathbf{m}'\mathbf{\Sigma}^{-1}\mathbf{x})}{(2\pi)^{d/2}|\mathbf{\Sigma}|^{1/2}}\left(\frac{Q(\mathbf{x})}{C(\mathbf{\Sigma},\mathbf{m})}\right)^{-d/2} K_{d/2}(Q(\mathbf{x})C(\mathbf{\Sigma},\mathbf{m})),$$

where

$$Q(\mathbf{x}) = \sqrt{\mathbf{x}'\mathbf{\Sigma}^{-1}\mathbf{x}} \ \ and \ \ C(\mathbf{\Sigma},\mathbf{m}) = \sqrt{2 + \mathbf{m}'\mathbf{\Sigma}^{-1}\mathbf{m}}.$$

Proof. Apply Proposition 4.1 from Kozubowski and Rachev (1999b), which identifies the density of geometric stable Lévy measure to obtain

$$\frac{d\Lambda}{d\mathbf{x}}(\mathbf{x}) = \int_0^\infty f(z^{-1/2}\mathbf{x} - z^{1/2}\mathbf{m})z^{-d/2-1}e^{-z}dz,$$

where $f(\cdot)$ is the density of the multivariate normal $N_d(\mathbf{0}, \mathbf{\Sigma})$ distribution with respect to the d-dimensional Lebesgue measure. Next, proceed similarly to the computation of AL densities described in Section 6.5. Alternatively, use the representation of $\mathbf{Y}$ through subordinated Brownian motion and Lemma 7, VI.2 of Bertoin (1996) or use the fact that multivariate AL laws are mixtures of normal distributions by generalized gamma convolutions (cf. Exercise 2.7.61) and the corresponding results for the latter laws derived in Takano (1989). □

Remark 6.9.1 Note that for any d the density of an AL Lévy measure is unbounded at $\mathbf{x} = \mathbf{0}$.

Remark 6.9.2 In the one-dimensional case ($d = 1$), writing $\sigma^2 = \boldsymbol{\Sigma}$, $\mu = \mathbf{m}$, and $\kappa = \sqrt{2}\sigma/(\mu + \sqrt{\mu^2 + 2\sigma^2})$, we have

$$\frac{d\Lambda}{dx}(\pm x) = \frac{1}{x}\exp\left(-\frac{\sqrt{2}x}{\sigma}\kappa^{\pm 1}\right), \quad x > 0,$$

which is the density (3.4.6) of the Lévy measure of univariate AL laws (see Section 3.4 of Chapter 3).

6.9.2 Asymmetric Laplace motion. Since multivariate AL laws are infinitely divisible, similar to the one-dimensional case, one can define a Lévy process on $[0, \infty)$ with independent increments—the Laplace motion $\{\mathbf{Y}(s), s \geq 0\}$—so that $\mathbf{Y}(0) = \mathbf{0}$, $\mathbf{Y}(1)$ is given by (6.2.1), while for $s > 0$ the ch.f. of $\mathbf{Y}(s)$ is

$$\Psi(\mathbf{t}) = \left(\frac{1}{1 + \frac{1}{2}\mathbf{t}'\boldsymbol{\Sigma}\mathbf{t} - i\mathbf{m}'\mathbf{t}}\right)^s, \quad s > 0 \tag{6.9.1}$$

[see, e.g., Teichroew (1957)]. Distributions on $\mathbb{R}^d$ given by (6.9.1) will be called *generalized asymmetric Laplace* (GAL) and denoted as $\mathcal{GAL}_d(\mathbf{m}, \boldsymbol{\Sigma}, s)$. For $d = 1$ we obtain the Bessel function distribution studied in Section 4.1 of Chapter 4. A GAL r.v. admits mixture representation (6.3.1) where W has a gamma distribution with density

$$g(x) = \frac{x^{s-1}}{\Gamma(s)}e^{-x}. \tag{6.9.2}$$

The density corresponding to (6.9.1) can be expressed in terms of Bessel function as follows:

$$p(\mathbf{x}) = \frac{2\exp(\mathbf{m}'\boldsymbol{\Sigma}^{-1}\mathbf{x})}{(2\pi)^{d/2}\Gamma(s)|\boldsymbol{\Sigma}|^{1/2}}\left(\frac{Q(\mathbf{x})}{C(\boldsymbol{\Sigma}, \mathbf{m})}\right)^{s-d/2} K_{s-d/2}(Q(\mathbf{x})C(\boldsymbol{\Sigma}, \mathbf{m})), \tag{6.9.3}$$

where

$$Q(\mathbf{x}) = \sqrt{\mathbf{x}'\boldsymbol{\Sigma}^{-1}\mathbf{x}} \text{ and } C(\boldsymbol{\Sigma}, \mathbf{m}) = \sqrt{2 + \mathbf{m}'\boldsymbol{\Sigma}^{-1}\mathbf{m}}.$$

In the one dimensional case, Sichel (1973) utilized (6.9.1) for modeling size distributions of diamonds excavated from marine deposits in southwest Africa. In financial applications, this process is known as the *variance gamma process* (see Part III for more details on these and other applications).

Remark 6.9.3 If $\boldsymbol{\Sigma} = \mathbf{I}_d$ and $\mathbf{m} = \mathbf{0}$ we obtain the symmetric multivariate Bessel density

$$p(\mathbf{x}) = C_d(||\mathbf{x}||/\beta)^a K_a(||\mathbf{x}||/\beta), \tag{6.9.4}$$

where $\beta = \sqrt{2}$, $a = s - d/2 > -d/2$ and C_d is a normalizing constant independent of $\mathbf{x}$ [see Fang et al. (1990), p. 92]. In the special case $a = 0$ and $\beta = \sigma/\sqrt{2}$, Fang et al. (1990) call the distribution corresponding to (6.9.4) a *multivariate Laplace distribution*. Note that this distribution belongs to our class of Laplace distributions only in the bivariate case ($d = 2$) (Exercise 6.12.14).

Remark 6.9.4 If $\boldsymbol{\Sigma} = \mathbf{I}_d$ and $s = \frac{d+1}{2}$, the density (6.9.3) simplifies to

$$p(\mathbf{x}) = \frac{e^{-\sqrt{2+||\mathbf{m}||^2}+\mathbf{m}'\mathbf{x}}}{(2\pi)^{(d-1)/2}\Gamma(\frac{d+1}{2})\sqrt{2 + ||\mathbf{m}||^2}}, \tag{6.9.5}$$

which is a direct generalization of the one-dimensional AL *density* [see Takano (1989, 1990), and Exercise 6.12.12]. Takano (1989) derived the Lévy measure corresponding to density (6.9.5) and showed that for $d \geq 2$ these distributions are self-decomposable if and only if $\mathbf{m} = \mathbf{0}$ (in contrast with the case $d = 1$ since all one-dimensional AL laws are self-decomposable; cf. Proposition 3.2.3).

6.9.3 Geometric infinite divisibility. Like their one-dimensional counterparts, all multivariate AL laws are *geometric infinitely divisible* [see, e.g., Kotz et al. (2000b)].

Proposition 6.9.1 *Let* $\mathbf{Y} \sim \mathcal{AL}_d(\mathbf{m}, \mathbf{\Sigma})$. *Then* $\mathbf{Y}$ *is geometric infinitely divisible and the relation*

$$\mathbf{Y} \stackrel{d}{=} \sum_{i=1}^{\nu_p} \mathbf{Y}_p^{(i)} \tag{6.9.6}$$

holds for all $p \in (0, 1)$, *where the* $\mathbf{Y}_p^{(i)}$*'s are i.i.d. with the* $\mathcal{AL}_d(\mathbf{m}p, p\mathbf{\Sigma})$ *distribution,* ν_p *is geometrically distributed with mean* $1/p$, *and* ν_p *and* $(Y_p^{(i)})$ *are independent.*

Proof. Write (6.9.6) in terms of ch.f.'s. and follow the proof for the one-dimensional case (see Proposition 3.4.3). □

6.10 Stability properties

In this section we collect various characterizations of the multivariate AL laws that exhibit their stability properties under appropriate summation schemes. The results presented here, unlike the majority of the previous ones, cannot be derived from the theory of generalized hyperbolic distributions because the latter do not possess any general convolution properties except in some special cases (such as the normal inverse Gaussian case or the normal variance gamma models).

6.10.1 Limits of random sums. Analogous to the one-dimensional case, the multivariate AL laws are the only possible limits of geometric sums (6.0.1) of i.i.d. r.v.'s with finite second moments. Actually, the following result can serve as an alternative definition of this class of distributions.

Proposition 6.10.1 *Let* ν_p *be a geometrically distributed r.v. with mean* $1/p$, *where* $p \in (0, 1)$. *A random vector* $\mathbf{Y}$ *has an AL distribution in* $\mathbb{R}^d$ *if and only if there exists an independent of* ν_p *sequence* $\{\mathbf{X}^{(i)}\}$ *of i.i.d. random vectors in* $\mathbb{R}^d$ *with finite covariance matrix, and* $a_p > 0$, $\mathbf{b}_p \in \mathbb{R}^d$, *such that*

$$a_p \sum_{j=1}^{\nu_p} (\mathbf{X}^{(j)} + \mathbf{b}_p) \stackrel{d}{\longrightarrow} \mathbf{Y} \quad as \quad p \to 0. \tag{6.10.1}$$

Proof. The result follows from the so-called *transfer theorem* for random summation [see, e.g., Rosiński (1976)] and its converse [see Szasz (1972)], together with the Central Limit Theorem for i.i.d. r.v.'s with a finite covariance matrix. □

Our next result determines the type of normalization that produces convergence in (6.10.1).

Theorem 6.10.1 *Let* $\mathbf{X}^{(j)}$ *be i.i.d. random vectors in* $\mathbb{R}^d$ *with mean vector* $\mathbf{m}$ *and covariance matrix* $\mathbf{\Sigma}$. *For* $p \in (0, 1)$, *let* ν_p *be a geometric r.v. with mean* $1/p$ *and independent of the sequence* $(\mathbf{X}^{(j)})$. *Then as* $p \to 0$,

$$a_p \sum_{j=1}^{\nu_p} (\mathbf{X}^{(j)} + \mathbf{b}_p) \stackrel{d}{\to} \mathbf{Y} \sim \mathcal{AL}_d(\mathbf{m}, \mathbf{\Sigma}), \tag{6.10.2}$$

where $a_p = p^{1/2}$ *and* $\mathbf{b}_p = \mathbf{m}(p^{1/2} - 1)$.

Proof. By the Cramér–Wald device [see, e.g., Billingsley (1968)], the convergence (6.10.2) is equivalent to

$$\mathbf{c}' a_p \sum_{j=1}^{\nu_n} (\mathbf{X}^{(j)} + \mathbf{b}_p) \stackrel{d}{\to} \mathbf{c}'\mathbf{Y}$$

for all vectors $\mathbf{c}$ in $\mathbb{R}^d$. Writing $W_j = \mathbf{c}'\mathbf{X}^{(j)}$, $\mu = \mathbf{c}'\mathbf{m}$, $b_p = (p^{1/2} - 1)\mu$, and $Y = \mathbf{c}'\mathbf{Y}$, we have

$$a_p \sum_{j=1}^{\nu_p} (W_j + b_p) \stackrel{d}{\to} Y \sim \mathcal{AL}(\mu, \sigma) \quad \text{as } p \to 0. \tag{6.10.3}$$

Here the W_j's are i.i.d. variables with mean μ and variance $\sigma^2 = \mathbf{c}'\mathbf{\Sigma}\mathbf{c}$, and Y is a univariate AL variable with ch.f.

$$\psi(t) = \frac{1}{1 + \frac{1}{2}\sigma^2 t^2 - i\mu t}.$$

The convergence (6.10.3) now follows from Proposition 3.4.4 for the univariate AL case (cf. equation (3.4.15)). □

Next, we study stability properties of AL random vectors.

6.10.2 Stability under random summation. The following stability property is a well-known characterization of α-stable random vectors: $\mathbf{X}$ is α-stable if and only if for any $n \geq 2$ we have the following equality in distribution:

$$\mathbf{X}^{(1)} + \cdots + \mathbf{X}^{(n)} \stackrel{d}{=} n^{1/\alpha}\mathbf{X} + \mathbf{d}_n, \tag{6.10.4}$$

where the $\mathbf{X}^{(i)}$'s are i.i.d. copies of $\mathbf{X}$ and $\mathbf{d}_n$ is some vector in $\mathbb{R}^d$ [see, e.g., Samorodnitsky and Taqqu (1994)].

We have an analogous characterization of AL random vectors with respect to geometric summation [see Kotz et al. (2000b)].

Theorem 6.10.2 *Let* $\mathbf{Y}$, $\mathbf{Y}^{(1)}$, $\mathbf{Y}^{(2)}, \ldots$ *be i.i.d. r.v.'s in* $\mathbb{R}^d$ *with finite second moments, and let* ν_p *be a geometrically distributed random variable independent of the sequence* $\{\mathbf{Y}^{(i)}, i \geq 1\}$. *For each* $p \in (0, 1)$, *the r.v.* $\mathbf{Y}$ *has the stability property*

$$a_p \sum_{i=1}^{\nu_p} (\mathbf{Y}^{(i)} + \mathbf{b}_p) \stackrel{d}{=} \mathbf{Y}, \tag{6.10.5}$$

with $a_p > 0$ *and* $\mathbf{b}_p \in \mathbb{R}^d$ *if and only if* $\mathbf{Y}$ *is* $\mathcal{AL}_d(\mathbf{m}, \mathbf{\Sigma})$ *with either* $\mathbf{\Sigma} = \mathbf{0}$ *or* $\mathbf{m} = \mathbf{0}$. *The normalizing constants are necessarily of the form*

$$a_p = p^{1/2}, \quad \mathbf{b}_p = \mathbf{0}.$$

This result follows from the characterization of strictly geometric stable laws given in Theorem 3.1 of Kozubowski (1997) and the fact that the only strictly geometric stable laws with finite second moments are $\mathcal{AL}_d(\mathbf{m}, \mathbf{\Sigma})$ laws with either $\mathbf{\Sigma} = \mathbf{0}$ or $\mathbf{m} = \mathbf{0}$.

Remark 6.10.1 Since in general multivariate AL r.v.'s do not satisfy relation (6.10.5), as is the case in the univariate case, the question arises as to whether

$$\mathbf{S}^{(p)} = a_p \sum_{i=1}^{\nu_p} (\mathbf{Y}^{(i)} + \mathbf{b}_p) \stackrel{d}{\to} \mathbf{Y}, \text{ as } p \to 0, \tag{6.10.6}$$

where $\mathbf{Y}^{(i)}$ are i.i.d. copies of $\mathbf{Y}$, ν_p is independent of $\{\mathbf{Y}^{(i)}, i \geq 1\}$ geometrically distributed, and $a_p > 0$ and $\mathbf{b}_p \in \mathbb{R}^d$. Note that the convergence (6.10.6) holds for all univariate AL laws (see Proposition 3.4.5), as well as for general geometric stable laws with index α less than two [see Kozubowski (1997)]. It is quite surprising that for $d > 1$, as noted by Kozubowski (1997), in general AL r.v.'s do not satisfy (6.10.6) unless $\mathbf{m} = \mathbf{0}$ or $\boldsymbol{\Sigma} = \mathbf{0}$. Indeed, if either $\boldsymbol{\Sigma} = \mathbf{0}$ or $\mathbf{m} = \mathbf{0}$, then (6.10.5) is satisfied and so is (6.10.6). Assume $\boldsymbol{\Sigma} \neq \mathbf{0}$ and suppose that $\mathbf{Y} \sim \mathcal{AL}_d(\mathbf{m}, \boldsymbol{\Sigma})$ satisfies (6.10.6). Then for any $\mathbf{c} \in \mathbb{R}^d$, we have

$$\mathbf{c}'\mathbf{S}^{(p)} = a_p \sum_{i=1}^{\nu_p} \left[\mathbf{c}'\mathbf{Y}^{(i)} + \mathbf{c}'\mathbf{b}_p\right] \stackrel{d}{\to} Y_{\mathbf{c}} = \mathbf{c}'\mathbf{Y} \text{ as } p \to 0. \tag{6.10.7}$$

By Corollary 6.8.1, part (ii), the r.v. $Y_{\mathbf{c}} = \mathbf{c}'\mathbf{Y}$ is univariate AL with $\sigma = (\mathbf{c}'\boldsymbol{\Sigma}\mathbf{c})^{1/2}$ and $\mu = \mathbf{c}'\mathbf{m}$. After the application of Proposition 3.4.5, we find that (6.10.7) holds with

$$a_p = Cp^{1/2}(1 + o(1)), \quad \text{where } C = \left[\sigma^2/(\mu^2 + \sigma^2)\right]^{1/2}. \tag{6.10.8}$$

Since the normalizing constant $a(p)$ in (6.10.7) should be independent of $\mathbf{c}$, (6.10.8) implies that $\mu = \mathbf{c}'\mathbf{m} = 0$ for every $\mathbf{c}$, and thus $\mathbf{m} = \mathbf{0}$. In the latter case $C = 1$ and (6.10.7) holds with $a_p = p^{1/2}$ and $\mathbf{b}_p = \mathbf{0}$.

6.10.3 Stability of deterministic sums. In the next result, taken from Kotz et al. (2000b), we show that a deterministic sum of i.i.d. AL r.v.'s, scaled by an appropriate *random variable*, has the same distribution as each component of the sum. It is a generalization of a similar characterization of the univariate Laplace distributions; see Proposition 2.2.11 in Chapter 2.

Theorem 6.10.3 *Let B_m, where $m > 0$, have a* Beta$(1, m)$ *distribution. Let $\{\mathbf{X}^{(i)}\}$ be a sequence of i.i.d. random vectors with finite second moment. Then the following statements are equivalent:*

(i) *For all $n \geq 2$, $\mathbf{X}^{(1)} \stackrel{d}{=} B_{n-1}^{1/2}(\mathbf{X}^{(1)} + \cdots + \mathbf{X}^{(n)})$.*

(ii) *$\mathbf{X}^{(1)}$ is $\mathcal{AL}_d(\mathbf{m}, \boldsymbol{\Sigma})$ with either $\boldsymbol{\Sigma} = \mathbf{0}$ or $\mathbf{m} = \mathbf{0}$.*

Proof. This result follows from the corresponding result for GS laws [see Kozubowski and Rachev (1999b)] and the fact that $\mathcal{AL}_d(\mathbf{m}, \boldsymbol{\Sigma})$ distributions with either $\boldsymbol{\Sigma} = \mathbf{0}$ or $\mathbf{m} = \mathbf{0}$ are strictly GS. The result for GS laws follows from the results of Pakes (1992ab). □

We conclude our discussion with yet another stability property of AL laws, that for the one-dimensional case was given in (2.2.28) [and noted by Pillai (1985)].

Proposition 6.10.2 *Let $\mathbf{Y}, \mathbf{Y}^{(1)}, \mathbf{Y}^{(2)}$, and $\mathbf{Y}^{(3)}$ be $\mathcal{AL}_d(\mathbf{m}, \boldsymbol{\Sigma})$ r.v.'s with either $\boldsymbol{\Sigma} = \mathbf{0}$ or $\mathbf{m} = \mathbf{0}$. Let $p \in (0, 1)$, and let I be an indicator random variable, independent of the $\mathbf{Y}^{(i)}$'s, with $P(I = 1) = p$, and $P(I = 0) = 1 - p$. Then the following equality in distribution is valid for any $p \in (0, 1)$:*

$$\mathbf{Y} \stackrel{d}{=} p^{1/2} I \mathbf{Y}^{(1)} + (1 - I)(\mathbf{Y}^{(2)} + p^{1/2}\mathbf{Y}^{(3)}). \tag{6.10.9}$$

Proof. Let $\mathbf{c} \in \mathbb{R}^d$. Since $\mathbf{c}'\mathbf{Y}$, $\mathbf{c}'\mathbf{Y}^{(1)}$, $\mathbf{c}'\mathbf{Y}^{(2)}$, and $\mathbf{c}'\mathbf{Y}^{(3)}$ are univariate $\mathcal{AL}(\mu, \sigma)$ with either $\mu = 0$ or $\sigma = 0$ (see Corollary 6.8.1), the result in the one-dimensional case [see equation (2.2.28) and Pillai (1985)] produces

$$\mathbf{c}'\mathbf{Y} \stackrel{d}{=} p^{1/2} I \mathbf{c}'\mathbf{Y}^{(1)} + (1 - I)(\mathbf{c}'\mathbf{Y}^{(2)} + p^{1/2}\mathbf{c}'\mathbf{Y}^{(3)}),$$

or equivalently,

$$\mathbf{c}'\mathbf{Y} \stackrel{d}{=} \mathbf{c}'(p^{1/2} I \mathbf{Y}^{(1)} + (1 - I)(\mathbf{Y}^{(2)} + p^{1/2}\mathbf{Y}^{(3)})).$$

The last relation implies (6.10.9). □

6.11 Linear regression with Laplace errors

In this final section we study a regression model with Laplace distributed error term. Consider the multiple linear regression model

$$\mathbf{Y} = \mathbf{X}\mathbf{b} + \mathbf{e}, \tag{6.11.1}$$

where $\mathbf{Y}$ is a $d \times 1$ random vector of observations, $\mathbf{X}$ is a $d \times k$ nonstochastic matrix of rank k, $\mathbf{b}$ is a $k \times 1$ vector of regression parameters with unknown values, and $\mathbf{e}$ is a $d \times 1$ random error term. Assume that $\mathbf{e} \sim \mathcal{AL}_d(\mathbf{0}, \sigma^2\mathbf{I}_d)$, where $\mathbf{I}_d$ is a $d \times d$ identity matrix (so that the mean vector and covariance matrix of $\mathbf{e}$ are, respectively, $\mathbf{0}$ and $\sigma^2\mathbf{I}_d$). Although the elements of $\mathbf{e}$ are uncorrelated, they are not independent. According to Theorem 6.3.1, $\mathbf{e}$ has the representation

$$\mathbf{e} \stackrel{d}{=} W^{1/2}\mathbf{N}, \tag{6.11.2}$$

where $\mathbf{N} \sim N_d(\mathbf{0}, \sigma^2\mathbf{I}_d)$ (multivariate normal with mean $\mathbf{0}$ and covariance matrix $\sigma^2\mathbf{I}_d$), while W is a standard exponential variable (independent of $\mathbf{N}$).

6.11.1 Least-squares estimation. The least-squares estimator (LSE) $\widehat{\mathbf{b}}$ of $\mathbf{b}$ satisfies the normal equations

$$(\mathbf{X}'\mathbf{X})\widehat{\mathbf{b}} = \mathbf{X}'\mathbf{Y}.$$

If $\mathbf{X}$ has full rank, the inverse of $\mathbf{X}'\mathbf{X}$ exists and $\widehat{\mathbf{b}}$ can be expressed as

$$\widehat{\mathbf{b}} = (\mathbf{X}'\mathbf{X})^{-1}\mathbf{X}'\mathbf{Y}, \tag{6.11.3}$$

which is the same as in the normal case.

Next, we consider the joint distribution of $\widehat{\mathbf{b}}$ and the vector of residuals $\widehat{\mathbf{e}} = \mathbf{Y} - \mathbf{X}\widehat{\mathbf{b}}$. In view of (6.11.1) and (6.11.3), we have

$$\begin{bmatrix} \widehat{\mathbf{b}} \\ \widehat{\mathbf{e}} \end{bmatrix} = \begin{bmatrix} (\mathbf{X}'\mathbf{X})^{-1}\mathbf{X}' \\ \mathbf{I}_d - (\mathbf{X}'\mathbf{X})^{-1}\mathbf{X}' \end{bmatrix} \mathbf{Y} = \begin{bmatrix} \mathbf{b} \\ \mathbf{0} \end{bmatrix} + \begin{bmatrix} (\mathbf{X}'\mathbf{X})^{-1}\mathbf{X}' \\ \mathbf{I}_d - (\mathbf{X}'\mathbf{X})^{-1}\mathbf{X}' \end{bmatrix} \mathbf{e},$$

where $\mathbf{e} \sim \mathcal{AL}_d(\mathbf{0}, \sigma^2\mathbf{I}_d)$. Now, since

$$\begin{bmatrix} \widehat{\mathbf{b}} \\ \widehat{\mathbf{e}} \end{bmatrix} - \begin{bmatrix} \mathbf{b} \\ \mathbf{0} \end{bmatrix}$$

is a linear function of $\mathbf{e}$, its distribution is AL according to Proposition 6.8.1.

Proposition 6.11.1 *Under the model* (6.11.1), *the least-squares estimator* $\widehat{\mathbf{b}}$ *and the vector of residuals* $\widehat{\mathbf{e}} = \mathbf{Y} - \mathbf{X}\widehat{\mathbf{b}}$ *have the following joint distribution:*

$$\begin{bmatrix} \widehat{\mathbf{b}} \\ \widehat{\mathbf{e}} \end{bmatrix} - \begin{bmatrix} \mathbf{b} \\ \mathbf{0} \end{bmatrix} \sim \mathcal{AL}_{k+d}(\mathbf{0}, \mathbf{\Sigma}) \quad \mathbf{\Sigma} = \sigma^2 \begin{bmatrix} (\mathbf{X}'\mathbf{X})^{-1} & \mathbf{0} \\ \mathbf{0} & \mathbf{I}_d - \mathbf{X}(\mathbf{X}'\mathbf{X})^{-1}\mathbf{X}' \end{bmatrix}. \tag{6.11.4}$$

Remark 6.11.1 As in the normal case, it follows that $E(\widehat{\mathbf{b}}) = \mathbf{b}$ (so that $\widehat{\mathbf{b}}$ is unbiased), $E(\widehat{\mathbf{e}}) = \mathbf{0}$, $\mathbf{Cov}(\widehat{\mathbf{b}}) = \sigma^2(\mathbf{X}'\mathbf{X})^{-1}$, and $\mathbf{Cov}(\widehat{\mathbf{e}}) = \sigma^2(\mathbf{I}_d - \mathbf{X}(\mathbf{X}'\mathbf{X})^{-1}\mathbf{X}')$. However, $\widehat{\mathbf{b}}$ and $\widehat{\mathbf{e}}$ are uncorrelated but not independent.

Remark 6.11.2 Note that since $Y_1, \dots Y_d$ are uncorrelated, $\mathrm{Var}(Y_i) = \sigma^2$, and $\widehat{\mathbf{b}}$ is unbiased for $\mathbf{b}$, the conditions of the Gauss–Markov theorem are fulfilled. Thus for any $\mathbf{c} \in \mathbb{R}^d$, the estimator $\mathbf{c}'\widehat{\mathbf{b}}$ of $\mathbf{c}'\mathbf{b}$ has the smallest possible variance among all linear estimators of the form $\mathbf{c}'\mathbf{Y}$ which are unbiased for $\mathbf{c}'\mathbf{b}$. In particular, for $j = 1, \dots d$, $\widehat{b}_j$ will have the smallest variance among all linear unbiased estimators of b_j.

6.11.2 Estimation of σ^2. As in the normal case, the estimator $\widehat{\mathbf{e}}'\widehat{\mathbf{e}}/(d-k)$ is unbiased for σ^2, which follows from the following result.

Proposition 6.11.2 *Under the model* (6.11.1), *the statistic* $\widehat{\mathbf{e}}'\widehat{\mathbf{e}}$ *is distributed as*

$$\sigma^2 W V,$$

where W and V are independent, W is standard exponential, and V has a chi-square distribution with $d-k$ degrees of freedom. Moreover, the r.v. $\widehat{\mathbf{e}}'\widehat{\mathbf{e}}/\sigma^2$ has the following density function:

$$p(x) = \left(\sqrt{x/2}\right)^{(d-k)/2-1} K_{(d-k)/2-1}(\sqrt{2x}) / \Gamma\left(\frac{d-k}{2}\right), \quad x > 0. \tag{6.11.5}$$

Proof. First, write $\widehat{\mathbf{e}} = (\mathbf{I}_d - \mathbf{X}(\mathbf{X}'\mathbf{X})^{-1}\mathbf{X}')(\mathbf{X}\mathbf{b} + \mathbf{e})$, note that the matrix $\mathbf{I}_d - \mathbf{X}(\mathbf{X}'\mathbf{X})^{-1}\mathbf{X}'$ is indepotent, and utilize the representation (6.11.2) to obtain

$$\widehat{\mathbf{e}}'\widehat{\mathbf{e}} = Z\mathbf{N}'(\mathbf{I}_d - \mathbf{X}(\mathbf{X}'\mathbf{X})^{-1}\mathbf{X}')\mathbf{N},$$

where $\mathbf{N}$ has a multivariate normal distribution with mean zero and covariance matrix $\sigma^2\mathbf{I}_d$. Now the first part of the proposition follows, since $\mathbf{N}'(\mathbf{I}_d - \mathbf{X}(\mathbf{X}'\mathbf{X})^{-1}\mathbf{X}')\mathbf{N}/\sigma^2$ has a chi-square distribution with $d-k$ degrees of freedom (a standard fact for the regression model (6.11.1) with normally distributed error term).

Next, apply the standard transformation theorem for random variables to obtain the density of WV in the form

$$p(x) = \frac{x^{(d-k)/2-1}}{2^{(d-k)/2}\Gamma\left(\frac{d-k}{2}\right)} \int_0^\infty y^{1-(d-k)/2-1} e^{-\frac{1}{2}(x/y+2y)} dy.$$

Finally, utilize the fact that the generalized inverse Gaussian density (6.3.2) with $\chi = x$, $\psi = 2$, and $\lambda = 1 - (d-k)/2$ integrates to one on $(0, \infty)$ so that

$$\int_0^\infty y^{1-(d-k)/2-1} e^{-\frac{1}{2}(x/y+2y)} dy = \frac{2K_{(d-k)/2-1}(\sqrt{2x})}{(2/x)^{1/2-(d-k)/4}},$$

which produces (6.11.5). □

Remark 6.11.3 This result may be used to obtain confidence intervals for σ.

Next, we derive the minimal mean squared error estimator for σ^2. Consider the class of estimators for σ^2 of the form $\delta_c = c\widehat{\mathbf{e}}'\widehat{\mathbf{e}}$. We know from Proposition 6.11.2 that for $c = 1/(d-k)$ we obtain an unbiased estimator. However, this estimator does not minimize the mean squared error (MSE) defined as

$$MSE = E(\delta_c - \sigma^2)^2 = \mathrm{Var}\delta_c + (E\delta_c - \sigma^2)^2.$$

To find c that minimizes the MSE, write

$$MSE = c^2\sigma^4 \mathrm{Var}(\widehat{\mathbf{e}}'\widehat{\mathbf{e}}/\sigma^2) + (c\sigma^2 E(\widehat{\mathbf{e}}'\widehat{\mathbf{e}}/\sigma^2) - \sigma^2)^2, \tag{6.11.6}$$

and compute the mean and variance of $\widehat{\mathbf{e}}'\widehat{\mathbf{e}}/\sigma^2$ that appear in (6.11.6) utilizing Proposition 6.11.2. Namely, we have

$$E(\widehat{\mathbf{e}}'\widehat{\mathbf{e}}/\sigma^2) = E(WV) = E(W)E(V) = 1 \cdot n$$

and

$$E(\widehat{\mathbf{e}}'\widehat{\mathbf{e}}/\sigma^2)^2 = E(W^2)E(V^2) = 2 \cdot (2n + n^2),$$

where $n = d - k$, so that

$$\mathrm{Var}(\widehat{\mathbf{e}}'\widehat{\mathbf{e}}/\sigma^2) = E(W^2V^2) - [E(WV)]^2 = 4n + n^2.$$

Consequently, (6.11.6) produces

$$MSE = \sigma^4[c^2(2n^2 + 4n) - 2cn + 1].$$

The minimum value is easily found to be $c^* = 1/(2(n+2))$. We summarize our discussion below.

Proposition 6.11.3 *Consider the model* (6.11.1) *and the class of estimators of* σ^2 *of the form* $c\widehat{\mathbf{e}}'\widehat{\mathbf{e}}$, *where* $c \in \mathbb{R}$. *Then the estimator*

$$\frac{\widehat{\mathbf{e}}'\widehat{\mathbf{e}}}{2(d-k+2)}$$

minimizes the MSE.

6.11.3 The distributions of standard t and F statistics. When studying the regression model (6.11.1) with multivariate student-t error term $\mathbf{e}$, Zellner (1976) noticed that tests and intervals based on the usual t and F statistics remain valid. He also remarked that his argument with conditioning holds for models (6.11.1) whenever the error term is a normal mixture (6.11.2) with a proper distribution of W, establishing the validity of the usual t and F statistics.

Proposition 6.11.4 *Consider the regression model* (6.11.1), *where* $\mathbf{e} \sim \mathcal{AL}_d(\mathbf{0}, \sigma^2\mathbf{I}_d)$ *and* $\mathbf{X}$ *is of full rank. Let* $\widehat{\mathbf{b}} = (\widehat{b}_1, \ldots \widehat{b}_k)'$ *be the least-squares estimator of* $\mathbf{b} = (b_1, \ldots b_k)'$, *and let* $s^2 = \widehat{\mathbf{e}}'\widehat{\mathbf{e}}/(d-k)$. *Then*

(i) *the statistic*

$$T_i = \frac{\widehat{b}_i - b_i}{s\sqrt{c_{ii}}}, \tag{6.11.7}$$

where c_{ii} *is the* i*th diagonal element in* $(\mathbf{X}'\mathbf{X})^{-1}$, *has a* t*-distribution with* $d-k$ *degrees of freedom;*

(ii) *if* $\mathbf{b} = \mathbf{0}$, *then the statistic*

$$F = \frac{(\widehat{\mathbf{b}}'\mathbf{X}'\mathbf{Y} - d\overline{\mathbf{Y}}^2)/(k-1)}{\widehat{\mathbf{e}}'\widehat{\mathbf{e}}/(d-k)} \tag{6.11.8}$$

has an F*-distribution with* $k-1$ *and* $d-k$ *degrees of freedom;*

(iii) *the statistic*

$$\frac{(\widehat{\mathbf{b}} - \mathbf{b})'\mathbf{X}'\mathbf{X}(\widehat{\mathbf{b}} - \mathbf{b})/k}{\widehat{\mathbf{e}}'\widehat{\mathbf{e}}/(d-k)}, \tag{6.11.9}$$

which is used in deriving confidence ellipsoids for $\mathbf{b}$, *has an* F*-distribution with* k *and* $d-k$ *degrees of freedom. Moreover, a* $100(1-\alpha)\%$ *confidence region for* $\mathbf{b}$ *is given by*

$$(\mathbf{b} - \widehat{\mathbf{b}})'\mathbf{X}'\mathbf{X}(\mathbf{b} - \widehat{\mathbf{b}}) \le k\frac{\widehat{\mathbf{e}}'\widehat{\mathbf{e}}}{d-k}F_{k,n-k}(\alpha), \tag{6.11.10}$$

where $F_{k,n-k}(\alpha)$ *is the upper* (100α)*th percentile of an* F*-distribution with* k *and* $d-k$ *degrees of freedom.*

Remark 6.11.4 Improved confidence ellipsoids were derived in Hwang and Chen (1986).

6.11.4 Inference from the estimated regression function. After fitting, a regression model can be used for predictions. Let $\mathbf{x}_0$ be a $k \times 1$ vector of predictor variables. Then $\mathbf{x}_0$ coupled with $\widehat{\mathbf{b}}$ can be used to estimate the regression function $\mathbf{x}_0'\mathbf{b}$ as well as the value of the response Y_0 at $\mathbf{x}_0$. It turns out that the confidence intervals for these predictions coincide with those for the normal case.

6.11.4.1 *Estimating the regression function at* $\mathbf{x}_0$. Note that since $\mathbf{x}_0'\mathbf{b}$ is a linear function of $\mathbf{b}$, the Gauss–Markov theorem implies that $\mathbf{x}_0'\widehat{\mathbf{b}}$ is BLUE for $\mathbf{x}_0'\mathbf{b}$, with variance of $\mathbf{x}_0'(\mathbf{X}'\mathbf{X})^{-1}\mathbf{x}_0\sigma^2$. Moreover, as in the normal case, the statistic

$$\frac{\mathbf{x}_0'\widehat{\mathbf{b}} - \mathbf{x}_0'\mathbf{b}}{s\sqrt{\mathbf{x}_0'(\mathbf{X}'\mathbf{X})^{-1}\mathbf{x}_0}}, \tag{6.11.11}$$

where $s^2 = \widehat{\mathbf{e}}'\widehat{\mathbf{e}}/(d-k)$, has a t-distribution with $d-k$ degrees of freedom.

6.11.4.2 *Forecasting a new observation at* $\mathbf{x}_0$. As in the normal model, a new observation Y_0 has an unbiased predictor $\mathbf{x}_0'\widehat{\mathbf{b}}$. According to model (6.11.1), we now have

$$\begin{bmatrix} \mathbf{Y} \\ Y_0 \end{bmatrix} = \begin{bmatrix} \mathbf{X} \\ \mathbf{x}_0' \end{bmatrix}\mathbf{b} + \begin{bmatrix} \mathbf{e} \\ e_0 \end{bmatrix},$$

where $[\mathbf{e}'e_0]' \sim \mathcal{AL}_{d+1}(\mathbf{0}, \sigma^2\mathbf{I}_{d+1})$. Note that the forecast error, $Y_0 - \mathbf{x}_0'\widehat{\mathbf{b}}$, can be expressed as

$$Y_0 - \mathbf{x}_0'\widehat{\mathbf{b}} = [\ -\mathbf{x}_0'(\mathbf{X}'\mathbf{X})^{-1}\mathbf{X}' \quad 1\ ]\begin{bmatrix} \mathbf{e} \\ e_0 \end{bmatrix}$$

so that it has a univariate AL distribution with mean zero and variance $\sigma^2(1 + \mathbf{x}_0'(\mathbf{X}'\mathbf{X})^{-1}\mathbf{x}_0)$ (see Corollary 6.8.1). It follows that the statistic

$$\frac{Y_0 - \mathbf{x}_0'\widehat{\mathbf{b}}}{s\sqrt{1 + \mathbf{x}_0'(\mathbf{X}'\mathbf{X})^{-1}\mathbf{x}_0}}$$

has t-distribution with $d-k$ degrees of freedom.

6.11.5 Maximum likelihood estimation. By (6.5.3) and (6.11.1), the likelihood function for the regression model has the form

$$p(\mathbf{y}|\mathbf{b},\sigma) = \frac{K_{d/2-1}(\sqrt{2}||\mathbf{y}-\mathbf{Xb}||/\sigma)}{2^{d/4-1/2}\pi^{d/2}\sigma^{1+d/2}||\mathbf{y}-\mathbf{Xb}||^{d/2-1}}, \tag{6.11.12}$$

where K_λ denotes the modified Bessel function of the third kind. Note that for any fixed value of σ, the functions $K_{d/2-1}(\sqrt{2}||\mathbf{y}-\mathbf{Xb}||/\sigma)$ and $||\mathbf{y}-\mathbf{Xb}||^{1-d/2}$ are both decreasing in $||\mathbf{y}-\mathbf{Xb}||$ (for $d=2$, which is the smallest value for d, the latter function is constant). Thus the maximum occurs whenever $||\mathbf{y}-\mathbf{Xb}||$ is minimized. Consequently, the maximum likelihood estimator (MLE) for $\mathbf{b}$ coincides with the least-squares estimator (LSE) for $\mathbf{b}$. To find the MLE for σ, we need to maximize the function

$$L(\sigma) = \frac{K_{d/2-1}(a/\sigma)}{\pi^{d/2}\sigma^{1+d/2}a^{d/2-1}}$$

with respect to $\sigma \in (0,\infty)$, where $a = \sqrt{2}||\mathbf{y}-\mathbf{X}\widehat{\mathbf{b}}||$ and $\widehat{\mathbf{b}}$ is the LSE (and MLE) for $\mathbf{b}$. The logarithmic derivative of L equals

$$\frac{d}{d\sigma}\log L(\sigma) = -\frac{1+d/2}{\sigma} - \frac{a}{\sigma^2}\frac{K'_{d/2-1}(a/\sigma)}{K_{d/2-1}(a/\sigma)}.$$

Using Property 4 of Bessel functions from the appendix, we have

$$\frac{d}{d\sigma}\log L(\sigma) = \frac{1}{\sigma}\frac{a}{\sigma}\left[R_{d/2-1}(a/\sigma) - d/(a/\sigma)\right], \tag{6.11.13}$$

where the function R_λ is defined by (A.0.15) in the appendix. In view of (6.11.13), Lemma 6.11.1 below implies the existence of a unique number $\widehat{\sigma} \in (0,\infty)$ such that the function $\log L(\sigma)$ is strictly increasing on $(0,\widehat{\sigma})$ and strictly decreasing on $(\widehat{\sigma},\infty)$. This number, which is the MLE of σ, is a unique solution of the equation

$$R_{d/2-1}(a/\sigma) = d/(a/\sigma). \tag{6.11.14}$$

Lemma 6.11.1 *Let d be an integer greater than or equal to two.*

(i) *If $d=2$, then the function $h_d(x) = xR_{d/2-1}(x)$ is strictly increasing for $x \in (0,\infty)$ with $\lim_{x\to\infty} h_d(x) = \infty$ and $\lim_{x\to 0^+} h_d(x) = 0$.*

(ii) *If $d>2$, then the function $h_d(x) = R_{d/2-1}(x) - d/x$ is strictly increasing for $x \in (0,\infty)$ with $\lim_{x\to\infty} h_d(x) = 1$ and $\lim_{x\to 0^+} h_d(x) = -\infty$.*

Proof. First, consider the case $d=2$. By Property 13 of Bessel functions (see the appendix), we have $\frac{d}{dx}xR_0(x) = x(R_0^2(x)-1)$. By Property 11 (see the appendix), $R_0(x) > 1$ so that $\frac{d}{dx}xR_0(x) > 0$, showing that the function $xR_0(x)$ is strictly increasing. The same property also produces $\lim_{x\to\infty} h_d(x) = \infty$. Finally, the limit $\lim_{x\to 0^+} h_d(x) = 0$ follows from the asymptotic behavior of the Bessel function (Property 6 in the appendix).

Next, consider $d>2$. Apply Property 12 (see the appendix) with $\lambda = d/2-1$ to obtain the following expression for the derivative of h_d:

$$\frac{d}{dx}h_d(x) = \frac{d}{dx}\left(-2/x + 1/R_{(d-2)/2-1}(x)\right). \tag{6.11.15}$$

Note that for $d>3$ the function $R_{(d-2)/2-1}(x)$ is decreasing (Property 11 in the appendix), while for $d=3$ we have $R_{-1/2}(x) = 1$ (by Property 4 in the appendix). In either case, the derivative (6.11.15) is positive (as the expression in parenthesis is a strictly increasing function), so that the function h_d is strictly increasing. The rest of (ii) follows from Properties 6 and 11 (see the appendix). □

Note that since $R_{d/2-1}(a/\sigma) > 1$ (see the appendix, Property 11), we must have $d/(a/\widehat{\sigma}) > 1$, so that the MLE of σ satisfies the inequality

$$\widehat{\sigma} > a/d = \sqrt{2}||\mathbf{y} - \mathbf{X}\widehat{\mathbf{b}}||/d.$$

Remark 6.11.5 Recall that the MLE for σ under normally distributed error term is given by $\tilde{\sigma} = ||\mathbf{y} - \mathbf{X}\widehat{\mathbf{b}}||/\sqrt{d}$. Consequently, in case $d = 2$, the MLE of σ under model (6.11.1) with AL distributed error term is greater than the one under the model with normally distributed error term.

Remark 6.11.6 The solution to (6.11.14) must be obtained numerically, except for a few special cases described below.

- Special case $d = 3$. Here the Bessel function has a closed form (see Property 5 in the appendix), and we have

$$R_{d/2-1}(x) = R_{1/2}(x) = 1 + 1/x.$$

  Consequently, equation (6.11.14) yields the solution

$$\widehat{\sigma} = a/2 = ||\mathbf{y} - \mathbf{X}\widehat{\mathbf{b}}||/\sqrt{2},$$

  which is greater than $\tilde{\sigma}$.

- Special case $d = 5$. Here we use the iterative property (A.0.16) of R_λ to write equation (6.11.14) as

$$1/R_{1/2}(a/\sigma) = 2/(a/\sigma).$$

  Since $R_{1/2}(x) = 1 + 1/x$, we obtain the following quadratic equation for $\widehat{\sigma}$:

$$2\widehat{\sigma}^2 + 2\widehat{\sigma}a - a^2 = 0,$$

  whose positive solution is $\widehat{\sigma} = \frac{\sqrt{3}-1}{2}a$. Again, we see that $\widehat{\sigma} \approx 0.366a$ is greater than $\tilde{\sigma} = a/\sqrt{10} \approx 0.316a$.

- Special case $d = 7$. Here we use the iterative property (A.0.16) of R_λ twice to write equation (6.11.14) as

$$3\sigma/a + 1/R_{1/2}(a/\sigma) = a/(2\sigma).$$

  Since $R_{1/2}(x) = 1 + 1/x$, we obtain the following cubic equation for $y = \widehat{\sigma}/a$:

$$y^3 + y^2 + y/6 - 1/6 = 0,$$

  whose real solution is

$$y = \frac{1}{3}\left((2 + \sqrt{31/8})^{1/3} + (2 - \sqrt{31/8})^{1/3} - 1\right).$$

  Consequently, the MLE of σ is

$$\widehat{\sigma} = \frac{a}{3}\left((2 + \sqrt{31/8})^{1/3} + (2 - \sqrt{31/8})^{1/3} - 1\right) \approx \frac{a}{2.47}.$$

  Again, we see that $\widehat{\sigma}$ is greater than $\tilde{\sigma} = a/\sqrt{14} \approx a/3.74$.

6.11.6 Bayesian estimation. Here we analyze the regression model (6.11.1) with an AL error term and likelihood function (6.11.12) from the Bayesian point of view. We assume a diffuse prior distribution for the parameters $\mathbf{b}$ and σ^2,

$$p(\mathbf{b}, \sigma^2) \propto 1/\sigma^2, \quad \mathbf{b} \in \mathbb{R}^k, 0 < \sigma^2 < \infty.$$

This standard improper distribution assumes that $\mathbf{b}$ and $\log \sigma^2$ are uniformly and independently distributed. Taking into account the likelihood function (6.11.12), we obtain the posterior joint distribution of $\mathbf{b}$ and σ^2,

$$p(\mathbf{b}, \sigma^2|\mathbf{y}) \propto \frac{K_{d/2-1}(\sqrt{2}||\mathbf{y} - \mathbf{X}\mathbf{b}||/\sqrt{\sigma^2})}{(\sigma^2)^{3/2+d/4}||\mathbf{y} - \mathbf{X}\mathbf{b}||^{d/2-1}}. \tag{6.11.16}$$

To obtain the marginal posterior p.d.f. of $\mathbf{b}$, we integrate (6.11.16) with respect to $u = \sigma^2$:

$$p(\mathbf{b}|\mathbf{y}) \propto ||\mathbf{y} - \mathbf{X}\mathbf{b}||^{1-d/2} \int_0^\infty \frac{K_{d/2-1}(\sqrt{2}||\mathbf{y} - \mathbf{X}\mathbf{b}||/\sqrt{u})}{(u)^{3/2+d/4}} du. \tag{6.11.17}$$

The change of variable $z = ||\mathbf{y} - \mathbf{X}\mathbf{b}||/\sqrt{u}$ in (6.11.17) leads to

$$p(\mathbf{b}|\mathbf{y}) \propto \left(\frac{2}{||\mathbf{y} - \mathbf{X}\mathbf{b}||^2}\right)^{d/2} \int_0^\infty z^{d/2+4} K_{d/2-1}(\sqrt{2}z) dz \propto \left(\frac{1}{||\mathbf{y} - \mathbf{X}\mathbf{b}||^2}\right)^{d/2}, \tag{6.11.18}$$

as the integral in (6.11.18) is a constant independent of $\mathbf{b}$ (the finiteness of the integral follows from relation (A.0.13); see the appendix). Since

$$||\mathbf{y} - \mathbf{X}\mathbf{b}||^2 = s^2(d-k) + (\mathbf{b} - \widehat{\mathbf{b}})'\mathbf{X}'\mathbf{X}(\mathbf{b} - \widehat{\mathbf{b}}), \tag{6.11.19}$$

where

$$s^2 = \widehat{\mathbf{e}}'\widehat{\mathbf{e}}/(d-k) = (\mathbf{Y} - \mathbf{X}\widehat{\mathbf{b}})'(\mathbf{Y} - \mathbf{X}\widehat{\mathbf{b}})/(d-k),$$

we recognize (6.11.18) as a k-dimensional Student-t p.d.f. with $v = d - k$ degrees of freedom [see, e.g., Zellner (1976), Johnson and Kotz (1972)]. The posterior density of $\mathbf{b}$ has the form

$$p(\mathbf{b}|\mathbf{y}) = \frac{\Gamma((v+k)/2)(1 + v^{-1}(\mathbf{b} - \widehat{\mathbf{b}})'\mathbf{R}^{-1}(\mathbf{b} - \widehat{\mathbf{b}}))^{(v+k)/2}}{(\pi v)^{k/2}\Gamma(v/2)|\mathbf{R}|^{1/2}}, \tag{6.11.20}$$

where $\mathbf{R} = (\mathbf{X}'\mathbf{X})^{-1}s^2$ is a positive-definite matrix. Note that the same posterior distribution results under the model (6.11.1) with multivariate normal and Student-t error terms [see Zellner (1976) for the latter]. We also see that whenever $v = d - k > 1$, the mean of the posterior distribution of $\mathbf{b}$ exists and equals $\widehat{\mathbf{b}}$. Consequently, the Bayesian estimator of $\mathbf{b}$ (under the squared error loss function and diffuse prior distribution) coincides with MLE and LSE for $\mathbf{b}$.

Next, we derive the marginal posterior p.d.f. of σ^2 by integrating (6.11.16) with respect to $\mathbf{b}$. Setting $u = \sigma^2$, $\delta^2 = s^2(d-k)/u$, $\lambda = 1 - (d-k)/2$, and using (6.11.19), we obtain after some algebra

$$p(u|\mathbf{y}) \propto \frac{\sqrt{2}^{1-d/2}}{u^{d/2+1}} \int_{\mathbb{R}^k} \frac{K_{k/2-\lambda}(\sqrt{2}\sqrt{\delta^2 + (\mathbf{b} - \widehat{\mathbf{b}})'(\mathbf{X}'\mathbf{X}/u)(\mathbf{b} - \widehat{\mathbf{b}})})}{(\sqrt{\delta^2 + (\mathbf{b} - \widehat{\mathbf{b}})'(\mathbf{X}'\mathbf{X}/u)(\mathbf{b} - \widehat{\mathbf{b}})}/\sqrt{2})^{k/2-\lambda}} d\mathbf{b}. \tag{6.11.21}$$

We now recognize the integrand in (6.11.21) as the main factor of a k-dimensional generalized hyperbolic density (6.5.4) with parameters λ, δ, $\xi = \alpha = \sqrt{2}$, $\boldsymbol{\mu} = \widehat{\mathbf{b}}$, $\boldsymbol{\beta} = \mathbf{0}$, and $\boldsymbol{\Sigma} = (\mathbf{X}'\mathbf{X})^{-1}u$.

Since the latter density integrates to one over $\mathbb{R}^k$, we evaluate the integral in (6.11.21) and obtain the following expression after some algebraic manipulations:

$$p(u|\mathbf{y}) \propto \left(\frac{1}{u}\right)^{(d-k)/4+3/2} K_{(d-k)/2-1}(\sqrt{2s^2(d-k)/u}). \tag{6.11.22}$$

Taking into consideration the integration formula (A.0.13), we finally obtain an exact expression for the posterior density of $u = \sigma^2$:

$$p(u|\mathbf{y}) = \frac{(\sqrt{s^2(d-k)})^{(d-k)/2+1} K_{(d-k)/2-1}(\sqrt{2s^2(d-k)/u})}{(\sqrt{2})^{(d-k)/2-1}(\sqrt{u})^{(d-k)/2+3}\Gamma((d-k)/2)}. \tag{6.11.23}$$

It can be shown that the r.v. with this density has the same distribution as $s^2(d-k)/X$, where X is an r.v. with density (6.11.5). The mean of this posterior distribution generally does not exist.

6.12 Exercises

Exercise 6.12.1 Let $\mathbf{X} \sim \mathcal{AL}_d(\mathbf{m}, \mathbf{\Sigma})$.

(a) Show that if $\mathbf{m} = \mathbf{0}$ (so that $\mathbf{X}$ is actually symmetric Laplace), then any one-dimensional marginal distribution of $\mathbf{X}$ is symmetric Laplace.

(b) Show that if every one-dimensional marginal distribution of $\mathbf{X}$ is symmetric Laplace, then $\mathbf{X}$ is symmetric Laplace, $\mathbf{X} \sim \mathcal{L}_d(\mathbf{\Sigma})$.

Thus for multivariate AL laws, the symmetry is a componentwise property, which is in contrast with geometric stable laws with index less than 2.

Exercise 6.12.2* Let $\mathbf{X} = (X_1, \ldots, X_d)'$ have a multivariate asymmetric Laplace distribution $\mathcal{AL}_d(\mathbf{m}, \mathbf{\Sigma})$, and let Ψ be the ch.f. of $\mathbf{X}$. Using the cumulant formula (5.3.2), show that $c_1(\mathbf{X}) = \mathbf{m}$, $c_2(\mathbf{X}) = \mathbf{\Sigma} + \mathbf{mm}'$, and

$$c_3(\mathbf{X}) = \mathbf{\Sigma} \otimes \mathbf{m} + \mathbf{m} \otimes \mathbf{\Sigma} + \text{vec}\,\mathbf{\Sigma}\mathbf{m}' + 2\mathbf{m}^{\otimes 2}\mathbf{m}' \tag{6.12.1}$$

[Kollo (2000)].

Exercise 6.12.3 Let $\mathbf{X} = (X_1, X_2)' \sim \mathcal{BAL}(m_1, m_2, \sigma_1, \sigma_2, \rho)$.

(a) Assuming that $m_1 = m_2 = m$, $\sigma_1 = \sigma_2 = \sigma$, and $\rho = 0$, find the p.d.f.'s of X_1, $X_1 + X_2$, $X_1 - X_2$, and X_2 given $X_1 = x_1$. What are the conditional mean and variance of the latter distribution?

(b) Repeat part (a) for a general BAL r.v. $\mathbf{X}$.

Exercise 6.12.4 By considering the appropriate characteristic functions, prove the "if" part of Theorem 6.10.2. Namely, show that if $\mathbf{X}^{(i)}$ are i.i.d. with the $\mathcal{L}_d(\mathbf{\Sigma})$ distribution and ν_p is an independent geometric variable with mean $1/p$, then the equality in distribution

$$p^{1/\alpha}\sum_{I=1}^{\nu_p}\mathbf{X}^{(i)} \stackrel{d}{=} \mathbf{X}^{(1)} \tag{6.12.2}$$

holds with $\alpha = 2$.

Exercise 6.12.5 Establish the implication (ii) $\to$ (i) of Theorem 6.10.3.

Exercise 6.12.6 Show that if $\mathbf{X} \sim N_d(\mathbf{0}, \boldsymbol{\Sigma})$ and W is an independent standard exponential variable, then the r.v.

$$\mathbf{Y} = \mathbf{m}W + \sqrt{W}\mathbf{X},$$

where $\mathbf{m} \in \mathbb{R}^d$, has the $\mathcal{AL}_d(\mathbf{m}, \boldsymbol{\Sigma})$ distribution.

Hint: Use the characteristic functions.

Exercise 6.12.7 Consider the regression model (6.11.1) from the Bayesian point of view. Assuming the diffuse prior distribution for the parameters $\mathbf{b}$ and σ^2, derive the posterior density (6.11.16) of the parameters and show that the posterior marginal densities of $\mathbf{b}$ and σ^2 are given by (6.11.20) and (6.11.23), respectively.

Exercise 6.12.8 Let $\mathbf{X}$ be an r.v. in $\mathbb{R}^d$ with the ch.f.

$$\Phi(\mathbf{t}) = E^{i\mathbf{t}'\mathbf{X}} = u(\mathbf{t}) + iv(\mathbf{t}) = r(\mathbf{t})e^{i\theta(\mathbf{t})}.$$

Then the function

$$\theta(\mathbf{t}) = \tan^{-1}\{v(\mathbf{t})/u(\mathbf{t})\}, \quad |\mathbf{t}| \le |r_0|,$$

where r_0 is the zero of $u(\mathbf{t})$ closest to the origin, is called the *characteristic symmetric function* of $\mathbf{X}$ [see Heathcote et al. (1995)]. For an (elliptically) symmetric distribution about the point $\mathbf{m}$, the above function is linear in $\mathbf{t}$ and has been used in testing multivariate symmetry [see Heathcote et al. (1995)].

Derive the characteristic symmetric function for a r.v. $\mathbf{X}$ with the $\mathcal{AL}_d(\mathbf{m}, \boldsymbol{\Sigma})$ distribution. Under what conditions on $\mathbf{m}$ and $\boldsymbol{\Sigma}$ is the distribution of $\mathbf{X}$ symmetric? What is $\theta(\mathbf{t})$ in this case?

Exercise 6.12.9 Let $\mathbf{X} = (X_1, \ldots, X_d)'$ be a random vector in $\mathbb{R}^d$. The variables $X_1, \ldots, X_d$ (the components of $\mathbf{X}$) are said to be *associated* if the inequality

$$Cov[f(\mathbf{X}), g(\mathbf{X})] \ge 0$$

holds for all measurable functions f and g that are nondecreasing in each coordinate (whenever the covariance is finite). It is well known that if $\mathbf{X} \sim N_d(\mathbf{0}, \boldsymbol{\Sigma})$, then the components of $\mathbf{X}$ are associated if and only if they are positively correlated ($\boldsymbol{\Sigma} \ge \mathbf{0}$) [Pitt (1982)]. Let $\mathbf{X}$ have an $\mathcal{AL}_d(\mathbf{m}, \boldsymbol{\Sigma})$ distribution.

(a) Show that if the components of $\mathbf{X}$ are associated, then they must be positively correlated, that is,

$$\boldsymbol{\Sigma} + \mathbf{m}\mathbf{m}' \ge \mathbf{0}. \tag{6.12.3}$$

(b)** Investigate whether the condition (6.12.3) is also sufficient for the association of the components of $\mathbf{X}$.

Exercise 6.12.10 Let $\mathbf{X}$ have a multivariate normal $N_d(\mathbf{m}, \boldsymbol{\Sigma})$ distribution, where $\boldsymbol{\Sigma}$ is a nonnegative definite covariance matrix of rank $r \le d$.

(a) Using the well-known decomposition $\boldsymbol{\Sigma} = \mathbf{C}\mathbf{C}'$, where $\mathbf{C}$ is a $d \times r$ matrix of rank r, show that the random vector

$$\mathbf{C}\mathbf{Z} + \mathbf{m}, \tag{6.12.4}$$

where

$$\mathbf{Z} = (Z_1, \ldots, Z_r)' \tag{6.12.5}$$

is a random vector with the standard normal and independent components, have the same distribution as the vector $\mathbf{X}$.

(b) Now let the components of the r.v. (6.12.5) be i.i.d. standard Laplace variables. Show that the distribution of the r.v. (6.12.4), referred to by Kalashnikov (1997) as multivariate Laplace distribution, does not belong to the class of AL laws. In particular, show that in general the univariate marginal distributions of the resulting random vector will not be Laplace. Discuss the similarities and the differences of the resulting distributions with the AL laws.

Exercise 6.12.11 Let $\mathbf{m} \in \mathbb{R}^d$ and let $\mathbf{\Sigma}$ be a $d \times d$ positive-definite matrix. Consider an elliptically symmetric distribution in $\mathbb{R}^d$ with density (6.3.5), where

$$g(x) = e^{-x^{\lambda/2}}. \tag{6.12.6}$$

This distribution is known as the *multivariate exponential power distribution* [see, e.g., Fernández et al. (1995)] as well as the *multivariate generalized Laplace distribution* [Ernst (1998)]. Haro-Lopéz and Smith (1999) refer to the special case with $\lambda = 1$ as the *elliptical Laplace distribution* in their robustness studies and show that it can be obtained as a scale mixture of multivariate normal distributions. For $d = 1$ we obtain the generalized Laplace distribution (the exponential power distribution) with density

$$f(x) = \frac{\lambda}{2s\Gamma(1/\lambda)} \exp\left\{-\left|\frac{x-\mu}{s}\right|^{\lambda}\right\}. \tag{6.12.7}$$

(a) Determine the proportionality constant k_d [see (6.3.5)] in this case.

(b) Set $\lambda = 1$ [in which case (6.12.7) produces the classical symmetric Laplace distribution] and check whether the marginal distributions corresponding to (6.3.5) are Laplace.

Exercise 6.12.12 Let $\mathbf{X}$ have a general multivariate Bessel distribution with the ch.f. (6.9.1) and the density (6.9.3).

(a) Show that in case $\mathbf{\Sigma} = \mathbf{I}_d$ and $s = \frac{d+1}{2}$, we obtain the density (6.9.5), which leads to

$$p(\mathbf{x}) = \frac{e^{-\sqrt{2}||\mathbf{x}||}}{\sqrt{2}(2\pi)^{(d-1)/2}\Gamma(\frac{d+1}{2})} \tag{6.12.8}$$

if $\mathbf{m} = \mathbf{0}$. Compare the latter density with that of the multivariate exponential power distributions discussed in Exercise 6.12.11.

(b) Show that the densities (6.9.5) and (6.12.8) lead to the AL and Laplace densities if $d = 1$. What are the parameters in this case?

Exercise 6.12.13 Generalizing elliptically symmetric distributions, Fernández et al. (1995) introduced a class of v-spherical distributions given by the density

$$p(\mathbf{x}; \mathbf{m}, \tau) = \tau^d g[v\{\tau(\mathbf{x} - \mathbf{m})\}], \tag{6.12.9}$$

where $v(\cdot)$ is a scalar function such that

- $v(\cdot) > 0$ (with a possible exception on a set of Lebesgue measure zero),
- $v(k\mathbf{a}) = kv(\mathbf{a})$ for all $k \geq 0$ and $\mathbf{a} \in \mathbb{R}^d$,

g is a nonnegative function, and $\mathbf{m} \in \mathbb{R}^d$ and $\tau^{-1} > 0$ are the location and scale parameters, respectively. [The functions $v(\cdot)$ and $g(\cdot)$ must be chosen such that (6.12.9) is a genuine probability density function.] Note that by choosing

$$v(\mathbf{a}) = \mathbf{a}'\boldsymbol{\Sigma}^{-1}\mathbf{a}$$

we obtain the elliptically symmetric distributions, which with g given by (6.12.6) are the exponential power distributions [cf. Exercise 6.12.11]. Fernández et al. (1995) introduced a *skewed multivariate generalization of the Laplace distribution* as a special case with $q = 1$ of the *skewed multivariate exponential power distribution* that has density (6.12.9) with

$$v(a_1, \ldots, a_d) = \left[\sum_{i=1}^{d} \{(a_i^+/\gamma)^q + (\gamma a_i^-)^q\}\right]^{1/q} \tag{6.12.10}$$

and

$$g(x) = c^d e^{-\frac{1}{2}x^q}. \tag{6.12.11}$$

[As before, $x^+ = \max(x, 0)$ and $x^- = \max(0, -x)$.]

(a) Show that if $\mathbf{X} = (X_1, \ldots, X_d)'$, where the X_i's are i.i.d. variables with the *skewed exponential power distribution* with the density

$$f(x) = c \begin{cases} e^{-(x/\gamma)^q/2} & \text{for } x \geq 0 \\ e^{-(-\gamma x)^q/2} & \text{for } x \leq 0, \end{cases} \tag{6.12.12}$$

where $\gamma, q > 0$ and

$$c^{-1} = 2^{1/q}\Gamma(1 + 1/q)(\gamma + 1/\gamma), \tag{6.12.13}$$

then the r.v. $\mathbf{X}$ has v-spherical density (6.12.9) with v given by (6.12.10) and g given by (6.12.11) [Fernández et al. (1995)]. In particular, we see that the d-dimensional skewed Laplace r.v. of Fernández et al. (1995) is generated as an i.i.d. sample of size d from a univariate AL distribution.

(b) Derive the mean, the variance, the moments EX^k, and the coefficients of skewness and kurtosis for a random variable X with density (6.12.12).

Exercise 6.12.14 Let $\mathbf{X}$ have a symmetric multivariate Bessel distribution with density given by (6.9.4). In the special case $a = 0$, Fang et al. (1990) call it a multivariate Laplace distribution. Here the density of $\mathbf{X}$ is proportional to

$$f(\mathbf{x}) \propto K_0(||\mathbf{x}||/\beta), \tag{6.12.14}$$

where K_0 is the modified Bessel function of the third kind and order 0.

(a) Show that the distribution in $\mathbb{R}^d$ with the density as in (6.12.14) is $\mathcal{AL}(\mathbf{m}, \boldsymbol{\Sigma})$ only if $d = 2$. What are $\mathbf{m}$ and $\boldsymbol{\Sigma}$ in this case?

(b) Show that if $\mathbf{X} \stackrel{d}{=} R\mathbf{U}^{(d)}$ is the polar representation of a symmetric multivariate Bessel r.v. in $\mathbb{R}^d$ with density (6.9.4), then the density of the r.v. R is

$$g_R(r) = c_r r^{a+d-1} K_a(r/\beta), \tag{6.12.15}$$

where

$$c_r^{-1} = 2^{a+d-2}\beta^{a+d}\Gamma(d/2)\Gamma(a + d/2). \tag{6.12.16}$$

What is this representation if $\mathbf{X}$ has density (6.12.14)? How does it compare with that of a symmetric Laplace $\mathcal{L}(\mathbf{I}_d)$ distribution?

Exercise 6.12.15 A d-dimensional r.v. with the ch.f.

$$\Psi(\mathbf{t}) = \frac{1}{1 + \left(\frac{1}{2}\mathbf{t}'\boldsymbol{\Sigma}\mathbf{t}\right)^{\alpha/2}}, \quad \mathbf{t} \in \mathbb{R}^d, \tag{6.12.17}$$

where $0 < \alpha \leq 2$ and $\boldsymbol{\Sigma}$ is a nonnegative definite matrix, is said to have a *multivariate Linnik* distribution [see, e.g., Anderson (1992), Pakes (1992a), Ostrovskii (1995)]. For $\alpha = 2$ it reduces to the symmetric multivariate Laplace distribution.

(a) Show that all components of a multivariate Linnik r.v. have univariate Linnik distributions.

(b) Show that all linear combinations $\mathbf{c}'\mathbf{X}$, where $\mathbf{X}$ has a multivariate Linnik distribution and $\mathbf{c} \in \mathbb{R}^d$, are univariate Linnik.

Exercise 6.12.16 By considering the appropriate characteristic functions, show that if $\mathbf{X}^{(i)}$'s are i.i.d. with multivariate Linnik distribution (6.12.17) and ν_p is an independent geometric variable with mean $1/p$, then the relation (6.12.2) holds. Thus multivariate Linnik variables are stable with respect to geometric summation, as are univariate (symmetric) Linnik and Laplace as well as multivariate symmetric Laplace variables.

Part III

Applications

Introduction

Laplace distributions found and continue to find applications in a variety of disciplines that range from image and speech recognition (input distributions) and ocean engineering (distributions of navigation errors) to finance (distributions of log-returns of a commodity). Now they are rapidly becoming distributions of first choice whenever "something" with heavier than Gaussian tails is observed in the data. Consequently, a large number of papers in diverse journals and monographs mention Laplace laws as the "right" distribution, and it is a daunting task to find and report them all.

The asymmetric Laplace distribution as described in this book is quite a recent invention. It was motivated by similar probabilistic considerations as was the asymmetric (skewed) normal distribution developed by Azzalini (1985, 1986). It is our belief that natural applications will inevitably arise. In fact an application in modeling of foreign currency exchange has recently been suggested. Several other applications are described in subsequent chapters. Similar comments apply to the multivariate generalizations of Laplace distributions.

In this part of the book, we attempt to present those applications that we consider in our subjective judgment the most interesting and promising. In our choice we were also restricted by the fact that our book is addressed to a wide range of potential "clients" of the Laplace distributions. Our personal taste may have played an unavoidable but hopefully not a damaging role. To make the material readable for our intended audience we present some of the more specialized and narrowly focused applications in essay form. Readers interested in further details are directed to the literature cited in the references.

7

Engineering Sciences

This is the first chapter in the third part of the book that deals with applications of various versions of Laplace distributions in the sciences, business, and various branches of engineering. We start with an application in communication theory, in particular signal processing, which seems to have dominated earlier results in the 1960s and 1970s. Next, we mention applications in fracture problems discovered in the late 1940s before the appearance of the Weibull distribution which dominated this field in the second half of the 20th century. Applications in navigation problems conclude the chapter.

7.1 Detection in the presence of Laplace noise

Detection of a known constant signal that is distorted by the presence of a random noise was discussed in communication theory on various occasions [see Marks et al. (1978), Dadi and Marks (1987), and the references therein].

Using statistical terms, the goal is to test for the presence or absence of a positive constant signal s in additive random noise. The hypothesis-testing problem in this context is formulated as

$$\begin{aligned} H_0 &: x_i = n_i, \quad i = 1, 2, \ldots, N; \\ H_1 &: x_i = s + n_i, \quad s > 0, \end{aligned}$$

where based on the observations $\{x_i, i = 1, 2, \ldots, N\}$ we are to decide whether the signal s is absent or present. The quantity α is the probability of the first type error (incorrectly accepting H_1); it is also called the significance level. Similarly, β, the detection probability or the power function of the test, is the probability of correctly accepting H_1.

In statistical terminology, we are dealing here with a test for location in the case of a simple hypothesis H_0 vs. a simple alternative H_1. However, in communication theory, the problem receives a different formulation, which uses the notion of a detector. This is best represented by the scheme presented in Figure 7.1. The detector represented in this figure is defined through the form of g, which is called in this context a zero-memory nonlinearity. Also the distribution of the noise n_i influences the value of the threshold T for the test statistic $t = \sum_{i=1}^{N} g(x_i)$, since the latter has a distribution that depends on the distribution of the noise.

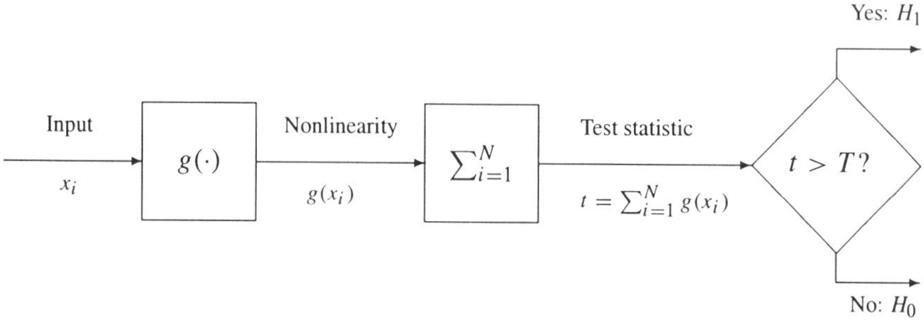

Figure 7.1: General scheme of a detector.

Various forms of detectors can be proposed by means of an appropriate definition of g. The well-known Neyman–Pearson optimal detector is defined if the density of the input is known. Its form (as well as its name) follows from the classical Neyman–Pearson lemma [Neyman and Pearson (1933)] which maximizes the power of the test. It is easy to observe that in general optimal nonlinearity should be of the form

$$g_{\text{opt}}(x) = \log \frac{f_n(x-s)}{f_n(x)},$$

where f_n is the distribution of n_i (which are assumed to be i.i.d. random variables) [see, e.g., Miller and Thomas (1972)].

In the analysis of detector performance, the noise is commonly assumed to be Gaussian. The assumption is often justified (for example for ultra-high-frequency signals — UHF) and results in a mathematically tractable analysis. However in many instances, as pointed out by Miller and Thomas (1972), a non-Gaussian noise assumption is necessary (for example for extremely low frequency — ELH).

One form of frequently encountered non-Gaussian noise is so-called impulsive noise. Such noise typically possesses much heavier tail behavior than Gaussian noise. Because of this, Laplace noise has been suggested as a model for some types of impulsive noise.

Indeed, models of noise based on Laplace distributions appear in various engineering studies in the last 40 years. Bernstein et al. (1974) comment on the non-Gaussian nature of ELF atmospheric noise, and they give a plot of a typical experimentally determined probability density function associated with such a noise which is very similar to a Laplace density. Mertz (1961) proposed a density for the amplitude of impulsive noise that in the limiting case results in the density of Laplace law. Kanefsky and Thomas (1965) considered a class of generalized Gaussian noises, obtained by generalizing the Gaussian density to arrive at a variable rate of exponential decay. The Laplace distribution is within this class of generalized Gaussian distributions. Also, Duttweiler and Messerschmitt (1976) refer to the Laplace distribution as a model for the distribution of speech.

For the case of Laplace noise given by the density

$$f(n) = \frac{\gamma}{2} e^{-\gamma |n|}, \quad n \in \mathbb{R}, \gamma > 0,$$

the Neyman–Pearson optimal detector found in Miller and Thomas (1972) is of the form

$$g_{\text{opt}}(x) = \begin{cases} \gamma s, & x > s, \\ 2\gamma x - \gamma s, & 0 \le x \le s, \\ -\gamma s, & x < 0. \end{cases}$$

See also Figure 7.2.

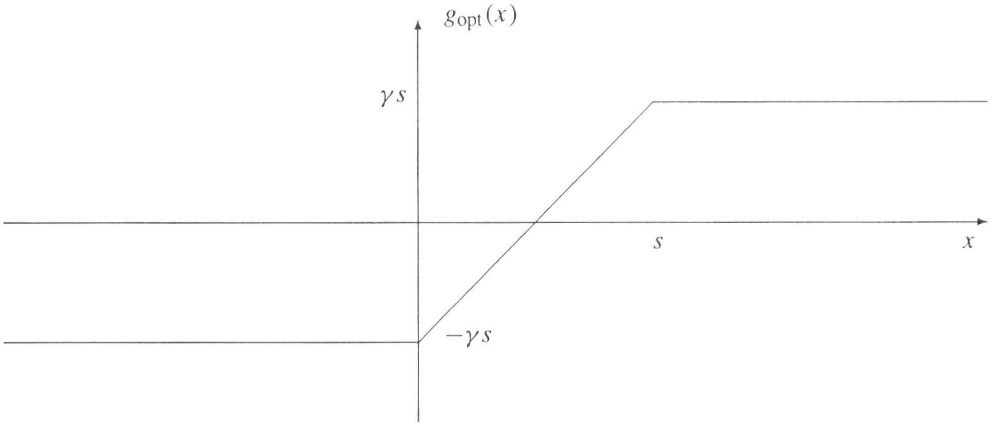

Figure 7.2: Nonlinearity in optimal detector for Laplace noise.

To solve the detection problem completely, it remains to find the distribution of the statistic

$$t = \sum_{i=1}^{N} g_{\text{opt}}(x_i).$$

This problem was solved in Marks et al. (1978) and results in the c.d.f.

$$\begin{aligned} F_N^{(0)}(x) = {} & \frac{1}{2^N} \sum_{k=1}^{N} \binom{N}{k} \sum_{r=0}^{k} (-1)^r \binom{k}{r} \sum_{l=0}^{N-k} \binom{N-k}{l} \\ & \cdot \left[e^{-(r+l)\gamma s} - e^{-\frac{x+N\gamma s}{2}} e_{k-1}\left(\frac{x + (N - 2(l+r))\gamma s}{2} \right) \right] u\left(x + (N - 2l - 2r)\gamma s\right) \\ & + \frac{1}{2^N} \sum_{m=0}^{N} \binom{N}{m} e^{-m\gamma s} u(x + (N - 2m)\gamma s), \end{aligned}$$

where $e_k(\cdot)$ is the incomplete exponential function

$$e_k(z) = \sum_{i=0}^{k} \frac{z^i}{i!}$$

and

$$u(x) = \begin{cases} 0 & \text{for } z < 0, \\ 1 & \text{for } z \geq 0. \end{cases}$$

For the proof of this result and further discussion of testing a hypothesis about the location parameters for the Laplace laws, see Part I, Chapter 2.6, Subsection 2.6.4.3. Note that in the above formulation, we use slightly different notation to be consistent with the original paper.

Since we are dealing here with the classical Laplace law which is symmetric, the distribution of the statistic $t(\cdot)$ under the alternative H_1 is given by

$$F_N^{(1)}(t) = 1 - F_N^{(0)}(-t).$$

The mean and variance of the test statistic are

$$\begin{aligned} \mathrm{E}_0 t &= -\mathrm{E}_1 t = N\left(1 - e^{-\gamma s} - \gamma s\right), \\ \mathrm{Var}_0 t &= \mathrm{Var}_1 t = N\left(3 - 2e^{-\gamma s} - e^{-2\gamma s} - 4\gamma s e^{-\gamma s}\right) \end{aligned}$$

(cf. Theorem 2.6.2 in Part I, Chapter 2.6).

In communication theory other detectors beside the optimal one are also considered. For example, the linear detector is given by $g_{lin}(x) = x$ and the sign detector is given by $g_{sign}(x) = \operatorname{sign} x$. We refer to Dadi and Marks (1987) and Marks et al. (1978) for a detailed discussion of the performance of these detectors under the Laplace noise and their limiting behavior when the sample size N increases without bound.

7.2 Encoding and decoding of analog signals

Another standard problem in communication theory is encoding and decoding of analog signals. The distribution of such signals depends on their nature. Among the most important ones are speech signals. It has been found that the Laplace distribution accurately models speech signals. Although it was also discovered that true speech signals are strongly correlated when measured in time, in many theoretical studies, in order to avoid complications it is often assumed that samples are independent. Theoretical findings have been compared to corresponding empirical properties observed in real speech samples. In one such a study, Duttweiler and Messerschmitt (1976) considered a reduced bit-rate wave form encoding of analog signals. A concise account of their findings is presented here (we emphasize that portion in which the Laplace distribution has played a prominent role). For additional details we refer the reader to the original paper.

The method considered in Duttweiler and Messerschmitt (1976) is called nearly instantaneous companding (NIC). NIC is distinguished among most other bit-rate reduction techniques by a performance that is largely insensitive to the statistic of the input signal. The analysis of this robustness was carried out in the paper by examining the method for sinusoidal signals, Gaussian independent samples, Laplace independent samples, and real speech samples (believed to be dependent Laplace samples). The method involves grouping of some standard encoding, in the study of the so-called $\mu 255$ (PCM) encoding (assuming n-bit quantization),[18] into groups consisting of N samples. Then it reencodes the groups, exploiting in a certain manner the information about the samples with the largest magnitude to reduce the bit size to $n - 2$. Next, the encoded signal is decoded in a complementary NIC decoder to obtain back the n-sized bit codes. Finally, in order to reobtain an analog signal, decoding through an appropriate decoder ($\mu 255$ PCM) is performed.

To verify the insensitivity of the technique to the initial distribution in the signal, the NIC signal-to-quantization noise ratio (SNR) with $n = 8$ and three sets of signal statistics (sinusoidal, Laplace, Gaussian) were discussed. In Figure 7.3, we present performance for Gaussian and Laplacian inputs (we should remember that Laplace inputs are believed to approximate better the true distribution of the speech data). The comparison to SNR is made with respect to the initial encoding (in our case $\mu 255$ PCM).

The performance of the decoder depends on the block size N. At $N = 8$ the degradation is about 7dB[19] with a Laplacian distribution and 6dB with a Gaussian. The Laplacian distribution is

---

[18] In a PCM encoding of analog-to-digital converter, each bit represents a fixed voltage level. So if the least significant bit corresponds to a level V volts, then the nth bit corresponds to a level $2^n V$ volts. To achieve recognizable voice quality sampling at rates of 8000 samples per second over a 13-bit range must often be used. To reduce the range requirement a logarithmic $\mu 255$ data compander can be used to compress speech into an 8-bit word according to the formula $y(x) = V \log(1 + \mu x/V)/\log(1 + \mu)$ with the value $\mu = 255$ most often used in telephone applications.

[19] A *decibel* is a dimensionless, logarithmic unit equal to one-tenth of the common logarithm of a number expressing a ratio of two powers. In the usual case for input and output quantities in telecommunications, the decibel is a very convenient unit to express signal-to-noise ratio.

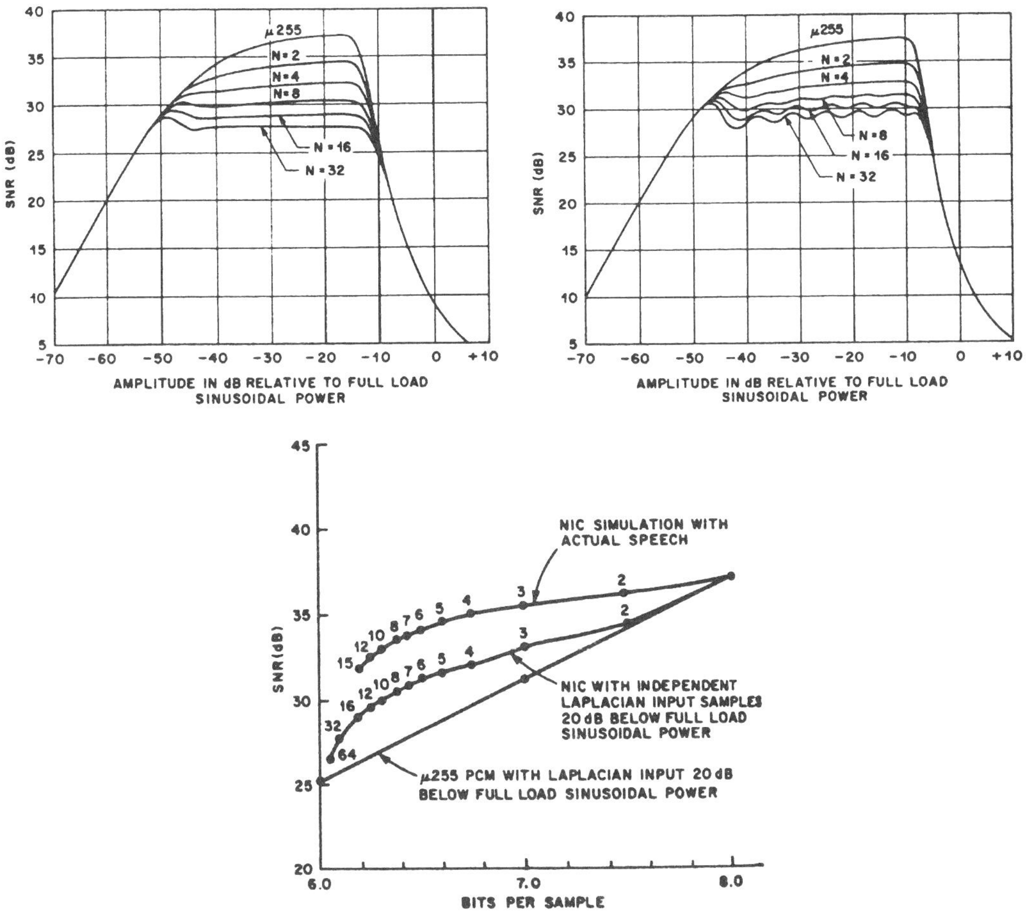

Figure 7.3: SNR vs. amplitude with independent Laplace (left) and Gaussian (right) samples. SNR of Laplacian samples vs. bits/sample (bottom). Graphs are reproduced from Duttweiler and Messerschmitt (1976) with permission of the IEEE (©1976 IEEE).

characteristic of speech, but speech samples are strongly correlated. For the simulated NIC with an actual speech input the degradation for $N = 8$ was 3.5dB.

Another interesting way of presenting SNR data consists of graphing the SNR vs. the average number of bits per sample as the block size N varies. Two of these plots appear in Figure 7.3 (*right*). One assumes independent Laplacian samples while the other is based on actual speech. The maximum advantage of NIC is 3dB with independent Laplacian samples and 6dB with actual speech. In both cases the maximum advantage occurs at about $N = 10$.

7.3 Optimal quantizer in image and speech compression

The Laplace distribution is commonly encountered in image and speech compression applications. One of the fundamental problems in this context consists of finding the so-called *optimal quantizer design*. Let us first explain the general idea of such a design.

Consider an analog signal that should be converted to a digital one. A quantizer is a method of analog-to-digital conversion. Specifically, a scalar quantizer maps each input (a continuous random variable) to its output approximation. The issue is to optimize the quantizer performance subject to some criteria. One criterion is to minimize the information rate of the quantizer as measured by its output entropy. In another approach, the mean square error of quantization is considered as a measure of performance.

Since Laplace distributions are commonly encountered in practical quantization problems, considerable attention was given to the problem of finding an optimal quantizer for Laplacian input sources. Here we shall discuss mostly the results of Sullivan (1996), but the works by Nitadori (1965), Lanfer (1978), Noll and Zielinski (1979), and Adams and Giesler (1978) are recommended to readers interested in the history of the problem.

Let an input variable, which will be subject to quantization, be modeled by a random variable X having a smooth p.d.f. $f(x)$. For convenience and without loss of generality, let us assume that $f(x)$ is zero for $x < 0$. An n-level scalar quantizer, where n is the number of possible values $\{y_i^{(n)}\}_{i=1}^n$ in the quantized output, is defined as

$$Q^{(n)}(X) = \sum_{i=0}^{n-1} y_i^{(n)} \mathbb{I}_{(t_{i+1}^{(n)}, t_i^{(n)}]}(X),$$

where $\mathbb{I}_{(a,b]}$ is an indicator function of an interval $(a, b]$ and $\{t_i^{(n)}\}_{i=0}^n$ are the $n+1$ decision thresholds for the quantizer given by

$$t_i^{(n)} = \sum_{j=i}^{n-1} \alpha_j, \quad i = 0, \dots, n-1, \quad t_n^{(n)} = 0.$$

The quantities $\{\alpha_i\}_{i=0}^{n-1}$ are positive steps ($\alpha_0 = \infty$) and the output values are defined through a set of n nonnegative reconstruction offsets $\{\delta_{i=0}^{n-1}\}$ by

$$y_i^{(n)} = t_{i+1}^{(n)} + \delta_i.$$

The distortion measure $d(\Delta)$ is any function of Δ that increases monotonically and smoothly (although not necessarily symmetrically) as its argument deviates from zero (for example the mean square error $d(\Delta) = |\Delta|^2$ is a distortion measure). The expected quantizer distortion is then defined by

$$\begin{aligned} D_f^{(n)} &= E[d(X - Q^{(n)}(X))] \\ &= \sum_{i=0}^{n-1} \int_{t_{i+1}^{(n)}}^{t_i^{(n)}} d(x - y_i^{(n)}) f(x) dx, \end{aligned}$$

and the probability of each output $y_i^{(n)}$ is

$$p_i^{(n)} = \int_{t_{i+1}^{(n)}}^{t_i^{(n)}} f(x) dx.$$

The output probabilities determine the output entropy of the quantizer, a lower bound on the expected bit rate required to encode the output, given by

$$H_f^{(n)} = -\sum_{i=0}^{n-1} p_i^{(n)} \log_2 p_i^{(n)} \quad \text{[bits per sample]}.$$

We are interested in a quantizer that minimizes the objective function

$$J_f^{(n)} = D_f^{(n)} + \lambda H_f^{(n)}$$

for some $\lambda \geq 0$. Such a quantizer is optimal in the sense that no other scalar quantizer can have lower distortion with equal or lower entropy.

In Sullivan (1996), the optimal quantizer as well a as fast algorithm for its computation for an exponentially distributed input were presented. In the case of mean squared-error distortion, the solution has an explicit form expressed by the Lambert function W, i.e., the inverse function to $f(W) = We^W$, which can be approximated by

$$W(z) = -1 + q - \frac{1}{3}q^2 + \frac{11}{72}q^3 - \frac{43}{540}q^4 + \dots,$$

where $q = \sqrt{2(ez+1)}$ [see Corless et al. (1996)]. This optimal solution α_i^* is given by

$$\alpha_{i+1}^* = \nu_i + W(-\nu_i e^{-\nu_i}),$$

where

$$\nu_i = 2 - \alpha_i^* \frac{e^{-\alpha_i^*}}{1 - e^{-\alpha_i^*}}.$$

The results on exponential source are then used to derive the optimal quantizer for a Laplace distribution. It is interesting to see how the exponential quantizer can be utilized in this case.

First, let us consider the quantizer that has an output value ϵ associated with the input value of $x = 0$. The boundaries of the step are defined by two nonnegative thresholds t_l and t_r, where $t_l + t_r > 0$, so if the input is between $-t_l$ and t_r, then the output value is equal to ϵ. The quantizer has the distortion

$$\eta(t_l, t_r, \epsilon) = \int_{-t_l}^{t_r} d(x - \epsilon)e^{-|x|}/2dx$$

and the entropy

$$T(e^{-t_l}/2, e^{-t_r}/2),$$

where

$$T(p, q) = B(p) + (1 - p)B(q/(1 - p)),$$
$$B(p) = -p\log_2 p - (1 - p)\log_2(1 - p).$$

The number of output levels on the right of t_r is n_r and on the left of t_l is n_l. Thus $n = n_r + n_l + 1$. Now we define the quantizer as the composition of three subquantizers. First, we have the one defined above for values around zero. Then for a Laplace random variable X, $X - t_r$ given that $X > t_r$ has exponential distribution and so does $-(X - t_l)$ given that $X < t_l$. Consequently, we can write

$$J_L^{(n)} = \eta(t_l, t_r, \epsilon) + \lambda T\left(\frac{1}{2}e^{-t_l}, \frac{1}{2}e^{-t_r}\right) + \frac{1}{2}(e^{-t_l}\hat{J}_e^{(n_l)} + e^{-t_r}J_e^{(n_r)}),$$

where $J_L^{(n)}$ stands for the objective function for the Laplace source, $J_e^{(n_r)}$ is the objective function for the exponential source, and $\hat{J}_e^{(n_l)}$ is the objective function of an n_l-level quantizer for an exponential source with distortion measure $\hat{d}(\Delta) = d(-\Delta)$. Using the results on the exponential source it is enough to find the minimizer for $\eta(t_l, t_r, \epsilon)$.

The method of computing these quantizers presented in Sullivan (1996) is noniterative, which is an improvement over some previous iterative refinement techniques. In addition, it is extremely fast and optimal for a general difference-based distortion measure, as well as for a restricted and unrestricted (asymptotic) number of quantization levels.

7.4 Fracture problems

In Epstein (1947, 1948), a potential application of the Laplace distribution is discussed in relation to the fracturing of materials under applied forces. The considered statistical models assume that the difference between the ideal model and observed values is due to randomly distributed flaws in the body that will weaken it. The simplest theory is based on the weakest link concept. It assumes that the strength of a given specimen is determined by the weakest point or, in other words, by the smallest values found in a sample of n, where n is the number of flaws in the considered material. This relates the problem to extreme value theory. For applications, the term *strength* can be interpreted in different ways: mechanical strength, electrical strength, resistance of painted specimens to the corrosive effects of the atmosphere, ability to stop the passage of light rays, or the life span of a device that ceases to function when any of a number of vital parts breaks down.

There is a dispute as to which distributions of the strength of a flaw are correct ones. Based on experimental data the following characteristics of the distribution should be accounted for: some experimenters have observed that the mode of the strength decreases as a function of the logarithm of the size of specimen; the distribution of strengths of specimens all of the same size appears to be negatively skewed; in the breakdown of capacitors the sizes of conducting particles (flaws) are distributed according to an exponential law. In the last example, it can be easily shown that the most probable value of the breakdown voltage depends linearly on the logarithm of the area.

Epstein (1947, 1948) considers several common distributions of the strength of a flaw given by a density $f(x)$ including Laplacian, Gaussian, and Weibull densities. Several issues are of interest in this context. First, one would like to know the asymptotic distribution of the smallest value in a sample of size n. Then it is important how specimen size (represented by n) affects the distribution of strengths. In particular, one would like to know how the mode, the mean, and variance of the smallest value depend on the size n. Rather standard arguments led Epstein to the results summarized in Table 7.1.

Distribution	**Smallest value distr. (large *n*)**	**Mode $\tilde{y}$**	**Mean**	**Variance**
Laplace $\frac{1}{2\lambda}e^{-\lvert x-\mu\rvert/\lambda}$	$\mu-\lambda\log\frac{wn}{2}$	$\mu-\lambda\log(n/2)$	$\tilde{y}-0.577\lambda$	$\frac{\lambda^2\pi^2}{6}$
Gaussian $\frac{1}{\sqrt{2\pi}\sigma}e^{-\frac{(x-\mu)^2}{2\sigma^2}}$	$\mu-\sigma(\sqrt{2\log n}-\frac{\log\log n+\log 4\pi}{2\sqrt{2\log n}}-\frac{\log w}{\sqrt{2\log n}})$	$\mu-\sigma(\sqrt{2\log n}-\frac{\log\log n+\log 4\pi}{2\sqrt{2\log n}})$	$\tilde{y}-\frac{0.577\sigma}{\sqrt{2\log n}}$	$\frac{\sigma^2\pi^2}{12\log n}$
Weibull $\alpha\beta x^{\beta-1}e^{-\alpha x^\beta}$	$\left(\frac{w}{n\alpha}\right)^{1/\beta}$	$\left(\frac{\beta-1}{\alpha n\beta}\right)^{1/\beta}$	$\frac{\Gamma\left(\frac{\beta+1}{\beta}\right)}{(n\alpha)^{1/\beta}}$	$\frac{\Gamma\left(\frac{\beta+2}{\beta}\right)-\Gamma^2\left(\frac{\beta+1}{\beta}\right)}{(n\alpha)^{2/\beta}}$

Table 7.1: Summary of the results from Epstein (1948) on the distribution of strength in the weakest link model depending on the distribution of the strength of a flaw (w stands for a standard exponential random variable).

From this summary we see that the assumption on the form of distribution affects in a significant way properties of the strength of a specimen. Moreover there are physical data available that follow each of the patterns exhibited by the distributions listed in the table. For example, breakdown voltages of capacitors have a distribution of the Laplace type. The derived properties when put in a physical context give information how the size depends on the distribution of strengths in the vicinity of flaws. Specifically, the specimens become weaker as the size increases. In the case of Laplace distribution it decreases linearly with $\log n$, for Gaussian distribution the dependence is through $\sqrt{\log n}$, and for Weibull distributions the dependence is through negative powers of n. The spread of the distribution remains unchanged in the Laplace case, and in two other cases it decreases with the specimen size.

Epstein's works carried out over 50 years ago generated vast literature on this subject related to extreme value distributions. For our purposes it is sufficient to note that the Laplace distribution appears on an equal footing with the more popular distributions at that period such as Gaussian and Weibull.

7.5 Wind shear data

Barndorff-Nielsen (1979) proposed the hyperbolic distributions for modeling turbulence[20] encountered by an aircraft. The model is quite complicated and difficult to handle when parameter estimation is considered. Kanji (1985), noticing that the Laplace and Gaussian distributions are limiting cases of the hyperbolic distributions, proposed the mixture of these two as a model for wind shear data.[21] Wind shears are encountered by an aircraft during the approach to landing and their distribution is critical for assessing the effectiveness and safety of aircraft and for training pilots to react correctly when they encounter a wind shear.

Kanji (1985) had worked with 24 sets of data on wind shear collected during the last two minutes of landing of a passenger aircraft. The measurement represents the gradient of airspeed change against its duration. The basic assumption is that a wind shear forms an individual gust that has a strictly defined form specified by its duration and the magnitude of change of the air velocity. The 120 seconds in flight before touchdown was split into four bands, the first two of 40 seconds length and the last two of 20 seconds length. The histograms of the data suggested that for the early stage (first 40 seconds) of landing the Laplace distribution seems to fit the data well, while for the last 20 seconds less peaky Gaussian distribution appears to be appropriate. Considering this, Kanji proposed the mixture model

$$p_1(x; \mu, \sigma, \alpha) = \alpha \frac{1}{\sqrt{2}\sigma} e^{-\sqrt{2}|x-\mu|/\sigma} + (1-\alpha) \frac{1}{\sqrt{2\pi}\sigma} e^{-(x-\mu)^2/(2\sigma^2)}, \tag{7.5.1}$$

a mixture of Laplace and Gaussian distributions having the same mean and variance. The proposed estimation procedure starts with the estimation of the mean and the variance for both components in the model and then employs the chi-square goodness-of-fit procedure to fit the mixing constant α. The inference led to the following approximate values of α in the four time bands: 0.9, 0.6, 0.5, 0.3, respectively, confirming that wind shear data lose their Laplacian character in the earlier stages to the Gaussian at the end of landing. The fit was significant in all but nine out of 24 cases.

In Jones and McLachlan (1990), a mixture of Laplace and Gauss distributions was studied in the same context. This time, however, the authors did not assume equal variance in the components and demonstrated an appropriately modified estimation procedure leading to even better fits than

[20]Random changes in wind velocity with insufficient duration to significantly affect an aircraft's flight path.

[21]A change in wind velocity of sufficient magnitude and duration to significantly affect an aircraft's flight path and require corrective action by the pilot or autopilot.

those obtained by Kanji (1985). Further discussion on justification and parameter estimation of the mixture model (7.5.1) can be found in Kapoor and Kanji (1990) and Scallan (1992), respectively.

7.6 Error distributions in navigation

In Anderson and Ellis (1971), we find an interesting discussion of error distributions reported in ocean engineering. The authors analyze 54 distributions and conclude that most of them have exponential or even heavier tails and only a few seem to follow Gaussian law. In heuristic fashion, they argue that only for equipment made of identical items do the collected data follow the Gaussian law. For example, it was observed that the frequency distributions of a single pilot operating the same set of facilities and under similar navigational environments appear to be well modeled by Gaussian laws. If the data are collected by instruments that, although nominally the same, are much more diverse and far from identical, then the data exhibit longer tails. This is due to variability of variance for different instruments. In the aircraft navigation data it was repeatedly observed that if data are collected from a fairly complex navigation system, there is a strong tendency to exhibit exponential tail behavior. The question is then how to rationalize the applicability of two so different distributions. The answer given in Anderson and Ellis (1971) suggests considering Gaussian distributions with random variances.

As an example, consider two gauges, one new and one older and worn out. The variance of the second one will lead to data far from their true value, and the distribution can be closer to a Laplace distribution than to a Gaussian one. The authors suggest the use of distributions with exponential or even heavier tails for navigation data. They derive such distributions by combining observations from a number of Gaussian distributions that cover a range of standard deviations. Of course, various distributions (or patterns, as the authors describe them) of standard deviations will lead to different distributions of the errors (we know that one such possibility is the Laplace distribution if the distribution of the standard deviation is Rayleigh; see 2.2.5). They note: "In the past, navigation statistics have tended to be a conglomeration of single observations from various origins and there has been no need to examine the range of standard deviations from each origin. Therefore, we do not know the pattern which these standard deviations are likely to follow."

Lack of information about the distribution of the standard deviation prevents the authors from making any strong recommendation on the type of error distribution, except that they strongly favor in some situations "log-tail" (in our terminology exponential-tail) distributions:

"The navigator will remember that the Gaussian distribution can arise if one observer (without blunders) operates one equipment (without integrators) under one set of stable conditions! If his information is based on a number of diverse sources (or even if it is based on one source and the navigator has a healthy pessimism) the log-tail distribution will be preferable within the limits in which he is likely to be interested."

In conclusion, after studying the difference in quantiles between the Gaussian distribution and an alternative distribution that is Gaussian with random variance (they do not consider Laplace distribution) they say: "If the Gaussian distribution is assumed for errors, and if the standard deviation is deduced from observations based on a large number of equipments and operators, there will in fact be considerably more extreme results than predicted by the assumption."

The argument they provide in favor of the models based on Gaussian mixtures with stochastic variance can be easily extended to other areas of applied research. For this reason Laplace distributions can serve as valuable models in situations heuristically described above.

In Hsu (1979), the model with Laplace distribution was investigated and compared with the real-life data on navigation errors for aircraft position. The data were collected by the U.S. Federal Aviation Administration over the Central East Pacific Track System. The position errors in the lateral direction (along the tracks) were recorded for the traffic heading to Oakland (3435 data points) and Los Angeles (4147 data points). The following five models were fitted to the data: Gaussian,

Laplace, t-distribution, a mixture of two Laplace, and a mixture of two t-distributions. The best fit, particularly in the tail region, was obtained by a mixture of two Laplace distributions. On the other hand, the Gaussian distribution performed rather poorly. It is worth emphasizing that a model adequately describing the tail behavior is of paramount importance in this application. The simplicity of the models based on Laplace distributions and their empirical adequacy adds much to its practical applicability, as illustrated by Hsu (1979) in the application of the proposed Laplace model for the calculation of aircraft midair collision risk. This risk is based on the probability of track overlap by two aircraft that take adjacent parallel tracks with some nominal lateral separation in nautical miles. The computation of this distribution (which is the convolution of the navigation error distributions for the considered two tracks) is possible for all models and it was found that the models other than the mixture of Laplace distributions tend to underestimate the overlap for most of the range of the nominal separation considered.

8
Financial Data

An area where the Laplace and related distributions can find most interesting and successful applications is modeling of financial data. This is due to the fact that traditional models based on Gaussian distribution are very often not supported by real-life data because of long tails and asymmetry present in these data. Since Laplace distributions can account for leptokurtic and skewed data they are natural candidates to replace Gaussian models and processes. In fact, some activity involving the Laplace distribution can already be observed in this area. Laplace motion and models based on multivariate Laplace laws have appeared in works on modeling stock market returns, currency exchange rates, and interest rates. In this chapter, we present several such applications.

It is important to mention that interesting materials exist on applications of hyperbolic and normal inverse Gaussian distributions to financial data [see, e.g., Eberlein and Keller (1995), Barndorff-Nielsen (1997)]. Since generalized Laplace distributions can be viewed as special cases of hyperbolic distributions, the mentioned work also supports their application to stochastic volatility modeling. In particular, the estimation based on German stock market data in Eberlein and Keller (1995) confirms most of claims in Section 8.4. We do not report these results as not directly related to the Laplace laws but we recommend the cited work to those interested in financial modeling.

8.1 Underreported data

Consider a Pareto random variable Y_* with p.d.f.

$$p_1(y_*) = \begin{cases} \frac{\gamma}{m}\left(\frac{m}{y_*}\right)^{\gamma+1}, & \text{for } y_* \geq m, \\ 0, & \text{for } 0 < y_* < m. \end{cases} \tag{8.1.1}$$

The Pareto distribution has been found useful for modeling a variety of phenomena, including distributions of incomes, property values, firm or city values, word frequencies, migration, etc. However, as remarked by Hartley and Revankar (1974), in many applications (particularly those dealing with income or property values) one may reasonably expect that the reported values *underestimate* the true values of a given variable of interest. To account for this, Hartley and Revankar (1974) consider Y_* with density (8.1.1) as an *unobservable* (true) variable, which is related to an *observable* variable

Y via the equation

$$Y = Y_* - U, \tag{8.1.2}$$

where the variable U $(0 \le U \le Y_*)$ is a *positive* underreporting error. The goal here is to make inference about the distribution of Y_* (that is, to estimate the parameters γ and m) based on a random sample from Y. To accomplish this, one needs to relate the p.d.f. of Y to the parameters γ and m of Y_*. Hartley and Revankar (1974) postulate that the proportion of Y_* that is underreported, denoted by

$$W_* = \frac{U}{Y_*}, \tag{8.1.3}$$

is distributed *independently* of Y_* with the p.d.f.

$$p_2(w_*) = \lambda(1 - w_*)^{\lambda - 1}, \quad 0 \le w_* \le 1, \quad \lambda > 0. \tag{8.1.4}$$

Then the observable r.v. Y given by (8.1.2) has the p.d.f.

$$g(y) = \frac{\gamma}{m}\frac{\lambda}{\lambda + \gamma}\begin{cases} \left(\frac{m}{y}\right)^{\gamma+1}, & \text{for } y \ge m, \\ \left(\frac{y}{m}\right)^{\lambda-1}, & \text{for } 0 < y < m. \end{cases} \tag{8.1.5}$$

We now recognize (8.1.5) as the p.d.f. of a *log-Laplace* distribution. Indeed, writing $X = \log Y$ and denoting

$$\sigma = \sqrt{\frac{1}{\lambda\gamma}}, \quad \kappa = \sqrt{\frac{\gamma}{\lambda}}, \quad \theta = \log m, \tag{8.1.6}$$

we find that the p.d.f. of X is

$$h(x) = \frac{1}{\sigma}\frac{1}{\kappa^{-1} + \kappa}\begin{cases} e^{-\kappa|x-\theta|/\sigma}, & \text{for } x \ge \theta, \\ e^{\frac{1}{\kappa}|x-\theta|/\sigma}, & \text{for } x < \theta, \end{cases} \tag{8.1.7}$$

which is a three-parameter $\mathcal{AL}^*(\theta, \kappa, \sigma)$ density [see also Hinkley and Revankar (1977)]. Thus AL laws have found applications in economics in connection with modeling (underreported) income and similar variables.

8.2 Interest rate data

In this section we present an application of AL distributions in modeling interest rates on 30-year Treasury bonds. Klein (1993) studied yield rates on average daily 30-year Treasury bonds from 1977 to 1990, finding that the empirical distribution is too "peaky" and "fat-tailed" to have been from a normal distribution. He rejected the traditional log-normal hypothesis and proposed the Paretian stable hypothesis, which would "account for the observed peaked middle and fat tails." The paper was followed by several discussions, where some researchers objected to the stable hypothesis and offered alternative models.

Kozubowski and Podgórski (1999a) suggested an AL model for interest rates, arguing that this relatively simple model is capable of capturing the peakedness, fat-tailedness, skewness, and high kurtosis observed in the data. These authors considered a data set consisting of interest rates on 30-year Treasury bonds on the last working day of the month [published in Huber's discussion of Klein's paper, p. 156]. The data cover the period of February 1977 through December 1993. Converting the

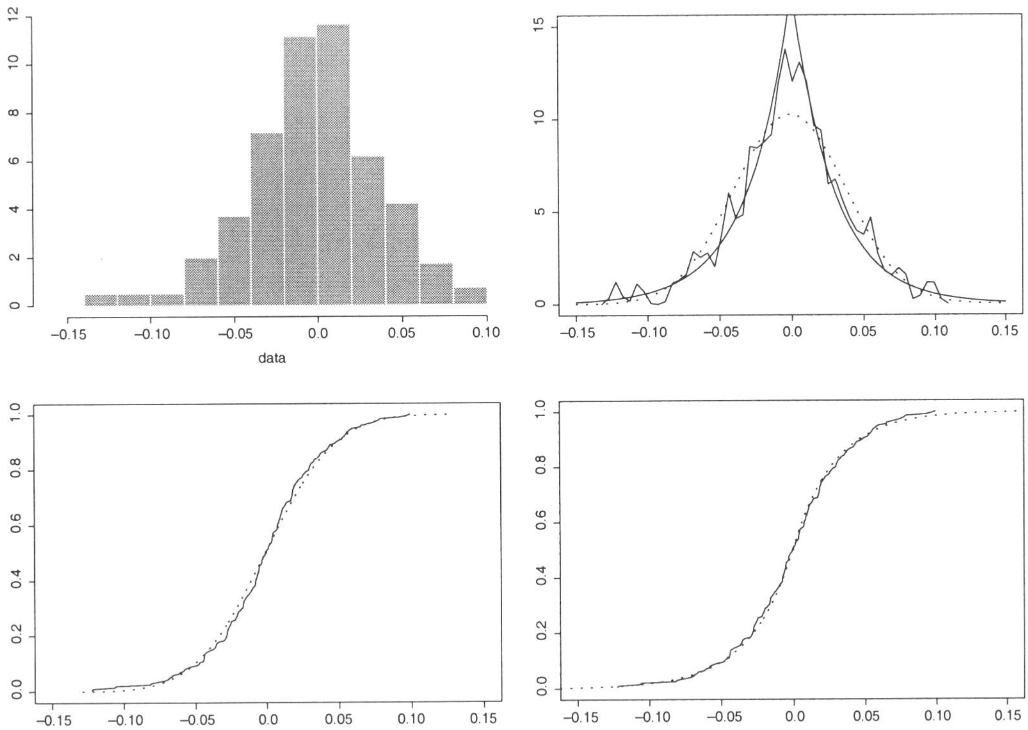

Figure 8.1: *Top left*: Histogram of interest rates on 30-year Treasury bonds. *Top right*: Nonparametric estimator of the density (thin solid line) vs. the theoretical ones (normal — dashed line; AL — thick solid line). *Bottom left*: Empirical c.d.f. vs. normal c.d.f. *Bottom right*: Empirical c.d.f. vs. AL c.d.f. (From Kozubowski and Podgórski (1999a).)

data to the logarithmic changes, $Y_t = \log(i_t/i_{t-1})$, where i_t is the is the interest rate on 30-year Treasury bonds on the last working day of the month t, the authors assume that the resulting 202 values of the logarithmic changes Y_i are i.i.d. observations from an AL distribution.

The histogram of the data set appears in Figure 8.1 (top left). The typical shape of an AL density is apparent: the distribution has high peak near zero and appears to have tails thicker than that of the normal distribution. Comparisons of the empirical c.d.f. with the normal c.d.f. (Figure 8.1, bottom left) and the empirical density with the normal density (Figure 8.1, top right) confirm these findings. We observe a disparity around the center of the distribution due to a high peak in the data. To fit an AL model, one needs to estimate the parameters μ and σ. Kozubowski and Podgórski (1999a) used the maximum likelihood estimators, obtaining

$$\widehat{\mu} = -0.007178218 \text{ and } \widehat{\sigma} = 0.294043202,$$

and then calculated the parameter κ and other related parameters. The resulting values are presented in Table 8.1, along with corresponding empirical counterparts:

1. Sample Mean: $\frac{1}{n}\sum Y_i$.
2. Sample Variance: $\frac{1}{n}\sum(Y_i - \overline{Y})^2$.
3. Sample Mean Deviation: $\frac{1}{n}\sum|Y_i - \overline{Y}|$.

Parameter	Theoretical value	Empirical value
Mean	−0.001018163	−0.001018163
Variance	0.001733809	001372467
Mean deviation	0.02944785	0.02945773
Mean dev./ Std. dev.	0.7072175	0.7582487
Coefficient of Skewness	−0.07334177	−0.2274964
Kurtosis (adjusted)	3.003586	3.599207

Table 8.1: Theoretical vs. empirical moments and related parameters of $Y \sim \mathcal{AL}(\widehat{\mu}, \widehat{\sigma})$.

4. Sample Coefficient of Skewness: $\widehat{\gamma}_1 = \frac{1}{n}\sum(Y_i - \overline{Y})^3/(\frac{1}{n}\sum(Y_i - \overline{Y})^2)^{3/2}$.
5. Sample Kurtosis (adjusted): $\widehat{\gamma}_2 = \frac{1}{n}\sum(Y_i - \overline{Y})^4/(\frac{1}{n}\sum(Y_i - \overline{Y})^2)^2 - 3$.

Except for a slight discrepancy in skewness, the match between empirical and theoretical values is remarkable. In Figure 8.1 the theoretical AL c.d.f. is compared with the empirical c.d.f. (bottom right) and the density kernel estimator based on the data is compared with the theoretical densities of normal and AL distributions with the estimated parameters. We observe a better agreement with the AL distribution than with the normal one.

8.3 Currency exchange rates

We present an application of AL distributions in modeling foreign currency exchange rates taken from Kozubowski and Podgórski (2000). Following the ideas of Mittnik and Rachev (1993), we may view an exchange rate change as a sum of a large number of small changes, where the sum is taken up to a random time ν_p (that has a geometric distribution):

$$\text{exchange rate change} = \sum_{i=1}^{\nu_p} \text{(small changes)}.$$

The random nature of time reflects the volatility and unpredictability of the factors that contribute to the establishment of a current exchange rate. Therefore, the AL laws (provided the small changes have finite variance) are very likely to approximate the distribution of the exchange rate change. We may think of ν_p as the moment when the probabilistic structure governing the exchange rates breaks down. This can be new information, political, economical, or other events that affect the fundamentals of the exchange market.

Kozubowski and Podgórski (2000) fitted AL laws to a bivariate data set on two currency commodities: the German Deutschmark vs. the U.S. Dollar (DMUS) and the Japanese Yen vs. the U.S. Dollar (YUS). The observations were daily exchange rates from 1/1/80 to 12/7/90 (2853 data points). (The standard change in the log(rate) from day t to day $t+1$ was used.)

The histograms of the data appear in Figure 8.2, where we observe a shape typical of AL density. The distributions have high peaks near zero and appear to have tails thicker than that of the normal distribution. The normal quantile plots (QQ plots) in Figure 8.3 (top) confirm these findings. Observe that the normal plots deviate from a straight line rather substantially. To fit an AL model, we need to estimate the parameters μ and σ. The maximum likelihood estimators produced

$$\widehat{\mu} = 0.0007558 \quad \text{and} \quad \widehat{\sigma} = 0.00521968$$

for the German Deutschmark data and

$$\widehat{\mu} = 0.0007272 \quad \text{and} \quad \widehat{\sigma} = 0.0049445$$

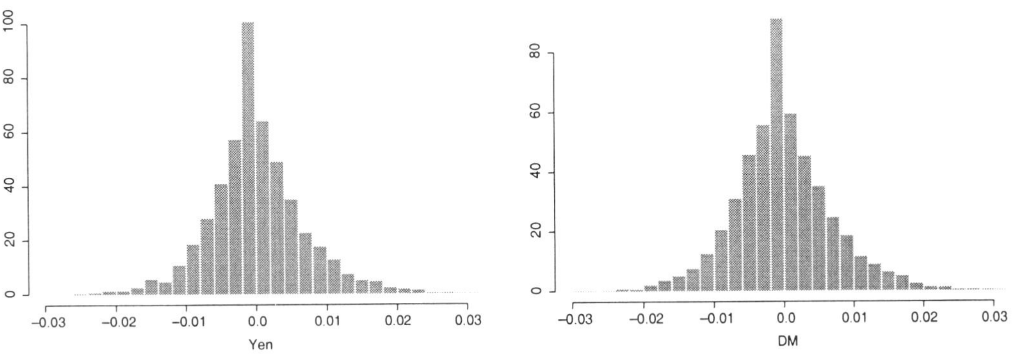

Figure 8.2: Japanese Yen (left) and German Deutschmark (right) daily exchange rates, 1/1/80 to 12/7/90.

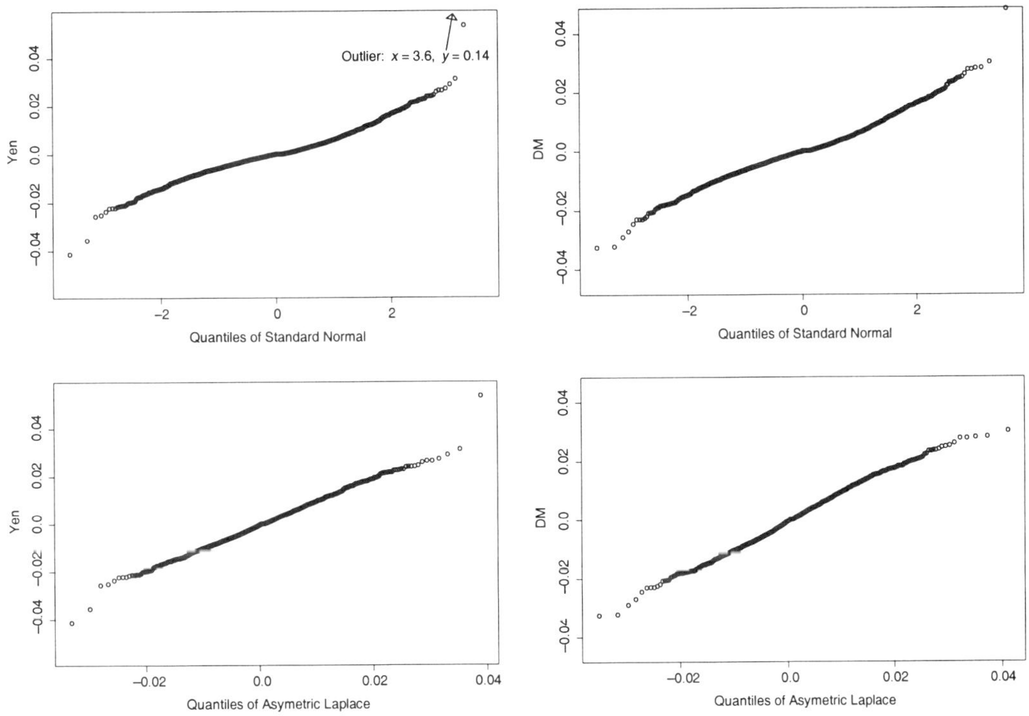

Figure 8.3: *Top*: Normal quantile plots of Japanese Yen (left) and German Deutschmark (right) exchange rate data. *Bottom*: Quantile plots of Japanese Yen (left) and German Deutschmark (right) exchange rate data vs. fitted AL distributions.

for the Japanese Yen data. The quantile plots of the two data sets with theoretical AL distributions are presented in Figure 8.3 (bottom). We see only a slight departures from the straight line. It is evident even to the naked eye that AL distributions model these data more appropriately than normal distributions. We refer the reader to Kozubowski and Podgórski (1999c) for a more in-depth study of modeling the distribution of currency exchange rates with AL laws.

8.4 Share market return models

8.4.1 Introduction. The application of Laplace motion as defined in Section 4.2, Chapter 4, to modeling share market returns has been investigated in many recent papers, starting with Clark (1973) (although indirectly) and during the last decade in Madan and Seneta (1990), Madan and Milne (1991), Longstaff (1994), Eberlein and Keller (1995), Barndorff-Nielsen (1996, 1997) (through more general models based on hyperbolic distributions), Madan et al. (1998), and Geman et al. (2000ab).

It is empirically evident that stock price changes do not follow normal distribution. In particular, sample excess kurtosis for many available financial data is significantly greater than zero (zero corresponds to normal distribution). This deviation from normality implies that the assumptions of the Central Limit Theorem may not be valid for individual random effects making up a price change. One solution, as postulated by Mandelbrot (1963), is to consider individual effects not having finite variance. The resulting distribution should then belong to the class of stable distributions (a.k.a. Paretian stable laws). An alternative solution, as suggested in Clark (1973), is to consider a subordinated Gaussian process. Considering cotton futures, he argues that their prices evolve at different rates during identical time intervals. This is presumably due to the fact that the number of individual effects that add together to give the price change during a fixed time unit, say a day, is random. Thus a version of the Central Limit Theorem with a random number of elements should be used to obtain an approximate distribution of a daily stock price. Clark (1973) describes the rationale behind these assumptions: "The different evolution of price series on different days is due to the fact that information is available to traders at a varying rate. On days when no new information is available, trading is slow, and the price process evolves slowly. On days when new information violates old expectations, trading is brisk, and the price process evolves much faster."

In economic literature, this argument is described through the assumption that the business (or economic) time runs randomly relative to physical time [see Madan and Seneta (1990), Geman et al. (2000ab)]. This sort of argument leads to the subordinated model of stock prices $S(t) = X(T(t))$, where $X(t)$ and $T(t)$ are two independent stochastic processes: $X(t)$ is the stock price in business time t, and $T(t)$ is business time at real time t.

If we assume that $B(t) = \log X(t)$ is a Brownian motion and that $T(t)$ is a gamma process, then the process $L(t) = \log S(t)$ is a Laplace motion. In the work of Madan et al. (1998) and some other works oriented toward applications in finance, this process is named the *variance gamma process*. This new model for security prices enjoys several major advantages when compared with other models discussed in the literature on the subject. In particular, it incorporates asymmetry, heavy-tailedness, continuous time specification, finite moments of all orders, and an elliptically contoured multivariate counterpart, and it provides adequate empirical fit. Additional features include approximation by a compound Poisson process and representation as a Brownian motion evaluated at random time governed by a gamma process. The last representation is interpreted as a mathematical interpretation of an economic clock ticking in a random fashion. All these features are direct consequences of the properties of the Laplace motion studied in Section 4.2, Chapter 4.

The following is a brief description of the model and its basic properties.

8.4.2 Stock market returns. We consider a particular commodity with stock price S_t at time t. We assume that $\{S_t\}_{t\geq 0}$ is a random process and the return over the time unit is given by

$$R = \frac{S_{t+1}}{S_t}.$$

Then the log-return is defined as

$$L = \log R. \tag{8.4.1}$$

In most models, it is assumed that the distribution of R does not depend on t, so the dependence of R on t is not exhibited in the notation.

More generally, the stochastic process

$$S(t) = S(0)\exp(L_t)$$

usually represents the stock price $S(t)$ at time t, where the process L_t has homogeneous increments, i.e., $L_{t+s} - L_t \stackrel{d}{=} L_s$. Note that (by (8.4.1)) we have $L \stackrel{d}{=} L_1$.

The literature on market returns includes a number of models for L_t: Brownian motion, symmetric stable processes, normally distributed jumps at Poisson jump times, models based on t-distribution, and generalized beta distributions. A model based on the Laplace motion (the variance gamma process) can be introduced by assuming that L_t has homogeneous and independent increments and that L_1 has a shifted generalized Laplace distribution. Thus

$$L_1 \stackrel{d}{=} \mathcal{GAL}(a, \mu, \sigma, \nu), \tag{8.4.2}$$

where the parameters of the generalized Laplace distribution (a, μ, σ, and ν), and the interest rate r are related through

$$a = r + \frac{1}{\nu}\log\left(1 - \mu - \frac{\sigma^2}{2}\right).$$

The additional shift $\log(1 - \mu - \sigma^2/2)/\nu$ is a result of the drift

$$E[\exp(L_t)] = 1/(1 - \mu t - \sigma^2 t/2)^{1/\nu}$$

and is added in order to have $E\exp(S(t)) = e^{rt}$.

Asymmetric generalized Laplace distribution (skewed Bessel K-function distribution) was probably, in this context, first considered in Longstaff (1994). He assumes that L_t is a conditional Brownian motion with the gamma stochastic variance and a shift in the mean proportional to this stochastic variance (without any substantiation of the gamma distribution for the variance). The stochastic process is not specified except for one-dimensional distributions, which allows for other than Laplace motion models for L_t (see Exercise 4.5.10 in Chapter 4).

Madan and Seneta (1990) considered the symmetric Laplace motion, showing that in this case ($\mu = 0$) the agreement of the Laplace model with real data is very good. Madan and Seneta (1990) compared the (symmetric) Laplace motion model with the normal, the stable, and the Press compound events model (ncp), using a chi-square goodness-of-fit test statistic on the data on 19 stocks quoted on the Sydney Stock Exchange. For 12 of the studied stocks, the minimum chi-square was attained by the Laplace motion model. The remaining seven cases were best characterized by the ncp for five cases and the stable for two cases (and none for the normal distribution). Thus the Laplace motion appears to be a good contender as a model of daily stock returns. The studies of Madan et al. (1998) confirm this opinion to an even greater extent for the asymmetric Laplace motion.

Madan et al. (1998) studied the empirical prices for the S&P 500 Index futures traded at the Chicago Mecantile Exchange (CME) obtained from the Financial Futures Institute in Washington, DC for the period from January 1992 to September 1994. Using the maximum likelihood approach, the authors fitted these data with the following models: Brownian motion (the popular Black–Sholes model), symmetric Laplace motion, and asymmetric Laplace motion. The three models were considered both for the statistical process of the stock price and for the risk neutral process that was obtained using the data on the three-month Treasury Bill rate obtained from the Federal Reserve Board in Washington, DC.

For the statistical process of the log-price, it was found that the log-normal process is strongly rejected in favor of the symmetric Laplace motion while the asymmetric Laplace motion makes no significant improvement in fit over the symmetric one.

The situation is essentially different for the risk neutral process where an enhancement of skewness is observed as a result of risk aversion in equilibrium. For example, the log-normal model is rejected in favor of the symmetric Laplace motion in 30.8% of the tests, while the analogous rate for asymmetric Laplace motion is 91.6%.

8.5 Option pricing

Once the model for the price change of a commodity is decided on, it is important to find an effective and operational formula for the price of an option. Probably the most important advantage of the Laplace model given by (8.4.2) is that it allows for a closed form of the price of a European option on the stock using the Black–Sholes formula for the Brownian motion model of price change. The results are shown in Longstaff (1994) and Madan et al. (1998).

The price of a European call option $C(S_0, K, t)$ for a strike of K and maturity t with the initial value of the stock $S(0) = S$ is given by

$$C(S, K, t) = e^{-rt} E[\max(S(t) - K, 0)], \tag{8.5.1}$$

where the expectation is taken with respect to the risk-neutral density. Evaluation of the option price (8.5.1) is based on the representation given in Theorem 4.2.1. Conditionally on the value of the random time, we have a standard Brownian motion model and the Black–Sholes formula can be applied. The European option price is then obtained by integrating out the gamma process.

Theorem 8.5.1 *The European call option price on a stock, when the stock price is given by the Laplace motion through the condition* (8.4.2), *is given by*

$$\begin{aligned} C(S, K, t) = {} & S \cdot \Psi\left(d\sqrt{\frac{1}{\nu} - \frac{(\alpha+s)^2}{2}},\, s(\xi+1)\Big/\sqrt{\frac{1}{\nu} - \frac{(\alpha+s)^2}{2}},\, \frac{t}{\nu}\right) \\ & - K \cdot e^{-rt} \cdot \Psi\left(d\sqrt{\frac{1}{\nu} - \frac{\alpha^2}{2}},\, \xi^2 s\Big/\sqrt{\frac{1}{\nu} - \frac{\alpha^2}{2}},\, \frac{t}{\nu}\right), \end{aligned}$$

where

$$d = \frac{1}{s}\left[\log\frac{S}{K} + rt + \frac{t}{\nu}\cdot\log\frac{2 - \nu(\alpha+s)^2}{2 - \nu\alpha^2}\right]$$

and Ψ *is the complementary Bessel function given by the following integral involving the standard normal distribution function* Φ:

$$\Psi(a, b, \gamma) = \int_0^\infty \Phi\left(\frac{a}{\sqrt{u}} + b\sqrt{u}\right)\frac{u^{\gamma-1}e^{-u}}{\Gamma(\gamma)}du.$$

The proof of this theorem can be found in Madan et al. (1998). This formula is similar to the one based on the Black–Sholes model. The only difference is that the Bessel function is used instead of the normal distribution. Computationally, the formula is more complex than the traditional Black–Sholes formula since it involves a double integral of elementary functions. It is nevertheless practical as used by Madan et al. (1998) in their numerical computations on the data discussed in the previous section.

Here for each fit to the three models the option price was computed for 143 weeks. Then the pricing error was computed. For a correct model the pricing errors should not exhibit any consistent pattern and they should not be predictable (orthogonality tests were used to determine whether the prices resulting from a given model were biased or not). From these studies it follows that the asymmetric Laplace motion provides acceptable pricing that removes the so-called volatility smile so often reported in the financial literature for the Black–Sholes prices. For a detailed description of the statistical analysis we refer readers to Madan et al. (1998).

8.6 Stochastic variance Value-at-Risk models

Research very closely related to modeling of stock market returns was presented in Levin and Albansese (1998) and Levin and Tchernitser (1999), where value-at-risk (VaR) models with multifactor gamma stochastic variance were recommended and supported by theoretical results and real-life data.

Let X be a random *risk factor*. Assume first that it is modeled by a one-dimensional random variable. An investment strategy is represented by a portfolio, say $\Pi(X)$, which depends on this factor and denotes the return of investment over some fixed period of time (e.g., one day or 10 days). The VaR at the level $p \in (0, 1)$ is then defined as the p-quantile of the distribution of $\Pi(X)$:

$$P(\Pi(X) \leq \text{VaR}) = p.$$

If the portfolio is a linear function of X, the distribution of the risk factor X determines the value of VaR. Usually, the assumption of a normality of X is not supported by real-life data. Figure 8.4 shows that the data are not well modeled by normal density. Assuming Gaussian distribution may lead to misleading values of VaR (they would be too small in absolute value when compared to the actual VaR's). The real data exhibit more peakedness, heavier tails, and often skewness. None of these features can be modeled accurately by a Gaussian density. (See also Figures 8.5 and 8.6.) For example, the returns of three-month FIBOR presented in Figure 8.4 show

- skewness equal to -0.98 and kurtosis equal to 49.0 for daily returns;
- skewness equal to -0.46 and kurtosis equal to 5.6 for 10-day returns.

In Figure 8.5, we see that the symmetric Laplace distribution fits currency exchange data by far better than does the Gaussian distribution.

It is not uncommon in financial research to consider a modification of the normality assumption by allowing for random variance in the normal model (see also Section 8.4). In addition, in the work discussed therein, the maximum entropy principle was evoked to determine the distribution of such a random variance of the risk factor. More precisely, consider the following assumptions on the distribution of the risk factor X.

Assumption 8.6.1 *Conditionally on V, the distribution of the risk factor X is normal with the mean θ and variance V, i.e.,*

$$X = \sqrt{V}Z + \theta,$$

where Z is a standard normal variable independent of a positive random variable V having the mean $V_0 = \sigma_0^2$.

Assumption 8.6.2 *The distribution of the variance $V \geq 0$ has to satisfy the maximum entropy principle under the constraint*

$$E(V) = V_0.$$

As we already know (see Section 2.4.5) these assumptions lead to the model with variance V distributed according to exponential law and thus by representation 2.2.3 the unconditional distribution of risk factor is given by the Laplace law $\mathcal{L}(\theta, \sigma_0)$. Of course, this allows for explicit computation of the VaR values via the formulas for the quantiles of the Laplace distribution.

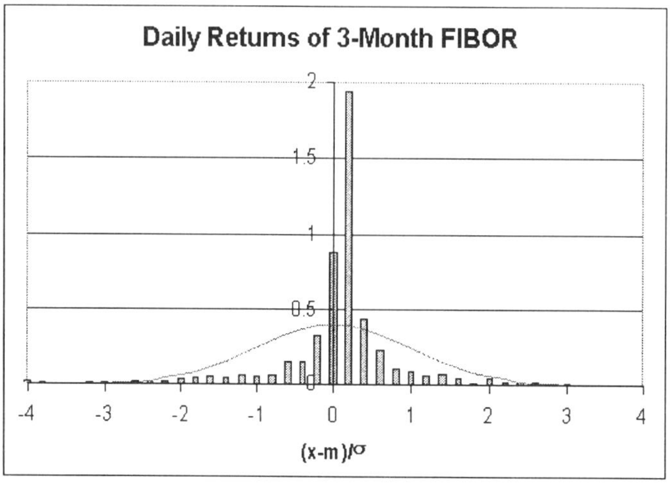

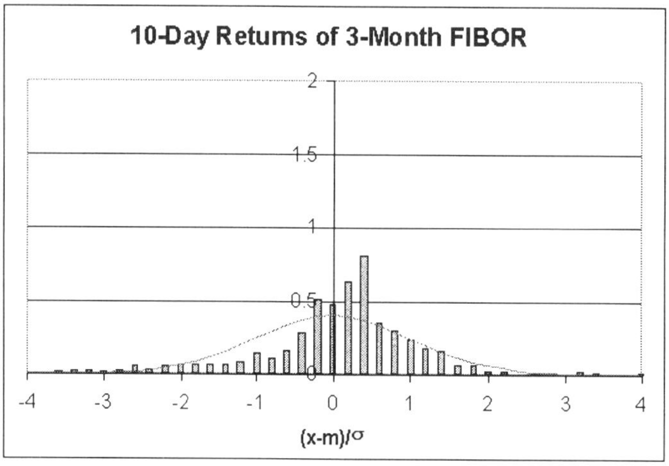

Figure 8.4: Comparing histograms of risk factors with Gaussian model: Daily and 10-day returns of three-month FIBOR. (Courtesy of Alexander Levin.)

Remark 8.6.1 In finance, the notion of *volatility* is commonly used to describe the square root of variance (the standard deviation). Note that if V is exponential, then the volatility $\sqrt{V}$ is distributed according to the Rayleigh distribution.

It may be reasonable to replace Assumption 8.6.1 by the following one.

Assumption 8.6.3 *Conditionally on V, the distribution of the risk factor X is normal with the mean $\theta - \gamma V$ and variance V, i.e.,*

$$X = \sqrt{V} Z - \gamma V + \theta,$$

where Z is a standard normal variable independent of a positive random variable V having the mean $V_0 = \sigma_0^2$. The parameter γ controls the correlation between the risk factor X and the stochastic variance V.

Then the distribution of the risk factor becomes asymmetric Laplace $\mathcal{AL}(\theta, -\sigma_0^2\gamma, \sigma_0)$. Again the VaR can be explicitly computed as the quantiles of asymmetric Laplace laws are readily available.

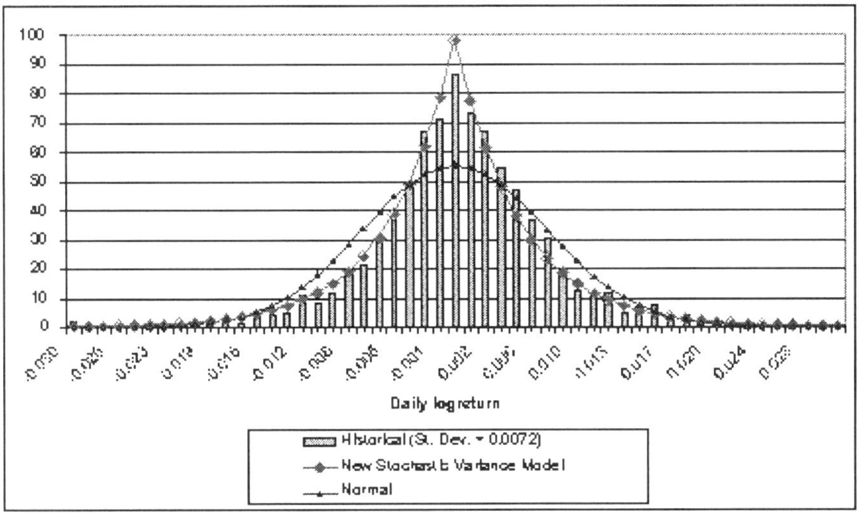

Historical and Model Distributions for JPY/USD Fx Rate (1988-1998)

Figure 8.5: Comparison of historical data and their fit by Gaussian and Laplace densities. (Courtesy of Alexander Levin.)

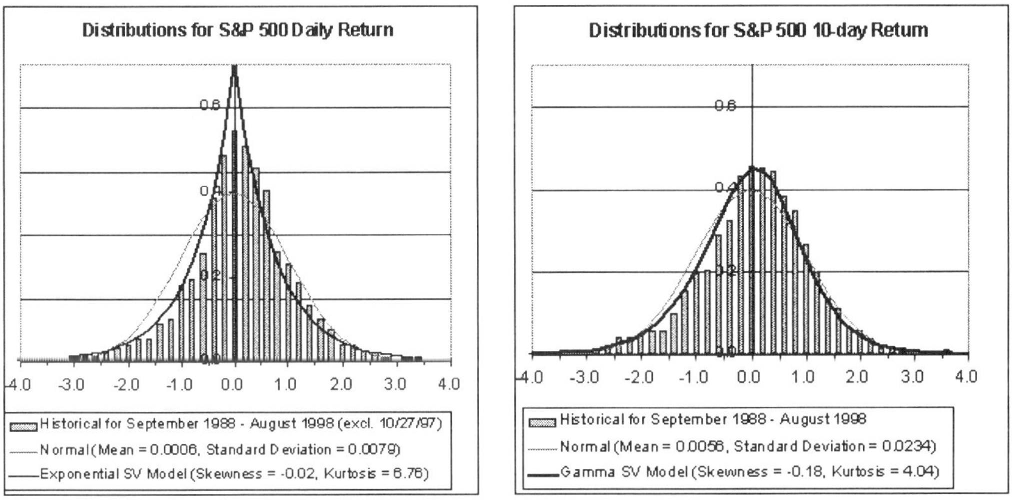

Figure 8.6: Comparing histograms of risk factors with Gaussian and Laplace models. *Left*: Daily returns of the S&P 500 Index showing that the asymmetric Laplace distribution fits the data quite well. *Right*: 10-day returns of the S&P 500 Index are fitted well by a generalized Laplace distribution (stochastic variance gamma model). (Courtesy of Alexander Levin.)

We see in Figure 8.6 that the data on returns of the S&P 500 Index are clearly skewed to the right. The fit of asymmetric Laplace on the left graph is far better than the Gaussian providing a sound empirical justification of the above model for risk factor distributions.

So far we have considered a fixed period within which we are modeling the return of our portfolio. A natural extension is to consider a stochastic variance model that depends on time. Our previous considerations, which lead to exponential distribution for stochastic variance over a fixed period, should naturally introduce a time factor into the model by considering a gamma process.

Assumption 8.6.4 *The total stochastic variance $V(t)$ follows a gamma process.*

As a consequence of this assumption, the stochastic variance over an arbitrary time interval is distributed according to the gamma law and the stochastic volatility is distributed according to the Nakagami distribution [the distribution of the square root of a gamma distributed variable; see, e.g., Nakagami (1964)]. We know from Chapter 4 that this leads to risk factors distributed according to generalized Laplace distributions (Bessel function distributions). In this case, the VaR is no longer expressed in terms of elementary functions as the Bessel function distributions involve a modified Bessel function that needs to be inverted to obtain VaR defined as a quantile for this distribution. Numerical procedures have to be used for the computational purposes.

The available financial data seem to confirm such a model. From Figure 8.6, we observe that the distribution over a longer period of time (10-day vs. daily) has a relatively smaller peak in the center, which agrees with the model having gamma distributed stochastic variance. The same observation can be made for the data presented in Figure 8.4.

The above model poses a challenging inferential problem how to estimate the parameters of distributions based on generalized Laplace model by exploiting the time scale. For example, the question arises as to which period of time would lead to an asymmetric (but not generalized) Laplace distribution of the risk factor. This problem was partially addressed in Levin and Tchernitser (1999) where an interesting calibration procedure was proposed allowing for computing the parameters of the model by matching appropriate moments of the distributions for variance and for the risk factor. As a first step, the method of moments could be used to estimate the parameters.

The next challenge is to extend these models to the case of a multivariate portfolio. Let $\mathbf{X}$ be a vector of risk factors and let $\Pi(\mathbf{X})$ be a portfolio depending on these factors. To compute VaR, one needs to identify multidimensional distribution of $\mathbf{X}$. Following the successful fit of the univariate models, we are looking for distributions which in the one-dimensional case are reduced to asymmetric Laplace or generalized Laplace distributions. For bivariate currency exchange data, studied in Levin and Tchernitser (1999), three models were examined: the Gaussian, a linear combination of Laplace variables, and bivariate Laplace (the elliptically contoured Laplace distribution). The two-dimensional data on exchange rates of German Mark and Japanese Yen vs. U.S. Dollar were used to verify the proposed models. As seen in Figure 8.7, the most convincing fit is provided by the elliptically contoured Bessel function distribution, which suggests that multivariate Laplace distributions can be very useful for multivariate modeling in finance.

8.7 A jump diffusion model for asset pricing with Laplace distributed jump-sizes

Another model that is an alternative to the Gaussian for the price of an asset (a stock or a stock index) was proposed in Kou (2000). As opposed to the variance gamma models discussed in Sections 8.5 and 8.6, which are purely jump processes, it contains both a continuous part modeled by a geometric Brownian motion and a jump part with the logarithm of the jump sizes having a Laplace distribution and the jump times corresponding to the arrival times of a Poisson process. The asset price $S(t)$ is given by the stochastic differential equation

$$\frac{dS(t)}{S(t)} = \mu dt + \sigma dW(t) + d\left(\sum_{i=1}^{N(t)}(V_i - 1)\right),$$

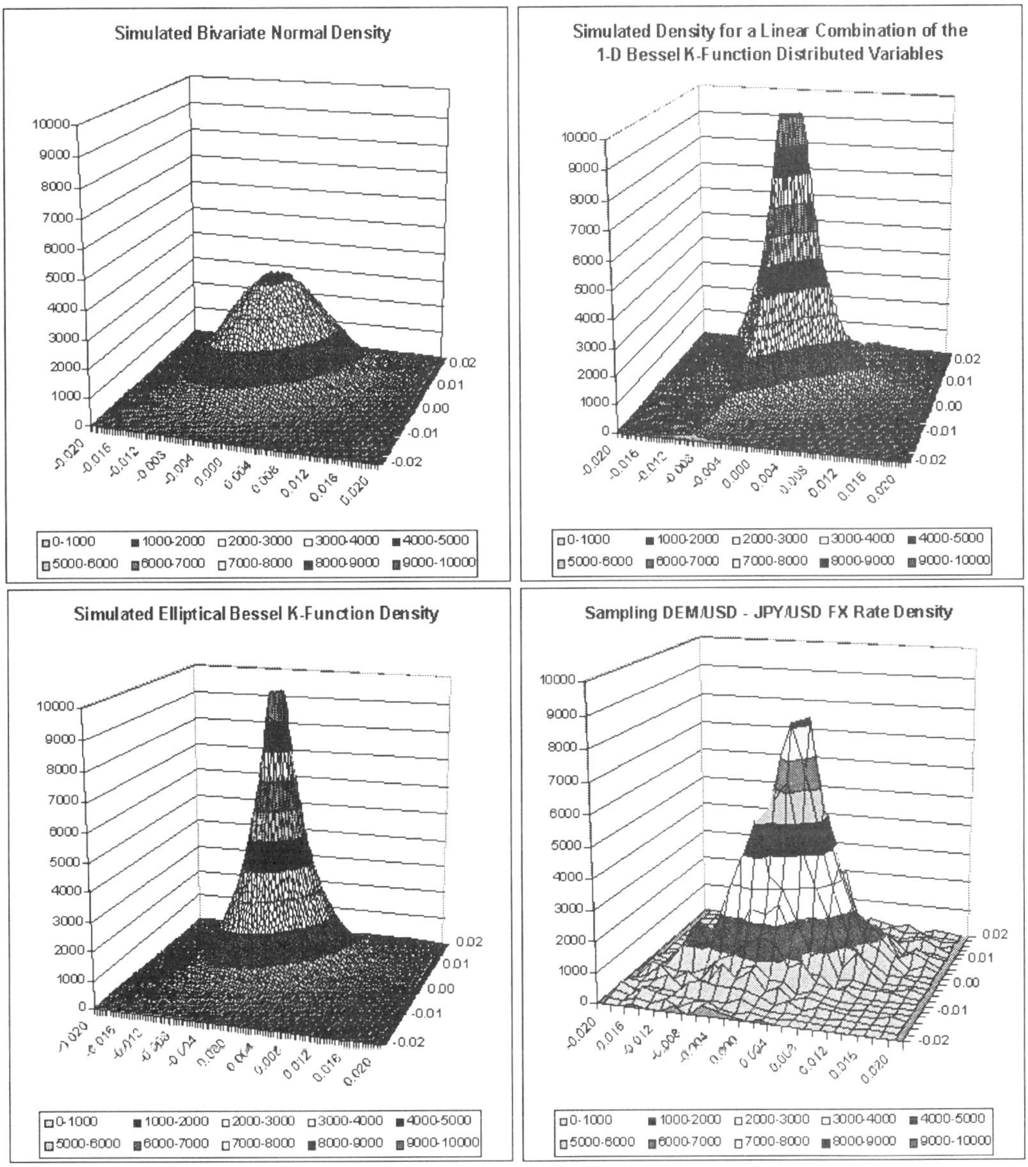

Figure 8.7: Bivariate distribution based on DEM/USD and JPY/USD data. *Top left*: Gaussian model. *Top right*: Model based on independent Laplace variables. *Bottom left*: Multivariate Laplace model. *Bottom right*: Historical distribution. (Courtesy of Alexander Levin.)

where $W(t)$ is a standard Wiener process, $N(t)$ is a Poisson process with rate λ, and $\{V_i\}$ is a sequence of independent identically distributed nonnegative random variables such that $X = \log V$ has the Laplace distribution $\mathcal{CL}(\theta, \eta)$. All the variables are assumed to be independent. The solution to the equation has the form

$$S(t) = S(0) \exp\left\{\left(\mu - \frac{1}{2}\sigma^2\right) t + \sigma W(t)\right\} \prod_{i=1}^{N(t)} V_i.$$

It is shown in Kou (2000) that this model has important features observed in the financial data (and nonexistent in the standard diffusion models), such as high peak and heavy tails, asymmetry, and volatility smile. Moreover, a closed formula for option pricing is available, although it is somewhat complicated and involves some special functions (the Hh function). We refer interested readers to the original work.

8.8 Price changes modeled by Laplace–Weibull mixtures

As we have mentioned, the ability to model heavy-tails as well as the center peak are important advantages of Laplace modeling in finance. Rachev and SenGupta (1993) propose *contaminated Laplace* distribution to accommodate the possibility of outliers. Namely, the following mixture model is discussed:

$$p(x; \pi, \lambda, \mu, \gamma) = \pi f_1(x; \lambda) + (1 - \pi) f_2(x; \mu, \gamma),$$

where f_1 is the $\mathcal{CL}(0, 1/\lambda)$ density,

$$f_1(x; \lambda) = (\lambda/2) \exp(-\lambda|x|),$$

and f_2 is the density of a symmetric Weibull distribution given by

$$f_2(x) = \frac{\gamma\mu}{2}|x|^{\gamma-1} \exp(-\mu|x|^{\gamma}),$$

where $\gamma > 1$, $\mu > 0$, $0 \leq \pi \leq 1$.

Obtaining maximum likelihood estimators for this multiparameter family of distributions is troublesome, mostly because of the presence of the Weibull component. However, the general E-M algorithm can be used for this purpose, and was successfully applied in Rachev and SenGupta (1993).

In the proposed model, the leading term is Laplace density with Weibull density being a possible contaminant. Therefore, it is of interest to test for the no mixture hypothesis: $\pi = 1$. Various cases, depending on which parameters are known, are discussed in Rachev and SenGupta (1993).

The model was then applied to price changes for real estate data from Paris. Mixture distributions are considered for such data because of a possibility of small changes in the corresponding buyers/investors population due to immigration or emigration. The data consisted of the average prices for one-bedroom apartments in Paris for 61 consecutive months. The data were transformed to $x_i = \log(\xi_{i+1}/\xi i)$ and then the E-M algorithm yielded the following estimates: $\hat{\pi} = 0.852$, $\hat{\gamma} = 5.070$, $\hat{\lambda} = 7.97$, and $\hat{\mu} = 45.39$. An initial Monte-Carlo study suggests rather good agreement of the estimated model with the observed data.

9

Inventory Management and Quality Control

Somewhat surprisingly, there are only a few and isolated applications of the Laplace distributions related to inventory management problems and quality control. The dominance of the gamma and exponential distributions in this field is still overwhelming. We have collected here a few results which hopefully will be elaborated by the researchers and practitioners in the not-too-distant future.

9.1 Demand during lead time

Distribution of demand during lead time in inventory control is essential for determining inventory decision variables such as *expected back order*, *lost sales*, *protection level*, and *stock out risk*.

Bagchi et al. (1983) show that based on theoretical considerations this distribution ought to be the Hermite distribution [see Johnson et al. (1992)] given by

$$P(W=0) = p_0 = e^{-a-b}$$

$$P(W=w) = p_w = p_0 \sum_{j=0}^{[[w/2]]} \frac{a^{w-2j}b^j}{(w-2j)!j!}, \qquad w = 1, 2, 3, \ldots,$$

where a and b are the parameters of the distribution such that $E(W) = a+2b$ and $\mathrm{Var}(W) = a+4b$. Indeed, this is the *exact* distribution of demand during lead time when unit demand is Poisson and lead time is normally distributed. However, in the applied literature [see, e.g., Peterson and Silver (1979)], the Laplace distributions are also recommended for this purpose, especially for slow-moving items or the universal normal approximation. We are thus interested in comparing normal and Laplace distributions as approximations to the (skewed) Hermite distribution. These approximations are based on the method of moments and the parameters are chosen by equating the means and variances. Bagchi et al. (1983) provide a table comparing

$$Q_R = 1 - P_R = \sum_{w=R+1}^{\infty} p_w$$

— the tails of Hermite distribution with mean 7 and variance 13 (corresponding to Poisson demand with mean equal to one and the normal lead with mean 7 and variance 6 — the relation being $E(W) = \lambda\mu$ and $\mathrm{Var}(W) = \lambda\mu + \lambda^2\sigma^2$, where $\lambda = 1$ is the mean of the Poisson demand and μ and σ^2 are the parameters of the normal lead time) with their normal and Laplace approximations. The results are summarized in Table 9.1.

Value	$Q_R = \sum_{w=R+1}^{\infty} p_w$	Q_R **approx.** (Q'_R)		**Percentage error** $100 \cdot (Q_R - Q'_R)/Q_R$	
(reorder points)	Hermite	Laplace	Normal	Laplace	Normal
7	.4163	.4110	.4449	1.27	−6.87
8	.3098	.2776	.3387	10.39	−9.33
9	.2335	.1875	.2440	12.70	−4.50
10	.1620	.1267	.1660	21.79	−2.47
11	.1140	.0856	.1060	24.91	7.08
12	.0741	.0578	.0636	21.98	14.17
13	.0484	.0391	.0358	19.21	26.03
14	.0300	.0264	.0188	12.00	37.33
15	.0186	.0178	.0092	4.30	50.54
16	.0108	.0120	.0043	−11.11	60.19
17	.0064	.0081	.0018	−26.56	71.88

Table 9.1: Approximations to the tail of Hermite demand during lead time (mean = 7, variance = 13). (Source: Bagchi et al. (1983).)

For the normal approximation, the maximum error decreases as the mean increases and increases as the variance increases. For the Laplace approximation, the maximum error seems to increase as the mean increases but decreases as the variance increases. The table indicates that the Laplace may approximate the Hermite well in the high percentage points of the right tail. The normal distribution yields better approximations in the middle percentage points. The percentage errors seem to move in opposite directions, with the normal distribution providing a better fit for moderate reorder points, and the Laplace is substantially dominating at high points. Further and more detailed investigations may be appropriate.

9.2 Acceptance sampling for Laplace distributed quality characteristics

In the theory of one-sided acceptance sampling we consider a measured quality characteristic, say X, which is compared to an upper specification limit, say U, to determine whether an item is classified as defective. The quality of all the items is then defined as the theoretical proportion p of its defective items; i.e., $p = P(X > U)$. If we have a sample of items from the lot for which quality is expressed in terms of $(X_1, \dots, X_n)$ and the estimated defective proportion is given by $\hat{p}$, then the decision rule to accept or reject the whole lot is given by

$$\text{if } \hat{p} \leq p^*, \text{ then accept the lot,}$$
$$\text{if } \hat{p} > p^*, \text{ then reject the lot,}$$

where p^* is a specified acceptance constant.

The theory is well developed if the distribution of X is normal. Sahli et al. (1997) pointed out that using the procedures based on the normality assumption when it is not valid could be quite misleading. The authors report that the procedure for which the sample size is $n = 45$ and under which the Gaussian assumption ensures an acceptance probability of 0.95 gives an acceptance probability of only 0.453 if we replace the Gaussian distribution by the Laplace distribution.

This demonstrates the importance of developing a theory for other than normal cases. Sahli et al. (1997) present an acceptance procedure for the symmetric Laplace distribution both in the case when only the center parameter is unknown and in the case when the center and scale parameters are unknown. The following is a summary of their findings.

In general, we assume that the distribution of X depends on a parameter θ and we define the lot acceptance probability based on our decision rule by

$$P_a(\theta) = P(\hat{p} \leq p^*).$$

The quality of the lot p is a function of θ. In the cases when all θ that give the same p also produce the same value of P_a, P_a can be treated as a function of p and the graph of P_a as a function of p is called an operating characteristic (OC) curve. The standard acceptance sampling plan design problem is to give a decision rule with corresponding OC curve passing through two given points (p_1, P_{a_1}) and (p_2, P_{a_2}). The problem is solved under the normal assumption by the following acceptance rules:

$$\bar{X} \leq U - \sigma z_{p^*} \quad \text{if the standard deviation } \sigma \text{ is known,}$$
$$\bar{X} \leq U - S z_{p^*} \quad \text{if } \sigma \text{ is unknown and } S^2 \text{ is the sample variance.}$$

Practical ways to choose the sample size n and p^* such that the OC curve passes through the two points (p_1, P_{a_1}) and (p_2, P_{a_2}) are provided by the International Organization for Standardization (1989).

For the Laplace distribution $\mathcal{CL}(\theta, \phi)$, we have the following relations between the parameters and the proportion p of the defective items:

$$\theta = U + \phi \log(2p).$$

The case of ϕ known. Let us take the decision rule using the median $\hat{\theta}$ (which is the MLE of θ):

$$\text{if } \hat{\theta} \leq \tilde{X}_U, \text{ then accept the lot,}$$
$$\text{if } \hat{\theta} > \tilde{X}_U, \text{ then reject the lot.}$$

The issue is to determine the acceptance constant $\tilde{X}_U$ and the sample size n such that the OC curve passes through two given points. Note that the function $P_a(p)$ is equal to the cumulative distribution function of the median, which in principle can be explicitely computed even though numerical algorithms have to be used. For example, if $\phi = 1$ and $U = 3$, then to ensure $P_a(0.0068) = 0.95$ and $P_a(0.0106) = .1$, we obtain $n = 51$ and $\tilde{X}_U = -1.0360$.

The case of ϕ unknown. A reasonable acceptance rule would be

$$\text{if } \hat{p} \leq p^*, \text{ then accept the lot,}$$
$$\text{if } \hat{p} > p^*, \text{ then reject the lot,}$$

where

$$\hat{p} = \frac{e^{(\hat{\theta}-U)/\hat{\phi}}}{2},$$

$\hat{\phi}$ is the sample mean absolute deviation (the MLE of ϕ), and p^* is to be determined. This is equivalent to

$$\text{if } \hat{\theta} \leq U - k\hat{\phi}, \text{ then accept the lot,}$$
$$\text{if } \hat{\theta} > U - k\hat{\phi}, \text{ then reject the lot,}$$

where k has to be determined. In order to determine the OC curve in this case one can either consider the exact distribution of the statistics $\hat{\phi}$ and $\hat{\theta}$ or apply some asymptotic results (see also Section 2.6). The complexity of the problem was partially analyzed in Sahli et al. (1997).

9.3 Steam generator inspection

The exponential distribution is found in applications in a variety of fields. Easterling (1978) notices that for heavy-tailed data the model consisting of the sum of an exponential variable and a Laplace-distributed independent measurement error can be utilized. In this paper the model is applied to measurements of tube degradation in a steam generator.

The steam generators in pressurized water reactors contain thousands of tubes through which heated water from the reactor flows to be converted into steam. The tubes can erode over time and, if the generator is not inspected and maintained properly, it can lead to leaks that require the plant to be shut down. To develop an appropriate inspection plan, an adequate statistical model for the degradation of the tubes has to be developed. In Easterling (1978), the actual degradation (extent of thinning) of a tube, D, expressed as a percentage of the initial tube wall thickness, is a random variable having an exponential distribution with mean θ:

$$h(d) = \frac{1}{\theta} e^{-d/\theta}.$$

The degradation is measured by a device called an eddy current tester and it is clear from the available experimental data that the measurements are made with heavy-tailed and biased errors E. A Laplace distribution with density

$$g(e) = \frac{1}{2\phi} e^{-|e-\mu|/\phi}$$

seems to be well fitted for these data. The measured degradation is then modeled as

$$M = D + E,$$

where E and D are independent and distributed according to these densities. Then the cumulative distribution function of M is given by

$$P(M \le m) = \begin{cases} \frac{\phi}{2(\phi+\theta)} e^{(m-\mu)/\phi}, & \text{if } m \le \mu, \\ 1 - \frac{\theta^2}{\theta^2-\phi^2} e^{-(m-\mu)/\theta} + \frac{\theta}{2(\theta-\phi)} e^{-(m-\mu)/\phi}, & \text{if } m > \mu. \end{cases}$$

From this explicit formula one can derive conditional moments of M and D [see Easterling (1978)].

The goodness-of-fit analysis of the above model was performed on some experimental data. The model appears to provide an adequate fit. However, as pointed out by Easterling (1978), correctly estimating the variances represented by θ and ϕ is a problem. Both represent variability in the model, and it is hard to discern if the variability comes from variance of the error or the variance of degradation.

9.4 Adjustment of statistical process control

The majority of applications of the Laplace distributions are due to inadequacy of Gaussian modeling. Along these lines, Gonzáles et al. (1999) present a rather surprising application of the Laplace distribution by finding an approximate solution to a Gaussian model (considered accurate) through exact solutions available for a corresponding Laplace model. Namely, an analytical solution to the average adjustment interval and the mean squared deviation from the target of the "bounded adjustment" schemes are found under the assumption that the disturbances are generated from a Laplace distribution. Then robustness of the solution on the distributional assumptions is demonstrated and used to derive the approximate results for the Gaussian case.

Feedback control schemes used in the parts and hybrid industries must often account for the cost of being off target, and the costs of adjustment and/or the sampling process. In such a case, feedback adjustment may be implemented by using *bounded* (*dead band*) adjustment schemes. In these schemes the disturbances are represented by an integrated moving average (IMA) time series model

$$z_{t+1} - z_t = a_{t+1} - \theta a_t,$$

where $z_0 = a_0 = 0$, the innovations a_t are independent and identically distributed (i.i.d.) normal random variables with mean zero and standard deviation σ_a, and $0 < \lambda = 1-\theta \leq 1$. The adjustments are given by $x_t = X_t - X_{t-1}$ and their effect is realized at time $t+1$. The possibility of sampling and adjusting the process occurs only at times tm, $m \in \mathbb{N}$. The corresponding disturbances are given by

$$z_{mt+m} - z_{mt} = u_{mt+m} - \theta_m u_{mt},$$

where u_{tm} are i.i.d. normal random variables with mean zero and standard deviation σ_m, and θ_m, σ_m, and $\lambda_m = 1 - \theta_m$ satisfy $\lambda_m^2\sigma_m^2 = m\lambda^2\sigma_a^2$ and $\theta_m\sigma_m^2 = \theta\sigma_a^2$. Optimal bounded adjustment schemes require that an action X_{tm} needed to bring the process back to target is taken every time the minimum mean squared error of forecasted deviation from target exceeds some threshold values $\pm L$. Important parameters for these schemes are the sampling interval m, the action limits $\pm L$, and the amount of adjustment required (which depends on the overcompensation s to be produced). Once these parameters are chosen, the average adjustment interval (AAI) and mean squared deviation (MSD) may be computed by solving certain integral equations. Under the disturbances described, the equations have the form

$$\begin{aligned} AAI(x) &= mh_0(x), \\ MSD(x) &= \sigma_m^2 + \lambda_m^2\sigma_m^2\{(1-m)/(2m) + g_2(x)\}, \end{aligned}$$

where $x = s/(\lambda_m\sigma_m)$, $g_2(x) = h_2(x)/h_0(x)$. The functions $h_k(x)$ for $k = 0$ and 2 are the solutions of the Fredholm integral equation

$$h_k(x) = x^k + \sigma_m \int_{-\Lambda}^{\Lambda} h_k(w)\phi\{\sigma_m(w-x)\}dw, \tag{9.4.1}$$

where $\Lambda = L/(\lambda_m\sigma_m)$ and $\phi(\cdot)$ is the density function of the innovations u_{tm}. See Gonzáles et al. (1999) and the references therein.

When innovations are Gaussian there is no analytic solution to (9.4.1). However, as shown in Gonzáles et al. (1999), analytical solution can be written explicitly if the innovations follow Laplace distribution. Namely, in the Laplacian case the solutions are

$$h_0(x) = \begin{cases} \Lambda^2 + \Lambda\sqrt{2} + 1 - x^2, & |x| \leq \Lambda, \\ 1 + \Lambda\sqrt{2}e^{-\sqrt{2}(|x|-\Lambda)}, & |x| > \Lambda, \end{cases}$$

$$h_2(x) = \begin{cases} \Lambda^4/6 + \Lambda^3\sqrt{2}/3 - x^4/6 + x^2, & |x| \leq \Lambda, \\ x^2 + \Lambda^3\frac{\sqrt{2}}{3}e^{-\sqrt{2}(|x|-\Lambda)}, & |x| > \Lambda. \end{cases}$$

These solutions can be used to obtain exact values of the AAI and MSD. The Fredholm equation can also be solved for the convolutions of Laplace distributions. Then the solutions can be used to approximate solutions for normal innovations by the Central Limit Theorem. However, as shown in Gonzáles et al. (1999), the limiting distribution can be approximated quite accurately by simply extrapolating the solution in the cases of the Laplace distribution and the twofold convolution of the Laplace distribution. For the corresponding results, we refer readers to the original paper.

9.5 Duplicate check-sampling of the metallic content

An application of generalized Laplace (Bessel function) distributions was obtained some 40 years ago by Rowland and Sichel (1960) in modeling duplicate measurements of the metallic content in the gold mines of South Africa (but to the best of our knowledge no more recent results are available at least in the probabilistic and statistical literature). Because such duplicate check-sampling is a common practice in industrial analysis this approach could be valuable for quality controllers working in other areas as well. In our presentation, we restrict ourselves to a description of the model, referring readers interested in quality control to the original paper.

The check measuring is based on duplicate measurements of a specimen in order to gauge the accuracy of qualitative determinations. The two measurements, called the original sample and the check sample, can be used to assess the quality of measurements. In standard applications, it is often reasonable to assume that the difference of measurements is normally distributed. However, in cases when the variance of the error is dependent on the level of specimen in a measurement, the use of the normal distribution is not appropriate.

This seems to be the case in duplicate measurements of the gold content in gold mines. Namely, the higher the level of the gold content in samples taken in a groove the larger the variance of the measured content. It was verified in various studies that for the double check sampling in the gold mines the ratios of two measurements have stabilized standard deviations and thus they should be used for statistical purposes instead of the differences.

Let X and Y represent the original and check sample. From the data collected from mines in South Africa, it was inferred that the distributions of X and Y are identical and thus the ratio $R = X/Y$ has a distribution that is asymmetric around one. As it is more convenient to use symmetric distributions in deriving control chart limits, the logarithm of the ratio, $L = \log R$, which is distributed symmetrically around zero, is a more suitable variable. The log-normal distribution has a prominent position in mine valuation, and is often used to model the distribution of R if all samples are taken in a small reef area (so the variance can be assumed constant). If variances of all such small reef areas were constant, all the ratios obtained in check sampling could be pooled together and would conform to the log-normal law. Unfortunately, the observed data reject such a model. It was observed that the logarithms of the observed ratios, which under the log-normal model should be normally distributed, reveal strongly leptokurtic features. According to Rowland and Sichel (1960) leptokurtosis is due to the "instability" of the logarithmic variances that is observed even for samples taken in two neighboring reef areas. Since standard statistical densities used for symmetric leptokurtic distributions, such as Pearson Type VII distribution (a t-distribution with not necessarily integer-valued degrees of freedom), were rejected by the χ^2-test, the authors resorted to a model which in the terminology of this book is represented by generalized symmetric Laplace (symmetric Bessel function) distributions.

The basis for the model follows the same scheme that was presented earlier in this book: the variable L is normally distributed with a stochastic variance (corresponding to the random choice of the location). The standard deviation is assumed to have a gamma distribution, and L is a product of the random standard deviation and a normal random variable, assumed to be independent. As a result of these assumptions we obtain the following density of L:

$$\gamma(l) = \frac{\sqrt{a/\pi}}{2^{\nu-1/2}\Gamma(\nu+1/2)}(\sqrt{2a}|l|)^{\nu} K_{\nu}(\sqrt{2a}|l|),$$

where a and ν are some positive parameters. This distribution corresponds to the density given by equation (4.1.32), if we take $a = 1/\sigma^2$ and $\nu = \tau - 1/2$. One should notice that the above density is also well defined for $\nu \in (-1/2, 0]$ although this case was not discussed in the original paper.

The derived model has fitted the data from various gold mines quite well. The formal derivation of the quality control charts based on this model and a discussion of their implementation in the mining practice can be found in Rowland and Sichel (1960).

10

Astronomy and the Biological and Environmental Sciences

In this short chapter, miscellaneous applications of Laplace distributions are briefly surveyed. In the first section we report that Laplace distribution may in certain instances provide a better fit than the more complicated hyperbolic distribution. The central part of this chapter is devoted to an important application to the area of dose response curves studied by Uppuluri (1981), which unfortunately has not been investigated further due to the untimely death of the author.

10.1 Sizes of sand particles, diamonds, and beans

Laplace distributions and, more generally, hyperbolic distributions were considered for modeling sizes of sand particles, diamonds, and beans.

Barndorff-Nielsen (1977) studied the distribution of the logarithm of particle size of wind-blown sands. The distribution for which the logarithm of the density function is a hyperbola (or, in higher dimensions, a hyperboloid) is proposed as a model. It was the first time when the class of hyperbolic distributions was introduced. It was also noted that the Laplace distribution is a limiting distribution with an appropriate passage to the limit of the corresponding parameters. For the Laplace distribution, the log-probability function is not a hyperbola but rather two straight half-lines attached at a single point.

The standard distribution in size statistics is the log-normal distribution. However, quite often *mixtures* of log-normal distribution seem to account better for long tails of observed data. Log-hyperbolic distributions (and in particular log-Laplace distributions) are mixtures of log-normal distributions, and both of them have asymptotically linear tails. These two features makes them particularly suitable for modeling size data.

The class of one-dimensional hyperbolic distributions introduced in Barndorff-Nielsen (1977) can be described in terms of the density

$$
\begin{aligned}
f(x;\phi,\gamma,\mu,\delta) = {} & \frac{1}{(\phi^{-1}+\gamma^{-1})\delta\sqrt{\phi\gamma}\,K_1(\delta\sqrt{\phi\gamma})} \\
& \cdot \exp\left(-\frac{1}{2}(\phi+\gamma)\sqrt{\delta^2+(x-\mu)^2}+\frac{1}{2}(\phi-\gamma)(x-\mu)\right).
\end{aligned}
$$

In the limiting case ($\delta \to 0$) we obtain an asymmetric Laplace distribution, while a Gaussian distribution is obtained when $\delta \to \infty$ and $\delta/\sqrt{\phi\gamma} \to \sigma^2$ (cf. Exercise 3.6.3).

Hyperbolic distributions provided an excellent fit to the data on sand particles from the studies by Bagnold (1954) as well as on samples of sand from the Danish west coast. It was also suggested that this class of distribution can be applied to other contexts when size data are considered. As an example, size distribution of diamonds from a large mining area in southwest Africa were discussed in Sichel (1973). He noticed that "diamond sizes in the marine deposit of southwest Africa are well represented by a two-parameter log-normal distribution provided the stones originate from a small compact mining block, on one and the same beach horizon." However, for larger mining areas deviations from the log-normal distributions are observed. Sichel (1973) introduced the mixture of log-normal distributions that in our terminology would be called generalized asymmetric log-Laplace distributions.

In Blaesild (1981), the bivariate hyperbolic distributions are proposed to fit W. Johannsen's bivariate data on the length and breadth of beans. These now classical sets of two-dimensional data showing nonnormal variations were fit by a bivariate hyperbolic distribution providing a reasonable agreement with the data. As the bivariate Laplace distributions constitute a subclass of hyperbolic distributions, it would be of interest to compare the Laplace fit to the more general but also more complicated hyperbolic fit. This was actually done in Fieller (1993), who studied the distribution of sizes of sand particles in relation to archaeological research. Fieller (1993) reported that "attempts to fit the log-hyperbolic models of Barndorff-Nielsen (1977) proved computationally impossible. Instead, a simpler version, based on the log skew Laplace distribution, proved computationally tractable and most satisfyingly answered the questions quite conclusively."

Similar comments apply to many other investigations of fitting the hyperbolic distribution to empirical data. Barndorff-Nielsen and Blaesild (1982) apply the hyperbolic model to the following six data sets:

1. grain sizes, acolian sand deposits;
2. grain sizes, river bed sediment;
3. differences between logarithms of duplicate determinations of content of gold per ore;
4. differences of streamwise velocity components in a turbulent atmospheric field of large Reynold numbers;
5. the lengths of beans whose breadths lie in a fixed interval;
6. personal incomes in Australia 1962–1963.

In four of these cases (data sets 1, 2, 4, and 6) the resulting distribution is close to the Laplace distribution (in the logarithmic scale we observe almost two straight half-lines instead of a hyperbola), while in two other cases the data seem to be "more" Gaussian (parabolic log-probability function).

10.2 Pulses in long bright gamma-ray bursts

A somewhat unusual application of an asymmetric Laplace distribution was found in the modeling of the shapes of long bright gamma-ray bursts discussed by Norris et al. (1996). The paper examines the temporal profiles of bursts detected by the burst and transient source experiment at the Compton Gamma Ray Observatory. The most frequently observed pulses are intermediate between asymmetric Laplace and asymmetric Gaussian. The general functional form of the pulse intensity is given by

$$I(t) = \begin{cases} A\exp(-(|t - t_{\max}|/\sigma_r)^\nu) & \text{for } t < t_{\max}, \\ A\exp(-(|t - t_{\max}|/\sigma_d)^\nu) & \text{for } t > t_{\max}, \end{cases}$$

where $t_{\max}$ is the time of the pulse's maximum intensity A, σ_r and σ_d are the rise ($t < t_{\max}$) and decay ($t > t_{\max}$) time constants, respectively, and ν is a measure of peakedness. For $\nu = 1$ we obtain an asymmetric Laplace shape, and for $\nu = 2$ the corresponding shape can be described by an asymmetric Gaussian distribution.

The paper focuses on deconvoluting the above shapes from the temporal data of the observed gamma ray bursts. The interactive numerical routine is used to fit pulses in bursts. The most frequently occurring peakedness lies approximately halfway between Gaussian and Laplacian distributions.

10.3 Random fluctuations of response rate

In many behavioral systems, one can observe pulse-like responses that recur regularly in time with a very low variation. The constant beating of the heart and the responses of the optic nerve of the horseshoe crab, *limulus* (which is famous for the long trains of action potentials produced when its visual receptor is subject to a steady light), are just two of many examples observed in nature. These responses, although random, are quite periodic, and their fluctuations are not modeled well by a Poisson process.

McGill (1962) proposes a stochastic model for such responses, which accommodates both periodic and random components. This model involves a mechanism that generates regularly spaced excitations that can initiate a response after a random delay. The excitations are not observed but their periodicity is indirectly seen in a regular pattern of responses.

The general model is rather simple. Suppose that excitations occur in equal nonrandom time intervals of length τ. At excitation $k\tau$, $k = 1, 2, \ldots$, we have a positive random variable S_k that represents a random delay between an excitation at $k\tau$ and the response that occurs at $k\tau + S_k$. We assume that the S_k's are i.i.d. random variables having exponential distribution with parameter λ. The goal is to find the distribution of the time between responses. In McGill (1962), this distribution is shown to have the form

$$f(t) = \begin{cases} \frac{\lambda\nu}{1-\nu} \sinh \lambda t \le \tau, \\ \frac{1+\nu}{2\nu} \lambda e^{-\lambda t}, & t \ge \tau, \end{cases}$$

where ν is a constant given by $\nu = e^{-\lambda\tau}$. This distribution is skewed and has the mode at $t = \tau$. Moreover, as $\lambda\tau$ increases without a bound, ν converges to zero and the distribution becomes asymptotically

$$f(t-\tau) = \frac{\lambda}{2} e^{-\lambda|t-\tau|},$$

which is the symmetric Laplace distribution. This asymptotic distribution applies to the case when the random component ("noise") is small relatively to the periodic component represented by τ.

The fact that the Laplace distribution arises as the limiting distribution is not surprising. By independence, we see that for large τ, $\tau + r = \tau + S_2 - \tilde{S}_1$, where $\tilde{S}_1 = \tau - S_1$ and the latter is approximately exponentially distributed for large τ. Now the limiting distribution follows from the representation of Laplace distribution as a difference of two exponential random variables. As noted by McGill (1962): "This simple point (that Laplace is a difference of exponentials) is ignored in most texts on statistics because, perhaps, no one imagines why anyone else would be interested. Our argument establishes a very good reason for being interested. The difference, and hence the Laplace distribution, provides a characterization of the error in a timing device that is under periodic excitation."

The model is then tested on two sets of real-life data: responses of a single fiber of the optic nerve of the horseshoe crab and interresponse times produced by a bar-pressing rat after a long conditioning period. The data are more leptokurtic than normal distribution and Laplace distribution fit the data quite well.

10.4 Modeling low dose responses

If a random variable Y has the Laplace distribution, then e^Y has the log-Laplace distribution. This distribution was considered in Uppuluri (1981) as a model in the study of the behavior of dose response curves at low doses. One of the problems in this context is linearity vs. nonlinearity of dose response for radiation carcinogenesis. Since animal experiments can only be performed at reasonably high doses, the problem of extrapolation to low doses becomes viable only under a suitable mathematical model. The following axiomatic approach leads to the model given by log-Laplace distribution.

Axiom 1 At small doses, the percent increase in the cumulative proportion of deaths is proportional to the percent increase in the dose.

Axiom 2 At larger doses, the percent increase in the cumulative proportion of survivors is proportional to the percent decrease in the dose.

Axiom 3 At zero dose, no deaths, and when the dose is infinite, no survivors, and the cumulative proportion of deaths $F(x)$ is a monotonic, nondecreasing function of the dose x.

Under these axioms we obtain that the cumulative distribution function of the dose response has the form

$$F(x) = F(1)x^{\mu}, \quad 0 \le x \le 1, \qquad 1 - F(x) = (1 - F(1))/x^{\lambda}, \quad x \ge 1,$$

for some positive μ and λ.

The log-Laplace distribution corresponding to the classical Laplace distribution is obtained if we additionally assume that $\lambda = \mu$ and $F(1) = 1/2$. Of course, the log-Laplace distribution for asymmetric Laplace distributions is also included in the above model.

10.5 Multivariate elliptically contoured distributions for repeated measurements

Lindsey (1999) discusses the need for other than normal multivariate distributions in the analysis of repeated measurements. The main deficiency of normal distributions is their inability to model heavier tails. As an alternative Lindsey (1999) proposes multivariate exponential power distributions given by the density

$$f(\mathbf{y}; \boldsymbol{\mu}, \boldsymbol{\Sigma}, \beta) = \frac{n\Gamma(n/2)}{\pi^{n/2}\sqrt{|\boldsymbol{\Sigma}|}\Gamma\left(1 + \frac{n}{2\beta}\right)2^{1+n/(2\beta)}} e^{-\frac{1}{2}[(\mathbf{y}-\boldsymbol{\mu})'\Sigma^{-1}(\mathbf{y}-\boldsymbol{\mu})]^{\beta}},$$

also known as the Kotz-type multivariate distribution [cf. Exercise 6.12.11 and Fang et al. (1990)].

For $\beta = 1/2$, this represents a certain generalization of the Laplace distribution. However, it is not a multivariate Laplace distribution as discussed in this book.

As an example, Lindsey (1999) considers blood sugar level for two treatments of rabbits involving two neutral protamine Hagedorn insulin mixtures. The estimate of β (around 0.40) strongly suggests nonnormality. The main reason is heavy tails exhibited by the data.

This example illustrates the ability of multivariate exponential power distribution to fit heavy-tailed data. However, as pointed out by Lindsey (1999), it has several unpleasant properties:

- The marginal and conditional distributions are more complex elliptically contoured distributions, not of the exponential power type.
- It seems to be difficult to introduce independence between observations.

In view of this, it would be interesting to compare the exponential power distributions with multivariate Laplace distributions as discussed in this book. To quote the author of the discussed paper: "The fact that the multivariate normal distribution is rejected in favor of a more heavily tailed distribution for these data does not imply that this (multivariate exponential power) is the most appropriate distribution for them."

10.6 ARMA models with Laplace noise in the environmental time series

An ARMA model with Laplace noise was used to fit the data on sulphate concentration in Damsleth and El-Shaarawi (1989). The data consisted of 147 weekly measurements of the sulphate concentration in the Turkey Lakes Watershed in Ontario, Canada, from early March 1982 to the end of 1984. The data exhibit some extreme values and thus there is a reasonable doubt about normality of the underlying time series. A standard time series analysis of the data suggests that an AR(1) model may be appropriate. Thus the model considered is

$$X_t = \phi X_{t-1} + a_t,$$

where a_t is a random noise. In the classical time series theory the model with a_t being Gaussian is typically being considered. The Laplace distribution is an alternative, which is distinct from the normal distribution. Computationally, the Laplace case is still straightforward, though sometimes cumbersome. The probability density function of X_t is given by

$$f(x) = \frac{1}{2}\sum_{j=0}^{\infty} \alpha_i |\phi|^{-i} e^{-|x|/|\phi|^i}$$

where

$$\alpha_i = (-1)^i \prod_{t=1}^{i} [\phi^{2t}/(1-\phi^{2i})] / \prod_{t=1}^{\infty} (1-\phi^{2t}).$$

The shape of this distribution exhibits "Laplacian features" (peak and heavy tails) for ϕ close to zero, and "Gaussian features" for ϕ close to unity. It is interesting that this density has all derivatives at zero, provided $\phi \neq 0$. In Damsleth and El-Shaarawi (1989), bivariate distribution of (X_t, X_{t-1}) is also computed in an explicit form.

Using the maximum likelihood method, the fit of both models (Gaussian and Laplacian) was made. The Laplace model fits the data better than the Gaussian one, both before and after logarithmic transformation of the data. Details are presented in the cited paper.

Appendix
Bessel Functions

The Bessel function of the first kind of order λ is given by the convergent series

$$J_\lambda(z) = z^\lambda \sum_{k=0}^{\infty} \frac{(-1)^k z^{2k}}{2^{2k+\lambda} k! \Gamma(\lambda + k + 1)}. \tag{A.0.1}$$

In particular,

$$J_0(z) = \sum_{k=0}^{\infty} \frac{(-1)^k z^{2k}}{2^{2k}(k!)^2} = \frac{1}{\pi} \int_0^\pi \cos(z \cos\theta) d\theta \tag{A.0.2}$$

and

$$J_1(z) = \sum_{k=0}^{\infty} \frac{(-1)^k z^{2k+1}}{2^{2k+1} k!(k+1)!} = \frac{1}{\pi} \int_0^\pi \cos(z \sin\theta - \theta) d\theta \tag{A.0.3}$$

[see, e.g., Abramowitz and Stegun (1965)].

We collect some results for the *modified Bessel function of the third kind* with index $\lambda \in \mathbb{R}$, denoted $K_\lambda(\cdot)$. We refer the reader to Abramowitz and Stegun (1965), Olver (1974), and Watson (1962) for definitions and further properties of these and related special functions.

There are many integral representations of $K_\lambda(u)$ in the literature. The following representations are relevant to our work. The first can be found in Watson (1962, p. 183), the second appears in Abramowitz and Stegun (1965, p. 376), while the third is given in Olver (1974).

$$K_\lambda(u) = \frac{1}{2}\left(\frac{u}{2}\right)^\lambda \int_0^\infty t^{-\lambda-1} \exp\left(-t - \frac{u^2}{4t}\right) dt, \quad u > 0, \tag{A.0.4}$$

$$K_\lambda(u) = \frac{(u/2)^\lambda \Gamma(1/2)}{\Gamma(\lambda + 1/2)} \int_1^\infty e^{-ut}(t^2 - 1)^{\lambda - 1/2} dt, \quad \lambda \geq -1/2, \tag{A.0.5}$$

$$K_\lambda(u) = \int_0^\infty e^{-u \cosh t} \cosh(\lambda t) dt, \quad \lambda \in \mathbb{R}. \tag{A.0.6}$$

Property 1 *The Bessel function $K_\lambda(u)$ is continuous and positive function of $\lambda \geq 0$ and $u > 0$.*

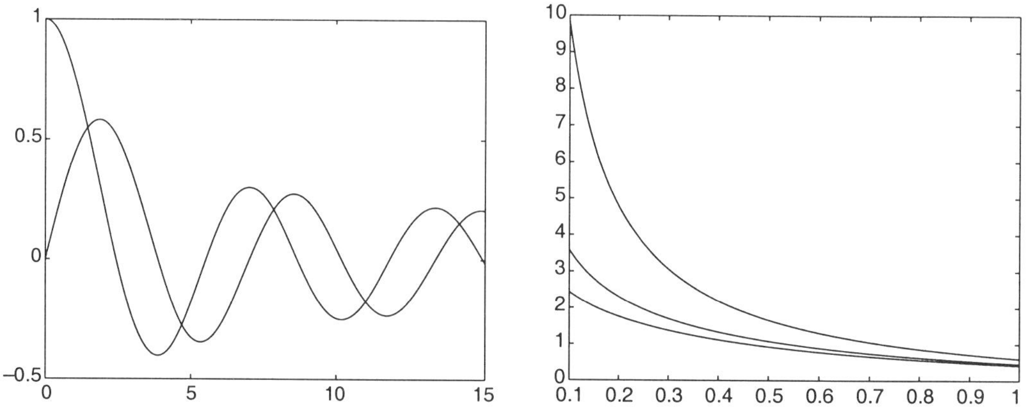

Figure A.1: Graphs of Bessel functions. *Left*: J_0 (starting at the origin) and J_1 (starting at one). *Right*: K_0 (the lowest), $K_{1/2}$, and K_1 (the highest).

Property 2 *If* $\lambda \geq 0$ *is fixed, then throughout the* u *interval* $(0, \infty)$, *the function* $K_\lambda(u)$ *is positive and decreasing.*

Property 3 *If* $u > 0$ *is fixed, then throughout the* λ *interval* $(0, \infty)$, *the function* $K_\lambda(u)$ *is positive and increasing.*

Property 4 *For any* $\lambda \geq 0$ *and* $u > 0$, *the Bessel function* K_λ *satisfies the relations*

$$K_\lambda(u) = K_\lambda(-u), \tag{A.0.7}$$

$$K_{\lambda+1}(u) = \frac{2\lambda}{u} K_\lambda(u) + K_{\lambda-1}(u), \tag{A.0.8}$$

$$K_{\lambda-1}(u) + K_{\lambda+1}(u) = -2K'_\lambda(u). \tag{A.0.9}$$

Property 5 *For* $\lambda = r + 1/2$, *where* r *is a nonnegative integer, the Bessel function* K_λ *has the closed form*

$$K_{r+1/2}(u) = \sqrt{\frac{\pi}{2u}} e^{-u} \sum_{k=0}^{r} \frac{(r+k)!}{(r-k)!k!} (2u)^{-k}. \tag{A.0.10}$$

In particular, for $r = 0$, *we obtain*

$$K_{1/2}(u) = \sqrt{\frac{\pi}{2u}} e^{-u}. \tag{A.0.11}$$

Property 6 *If* λ *is fixed, then*

$$\text{as } x \to 0^+, \quad K_\lambda(x) \sim \Gamma(\lambda) 2^{\lambda-1} x^{-\lambda} \quad (\lambda > 0), \quad K_0(x) \sim \log(1/x). \tag{A.0.12}$$

Property 7 *For any* $a > 0$ *and* μ, λ *such that* $\mu + 1 \pm \lambda > 0$, *we have*

$$\int_0^\infty x^\mu K_\lambda(ax)dx = \frac{2^{\mu-1}}{a^{\mu+1}} \Gamma\left(\frac{1+\mu+\lambda}{2}\right) \Gamma\left(\frac{1+\mu-\lambda}{2}\right) \tag{A.0.13}$$

[see Gradshteyn and Ryzhik (1980).]

Property 8 *For any $\mu > 0$ and $\beta u > 0$, we have*

$$\int_u^{\infty} x^{\mu-1}(x-u)^{\mu-1}e^{-\beta x}dx = \frac{\Gamma(\mu)}{\sqrt{\pi}}\left(\frac{u}{\beta}\right)^{\mu-\frac{1}{2}} e^{-\frac{\beta u}{2}} K_{\mu-\frac{1}{2}}\left(\frac{\beta u}{2}\right) \tag{A.0.14}$$

[see Gradshteyn and Ryzhik (1980).]

Property 9 *For any $\nu > 0$, we have*

$$[x^{\nu}K_{\nu}(x)]' = -x^{\nu-1}K_{\nu-1}(x)$$

[see Olver (1974), (8.05), p. 251, and (10.05), p. 60.]

Property 10 *For any $\nu > 0$, we have*

$$K_{\nu}(x) = K_{-\nu}(x)$$

[see Olver (1974), (8.05), p. 251.]

Consider the function

$$R_{\lambda}(x) = \frac{K_{\lambda+1}(x)}{K_{\lambda}(x)}. \tag{A.0.15}$$

The function R_{λ} has a number of important properties.

Property 11 *For $\lambda \geq 0$ the function $R_{\lambda}(x)$ is strictly decreasing in x with $\lim_{x\to\infty} R_{\lambda}(x) = 1$ and $\lim_{x\to 0^+} R_{\lambda}(x) = \infty$.*

Property 12 *Property 4 of Bessel functions produces the recursive relation*

$$R_{\lambda}(x) = \frac{2\lambda}{x} + \frac{1}{R_{\lambda-1}(x)}. \tag{A.0.16}$$

Property 13 *Property 4 of Bessel functions produces the following expression for the derivative of R_{λ}:*

$$\frac{d}{dx}R_{\lambda}(x) = R_{\lambda}^2(x) - \frac{2\lambda+1}{x}R_{\lambda}(x) - 1. \tag{A.0.17}$$

See Jorgensen (1982) for these and other properties of the function R_{λ} (and Bessel function).

References

[1] Abramowitz, M. and Stegun, I.A. (1965). *Handbook of Mathematical Functions*, Dover Publications, New York.

[2] Adams, Jr., W.C. and Giesler, C.E. (1978). Quantizing characteristics for signals having Laplacian amplitude probability density function, *IEEE Trans. Comm.* **COM-26**, 1295–1297.

[3] Agrò, G. (1995). Maximum likelihood estimation for the exponential power function parameters, *Comm. Statist. Simulation Comput.* **24**(2), 523–536.

[4] Ahsanullah, M. and Rahim, M.A. (1973). Simplified estimates of the parameters of the double-exponential distribution based on optimum order statistics from a middle-censored sample, *Naval Res. Logist. Quarterly* **20**, 745–751.

[5] Akahira, M. (1986). On the loss of information for the estimators in the double exponential case, Reported at the *Symposium on Inference Based on Incomplete Information with Applications*, University of Tsukuba, Kyushu University.

[6] Akahira, M. (1987). Second order asymptotic comparison of estimators of a common parameter in the double exponential case, *Ann. Inst. Statist. Math.* **39**, 25–36.

[7] Akahira, M. (1990). Second order asymptotic comparison of the discretized likelihood estimator with asymptotically efficient estimators in the double exponential case, *Metron* **48**, 5–17.

[8] Akahira, M. and Takeuchi, K. (1990). Loss of information associated with the order statistics and related estimators in the double exponential distribution case, *Austral. J. Statist.* **32**, 281–291.

[9] Akahira, M. and Takeuchi, K. (1993). Second order asymptotic bound for the variance of estimators for the double exponential distribution, in *Statistical Science & Data Analysis* (eds., K. Matusita et al.), VSP Publishers, Amsterdam, pp. 375–382.

[10] Alamatsaz, M.H. (1993). On characterizations of exponential and gamma distributions, *Statist. Probab. Lett.* **17**, 315–319.

[11] Ali, M.M., Umbach, D. and Hassanein, K.M. (1981). Estimation of quantiles of exponential and double exponential distributions based on two order statistics, *Comm. Statist. Theory Methods* **A10**(19), 1921–1932.

[12] Anderson, D.N. (1992). A multivariate Linnik distribution, *Statist. Probab. Lett.* **14**, 333–336.

[13] Anderson, D.N. and Arnold, B.C. (1993). Linnik distributions and processes, *J. Appl. Probab.* **30**, 330–340.

[14] Anderson, E.W. and Ellis, D.M. (1971). Error distributions in navigation, *J. Inst. of Navigation* **24**, 429–442.

[15] Andrews, D.F., Bickel, P.J., Hampel, F.R., Huber, P.J., Rogers, W.H. and Tukey, J.W. (1972). *Robust Estimates of Location*, Princeton University Press, Princeton.

[16] Antle, C.E. and Bain, L.J. (1969). A property of maximum likelihood estimators of location and scale parameters, *SIAM Rev.* **11**(2), 251–253.

[17] Arnold, B.C. (1973). Some characterizations of the exponential distribution by geometric compounding, *SIAM J. Appl. Math.* **24**(2), 242–244.

[18] Asrabadi, B.R. (1985). The exact confidence interval for the scale parameter and the MVUE of the Laplace distribution, *Comm. Statist. Theory Methods* **14**(3), 713–733.

[19] Atkinson, A.C. (1982). The simulation of generalized inverse Gaussian and hyperbolic random variables, *SIAM J. Sci. Statist. Comput.* **3**(4), 502–515.

[20] Azzalini, A. (1985). A class of distributions that includes the normal ones, *Scand. J. Statist.* **12**, 171–178.

[21] Azzalini, A. (1986). Further results on a class of distributions which includes the normal ones, *Statistica* **46**(2), 199–208.

[22] Azzalini, A. and Capitanio, A. (1999). Statistical applications of the multivariate skew normal distribution, *J. Roy. Statist. Soc. Ser. B* **61**(3), 579–602.

[23] Azzalini, A. and Dalla Valle, D. (1996). The multivariate skew-normal distribution, *Biometrika* **83**(4), 715–726.

[24] Bagchi, U., Hayya, J.C. and Ord, J.K. (1983). The Hermite distribution as a model of demand during lead time for slow-moving items, *Decision Sciences* **14**, 447–466.

[25] Bagnold, R.A. (1954). *The Physics of Blown Sand and Desert Dunes*, Methuen, London.

[26] Bain, L.J. and Engelhardt, M. (1973). Interval estimation for the two-parameter double-exponential distribution, *Technometrics* **15**(4), 875–887.

[27] Balakrishnan, N. (1988). Recurrence relations among moments of order statistics from two related outlier moments, *Biometrical J.* **30**, 741–746.

[28] Balakrishnan, N. (1989). Recurrence relations among moments of order statistics from two related sets of independent and not-identically distributed random variables, *Ann. Inst. Statist. Math.* **41**, 323–329.

[29] Balakrishnan, N. and Ambagaspitiya, R.S. (1988). Relationships among moments of order statistics in samples from two related outlier moments and some applications, *Comm. Statist. Theory Methods* **17**(7), 2327–2341.

[30] Balakrishnan, N. and Ambagaspitiya, R.S. (1994). On skewed-Laplace distributions, Report, McMaster University, Hamilton, Ontario, Canada.

[31] Balakrishnan, N. and Chandramouleeswaran, M.P. (1994a). Reliability estimation and tolerance limits for Laplace distribution based on censored samples, Report, McMaster University, Hamilton, Ontario, Canada.

[32] Balakrishnan, N. and Chandramouleeswaran, M.P. (1994b). Prediction intervals for Laplace distribution based on censored samples, Report, McMaster University, Hamilton, Ontario, Canada.

[33] Balakrishnan, N., Chandramouleeswaran, M.P. and Ambagaspitiya, R.S. (1994). BLUE's of location and scale parameters of Laplace distribution based on Type-II censored samples and associated inference, Report, McMaster University, Hamilton, Ontario, Canada.

[34] Balakrishnan, N., Chandramouleeswaran, M.P. and Govindarajulu, Z. (1994). Inference on parameters of the Laplace distribution based on Type-II censored samples using Edgeworth approximation, Report, McMaster University, Hamilton, Ontario, Canada.

[35] Balakrishnan, N. and Cohen, A.C. (1991). *Order Statistics and Inference: Estimation Methods*, Academic Press, San Diego.

[36] Balakrishnan, N. and Cutler, C.D. (1994). Maximum likelihood estimation of the Laplace parameters based on Type-II censored samples, in *H.A. David Festschrift Volume* (eds., D.F. Morrison, H.N. Nagaraja, and P.K. Sen), Springer-Verlag, New York, pp. 145–151.

[37] Balakrishnan, N., Govindarajulu, Z. and Balasubramanian, K. (1993). Relationships between moments of two related sets of order statistics and some extensions, *Ann. Inst. Statist. Math.* **45**, 243–247.

[38] Balakrishnan, N. and Kocherlakota, S. (1986). Effects of nonnormality on $\overline{X}$ charts: Single assignable cause model, *Sankhyā Ser. B* **48**, 439–444.

[39] Balanda, K.P. (1987). Kurtosis comparisons of the Cauchy and double exponential distributions, *Comm. Statist. Theory Methods* **16**(2), 579–59.

[40] Baringhaus, L. and Grübel, R. (1997). On a class of characterization problems for random convex combinations, *Ann. Inst. Statist. Math.* **49**(3), 555–567.

[41] Barndorff-Nielsen, O.E. (1977). Exponentially decreasing distributions for the logarithm of particle size, *Proc. Roy. Soc. London Ser. A* **353**, 401–419.

[42] Barndorff-Nielsen, O.E. (1979). Models for non-Gaussian variation, with application to turbulence, *Proc. Roy. Soc. London Ser. A* **368**, 501–520.

[43] Barndorff-Nielsen, O.E. (1996). Processes of Normal Inverse Gaussian Type, Research Report **339**, Dept. Theor. Statist., Aarhus University.

[44] Barndorff-Nielsen, O.E. (1997). Normal inverse Gaussian distributions and stochastic volatility modelling, *Scand. J. Statist.* **24**, 1–13.

[45] Barndorff-Nielsen, O.E. and Blaesild, P. (1981). Hyperbolic distributions and ramifications: contributions to theory and application, in *Statistical Distributions in Scientific Work*, Vol 4 (eds., C. Taillie et al.), Reidel, Dordrecht, pp. 19–44.

[46] Barndorff-Nielsen, O.E. and Blaesild, P. (1982). Hyperbolic distributions, in *The Encyclopedia of Statistical Sciences*, Vol. 3 (eds., S. Kotz and N.L Johnson), Wiley, New York, pp. 700–707.

[47] Belinskiy, B.P. and Kozubowski, T.J. (2000). Exponential mixture representation of geometric stable densities, *J. Math. Anal. Appl.* **246**, 465–479.

[48] Bernstein, S.L., Burrows, M.L., Evans, J.E., Griffiths, A.S., McNeill, D.A., Niessen, C.W., Richer, I., White, D.P. and Willim, D.K. (1974). Long-range communications at extremely low frequencies, *Proc. IEEE* **62**, 292–312.

[49] Bertoin, J. (1996). *Lévy Processes*, University Press, Cambridge.

[50] Bhattacharyya, B. C. (1942). The use of McKay's Bessel function curves for graduating frequency distributions, *Sankhyā* **6**, 175–182.

[51] Billingsley, P. (1968). *Convergence of Probability Measures*, Wiley, New York.

[52] Biswas, S. and Sehgal, V.K. (1991). *Topics in Statistical Methodology*, Wiley, New York.

[53] Black, C.M., Durham, S.D., Lynch, J.D. and Padgett, W.J. (1989). A new probability distribution for the strength of brittle fibers, *Fiber-Tex 1989*, The Third Conference on Advanced Engineering Fibers and Textile Structures for Composites, *NASA Conference Publication 3082*, 363–374.

[54] Blaesild, P. (1981). The two-dimensional hyperbolic distribution and related distributions, with an application to Johannsen's bean data, *Biometrika* **68**, 251-63.

[55] Bondesson, L. (1992). *Generalized Gamma Convolutions and Related Classes of Distributions and Densities*, Springer, New York.

[56] Bouzar, N. (1999). On geometric stability and Poisson mixtures, *Illinois J. Math.* **43**(3), 520–527.

[57] Box, G.E.P. and Tiao, G.C. (1962). A further look at robustness via Bayes's theorem, *Biometrika* **49**, 419–432.

[58] Breiman, L. (1993). *Probability* (2nd ed.), SIAM, Philadelphia.

[59] Brunk, H.D. (1955). Maximum likelihood estimation of monotone parameters, *Ann. Math. Statist.* **26**, 607–616.

[60] Brunk, H.D. (1965). Conditional expectation given a σ-lattice and applications, *Ann. Math. Statist.* **36**, 1339–1350.

[61] Buczolich, Z. and Székely, G. (1989). When is a weighted average of ordered sample elements a maximum likelihood estimator of the location parameter? *Adv. Appl. Math.* **10**, 439–456.

[62] Bunge, J. (1993). Some stability classes for random numbers of random vectors, *Comm. Statist. Stochastic Models* **9**, 247–254.

[63] Bunge, J. (1996). Composition semigroups and random stability, *Ann. Probab.* **24**(3), 1476–1489.

[64] Carlton, A.G. (1946). Estimating the parameters of a rectangular distribution, *Ann. Math. Statist.* **17**, 355–358.

[65] Chan, L.K. (1970). Linear estimation of the location and scale parameters of the Cauchy distribution based on sample quantiles, *J. Amer. Statist. Assoc.* **65**, 851–859.

[66] Chan, L.K. and Chan, N.N. (1969). Estimates of the parameters of the double exponential distribution based on selected order statistics, *Bulletin of the Institute of Statistical Research and Training* **3**, 21–40.

[67] Cheng, S.W. (1978). Linear quantile estimation of parameters of the double exponential distribution, *Soochow J. Math.* **4**, 39–49.

[68] Chew, V. (1968). Some useful alternatives to the normal distribution, *Amer. Statist.* **22** (3), 22–24.

[69] Childs, A. and Balakrishnan, N. (1996). Conditional inference procedures for the Laplace distribution based on type-II right censored samples, *Statist. Probab. Lett.* **31**(1), 31–39.

[70] Childs, A. and Balakrishnan, N. (1997a). Some extensions in the robust estimation of parameters of exponential and double exponential distributions in the presence of multiple outliers, in *Handbook of Statistics — Robust Methods*, Vol. 15 (eds., C.R. Rao and G.S. Maddala), North-Holland, Amsterdam, pp. 201–235.

[71] Childs, A. and Balakrishnan, N. (1997b). Maximum likelihood estimation of Laplace parameters based on general type-II censored examples, *Statist. Papers* **38**(3), 343–348.

[72] Chipman, J.S. (1985). Theory and measurement of income distribution, in *Advances in Econometrics*, Vol. 4 (eds., R.L. Basmann and G.F. Rhodes, Jr.), JAI Press, Greenwich, pp. 135–165.

[73] Christoph, G. and Schreiber, K. (1998a). Discrete stable random variables, *Statist. Probab. Lett.* **37**, 243–247.

[74] Christoph, G. and Schreiber, K. (1998b). The generalized discrete Linnik distributions, in *Advances in Stochastic Models for Reliability, Quality, and Safety* (eds., W. Kahle et al.), Birkhäuser, Boston, pp. 3–18.

[75] Christoph, G. and Schreiber, K. (1998c). Positive Linnik and discrete Linnik distributions, in *Asymptotic Methods in Probability and Statistics with Applications*, Proceedings of the Conference in St-Petersburg, June 1998, Birkhäuser, Boston.

[76] Chu, J.T. and Hotelling, H. (1955). The moments of the sample median, *Ann. Math. Statist.* **26**, 593–606.

[77] Christensen, R. (2000). Unpublished Notes, Department of Statistics, University of New Mexico, Albuquerque, NM.

[78] Cifarelli, D.M and Regazzini, E. (1976). On a characterization of a class of distributions based on efficiency of certain estimator, *Rend. Sem. Mat. Univers. Politecn. Torino* (1974–1975) **33**, 299–311 (in Italian).

[79] Clark, P.K. (1973) A subordinated stochastic process model with finite variance for speculative prices, *Econometrica* **41**, 135–155.

[80] Conover, W.J., Wehmanen, O. and Ramsey, F.L. (1978). A note on the small-sample power functions for nonparametric tests of location in the double exponential family, *J. Amer. Statist. Assoc.* **73**, 188–190.

[81] Corless, R.M., Gonnet, G.H., Hare, D.E.G., Jeffrey, D.J. and Knuth, D.E. (1996). On the Lambert W function, *Adv. Comput. Math.* **5**, 329–359.

[82] Craig, C.C. (1932). On the distributions of certain statistics, *Amer. J. Math.* **54**, 353–366.

[83] Craig, C.C. (1936). On the frequency function of xy, *Ann. Math. Statist.* **7**, 1–15.

[84] Crum, W.L. (1923). The use of the median in determining seasonal variation, *J. Amer. Statist. Assoc.* **18**, 607–614.

[85] Dadi, M.I. and Marks, R.J., II (1987). Detector relative efficiencies in the presence of Laplace noise, *IEEE Trans. Aerospace Electron. Systems* **23**, 568–582.

[86] Damsleth, E. and El-Shaarawi, A.H. (1989). ARMA models with double-exponentially distributed noise, *J. Roy. Statist. Soc. Ser. B* **51**(1), 61–69.

[87] David, H.A. (1981). *Order Statistics* (2nd ed.), Wiley, New York.

[88] Devroye, L. (1986). *Non-Uniform Random Variate Generation*, Springer-Verlag, New York.

[89] Devroye, L. (1990). A note on Linnik distribution, *Statist. Probab. Lett.* **9**, 305–306.

[90] Devroye, L. (1993). A triptych of discrete distributions related to the stable law, *Statist. Probab. Lett.* **18**, 349–351.

[91] Dewald, L.S. and Lewis, P.A.W. (1985). A new Laplace second-order autoregressive time series model-NLAR(2), *IEEE Trans. Inform. Theory* **31**(5), 645–651.

[92] Dharmadhikari, S. and Joag-Dev, K. (1988). *Unimodality, Convexity, and Applications*, Academic Press, San Diego.

[93] Divanji, G. (1988). On Semi-α-Laplace distributions, *J. Indian Statist. Assoc.* **26**, 31–38.

[94] Dixon, W.J. (1957). Estimates of the mean and standard deviation of a normal population, *Ann. Math. Statist.* **28**, 806–809.

[95] Dixon, W.J. (1960). Simplified estimation from censored normal samples, *Ann. Math. Statist.* **31**, 385–391.

[96] Dreier, I. (1999). Inequalities for real characteristic functions and their moments, *Ph.D. Dissertation*, Technical University of Dresden, Germany (in German).

[97] Dugué, D. (1951). Sur certains exemples de décompositions en arithmétique des lois de probabilité, *Ann. Inst. H. Poincaré* **12**, 159–169.

[98] Duttweiler, D.L. and Messerschmitt, D.G. (1976). Nearly instantaneous companding for nonuniformly quantized PCM, *IEEE Trans. Comm.* **COM-24**(8), 864–873.

[99] Easterling, R.G. (1978). Exponential responses with double exponential measurement error — A model for steam generator inspection, *Proceedings of the DOE Statistical Symposium, US Department of Energy*, pp. 90–110.

[100] Eberlein, E. and Keller, U. (1995). Hyperbolic distributions in finance, *Bernoulli* **1**(3), 281–299.

[101] Edwards, A.W.F. (1974). Letter to the Editor, *Technometrics* **16**(4), 641–642.

[102] Edwards, L. (1948). The use of normal significance limits when the parent population is of Laplace form, *J. Institute of Actuaries Students' Society* **8**, 87–99.

[103] Efron, B. (1986). Double exponential families and their use in generalized linear regression, *J. Amer. Statist. Assoc.* **81**, 709–721.

[104] Eisenhart, Ch. (1983). Laws of Error I-III, in *Encyclopedia of Statistical Sciences*, Vol 4 (eds., S. Kotz et al.), Wiley, New York, pp. 530–562.

[105] Elderton, W.P. (1938). *Frequency Curves and Correlation*, Cambridge University Press, Cambridge.

[106] Epstein, B. (1947). Application of the theory of extreme values in fracture problems, *J. Amer. Statist. Assoc.* **43**, 403–412.

[107] Epstein, B. (1948). Statistical aspects of fracture problems, *J. Appl. Physics* **19**, 140–147.

[108] Erdogan, M.B. (1995). Analytic and asymptotic properties of non-symmetric Linnik's probability densities, *PhD Thesis*, Bilkent University, Ankara; appeared in *J. Fourier Anal. Appl.* **5**(6), 523–544, 1999.

[109] Erdogan, M.B. and Ostrovskii, I.V. (1997). Non-symmetric Linnik distributions, *C. R. Acad. Sci. Paris* t. 325, Série I, 511–516.

[110] Erdogan, M.B. and Ostrovskii, I.V. (1998a). On mixture representation of the Linnik density, *J. Austral. Math. Soc. Ser. A* **64**, 317–326.

[111] Erdogan, M.B. and Ostrovskii, I.V. (1998b). Analytic and asymptotic properties of generalized Linnik probability densities, *J. Math. Anal. Appl.* **217**, 555–579.

[112] Ernst, M.D. (1998). A multivariate generalized Laplace distribution, *Comput. Statist.* **13**, 227–232.

[113] Fang, K.T., Kotz, S. and Ng, K.W. (1990). *Symmetric Multivariate and Related Distributions*, Monographs on Statistics and Probability **36**, Chapman-Hall, London.

[114] Farebrother, R.W. (1986). Pitman's measure of closeness, *Amer. Statist.* **40**, 179–180.

[115] Farison, J.B. (1965). On calculating moments for some common probability laws, *IEEE Trans. Inform. Theory* **11**(4), 586–589.

[116] Fechner, G.T. (1897). *Kollektivmasslehre*, W. Engelmann, Leipzig.

[117] Feller, W. (1971). *Introduction to the Theory of Probability and its Applications*, Vol. 2 (2nd ed.), Wiley, New York.

[118] Ferguson, T.S. and Klass, M.J. (1972). A representation of independent increment processes without Gaussian components, *Ann. Math. Statist.* **43**, 1634–1643.

[119] Fernández, C., Osiewalski, J. and Steel, M.F.J. (1995). Modeling and inference with v-distributions, *J. Amer. Statist. Assoc.* **90**, 1331–1340.

[120] Fernández, C. and Steel, M.F.J. (1998). On Bayesian modeling of fat tails and skewness, *J. Amer. Statist. Assoc.* **93**, 359–371.

[121] Fieller, N.R.J. (1993). Archaeostatistics: Old statistics in ancient contexts, *Statistician* **42**, 279–295.

[122] Findeisen, P. (1982). Characterization of the bilateral exponential distribution, *Metrika* **29**, 95–102 (in German).

[123] Fisher, R.A. (1922). On the mathematical foundations of theoretical statistics, *Philos. Trans. Roy. Soc. London Ser. A* **222**, 309–368.

[124] Fisher, R.A. (1925). Theory of statistical estimation, *Proc. Camb. Philos. Soc.* **22**, 700–725.

[125] Fisher, R.A. (1934). Two new properties of mathematical likelihood, *Proc. Roy. Soc. London Ser. A* **147**, 285–307.

[126] Fuita, Y. (1993). A generalization of the results of Pillai, *Ann. Inst. Statist. Math.* **45**(2), 361–365.

[127] Galambos, J. and Kotz, S. (1978). *Characterizations of Probability Distributions. A Unified Approach with an Emphasis on Exponential and Related Models*, *Lecture Notes in Math.* **675**, Springer, Heidelberg.

[128] Gallo, F. (1979). On the Laplace first law: Sample distribution of the sum of values and the sum of absolute values of the errors, distribution of the related T, *Statistica* **39**, 443–454.

[129] Geary, R.C. (1935). The ratio of the mean deviation to the standard deviation as a test of normality, *Biometrika* **27**, 310–332.

[130] Geman, H., Madan, D.B. and Yor, M. (2000a). Asset prices are Brownian motion: Only in business time, in *Quantitative Analysis in Financial Markets*, Volume II (ed., Marco Avellaneda), World Scientific Publishing Company, in press.

[131] Geman, H., Madan, D.B. and Yor, M. (2000b). Time changes for Lévy processes, *Mathematical Finance*, in press.

[132] George, E.O. and Mudholkar, G.S. (1981). A characterization of the logistic distribution by a sample median, *Ann. Inst. Statist. Math.* **33**, Part A, 125–129.

[133] George, E.O. and Rousseau, C.C. (1987). On the logistic midrange, *Ann. Inst. Statist. Math.* **39**, Part A, 627–635.

[134] George, Sabu and Pillai, R.N. (1988). A characterization of spherical distributions and some applications of Meijer's G-function, *Proceedings of the Symposium on Special Functions and Problem-oriented Research (Trivandrum, 1988)*, Publication **14**, Centre Math. Sci., Trivandrum, Kerala, India, pp. 61–71.

[135] Gnedenko, B.V. (1970). Limit theorems for sums of random number of positive independent random variables, *Proc. 6th Berkeley Symp. on Math. Statist. Probab.*, Vol. 2, pp. 537–549.

[136] Gnedenko, B.V. and Janjić, S. (1983). A characteristic property of one class of limit distributions, *Math. Nachr.* **113**, 145–149.

[137] Gnedenko, B.V. and Korolev, V.Yu. (1996). *Random Summation: Limit Theorems and Application*, CRC Press, Boca Raton.

[138] Gokhale, D.V. (1975). Maximum entropy characterizations of some distributions, in *Statistical Distributions in Scientific Work*, Vol. 3 (eds., G.P. Patil, S. Kotz and J.K. Ord), Reidel, pp. 299–305.

[139] Goldfeld, S.M and Quandt, R.E. (1981). Econometric modelling with non-normal disturbances, *J. Econometrics* **17**, 141–155.

[140] González, F.J., Puig-Pey, J. and Luceño, A. (1999). Analytical expressions for the average adjustment interval and mean squared deviation for bounded adjustment schemes, *Comm. Statist. Simulation Comput.* **28**, 623–635.

[141] Govindarajulu, Z. (1963). Relationships among moments of order statistics in samples from two related populations, *Technometrics* **5**, 514–518.

[142] Govindarajulu, Z. (1966). Best linear estimates under symmetric censoring of the parameters of a double exponential population, *J. Amer. Statist. Assoc.* **61**, 248–258. (Correction: **71**, 255.)

[143] Gradshteyn, I.S. and Ryzhik, I.M. (1980). *Tables of Integrals, Series, and Products*, Academic Press, New York.

[144] Greenwood, J.A., Olkin, I. and Savage, I.R. (1962). Index to *Annals of Mathematical Statistics*, Volumes 1–31, 1930–1960. University of Minnesota, Minneapolis St. Paul: North Central Publishing.

[145] Grice, J.V., Bain, L.J. and Engelhardt, M. (1978). Comparison of conditional and unconditional confidence intervals for the double exponential distribution, *Comm. Statist. Simulation Comput.* **7**(5), 515–524.

[146] Gumbel, E.J. (1944). Ranges and midranges, *Ann. Math. Statist.* **15**, 414–422.

[147] Hájek, J. (1969). *Nonparametric Statistics*, Holden-Day, San Francisco.

[148] Hald, A. (1995). *History of Mathematical Statistics* 1750–1930, Wiley, New York.

[149] Hall, D.L. and Joiner, B.L. (1983). Asymptotic relative efficiencies of R-estimators of location, *Comm. Statist. Theory Methods* **12**(7), 739–763.

[150] Haro-López, R.A. and Smith, A.F.M. (1999). On robust Bayesian analysis for location and scale parameters, *J. Multivariate Anal.* **70**, 30–56.

[151] Harris, B. (1966). *Theory of Probability*, Addison-Wesley, Reading.

[152] Harter, H.L., Moore, A.H. and Curry, T.F. (1979). Adaptive robust estimation of location and scale parameters of symmetric populations, *Comm. Statist. Theory Methods* **A8**(15), 1473–1491.

[153] Hartley, M.J and Revankar, N.S. (1974). On the estimation of the Pareto law from underreported data, *J. Econometrics* **2**, 327–341.

[154] Harville, D.A. (1997). *Matrix Algebra From a Statistician's Perspective*, Springer, New York.

[155] Hausdorff, F. (1901). Beitrage zur Wahrscheinlichkeitsrechnung, *Verhandlungen der Konigliche Sachsischen Gesellschaft der Wissenschaften, Leipzig, Mathematisch-Physische Classe* **53**, 152–178.

[156] Hayfavi, A. (1998). An improper integral representation of Linnik's probability densities, *Tr. J. Math.* **22**, 235–242.

[157] Heathcote, C.R., Rachev, S.T. and Cheng, B. (1995). Testing multivariate symmetry, *J. Multivariate Anal.* **54**, 91–112.

[158] Henze, N. (1986). A probabilistic representation of the skew-normal distribution, *Scand. J. Statist.* **13**, 271–275.

[159] Hettmansperger, T.P. and Keenan, M.A. (1975). Tailweight, statistical inference and families of distributions — a brief survey, in *Statistical Distributions in Scientific Work*, Vol. 1 (eds., G.P. Patil, S. Kotz and J.K. Ord), Reidel, Dordrecht, pp. 161–172.

[160] Hinkley, D.V. and Revankar, N.S. (1977). Estimation of the Pareto law from underreported data, *J. Econometrics* **5**, 1–11.

[161] Hirschberg, J.G., Molina, D.J. and Slottje, D.J. (1989). A selection criterion for choosing between functional forms of income, *Econometric Rev.* **7**(2), 183–197.

[162] Hoaglin, D.C., Mosteller, F. and Tukey, J.W. (eds.) (1983). *Understanding Robust and Exploratory Data Analysis*, Wiley, New York.

[163] Hogg, R.V. (1972). More light on the kurtosis and related statistics, *J. Amer. Statist. Assoc.* **67**, 422–424.

[164] Holla, M.S. and Bhattacharya, S.K. (1968). On a compound Gaussian distribution, *Ann. Inst. Statist. Math.* **20**, 331–336.

[165] Hombas, V.C. (1986). The double exponential distribution: Using calculus to find a maximum likelihood estimator, *Amer. Statist.* **40**(2), 178.

[166] Horn, P.S. (1983). A measure for peakedness, *Amer. Statist.* **37**(1), 55–56.

[167] Hsu, D.A. (1979). Long-tailed distributions for position errors in navigation, *Appl. Statist.* **28**, 62–72.

[168] Huang, W.J. and Chen, L.S. (1989). Note on a characterization of gamma distributions, *Statist. Probab. Lett.* **8**, 485–487.

[169] Huber, P.J. (1981). *Robust Statistics*, Wiley, New York.

[170] Hwang, J.T. and Chen, J. (1986). Improved confidence sets for the coefficients of a linear model with spherically symmetric errors, *Ann. Statist.* **14**(2), 444–460.

[171] Iliescu, D.V. and Vodă, V.Gh. (1973). Proportion-p estimators for certain distributions, *Statistica* **33**, 309–321.

[172] International Organization for Standardization (1989). Sampling procedures and charts for inspection by variable for percent nonconforming (ISO 3951), Geneva, Switzerland.

[173] Jacques, C., Rémillard, B. and Theodorescu, R. (1999). Estimation of Linnik law parameters, *Statist. Decisions* **17**, 213–235.

[174] Jakuszenkow, H. (1978). Estimation of variance in Laplace's distribution, *Zeszyty Nauk. Politech. Lódz. Matematyka* **10**, 29–36 (in Polish).

[175] Jakuszenkow, H. (1979). Estimation of the variance in the generalized Laplace distribution with quadratic loss function, *Demonstratio Math.* **12**(3), 581–591.

[176] Janicki, A. and Weron, A. (1994). *Simulation and Chaotic Behavior of α-Stable Stochastic Processes*, Marcel Dekker, New York.

[177] Janjić, S. (1984). On random variables with the same distribution type as their random sum, *Publications de l'Institut Mathématique, Nouvelle série, tome 35* **49**, 161–166.

[178] Janković, S. (1992). Preservation of type under mixing, *Teor. Veroyatnost. i Primenen.* **37**, 594–599.

[179] Janković, S. (1993a). Some properties of random variables which are stable with respect to the random sample size, *Stability Problems for Stochastic Models, Lecture Notes in Math.* **1546**, 68–75.

[180] Janković, S. (1993b). Enlargement of the class of geometrically infinitely divisible random variables, *Publ. Inst. Math. (Beograd) (N.S.)* **54**(68), 126–134.

[181] Jayakumar, K. and Pillai, R.N. (1993). The first-order autoregressive Mittag-Leffler process, *J. Appl. Probab.* **30**, 462–466.

[182] Jayakumar, K. and Pillai, R.N. (1995). Discrete Mittag-Leffler distribution, *Statist. Probab. Lett.* **23**, 271–274.

[183] Jaynes, E.T. (1957). Information theory and statistical mechanics I, *Phys. Rev.* **106**, 620–630.

[184] Johnson, M.E. (1987). *Multivariate Statistical Simulation*, Wiley, New York.

[185] Johnson, N.L. (1954). System of frequency curves derived from the first law of Laplace, *Trabajos de Estadistica* **5**, 283–291.

[186] Johnson, N.L. and Kotz, S. (1970). *Continuous Univariate Distributions* — 1, Wiley, New York.

[187] Johnson, N.L. and Kotz, S. (1972). *Distributions in Statistics: Continuous Multivariate Distributions*, Wiley, New York.

[188] Johnson, N.L., Kotz, S. and Balakrishnan, N. (1994). *Continuous Univariate Distributions* — 1 (2nd ed.), Wiley, New York.

[189] Johnson, N.L., Kotz, S. and Balakrishnan, N. (1995). *Continuous Univariate Distributions* — 2 (2nd ed.), Wiley, New York.

[190] Johnson, N.L., Kotz, S. and Kemp, A.W. (1992). *Univariate Discrete Distributions* (2nd ed.), Wiley, New York.

[191] Jones, P.N. and McLachlan, G.J. (1990). Laplace-normal mixtures fitted to wind shear data, *J. Appl. Statistics* **17**, 271–276.

[192] Jorgensen, B. (1982). *Statistical Properties of the Generalized Inverse Gaussian Distribution, Lecture Notes in Statist.* **9**, Springer-Verlag, New York.

[193] Joshi, S.N. (1984). Expansion of Bayes risk in the case of double exponential family, *Sankhyā Ser. A* **46**, 64–74.

[194] Kacki, E. (1965a). Absolute moments of the Laplace distribution, *Prace Matematyczne* **10**, 89–93 (in Polish).

[195] Kacki, E. (1965b). Certain special cases of the parameter estimation of a mixture of two Laplace distributions, *Zeszyty Naukowe Politechniki Lodzkiej, Elektryka* **20** (in Polish).

[196] Kacki, E. and Krysicki, W. (1967). Parameter estimation of a mixture of two Laplace distributions (general case), *Roczniki Polskiego Towarzystwa Matematycznego, Seria I: Prace Matematyczne* **11**, 23–31 (in German).

[197] Kafaei, M.-A. and Schmidt, P. (1985). On the adequacy of the "Sargan distribution" as an approximation to the normal, *Comm. Statist. Theory Methods* **14**(3), 509–526.

[198] Kagan, A.M., Linnik, Yu.V. and Rao, C.R. (1973). *Characterization Problems in Mathematical Statistics*, Wiley, New York.

[199] Kakosyan, A.V., Klebanov, L.B. and Melamed, I.A. (1984). *Characterization of Distributions by the Method of Intensively Monotone Operators*, *Lecture Notes in Math.* **1088**, Springer, Heidelberg.

[200] Kalashnikov, V. (1997). *Geometric Sums: Bounds for Rare Events with Applications*, Kluwer Acad. Publ., Dordrecht.

[201] Kanefsky, M. and Thomas, J.B. (1965). On polarity detection schemes with non-Gaussian inputs, *J. Franklin Institute* **280**, 120–138.

[202] Kanji, G.K. (1985). A mixture model for wind shear data. *J. Appl. Statistics* **12**, 49–58.

[203] Kapoor, S. and Kanji, G.K. (1990). Application of the characterization theory to the mixture model, *J. Appl. Statist.* **17**, 263–270.

[204] Kappenman, R.F. (1975). Conditional confidence intervals for double exponential distribution parameters, *Technometrics* **17**(2), 233–235.

[205] Kappenman, R.F. (1977). Tolerance intervals for the double exponential distribution, *J. Amer. Statist. Assoc.* **72**, 908–909.

[206] Kapteyn, J.C. (1903). *Skew Frequency Curves*, Astronomical Laboratory, Groningen.

[207] Kapur, J.N. (1993). *Maximum-Entropy Models in Science and Engineering* (revised ed.), Wiley, New York.

[208] Karst, O.J. and Polowy, H. (1963). Sampling properties of the median of a Laplace distribution, *Amer. Math. Monthly* **70**, 628–636.

[209] Kelker, D. (1971). Infinite divisibility and variance mixtures of the normal distribution, *Ann. Math. Statist.* **42**(2), 802–808.

[210] Kendall, M.G., Stuart, A. and Ord, J.K. (1994). *Kendall's Advanced Theory of Statistics, Vol. 1, Distribution theory* (6th edition), Halsted Press (Wiley, Inc.), New York.

[211] Keynes, J.M. (1911). The principal averages and the laws of error which lead to them, *J. Roy. Statist. Soc.* **74**, New Series, 322–331.

[212] Khan, A.H. and Khan, R.V. (1987). Relations among moments of order statistics in samples from doubly truncated Laplace and exponential distributions, *J. Statist. Res.* **21**, 35–44.

[213] Klebanov, L.B., Maniya, G.M. and Melamed, I.A. (1984). A problem of Zolotarev and analogs of infinitely divisible and stable distributions in a scheme for summing a random number of random variables, *Theory Probab. Appl.* **29**, 791–794.

[214] Klebanov, L.B., Melamed, J.A., Mittnik, S. and Rachev, S.T. (1996). Integral and asymptotic representations of geo-stable densities, *Appl. Math. Lett.* **9**, 37–40.

[215] Klebanov, L.B. and Rachev, S.T. (1996). Sums of random number of random variables and their approximations with ν-accompanying infinitely divisible laws, *Serdica Math. J.* **22**, 471–496.

[216] Klein, G.E. (1993). The sensitivity of cash-flow analysis to the choice of statistical model for interest rate changes (with discussions), *Transactions of the Society of Actuaries* XLV, 79–186.

[217] Kollo, T. (2000). Private communication (to S. Kotz).

[218] Kotz, S., Balakrishnan, N. and Johnson, N.L. (2000). *Continuous Multivariate Distributions* (2nd ed.), Wiley, New York.

[219] Kotz, S., Fang, K.T. and Liang, J.J. (1997). On multivariate vertical density representation and its application to random number generation, *Statistics* **30**, 163–180.

[220] Kotz, S. and Johnson, N.L. (1982). *Encyclopedia of Statistical Sciences*, Vol. 2, Wiley, New York.

[221] Kotz, S., Johnson, N.L. and Read, C.B. (1985). Log-Laplace distribution, in *Encyclopedia of Statistical Sciences*, Vol. 5 (eds., S. Kotz, N.L. Johnson, and C.B. Read), Wiley, New York, pp. 133–134.

[222] Kotz, S., Kozubowski, T.J. and Podgórski, K. (2000a). Maximum entropy characterization of asymmetric Laplace distribution, *Technical Report No. 361*, Department of Statistics and Applied Probability, University of California, Santa Barbara; to appear in *Int. Math. J.*

[223] Kotz, S., Kozubowski, T.J. and Podgórski, K. (2000b). An asymmetric multivariate Laplace distribution, *Technical Report No. 367*, Department of Statistics and Applied Probability, University of California, Santa Barbara.

[224] Kotz, S., Kozubowski, T.J. and Podgórski, K. (2000c). Maximum likelihood estimation of asymmetric Laplace parameters, Preprint.

[225] Kotz, S. and Ostrovskii, I.V. (1996). A mixture representation of the Linnik distribution, *Statist. Probab. Lett.* **26**, 61–64.

[226] Kotz, S., Ostrovskii, I.V. and Hayfavi, A. (1995). Analytic and asymptotic properties of Linnik's probability densities, I and II, *J. Math. Anal. Appl.* **193**, 353–371 and 497–521.

[227] Kotz, S. and Steutel, F.W. (1988). Note on a characterization of exponential distributions, *Statist. Probab. Lett.* **6**, 201–203.

[228] Kotz, S. and Trout, M.D. (1996). On the vertical density representation and ordering of distributions, *Statistics* **28**, 241–247.

[229] Kou, S.G. (2000). A jump diffusion model for option pricing with three properties: leptokurtic feature, volatility smile, and analytical tractability, Preprint, Columbia University.

[230] Koutrouvelis, I.A. (1980). Regression-type estimation of the parameters of stable laws, *J. Amer. Statist. Assoc.* **75**, 918–928.

[231] Kozubowski, T.J. (1993). Estimation of the parameters of geometric stable laws, *Technical Report No. 253*, Department of Statistics and Applied Probability, University of California, Santa Barbara; appeared in *Math. Comput. Modelling* **29**(10–12), 241–253, 1999.

[232] Kozubowski, T.J. (1994a). Representation and properties of geometric stable laws, in *Approximation, Probability, and Related Fields* (eds., G. Anastassiou and S.T. Rachev), Plenum, New York, pp. 321–337.

[233] Kozubowski, T.J. (1994b). The inner characterization of geometric stable laws, *Statist. Decisions* **12**, 307–321.

[234] Kozubowski, T.J. (1997). Characterization of multivariate geometric stable distributions, *Statist. Decisions* **15**, 397–416.

[235] Kozubowski, T.J. (1998). Mixture representation of Linnik distribution revisited, *Statist. Probab. Lett.* **38**, 157–160.

[236] Kozubowski, T.J. (1999). Fractional moment estimation of Linnik and Mittag-Leffler parameters, *Math. Comput. Modelling*, in press.

[237] Kozubowski, T.J. (2000a). Exponential mixture representation of geometric stable distributions, *Ann. Inst. Statist. Math.* **52**(2), 231–238.

[238] Kozubowski, T.J. (2000b). Computer simulation of geometric stable random variables, *J. Comp. Appl. Math.* **116**, 221–229.

[239] Kozubowski, T.J. and Panorska, A.K. (1996). On moments and tail behavior of ν-stable random variables, *Statist. Probab. Lett.* **29**, 307–315.

[240] Kozubowski, T.J. and Panorska, A.K. (1998). Weak limits for multivariate random sums, *J. Multivariate Anal.* **67**, 398–413.

[241] Kozubowski, T.J. and Panorska, A.K. (1999). Multivariate geometric stable distributions in financial applications, *Math. Comput. Modelling* **29**, 83–92.

[242] Kozubowski, T.J. and Podgórski, K. (1999a). A class of asymmetric distributions, *Actuarial Research Clearing House* **1**, 113–134.

[243] Kozubowski, T.J. and Podgórski, K. (1999b). A multivariate and asymmetric generalization of Laplace distribution, *Comput. Statist.*, in press.

[244] Kozubowski, T.J. and Podgórski, K. (1999c). Asymmetric Laplace laws and modeling financial data, *Math. Comput. Modelling*, in press.

[245] Kozubowski, T.J. and Podgórski, K. (2000). Asymmetric Laplace distributions, *Math. Sci.* **25**, 37–46.

[246] Kozubowski, T.J., Podgórski, K. and Samorodnitsky, G. (1998). Tails of Lévy measure of geometric stable random variables, *Extremes* **1**(3), 367–378.

[247] Kozubowski, T.J. and Rachev, S.T. (1994). The theory of geometric stable distributions and its use in modeling financial data, *European J. Oper. Res.* **74**, 310–324.

[248] Kozubowski, T.J. and Rachev, S.T. (1999a). Univariate geometric stable laws, *J. Comput. Anal. Appl.* **1**(2), 177–217.

[249] Kozubowski, T.J. and Rachev, S.T. (1999b). Multivariate geometric stable laws, *J. Comput. Anal. Appl.* **1**(4), 349–385.

[250] Krein, M. (1944). On the extrapolation problem of A.N. Kolmogorov, *Doklady Akad. Nauk SSSR* **46**(8), 339–342 (in Russian).

[251] Krysicki, W. (1966a). An application of the method of moments to the problem of parameter estimation of a mixture of two Laplace distributions, *Zeszyty Naukowe Politechniki Lódzkiej, Wlókiennictwo* **14**, 3–14 (in Polish).

[252] Krysicki, W. (1966b). Estimation of parameters in a mixture of two Laplace's distributions with use of the method of moments. *Zeszyty Naukowe Politechniki Lódzkiej* **77**, 5–13 (in Polish).

[253] Laha, R.G. (1961). On a class of unimodal distributions, *Proc. Amer. Math. Soc.* **12**, 181–184.

[254] Lanfer, H. (1978). Maximum signal-to-noise-ratio quantization for Laplacian-distributed signals, *Inform. Syst. Theory in Digital Comm.* NTG-Report, VDE-Verlag, Berlin, Germany **65**, 52.

[255] Laplace, P.S. (1774). Mémoire sur la probabilité des causes par les événements, *Mémoires de Mathématique et de Physique, Presentés à l'Académie Royale des Sciences, par divers Savans et lûs dans ses Assemblées* **6**, 621–656. English translation *Memoir on the Probability of the Causes of Events* in *Statistical Science* **1**(3), 1986, 364–378.

[256] Latta, R.B. (1979). Composition rules for probabilities from paired comparisons, *Ann. Statist.* **7**, 349–371.

[257] Lehmann, E.L. (1983). *Theory of Point Estimation*, Wiley, New York.

[258] Lehmann, E.L. and Casella, G. (1998). *Theory of Point Estimation* (2nd ed.), Springer, New York.

[259] Levin, A. and Albanese, C. (1998). Bayesian Value-at-Risk: Calibration and Simulation, Presentation at the SIAM Annual Meeting'98, Toronto, July 13–17, 1998.

[260] Levin, A. and Tchernitser, A. (1999). Multifactor gamma stochastic variance Value-at-Risk model, Presentation at the Conference *Applications of Heavy Tailed Distributions in Economics, Engineering, and Statistics*, American University, Washington, DC, June 3–5, 1999.

[261] Lien, D.H.D, Balakrishnan, N. and Balasubramanian, K. (1992). Moments of order statistics from a non-overlapping mixture model with applications to truncated Laplace distribution, *Comm. Statist. Theory Methods* **21**, 1909–1928.

[262] Lin, G.D. (1994). Characterizations of the Laplace and related distributions via geometric compounding, *Sankhyā Ser. A* **56**, 1–9.

[263] Lindsey, J.K. (1999). Multivariate elliptically contoured distributions for repeated measurements, *Biometrics* **55**, 1277–1280.

[264] Ling, K.D. (1977). Bayesian predictive distribution for sample from double exponential distribution, *Nanta Mathematica* **10**(1), 13–19.

[265] Ling, K.D. and Lim, S.K. (1978). On Bayesian predictive distributions for samples from double exponential distribution based on grouped data, *Nanta Mathematica* **11**(2), 191–201.

[266] Lingappaiah, G.S. (1988). On two-piece double exponential distribution, *J. Korean Statist. Soc.* **17**(1), 46–55.

[267] Linnik, Ju.V. (1953). Linear forms and statistical criteria, I, II, *Ukr. Mat. Zhurnal* **5**, 207–290 (in Russian); also in *Selected Translations in Math. Statist. Probab.* **3** (1963), 1–90.

[268] Liseo, B. (1990). The skew normal class of densities: inferential aspects from a Bayesian viewpoint, *Statistica* **50**, 59–70 (in Italian).

[269] Loh, W.-Y. (1984). Random quotients and robust estimation, *Comm. Statist. Theory Methods* **13**(22), 2757–2769.

[270] Longstaff, F. A. (1994). Stochastic volatility and option valuation: a pricing-density approach, Preprint, The Anderson Graduate School of Management, UCLA.

[271] Lukacs, E. (1955). A characterization of gamma distribution, *Ann. Math. Statist.* **26**, 88–95.

[272] Lukacs, E. (1957). Remarks concerning characteristic functions, *Ann. Math. Statist.* **28**, 717–723.

[273] Lukacs, E. (1970). *Characteristic Functions*, Griffin, London.

[274] Lukacs, E. and Laha, R.G. (1964). *Applications of Characteristic Functions*, Hafner Publishing Company, New York.

[275] Madan, D.B., Carr, P. and Chang, E.C. (1998). The variance gamma process and option pricing, *European Finance Review* **2**, 74–105.

[276] Madan, D.B. and Milne, F. (1991). Option pricing with VG martingale components, *Mathematical Finance* **1**, (4), 33–55.

[277] Madan, D.B. and Seneta, E. (1990). The variance gamma (V.G.) model for share markets returns, *J. Business* **63**, 511–524.

[278] Magnus, J.R. (2000). Estimation of the mean of a univariate normal distribution with known variance, Preprint, Center for Economic Research, Tilburg University.

[279] Magnus, J.R. and Neudecker, H. (1999). *Matrix Differential Calculus with Applications in Statistics and Economics*, Wiley, Chichester.

[280] Mandelbrot, B. (1963). The variations of certain speculative prices, *J. Business* **36**, 394–419.

[281] Manly, B.F.J. (1976). Some examples of double exponential fitness functions, *Heredity* **36**, 229–234.

[282] Mantel, N. (1973). A characteristic function exercise, *Amer. Statist.* **27**(1), 31.

[283] Mantel, N. (1987). The Laplace distribution and 2×2 unit normal determinants, *Amer. Statist.* **41**(1), 88.

[284] Mantel, N. and Pasternak, B.S. (1966). Light bulb statistics, *J. Amer. Statist. Assoc.* **61**, 633–639.

[285] Marks, R.J., Wise, G.L., Haldeman, D.G., and Whited, J.L. (1978). Detection in Laplace noise, *IEEE Trans. Aerospace Electron. Systems* **AES-14**(6), 866–871.

[286] Marshall, A.W. and Olkin, I. (1993). Maximum likelihood characterizations, *Statist. Sinica* **3**, 157–171.

[287] Mathai, A.M. (1993). Generalized Laplace distribution with applications, *J. Appl. Statist. Sci.* **1**(2), 169–178.

[288] McAlister, D. (1879). The law of the geometric mean, *Proceeding of the Royal Soc.* **29**, 367–376.

[289] McGill, W.J. (1962). Random fluctuations of response rate, *Psychometrika* **27**, 3–17.

[290] McGraw, D.K. and Wagner, J.F. (1968). Elliptically symmetric distributions, *IEEE Trans. Inform. Theory* **14**, 110–120.

[291] McKay, A.T. (1932). A Bessel function distribution, *Biometrika* **24**, 39–44.

[292] McKay, A.T. and Pearson, E.S. (1933). A note on the distribution of range in samples of n, *Biometrika* **25**, 415–420.

[293] McLeish, D.L. (1982). A robust alternative to the normal distribution, *Canad. J. Statist.* **10**(2), 89–102.

[294] Mendenhall, W. and Hader, R.J. (1958). Estimation of parameters of mixed exponentially distributed failure time distributions from censored life test data, *Biometrika* **45**, 504–520.

[295] Mertz, P. (1961). Model of impulsive noise for data transmission, *IEEE Trans. Comm.* **CS-9**, 130–137.

[296] Miller, G. (1978). Properties of certain symmetric stable distributions, *J. Multivariate Anal.* **8**(3), 346–360.

[297] Miller, J.H. and Thomas, J.B. (1972). Detectors for discrete-time signals in non-Gaussian noise, *IEEE Trans. Inform. Theory* **IT-18**(2), 241–250.

[298] Milne, R.K. and Yeo, G.F. (1989). Random sum characterizations, *Math. Sci.* **14**, 120–126.

[299] Missiakoulis, S. (1983). Sargan densities, which one? *J. Econometrics* **23**, 223–233.

[300] Missiakoulis, S. and Darton, R. (1985). The distribution of 2×2 unit normal determinants, *Amer. Statist.* **39**(3), 241.

[301] Mitchell, A.F.S. (1994). A note on posterior moments for a normal mean with double exponential prior, *J. Roy. Statist. Soc. Ser. B* **56**(4), 605–610.

[302] Mittnik, S. and Rachev, S.T. (1991). Alternative multivariate stable distributions and their applications to financial modelling, in *Stable Processes and Related Topics* (eds., S. Cambanis et al.), Birkhäuser, Boston, pp. 107–119.

[303] Mittnik, S. and Rachev, S.T. (1993). Modeling asset returns with alternative stable distributions, *Econometric Rev.* **12**(3), 261–330.

[304] Mohan, N.R., Vasudeva, R. and Hebbar, H.V. (1993). On geometrically infinitely divisible laws and geometric domains of attraction, *Sankhyā Ser. A* **55**, 171–179.

[305] Nakagami, M. (1964). On the intensity distribution and its applications to signal statistics, *Radio Sci. J. Res.* **68D**(9), 995–1003.

[306] Navarro, J. and Ruiz, J.M. (2000). Personal communication (to S. Kotz).

[307] Neyman, J. and Pearson, E.S. (1928). On the use and interpretation of certain test criteria for purposes of statistical inference, I, *Biometrika* **20A**, 175–240.

[308] Neyman, J. and Pearson, E.S. (1933). On the problem of the most efficient tests of statistical hypotheses, *Phil. Trans. Roy. Soc. London, Ser. A* **231**, 289–337.

[309] Nicholson, W.L. (1958). On the distribution of 2×2 random normal determinants, *Ann. Math. Statist.* **29**, 575–580.

[310] Nikias, C.L. and Shao, M. (1995). *Signal Processing with Alpha-Stable Distributions and Applications*, Wiley, New York.

[311] Nikitin, Y. (1995). *Asymptotic Efficiency of Nonparametric Tests*, Cambridge University Press, Cambridge.

[312] Nitadori, K. (1965). Statistical analysis of ΔPCM, *Electron. Comm. in Japan* **48**, 17–26.

[313] Noll, P. and Zelinski, R. (1979). Comments on"Quantizing characteristics for signals having Laplacian amplitude probability density function," *IEEE Trans. Comm.* **COM-27**, 1295–1297.

[314] Norris, J.P., Nemiroff, R.J., Bonnell, J.T., Scargle, J.D., Kouveliotou, C., Paciesas, W.S., Meegan, C.A. and Fishman, G.J. (1996). Attributes of pulses in long bright gamma-ray bursts, *Astrophysical Journal* **459**, 393–412.

[315] Norton, R.M. (1984). The double exponential distribution: Using calculus to find a maximum likelihood estimator, *Amer. Statist.* **38**(2), 135–136.

[316] Nyquist, H., Rice, S.O. and Riordan, J. (1954). The distribution of random determinants, *Quart. Appl. Math.* **12**(2), 97–104.

[317] Ogawa, J. (1951). Contribution to the theory of systematic statistics, I, *Osaka Math. J.* **3**(2), 175–213.

[318] Okubo, T. and Narita, N. (1980). On the distribution of extreme winds expected in Japan, *National Bureau of Standards Special publication 560-1*, 12 pp.

[319] Olver, A.K.P. (1974). *Asymptotics and Special Functions*, Academic Press, New York.

[320] Ord, J.K. (1983). Laplace distribution, in *Encyclopedia of Statistical Science*, Vol. 4 (eds., S. Kotz, N.L. Johnson, and C.B. Read), Wiley, New York, pp. 473–475.

[321] Osiewalski, J. and Steel, M.F.J. (1993). Robust Bayesian inference in l_q-spherical models, *Biometrika* **80**(2), 456–460.

[322] Ostrovskii, I.V. (1995). Analytic and asymptotic properties of multivariate Linnik's distribution, *Mathematical Physics, Analysis, Geometry* **2**(3/4), 436–455.

[323] Pace, L. and Salvan, A. (1997). *Principles of Statistical Inference*, World Scientific Press, Singapore.

[324] Pakes, A.G. (1992a). A characterization of gamma mixtures of stable laws motivated by limit theorems, *Statist. Neerlandica* **2–3**, 209–218.

[325] Pakes, A.G. (1992b). On characterizations through mixed sums, *Austral. J. Statist.* **34**(2), 323–339.

[326] Pakes, A.G. (1995). Characterization of discrete laws via mixed sums and Markov branching processes, *Stochastic Process. Appl.* **55**, 285–300.

[327] Pakes, A.G. (1997). The laws of some random series of independent summands, in *Advances in the Theory and Practice of Statistics: A Volume in Honor of Samuel Kotz* (eds., N.L. Johnson and N. Balakrishnan), Wiley, New York, pp. 499–516.

[328] Pakes, A.G. (1998). Mixture representations for symmetric generalized Linnik laws, *Statist. Probab. Lett.* **37**, 213–221.

[329] Pearson, E.S. (1935). A comparison of β_2 and Mr. Geary's w_n criteria, *Biometrika* **27**, 333–352.

[330] Pearson, E.S. and Hartley, H.O. (1942). The probability integral of the range in samples of n observations from a normal population, *Biometrika* **32**, 301–310.

[331] Pearson, K., Jefferey, G.B. and Elderton, E.M. (1929). On the distribution of the first product moment-coefficient, in samples drawn from an indefinitely large normal population, *Biometrika* **21**, 164–193.

[332] Pearson, K., Stouffer, S.A. and David, F.N. (1932). Further applications in statistics of the $T_m(x)$ Bessel function, *Biometrika* **24**, 293–350.

[333] Peterson, R. and Silver, E.A. (1979). *Decision Systems for Inventory Management and Production Planning*, Wiley, New York.

[334] Pillai, R.N. (1985). Semi-α-Laplace distributions, *Comm. Statist. Theory Methods* **14**(4), 991–1000.

[335] Pillai, R.N. (1990). On Mittag-Leffler functions and related distributions, *Ann. Inst. Statist. Math.* **42**(1), 157–161.

[336] Pitt, L. (1982). Positively correlated normal variables are associated, *Ann. Probab.* **10**(2), 496–499.

[337] Poiraud-Casanova, S. and Thomas-Agnan, C. (2000). About monotone regression quantiles, *Statist. Probab. Lett.* **48**, 101–104.

[338] Press, S.J. (1967). On the sample covariance from a bivariate normal distribution, *Ann. Inst. Statist. Math.* **19**, 355–361.

[339] Problem 64–13 (1966). *SIAM Review* **8**(1), 108–110.

[340] Rachev, S.T. and SenGupta, A. (1992). Geometric stable distributions and Laplace-Weibull mixtures, *Statist. Decisions* **10**, 251–271.

[341] Rachev, S.T. and SenGupta, A. (1993). Laplace-Weibull mixtures for modeling price changes, *Management Science* **39**(8), 1029–1038.

[342] Raghunandanan, K. and Srinivasan, R. (1971). Simplified estimation of parameters in a double exponential distribution, *Technometrics* **13**(3), 689–691.

[343] Ramachandran, B. (1997). On geometric stable laws, a related property of stable processes, and stable densities, *Ann. Inst. Statist. Math.* **49**(2), 299–313.

[344] Ramsey, F.L. (1971). Small sample power functions for nonparametric tests of location in the double exponential family, *J. Amer. Statist. Assoc.* **66**, 149–151.

[345] Rao, C.R. (1961). Asymptotic efficiency and limiting information, *Proc. Fourth Berkeley Symp. Math. Statist. Probab.* **1**, 531–545.

[346] Rao, C.R. (1965). *Linear Statistical Inference and Its Applications*, Wiley, New York.

[347] Rao, C.R and Ghosh, J.K. (1971). A note on some translation-parameter families of densities for which the median is an M.L.E., *Sankhyā Ser. A* **33**(1), 91–92.

[348] Rao, A.V., Rao, A.V.D. and Narasimhan, V.L. (1991). Optimum linear unbiased estimation of the scale parameter by absolute values of order statistics in the double exponential and double Weibull distributions, *Comm. Statist. Simulation Comput.* **20**(4), 1139–1158.

[349] Rényi, A. (1956). Posson-folyamat egy jemllemzëse, *Magyar Tud. Akad. Nat. Kutató Int. Közl.*, v. 1, 4, 519–527 (in Hungarian).

[350] Reza, F.M. (1961). *An Introduction to Information Theory*, McGraw-Hill, New York.

[351] Rider, P. (1961). The method of moments applied to a mixture of two exponential distributions, *Ann. Statist.* **32**, 143–147.

[352] Robertson, T. and Waltman, P. (1968). On estimating monotone parameters, *Ann. Math. Statist.* **39**(3), 1030–1039.

[353] Rohatgi, V.K. (1984). *Statistical Inference*, Wiley, New York.

[354] Rosenberger, J.L. and Gasko, M. (1983). Comparing location estimators: trimmed means, medians, and trimean, in *Understanding Robust and Exploratory Data Analysis* (eds., D.C. Hoaglin, F. Mosteller, and J.W. Tukey), Wiley, New York, pp. 297–338.

[355] Rosiński, J. (1976). Weak compactness of laws of random sums of identically distributed random vectors in Banach spaces, *Colloq. Math.* **35**, 313–325.

[356] Rowland, R.St.H. and Sichel, H.S. (1960). Statistical quality control of routine underground sampling, *J. S. Afr. Inst. Min. Metall.* **60**, 251–284.

[357] Sahli, A., Trecourt, P. and Robin, S. (1997). Acceptance sampling by Laplace distributed variables, *Comm. Statist. Theory Methods* **26**, 2817–2834.

[358] Saleh, A.K.Md., Ali, M.M. and Umbach, D. (1983). Estimation of the quantile function of a location-scale family of distributions based on a few selected order statistics, *J. Statist. Plann. Inference* **8**, 75–86.

[359] Salzer, H.E., Zucker, R. and Capuano, R. (1952). Table of the zeros and weight factors of the first twenty Hermite polynomials, *J. Research NBS* **48**, 111–116.

[360] Samorodnitsky, G. and Taqqu, M. (1994). *Stable Non-Gaussian Random Processes*, Chapman & Hall, New York.

[361] Sansing, R.C. (1976). The t-statistic for a double exponential distribution, *SIAM J. Appl. Math.* **31**, 634–645.

[362] Sansing, R.C. and Owen, D.B. (1974). The density of the t-statistic for non-normal distributions, *Comm. Statist.* **3**(2), 139–155.

[363] Sarabia, J.M. (1993). Problem 48, *Qüestiió* **17**(1) (in Spanish).

[364] Sarabia, J.M. (1994). Personal communication (to S. Kotz).

[365] Sarhan, A.E. (1954). Estimation of the mean and standard deviation by order statistics, *Ann. Math. Statist.* **25**, 317–328.

[366] Sarhan, A.E. (1955). Estimation of the mean and standard deviation by order statistics, Part III, *Ann. Math. Statist.* **26**, 576–592.

[367] Sarhan, A.E. and Greenberg, B. (1967). Linear estimates for doubly censored samples from the exponential distribution with observations also missing from the middle, *Bull. Int. Statist. Institute, 36th Session* **42**(2), 1195–1204.

[368] Sassa, H. (1968). The probability density of a certain statistic in one sample from the double exponential population, *Bull. Tokyo Gakugei Univ.* **19**, 85–89 (in Japanese).

[369] Scallan, A.J. (1992). Maximum likelihood estimation for a normal/Laplace mixture distribution, *The Statistician* **41**, 227–231.

[370] Serfling, R.J. (1980). *Approximation Theorems of Mathematical Statistics*, Wiley, New York.

[371] Sharma, D. (1984). On estimating the variance of a generalized Laplace distribution, *Metrika* **31**, 85–88.

[372] Shepp, L.A. (1962). Symmetric Random Walk, *Trans. Amer. Math. Soc.* **104**, 144–153.

[373] Shyu, J.-C. and Owen, D.B. (1986a). One-sided tolerance intervals for the two-parameter double exponential distribution, *Comm. Statist. Simulation Comput.* **15**(1), 101–119.

[374] Shyu, J.-C. and Owen, D.B. (1986b). Two-sided tolerance intervals for the two parameter double exponential distribution, *Comm. Statist. Simulation Comput.* **15**(2), 479–495.

[375] Shyu, J.-C. and Owen, D.B. (1987). β-expectation tolerance intervals for the double exponential distribution, *Comm. Statist. Simulation Comput.* **16**(1), 129–139.

[376] Sichel, H.S. (1973). Statistical valuation of diamondiferous deposits, *J. S. Afr. Inst. Min. Metall.* **73**, 235–243.

[377] Springer, M.D. (1979). *The Algebra of Random Variables*, Wiley, New York.

[378] Srinivasan, R. and Wharton, R.M. (1982). Confidence bands for the Laplace distribution, *J. Statist. Comput. Simulation* **14**, 89–99.

[379] Steutel, F.W. (1970). *Preservation of Infinite Divisibility Under Mixing and Related Topics*, Mathematisch Centrum, Amsterdam.

[380] Steutel, F.W. and van Harn, K. (1979). Discrete analogues of self-decomposability and stability, *Ann. Probab.* **7**, 497–501.

[381] Stigler, S.M. (1986a). Laplace's 1774 Memoir on Inverse Probability, *Statistical Science* **1**(3), 359–378.

[382] Stigler, S.M. (1986b). *The History of Statistics: The Measurement of Uncertainty before* 1900, Harvard Univ. Press, Cambridge.

[383] Stoyanov, J. (2000). Krein condition in probabilistic moment problems, *Bernoulli* **6**(5), 939–949.

[384] Subbotin, M.T. (1923). On the law of frequency of error, *Mathematicheskii Sbornik* **31**, 296–300.

[385] Sugiura, N. and Naing, M.T. (1989). Improved estimators for the location of double exponential distribution, *Comm. Statist. Theory Methods* **18**, 541–554.

[386] Sullivan, G.J. (1996). Efficient scalar quantization of exponential and Laplacian random variables, *IEEE Trans. Inform. Theory* **42**, 1265–1274.

[387] Szasz, D. (1972). On classes of limit distributions for sums of a random number of identically distributed independent random variables, *Theory Probab. Appl.* **17**, 401–415.

[388] Takano, K. (1988). On the Lévy representation of the characteristic function of the probability distribution $Ce^{-|x|}dx$, *Bull. Fac. Sci. Ibaraki Univ.* **20**, 61–65.

[389] Takano, K. (1989). On mixtures of the normal distribution by the generalized gamma convolutions, *Bull. Fac. Sci. Ibaraki Univ.* **21**, 29–41.

[390] Takano, K. (1990). Correction and addendum to "On mixtures of the normal distribution by the generalized gamma convolutions," *Bull. Fac. Sci. Ibaraki Univ.* **22**, 50–52.

[391] Takeuchi, K. and Akahira, M. (1976). On the second order asymptotic efficiencies of estimators, in *Proc. Third Japan-USSR Symp. Probab. Theory* (eds., G. Maruyama and J.V. Prokhorov), pp. 604–638; *Lecture Notes in Math.* **550**, Springer-Verlag, Berlin.

[392] Taylor, J.M.G. (1992). Properties of modelling the error distribution with an extra shape parameter, *Comput. Statist. Data Anal.* **13**, 33–46.

[393] Teicher, H. (1961). Maximum likelihood characterization of distributions, *Ann. Math. Statist.* **32**, 1214–1222.

[394] Teichroew, D. (1957). The mixture of normal distributions with different variances, *Ann. Math. Statist.* **28**, 510–512.

[395] Tiao, G.C. and Lund, D.R. (1970). The use of OLUMV estimators in inference robustness studies of the location parameter of a class of symmetric distributions, *J. Amer. Statist. Assoc.* **65**, 370–386.

[396] Tomkins, R.J. (1972). A generalization of Kolmogorov's law of the iterated logarithm, *Proc. Amer. Math. Soc.* **32**(1), 268–274.

[397] Tricker, A.R. (1984). Effects of rounding on the moments of a probability distribution, *Statistician* **33**(4), 381–390.

[398] Troutt, M.D. (1991). A theorem on the density of the density ordinate and an alternative interpretation of the Box-Muller method, *Statistics* **22**, 463–466.

[399] Tse, Y.K. (1987). A note on Sargan densities, *J. Econometrics* **34**, 349–354.

[400] Ulrich, G. and Chen, C.-C. (1987). A bivariate double exponential distribution and its generalization, *ASA Proceedings on Statistical Computing*, 127–129.

[401] Umbach, D., Ali, M.M. and Saleh, A.K.Md. (1984). Hypothesis testing for the double exponential distribution based on optimal spacing, *Soochow J. Math.* **10**, 133–143.

[402] Uppuluri, V.R.R. (1981). Some properties of log-Laplace distribution, in *Statistical Distributions in Scientific Work*, Vol 4 (eds., G.P. Patil, C. Taillie and B. Baldessari), Reidel, Dordrecht, pp. 105–110.

[403] Uthoff, V.A. (1973). The most powerful scale and location invariant test of the normal versus the double exponential, *Ann. Statist.* **1**, 170–174.

[404] van der Waerden, B.L. (1952). Order tests for the two-sample problem and their power, *Indag. Math.* **14**, 453–458. [Correction: *Indag. Math.* **15**, 80.]

[405] Van Eeden, C. (1957). Maximum likelihood estimation of partially or completely ordered parameters, I and II, *Indag. Math.* **19**, 128–136, 201–211.

[406] van Zwet, W.R. (1964). *Convex Transformations of Random Variables*, Math. Centre, Amsterdam.

[407] Watson, G.N. (1962). *A Treatise on the Theory of Bessel functions*, Cambridge University Press, London.

[408] Weida, F.M. (1935). On certain distribution functions when the law of the universe is Poisson's first law of error, *Ann. Math. Statist.* **6**, 102–110.

[409] Weron, R. (1996). On the Chambers-Mallows-Stuck method for simulating skewed stable random variables, *Statist. Probab. Lett.* **28**, 165–171.

[410] Weron, K. and Kotulski, M. (1996). On the Cole-Cole relaxation function and related Mittag-Leffler distributions, *Physica A* **232**, 180–188.

[411] Wilcoxon, F. (1945). Individual comparisons by ranking methods, *Biometrics* **1**, 80–83.

[412] Wilson, E.B. (1923). First and second laws of error, *J. Amer. Statist. Assoc.* **18**, 841–852.

[413] Wilson, E.B. and Hilferty, M.M. (1931). The distribution of chi-square, *Proc. Nat. Acad. Sci.* **17**, 684–688.

[414] Yamazato, M. (1978). Unimodality of infinitely divisible distribution functions of class L, *Ann. Probab.* **6**, 523–531.

[415] Yen, V.C. and Moore, A.H. (1988). Modified goodness-of-fit test for the Laplace distribution, *Comm. Statist. Simulation Comput.* **17**, 275–281.

[416] Yeo, G.F. and Milne, R.K. (1989). On characterizations of exponential distributions, *Statist. Probab. Lett.* **7**, 303–305.

[417] Younes, L. (2000). Orthogonal expansions for a continuous random variable with statistical applications, *Ph.D. Thesis*, University of Barcelona, Barcelona, Spain.

[418] Zeckhauser, R. and Thompson, M. (1970). Linear regression with non-normal error terms, *Rev. Econom. Statist.* **52**, 280–286.

[419] Zellner, A. (1976). Bayesian and non-Bayesian analysis of the regression model with multivariate student-*t* error terms, *J. Amer. Statist. Assoc.* **71**, 400–405.

[420] Zolotarev, V.M. (1986). *One-Dimensional Stable Distributions*, Volume 65 of Translations of Mathematical Monographs, American Mathematical Society.

Index

acceptance sampling, 304, 305, 338
α-Laplace distribution, 199, 203, 230, 324, 337
analogue of the t-distribution, 46, 118
Anderson–Darling test, 105
asset pricing, 300
association, 269
asymmetric Laplace distribution, 37, **133–174**, 179, 188, 197, 198, 221, 230, 331, 332
  absolute moments, 134, 135, 142, 156, 157
  alternative parametrization, 136
  characteristic function of, 136, 137, 140
  characterizations of, 134, 144, 160
  coefficient of skewness, 143
  coefficient of variation, 143
  cumulant generating function, 140, 141
  cumulants, 141, 142, 145
  density, 137, 138
  distribution function, 137, 138
  entropy, 155, 156
  estimation, 158
  Fisher information matrix, 158, 160, 161, 164, 166, 169, 170, 174
  generalized, *see* generalized asymmetric Laplace distribution
  geometric infinite divisibility, 151
  infinite divisibility, 135, 153
  interquartile range, 144
  kurtosis, 143
  Lévy density of, 151
  likelihood function, 158, 159, 163, 168
  mean, 135, 142
  mean deviation, 142
  median, 143, 144
  moment generating function, 140
  moments, 142, 144
  parametrizations, 136–140
  quantiles, 143
  simulation, 135, 149
  standard, 134, 149, 151
  variance, 136, 142
asymptotic relative efficiency, 80
asymptotically most powerful, 110–112
average adjustment interval, 306, 307

Bayes risk, 87
Bernoulli random variable, 29, 63, 147
Bessel function, 26, 36, 37, 39–41, 315, 322, 337, 341
Bessel function distribution, 36, 150, **179–222**, 257, 335
  characteristic function, 179
  definition, 180
  density, 180, 189
  mean, 192
  moments, 192

multivariate, *see* multivariate Bessel distribution
standard, 181
standard density, 189
variance, 192
best linear unbiased estimator, 63, 84
bilateral exponential distribution, 16, 326
bivariate asymmetric Laplace distribution, 240
characteristic function, 240, 241, 243
definition, 240
densities, 241, 242, 249
moments, 241
simulation, 243, 245
variance-covariance matrix, 241, 249
bivariate hyperbolic distribution, 310
bivariate Laplace distributions, 229, 231
bivariate normal distribution, 237
Black–Sholes formula, 296, 297
breakdown voltage, 284, 285
brittle fracture distribution, 23
broadened median, 67
Brownian motion, 193–196, 198, 222, 224, 294–296, 300, 326

canonical median, 66, 67
Carleman condition, 114
Cauchy distribution, 19, 41, 65, 106, 115, 119, 206, 225, 238, 321, 323
censored sample, 77, 84–90, 93, 99, 103, 122, 123, 319, 321, 323, 324, 335, 339
chi-square distribution, 13, 23–26, 92, 109, 146, 148, 184, 188, 248, 262, 285, 341
class L, 213, 214, 341
code modulation, 280
coefficient of skewness, 21, 143, 176, 222
coefficient of variation, 20, 143
communication theory, 277, 280
complementary Bessel function, 296
completely monotone density, 50
compound Laplace distribution, 124, 125
compound Poisson process, 193, 294
confidence ellipsoid, 264
contaminated Laplace, 302
control chart, 39, 308
convolution of exponential distributions, 135, 146
cosine distribution, 15–17
Cramér–Rao lower bound, 69, 70, 90, 91, 177, 223
Cramér–von Mises test, 105, 106
Cramér–Wald device, 259
currency exchange, 275, 289, 292, 293, 300

decibel, 280
demand during lead time, 303, 304
detector, 277–280
diffuse prior, 267
discrete Laplace distribution, 130
discrete Linnik distribution, 219, 323
discrete Mittag–Leffler distribution, 219, 329
discrete stable distribution, 219, 323
distortion measure, 282, 283
domain of attraction, 213
dose response for radiation carcinogenesis, 312
double exponential distribution, 6, 13, 16
double geometric distribution, 130
double Pareto distribution, 41
doubly exponential law, 16
duplicate check-sampling, 308

E-M algorithm, 302
economic clock, 294
Edgeworth expansion, 130
efficiency of an estimator, 85
efficient score function, 120
elliptically contoured K-Bessel distribution, 300
elliptically contoured distribution, 229, 312, 333
elliptically contoured Laplace distribution, 232, 270, 300
elliptically symmetric distribution, 234, 270, 271, 335
empirical characteristic function, 216, 217
empirical distribution, 215, 290
encoding of analog signals, 280
entropy, 21, 51, 119, 134, 135, 155–157, 177, 282, 283, 327, 330, 331
environmental sciences, 309
European option, 296
exponential family, 16, 64, 78
exponential integral, 119, 196
exponential mixture, 192, 199, 205, 206, 322, 332
exponential power distribution, 71, 220, 223, 270, 271, 319

multivariate, *see* multivariate exponential power distribution

F-distribution, 116, 264
feedback adjustment, 307
financial data, 134, 221, 289, 294, 300, 302, 332, 333, 335
first law of error, 3
first law of Laplace, 5, 6
Fisher information, 65, 70, 95, 119, 158, 160, 161, 164, 166, 169, 170, 174, 178
folded Cauchy density, 205
fractional moments, 42, 213, 218, 332
fracturing of materials, 284
Fredholm integral equation, 307
functions of order statistics, 56

gamma distribution, 20, 32, 33, 47, 124, 131, 180, 183, 184, 188, 190, 194, 257, 295, 308, 334
characterization of, 33
gamma process, 193, 194, 196, 198, 224, 294, 296, 300
gamma white noise, 224
gamma-ray bursts, 310
Gauss–Markov theorem, 262, 264
generalized asymmetric Laplace distribution, 180, 222, 257
generalized beta distribution, 295
generalized gamma convolution, 131, 256, 322, 340
generalized Gaussian distribution, 278
generalized hyperbolic distribution, 240, 246, 253, 254, 258
generalized inverse Gaussian distribution, 246, 320, 329
generalized Laplace distribution, 180, 189, 190, 219, 220, 222, 248, 270, 289, 295, 299, 300, 325, 329, 335, 339
generalized Linnik law, 216, 219, 337
generating function, 27, 219
geometric infinitely divisible, 48, 151, 153, 258
geometric mean, 9, 112, 335
geometric random variable, 22
geometric stable distribution, 32, 177, 179, 199, 200, 206, 208, 213–216, 218, 220, 240, 246, 256, 259, 260, 268, 322, 331–333, 337, 338
characterization of, 155
geometric summation, 22, 27, 30, 134, 155, 200, 202, 259, 272
Gini mean difference, 112
goodness-of-fit test, 105, 110, 295, 341

harmonic mean, 9, 114
Hermite distribution, 303, 320
Hermite polynomial, 217
homogeneous increments, 295
hyperbolic distribution, 174, 175, 179, 230, 239, 240, 247, 248, 253, 254, 285, 289, 294, 309, 310, 322, 325
generalized, *see* generalized hyperbolic distribution

image and speech compression, 281
impulsive noise, 278
incomplete exponential function, 108
incomplete gamma function, 46, 99
indicator function, 140, 196
infinite divisibility, 27, 46, 47, 117, 128, 135, 150, 256, 258, 330, 339
integrated moving average, 307
interest rates, 289–291, 331
inventory control, 303
inverse Gaussian distribution, 230, 246, 289, 321
generalized, *see* generalized inverse Gaussian distribution

jump diffusion model, 300
jump function, 197

κ-criterion, 176
Kolmogorov goodness-of-fit test, 110
Kolmogorov statistic, 110
Kolmogorov–Smirnov test, 110, 111
Kotz-type multivariate distribution, 312
Krein condition, 114, 340
Kronecker product, 237
kurtosis, 114, 125, 143, 176, 222, 271, 290, 294, 297, 321, 328

Lambert function, 283
Laplace distribution, 15
central absolute moments, 20
central moments, 20, 50
characteristic function, 19
classical, 22–26, 35, 38–41, 43

coefficient of skewness, 21
coefficient of variation, 20
conditional inference, 95–99
confidence intervals, 91, 92, 99
cumulant generating function, 19
cumulative distribution function, 18
density, 16, 18
discrete, *see* discrete Laplace distribution
entropy, 21, 52
Fisher information matrix, 65, 74
generalized, *see* generalized Laplace distribution
genesis of, 12
goodness-of-fit tests, 105, 110
joint distribution of order statistics, 55
kurtosis, 21
likelihood function, 66
linear estimation, 88
maximum likelihood estimation, *see also* maximum likelihood estimation, 71, 74, 77
mean, 19, 20
mean deviation, 20
mean of the sample median, 60, 66
median, 21
midrange, 56, 58
minimal sufficient statistic, 64
moment generating function, 19, 20
moments of order statistics, 60, 61, 63
multivariate, *see* multivariate Laplace distribution
orthogonal representation, 25
parameters, 16
quantile estimation, 91
quantiles, 21
sample median, 60
standard, 18
standard classical, 18–20, 22, 26, 47, 49, 54, 55, 61–63
test for location, 106
variance, 15, 18
variance of the sample median, 60, 61, 66

Laplace motion, 179, **193–199**, 223, 224, 257, 289, 294–297
asymmetric, 193, **197–199**
compound Poisson approximation, 193
covariance structure, 194, 195, 198
Lévy–Khinchine representation, 196
self-similarity property, 194
series representation, 196, 198
space scale parameter, 193, 197
standard, 194–196
symmetric, **193–197**
time scale parameter, 193, 197
trajectories, 193, 194, 197, 198
with drift, 193, 194, 197

Laplace noise, 277–279, 313, 324, 335
Laplace–Weibull mixture, 302, 337
law of the iterated logarithm, 131, 340
least-squares estimator, 216, 261–263, 265
leptokurtic, 15, 21, 125, 143, 289, 308, 311, 332
L-estimator, 84
Lévy measure, 47, 48, 150, 151, 196, 198, 214, 239, 256–258, 332
Lévy process, 46, 179, 193, 248, 257, 322, 326
Lévy–Khinchine representation, 47, 48, 150, 196, 214, 340
likelihood ratio test, 103–105
limits of geometric sums, 30, 152, 202, 258
linear combinations of order statistics, 166
linear unimodal, 252
Linnik distribution, 35, **199–206**, 215, 221, 225, 272, 320, 324, 325, 328, 331, 332, 336
absolute moments, 213
characteristic function, 199, 207, 208, 213
densities, 199, 205–208, 210–212
discrete, *see* discrete Linnik distribution
distribution function, 204, 206, 208, 209
estimation, **215–218**
exponential mixture representation, 192, 205, 206
generalized, *see* generalized Linnik law
infinite divisibility, 200
Lévy–Khinchine representation, 214
mixture representations, 204, 205
multivariate, *see* multivariate Linnik distribution
nonsymmetric, 199
positive, *see* positive Linnik law
scale parameter, 199, 217
series expansions, 210
simulations, 200, 204, 215
tail index, 212

Liouville number, 211, 212
locally most powerful, 111

log-gamma distribution, 184
log-Laplace distribution, 24, 172, 219, 222, 290, 309, 310, 312, 331, 341
log-normal distribution, 9, 222, 290, 308–310
log-return, 294
logistic distribution, 15–17, 60, 65, 115, 116, 119, 120, 126–128, 326
 characterization of, 126, 326
Lomax distribution, 41
loss of information, 62, 94–97, 319

Marshall–Olkin exponential distribution, 246
maximum entropy principle, 51, 135, 155, 297
maximum likelihood, 66, 69, 71, 74, 78, 121, 123, 158, 163, 168, 172, 177, 178, 265, 291, 292, 295, 302, 313, 319, 322, 323, 328, 331, 335, 336, 339–341
maximum likelihood estimation, **66–76**
 of location parameters, 78
 under censoring, 77
mean deviation, 4, 6, 20, 52, 71, 72, 142
mean squared deviation, 4, 307
mean squared error, 263
median law of error, 10, 11
median test, 110–112
method of moments, 79, 83, 158, 178, 216, 217, 222, 223, 300, 303, 333, 338
midmean, 67
midrange, 56, 58, 60, 85, 86, 88, 127, 128, 161, 326, 327
Mittag–Leffler distribution, 29, 34, 201, 219, 221, 332, 337
 discrete, *see* discrete Mittag–Leffler distribution
mixture of exponentially distributed random variables, 147, 338
mixture of Laplace and Gaussian distributions, 285, 339
mixture of log-normal distributions, 309
mixture of normal distributions, 22, 131, 134, 180, 181, 190, 199, 205, 246, 256, 330, 340
mixture of stable laws, 199, 337
mixture of two Laplace distributions, 83, 123, 287, 333
mixture of two Student t distributions, 287
mode-median-mean inequalities, 177
modified Bessel function, 174, 315
multivariate asymmetric Laplace distribution, **239–268**
 characteristic function, 241, 248
 conditional distributions, 253, 254
 covariance, 249, 254
 definition, 240, **243**, 258
 densities, 249, 251
 geometric infinite divisibility, 258
 infinite divisibility, 256
 Lévy measure, 239, 256, 257
 linear combination of, 240, 252, 254, 255
 linear regression, 254, 255
 marginals, 255
 mean, 249, 254
 simulation, 239, 243, 244
 unimodality, 251
multivariate Bessel distribution, 257, 270, 271
multivariate Cauchy distribution, 238
multivariate exponential power distribution, 270, 271, 312
multivariate Laplace distribution, **229–268**, 270–272, 312, 331, 332
 covariance, 232, 233, 235
 density, 229, 232, 233
 mean vector, 232
 polar representation, 247
 simulation, *see* simulation, multivariate asymmetric Laplace distribution
 symmetric, *see* multivariate symmetric Laplace distribution
multivariate Linnik distribution, 230, 236, 272, 320, 336
multivariate symmetric Laplace distribution, **231–236**

Nakagami distribution, 300
navigation, 275, 277, 286, 287, 320, 328
nearly instantaneous companding, 280
Neave–Tukey quick test, 111
neutral Laplace estimator, 126
Neyman–Pearson lemma, 278
Neyman–Pearson optimal detector, 278
non-Gaussian noise, 278, 335
nonparametric tests of location, 110, 111, 113
normal characteristic function, 154
normal distribution, 3
 characterization of, 70
ν-stable law, 222, 255, 332

ocean engineering, 275, 286

operating characteristic curve, 305
optimal quantizer, 281, 283
option pricing, 296, 297, 332, 334
order statistics, **53–63**, 122, 159, 319–321, 323, 324, 327, 331, 333, 338, 339
  joint distribution of two, *see* joint distribution of order statistics, Laplace distribution
orthogonal representation, 25

Paretian stable distribution, 27, 221, 289, 290, 294
Pareto distribution, 24, 26, 41, 117, 146, 149, 184, 225, 327, 328
Pascal distribution, 185
Pascal-stable distribution, 185
Pearson Type VII distribution, 124, 308
platykurtic, 21, 125
Poisson approximation, 196, 198
Poisson process, 196, 197, 199, 300, 311
Polya-type characteristic function, 224
positive Linnik law, 219, 323
power function distribution, 117
prediction interval, 87
prior, 269
product of two independent Laplace variates, 40
product of two independent normal variables, 41
product-moment coefficient, 186, 337
pure jumps process, 193

quality control, 303, 338

Rademacher sequence, 196, 199
random fluctuations of response rate, 311, 335
random summation, 134, 239, 240, 258, 259, 327
range, 56, 58
rank sum statistic, 110, 111
rank test, 120
ratio of two independent Laplace variates, 41
Rayleigh distribution, 23, 298
real estate data, 302
reciprocal property, 205, 225
regression, 239, 254–256, 261–265, 267, 325, 332, 337, 342
repeated measurements, 312
R-estimator, 120, 327
restricted maximum likelihood, 78
risk neutral process, 295, 296

sample coefficient of skewness, 292
sample kurtosis, 292
sample mean deviation, 291
sample median, 3, 9–11, 53, 55, 56, 60, 61, 118–123, 126, 127, 323, 326
sample quantile, 88
S&P 500 Index, 295, 299
Sargan distribution, 220, 225, 330
scale and location invariant test, 105, 341
second law of error, 3
self-decomposability, 49, 147, 184, 185, 200, 213, 214, 258, 339
share market returns, 294
sign test, 120
signal-to-quantization noise ratio, 280
simplified linear estimator, 88
size data, 309
skew-normal distribution, 134, 320, 328, 334
skewed exponential power distribution, 271
skewed Laplace distribution, 52, 68, 121, 133, 134, 140, 321
slow-moving item, 303
Smirnov one-sided statistic, 110
s-ordering, 115
spacings, 90, 91
speech recognition, 275
speech signals, 280
spherically symmetric distribution, 238
stability, 27, 134, 155, 185, 200, 203, 221, 258–260, 339
stable distribution, 27, 128, 204, 221, 240, 272, 294, 295, 341
  discrete, *see* discrete stable distribution
stable process, 295, 329, 338
standard bivariate Laplace distribution, 232, 236, 237
standard exponential distribution, 23
standard gamma distribution, 36
star unimodality, 251
statistical process control, 306
steam generator inspection, 306, 324
stochastic variance, 22, 52, 134, 180, 181, 190, 205, 286, 297–300, 308, 333
stock price, 294–296

strictly geometric stable distribution, 29, 35, 199–201, 203, 213, 218, 255, 259
subordinated Brownian motion, 198, 248, 256
subordinated Gaussian process, 294
subordinated model of stock prices, 294
sufficient statistic, 64, 94, 223
sum of two independent Laplace r.v.'s, 38

tails, 15, 200, 212
t-distribution, 118, 263, 264, 287, 295, 308
testing multivariate symmetry, 269, 328
time series, 307, 313, 324
tolerance factor, 100–103
tolerance interval, 99–103, 330, 339
tolerance limit, 99, 102
transfer theorem, 258
Treasury Bill, 295
Treasury bonds, 290, 291
triangular distribution, 15–17, 68
t-statistic, 43, 45, 118, 339
Tukey quick test, 111
two-piece double exponential, 133, 334
two-tailed exponential distribution, 16
two-tailed power distribution, 68

underground sampling, 338
underreported data, 289, 327, 328
uniform distribution, 16
uniformly minimum variance unbiased estimator, 70
uniformly most powerful, 105

Value-at-Risk, 297, 333
van der Waerden test, 111
variance gamma distribution, 180
variance gamma process, 180, 257, 258, 294, 295, 299, 334
υ-distribution, 325
vec operator, 237
vec-permutation matrix, 237
vertical density function, 116, 331
volatility, 289, 292, 298, 300, 321, 334
volatility smile, 297, 302, 332
υ-spherical distribution, 270

weakest link, 284
Weibull distribution, 106, 277, 285, 302, 338
Wilcoxon test, 111
wind shear, 285, 329, 330